Drinking Water Quality

This textbook provides a comprehensive review of the problems associated with the supply of drinking water in the developed world. Since the first edition of this book was published there have been enormous changes in the water industry, especially in the way drinking water is perceived and regulated. Water companies and regulators have been presented with numerous new challenges – global warming has seriously affected the sustainability of water supplies as well as impacting water quality; advances in chemical and microbial analysis have revealed many new contaminants in water that were previously undetectable or unknown; and recent terrorist attacks have demonstrated how vulnerable water supplies could be to contamination or disruption. This new edition is an overview of the current and emerging problems, and what can be done to solve them. It has been completely updated, and includes the new WHO Revised Drinking Water Guidelines.

Drinking Water Quality is an ideal textbook for courses in environmental science, hydrology, environmental health, and environmental engineering. It also provides an authoritative reference for practitioners and professionals in the water supply industry.

N. F. GRAY is a Professor at the Centre for the Environment at Trinity College, Dublin. He has worked in the area of water technology for 30 years, and is internationally known as a lecturer and author in water quality and pollution control. His research specializes in the operational problems associated with supplying drinking water and treating wastewaters.

Drinking Water Quality

Second Edition

N. F. Gray
University of Dublin

CAMBRIDGE
UNIVERSITY PRESS

CAMBRIDGE UNIVERSITY PRESS

Cambridge, New York, Melbourne, Madrid, Cape Town, Singapore, São Paulo, Delhi

Cambridge University Press
The Edinburgh Building, Cambridge CB2 8RU, UK

Published in the United States of America by Cambridge University Press, New York

www.cambridge.org
Information on this title: www.cambridge.org/9780521878258

First published by John Wiley & Sons, Ltd 1994
This edition published 2008.

Printed in the United Kingdom at the University Press, Cambridge

A catalogue record for this publication is available from the British Library

ISBN 978-0-521-87825-8 hardback
ISBN 978-0-521-70253-9 paperback

Contents

Preface to the second edition

Since writing the first edition there have been enormous changes in the water industry especially in the way drinking water quality is perceived and regulated. That first edition was written at the same time as the 1993 revision of the World Health Organization (WHO) guidelines as published, which has subsequently resulted in the revision of all the major drinking water standards, including those covering the European Union and the USA. That early edition reflected those changes. So the preparation of this new edition was timed to coincide with the publication, late in 2004, of the latest revision of the drinking water guidelines by the WHO, which has adopted a more rigorous health-based approach in setting guidelines. These new guidelines have been used as the basis of this new edition.

The problems associated with global warming leading to regional changes in climate and water availability are seriously affecting sustainability of supplies as well as seriously impacting on quality. Advances in chemical and microbial analysis have revealed that water contains many new contaminants that were previously undetectable or unknown, constantly presenting water utilities and regulators with new challenges. Also the recent terrorist attacks have demonstrated how vulnerable water supplies are to contamination or disruption. Thus, while the existing risks remain and need to be dealt with on a day-to-day basis, these new problems require innovative technical and management solutions. The aim of this new edition is to give an overview of the current and emerging problems and what can be done to solve them.

This new edition has been extensively updated and expanded using a different framework. It now comprises of 31 chapters clustered into 5 distinct parts, each dealing with a separate element of the water supply chain. *Part I. Introduction to water supply* comprises of three introductory chapters. The first deals with the fundamentals of the water industry: how much water is used; what is required by consumers in terms of quality; and the operation, management and regulation of the water utilities. The remainder of the chapter looks at the new management approaches to water supply, in particular water demand management, and how water conservation is becoming an integral part of sustaining future supplies. The second chapter reviews how drinking water standards are developed and the role of risk assessment in that process. Water safety plans are now the basis for achieving good quality and maintaining supplies and this is discussed in detail. The chapter also explores how guidelines and standards have changed over the past 15 years. Part I closes with a quick overview of the problems relating to drinking water quality, where those problems arise within the supply chain and where more information can be accessed within this text.

Part II. *Problems with the resource* examines in depth the water quality problems that arise within water resources due to natural and man-made influences. Chapter 4 is a brief overview of how quality varies due to land use and natural geology, comparing surface and ground waters. The remaining chapters look at each group of potential contaminants in turn examining the source, effects on consumers and appropriate solutions. There are separate chapters dealing with nitrate and nitrite; organic micro-pollutants including pesticides, industrial solvents and polycyclic aromatic hydrocarbons; endocrine-disrupting (oestrogen-mimicking) compounds, pharmaceutical and personal care products; odour and taste; metals including iron, manganese, arsenic and other heavy metals; hardness; algae and algal toxins; radon and non-radon radionuclides; and pathogens.

Part III. *Problems arising from water treatment* looks at how the very action of improving water quality can itself cause significant aesthetic and occasionally health-related problems. After a brief review of treatment technology and how problems arise, there are individual chapters dealing with the main issues: flocculants such as aluminium and acrylamide; odour and taste; the contentious issue of fluoridation; disinfection by-products including trihalomethanes; and pathogen removal. *Part* IV. *Problems arising in the distribution network* deals specifically with the transport of water from the treatment plant to the consumer, which can seriously affect water quality. The design and management of service reservoirs, the mains and the individual service pipe that connects individual households to the network are explained. Chapters dealing with aesthetic quality, asbestos, bitumen and coal-tar linings, the remarkable variety of animals, both large and small, that live in the network and occasionally pop out of the tap, and finally the problems of biofilm development and pathogens within the mains are all dealt with in detail. After all this the water is still very susceptible to contamination from our own household plumbing and storage system. *Part* V. *Problems in household plumbing systems* explains how household systems work and follows with specialist chapters on corrosion, including lead contamination, pathogens, such as *Legionella* and *Mycobacterium,* and other quality problems.

Part VI. *The water we drink* deals with a number of consumer-related issues. The first chapter deals with alternatives to tap water and includes detailed sections on bottled water, point-of-use and point-of-entry treatment systems and other sources of water such as rainwater harvesting and water reuse. I said at the beginning that much has changed in the water industry in the developed world. What hasn't changed is the number of people facing water scarcity in developing countries, leading to poverty, starvation, serious illness and frequently death. After decades of dedicated and life-saving work by agencies such as WaterAid, the situation continues to worsen, driven by the dual problems of climate change and conflict. Water security in the twenty-first century looks at the problem of terrorism and ensuring water is protected more effectively from deliberate or accidental contamination or interruption. The final chapter gives an overview of drinking water quality and how climate change will affect it over the coming decades.

Each chapter in the book concludes with a brief discussion of the relevance of the specific problems for consumers and suppliers alike. There are also a number of appendices giving drinking water standards in Europe, USA, as well as the new WHO guidelines and much else.

In this text I have attempted to provide a cohesive and comprehensive introduction to the water supply industry and the supply chain, which I hope will be equally useful to engineers, scientists, managers and even the general consumers who wants to know more about the water they drink. The text has been designed to give you an integrated overview of drinking water quality and to act as a reference guide. It should be used in conjunction with the Internet where very detailed information can be accessed and for that reason key URLs have been given where appropriate in the text.

The royalties from this text have been donated to WaterAid (www.wateraid. org.uk), which is an international non-governmental organization (NGO) dedicated to the provision of safe drinking water, sanitation and hygiene education to the world's poorest communities. As you turn on your tap and safe clean water pours out, remember that this really is something very special and that it has given us the wonderful society in which we live today. If you can help WaterAid in giving this gift to others then thank you.

Chapter 1
The water business

1.1 Introduction

The water supply industry is vitally important not only to maintain the health of the community, but for the sustainability of industry, business and agriculture. Without adequate water supplies our present society would never have evolved, and our lives today would be unrecognizable. Our dependence on treated water is now incalculable, and threats to that supply are comparable to the worst natural and man-made disasters. The volumes of water consumed each day by agriculture, industry and the public are vast, requiring an enormous infrastructure to satisfy the demand. Like the other service providers, electricity, telephone and gas, the water utilities deliver their product to the home, which requires a network of distribution pipes to service each household, but unlike the other utilities these are stand alone local or regional networks, rather than integrated national supply networks.

In England and Wales there are 26 private water companies that together supplied 52.7 million consumers in 2004/5 with 15 807 million litres (Ml d^{-1}) of water each day. Sixty-eight per cent of this came from surface waters and the remainder (32%) from groundwater. It requires 1344 plants to treat this volume of water, which is supplied to consumers via 326 471 km of distribution mains. When this is broken down by region, the greatest demand is in the south-east and north-west regions, which have the largest populations. However, the areas of highest demand do not normally correspond to the areas where adequate water resources are to be found, so shortages occur. The current demand for potable water in England and Wales has stabilized and is currently at 91% of the peak demand recorded in 1990/1 (Table 1.1).

1.2 Water consumption

Water demand varies significantly between countries due to differences in culture, climate and economic wealth (Smith and Ali, 2006). The demand for water also varies over the 24-hour period. This is known as the diurnal variation, with peak usage in the UK occurring between 07.00 and 12.00 and from

Table 1.1 *Water supplied to the public distribution system 1990/1–2004/5. Adapted from Defra (2006) with permission from Defra*

Ml d^{-1}

United Kingdom	1990/1	1991/2	1992/3	1993/4	1994/5	1995/6	1996/7	1997/8	1998/9	1999/ 2000	2000/1	2001/2	2002/3	2003/4	2004/5
England and Wales[1]															
metered[2]	4824	4785	4646	4572	4687	4676	4743	4791	4866	5076	5142	5464	5509	5704	5765
unmetered[3]	12 558	12 414	12 109	12 185	12 424	12 643	11 918	11 192	10 477	10 254	10 117	10 319	10 312	10 373	10 042
England and Wales total	**17 382**	**17 199**	**16 755**	**16 757**	**17 111**	**17 319**	**16 661**	**15 983**	**15 343**	**15 331**	**15 259**	**15 783**	**15 821**	**16 077**	**15 807**
Scotland															
metered[2]	656	643	603	610	612	574	625	555	553	531	438	532	454	478	470
unmetered[3]	1645	1596	1603	1662	1651	1748	1686	1782	1775	1832	1962	1876	1933	1919	1920
Scotland total	**2301**	**2239**	**2206**	**2272**	**2263**	**2322**	**2312**	**2336**	**2329**	**2363**	**2400**	**2408**	**2387**	**2397**	**2390**
Northern Ireland															
metered[2]	143	152	186	184	188	186	187	174	158	157	147	144	156	151	148
unmetered[3]	538	527	483	486	498	517	520	516	534	547	573	591	558	525	492
Northern Ireland total	**681**	**679**	**669**	**670**	**686**	**703**	**707**	**690**	**692**	**704**	**720**	**735**	**714**	**676**	**640**
United Kingdom[1]															
metered[2]	5623	5580	5435	5366	5487	5436	5555	5520	5577	5764	5727	6140	6119	6333	6383
unmetered[3]	14 741	14 537	14 195	14 333	14 573	14 908	14 124	13 490	12 786	12 633	12 652	12 786	12 803	12 817	12 454
United Kingdom total	**20 364**	**20 117**	**19 630**	**19 699**	**20 060**	**20 344**	**19 680**	**19 009**	**18 364**	**18 398**	**18 379**	**18 926**	**18 922**	**19 150**	**18 837**

[1] Includes water supplied by water supply companies.
[2] Metered water is water measured at the point of delivery to premises.
[3] Unmetered water includes leakage from the distribution system and water used for miscellaneous purposes such as fire-fighting, sewer cleaning, water mains flushing and temporary supplies for construction sites.

Table 1.2 *Typical current domestic water use in England and Wales*

Use	%	Use	%
Toilet	35	Washbasin	8
Kitchen sink	15	Outside use	6
Bath	15	Shower	5
Washing-machine	12	Dishwasher	4

Table 1.3 *Average water use of a range of activities and appliances*

Purpose of water use	Frequency	Litres
Cooking, drinking, washing-up and personal hygiene	per person per day	27
Bath	one	90
Shower	one	20
Toilet	one	6–9[a]
Automatic washing-machine	one	100
Dishwashing machine	one	50
Hosepipe/sprinkler	per minute	18

[a] Modern cisterns now use 4 litres, or more commonly 6 litres, compared to 9 litres in older systems.

18.00–20.00 each day (Figure 20.1). Demand is greater during weekends by about 12%, with demand being higher in the summer than in the winter. In the UK the typical household water consumption, typical here meaning a family of two adults and two children, is currently $510 \, l \, d^{-1}$. This is equivalent to a per capita water consumption rate of $150–180 \, l \, d^{-1}$. Less than 20% of the water supplied is consumed for drinking or food preparation, with toilet flushing the single major use of water (Table 1.2). Ownership of certain white goods, which has increased dramatically over the past 20 years, has an important influence on water usage. For example in the UK, 94% of households owned a washing-machine in 2003 compared to just 79% in 1983. A similar trend has been seen with dishwasher ownership, which is currently 31% in the UK compared to 5% over the same period. On average a dishwasher adds an extra $6 \, l \, d^{-1}$ to the per capita consumption, increasing the overall demand by about 4%. Table 1.3 gives some idea of the amount of water such appliances use. At the top of the list are automatic washing-machines, which can use a staggering 100 litres each time they are used. A bath uses on average 90 litres a time compared with a shower that can use as little as $5 \, l \, min^{-1}$, although this depends on the showerhead used. For example, a power shower can use in excess of 17 litres per minute. Garden sprinklers use about a $1000 \, l \, h^{-1}$, which is the average daily water usage for seven or eight people. Clearly not all the public supplies are utilized for

Table 1.4 *Estimated daily use of water supplied by the former Severn and Trent Water Authority during 1984–5. Adapted from Archibald (1986) with permission from the Economic and Social Research Council*

Type of use	Amount used (Ml d^{-1})	Purpose	Amount used (Ml d^{-1})
Domestic	840	Basic	288
		Toilet flushing	242
		Bathing	155
		Washing-machine	114
		External use	27
		Luxury appliances	14
Industrial and commercial	530	Processing	256
		Domestic	153
		Cooling: direct	77
		Cooling: recycled	44
Agricultural	50	Livestock	35
		Domestic	10
		Protected crops	3
		Outdoor irrigation	2
Unaccounted for	522	Distribution system	287
		Consumers' service pipes	167
		Trunk mains	52
		Service reservoirs	16
Total	1942		

domestic purposes. This is illustrated by the analysis of daily water usage in 1984–5 by the former Severn and Trent Water Authority. Of the total 1942 Ml d^{-1} supplied each day, 840 Ml was used for domestic purposes, 530 Ml for industrial, 50 Ml for agricultural purposes and a remarkable 522 Ml (26.9%) was lost every day through a leaky distribution system (Table 1.4).

Losses from leaks are a widespread problem as water mains not only deteriorate with age, but are often damaged by heavy vehicles, building work or subsidence. Leakage control is a vital method of conserving water. Detecting and repairing leaks is both labour intensive and time consuming, which means that it is very expensive. However, if leaks are not controlled then water demand will escalate, with most of the extra demand seeping away into the ground instead of making its way to the consumer. In England and Wales 3608 Ml of the 15 378 Ml of treated water supplied each day was lost during 2004/5 through leakage, 2584 Ml d^{-1} from the distribution mains (17% of total input) and 1024 Ml d^{-1} from supply pipes (7% of total input). The current leakage rate is

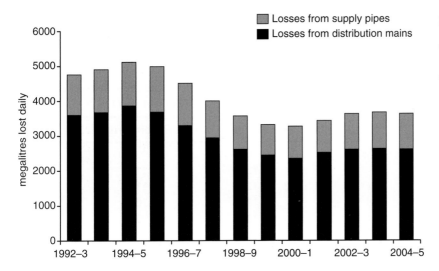

Figure 1.1 Water leakage reported in England and Wales during the period 1992/3 to 2004/5. Adapted from Defra (2006) with permission from Defra.

33% lower than the peak leakage rates reported in 1994/5, but has remained static each year since 1997/8 at between 22% and 25% despite a huge investment by the water companies in repairs and replacement to the distribution network each year (Figure 1.1). Currently, 24% of all water treated is being lost due to leakage.

Studies of water usage are very difficult to carry out, as individual use of water is so variable. Actual average consumption values may also hide water lost by leakage within the household plumbing system for example. The National Water Council (1982) carried out a detailed study of water usage and found that the actual average consumption levels were slightly lower than those calculated by the companies at that time (Table 1.5). This and subsequent studies on the pattern of domestic water usage have shown that per capita consumption decreases slightly with an increase in household size, and that social groupings also have an influence, with Social Group A using about $50 \, \mathrm{l} \, \mathrm{d}^{-1}$ more than Social Group E (Bailey *et al.*, 1986) (Table 1.5). It was also shown that the daily volume of water consumed per household for non-potable purposes is dependent on the size of the household. So on a national basis about 3% of the total volume of domestic water consumed each day is used for potable purposes, which is equivalent to about 10 litres for the average household. Interestingly, 25% of the first draws of water taken from the system each day are for potable purposes, a habit that may have significant consequences in those areas where the water is corrosive and lead or galvanized plumbing is used (Chapter 27).

1.3 Acceptable water quality

Water rapidly absorbs both natural and man-made substances, generally making the water unsuitable for drinking without some form of treatment.

Table 1.5 *Comparison of average water consumption per person in three water company areas with respect to household size and socio-economic group Adapted from National Water Council (1982) with permission from the National Water Council*

	Average water consumption ($1 d^{-1}$)		
	South West	Severn Trent	Thames
Household size			
1	126	116	136
2	124	118	151
3	118	109	123
4	110	92	116
5	103	96	103
6	97	75	97
7+	92	69	64
Social group			
A	138	126	134
B	124	117	126
C1	122	100	124
C2	111	93	113
D	103	93	102
E	86	76	94

Important categories of substances that can be considered undesirable in excess are:

1. *Colour.* This is due to the presence of dissolved organic matter from peaty soils, or the mineral salts of iron and manganese.
2. *Suspended matter.* This is fine mineral and plant material that is unable to settle out of solution under the prevailing conditions.
3. *Turbidity.* This is a measure of the clarity, or transparency, of the water. Cloudiness can be caused by numerous factors such as fine mineral particles in suspension, high bacteria concentrations, or even fine bubbles due to over-aeration of the water.
4. *Pathogens.* These can be viruses, bacteria, protozoa or other types of pathogenic organism that can adversely affect the health of the consumer. They can arise from animal or human wastes contaminating the water resource.
5. *Hardness.* Excessive and extremely low hardness are equally undesirable. Excessive hardness arises mainly from groundwater resources whereas very soft waters are characteristic of some upland catchments.
6. *Taste and odour.* Unpleasant tastes and odours are due to a variety of reasons such as contamination by wastewaters, excessive concentration of certain chemicals such as iron, manganese or aluminium, decaying vegetation, stagnant conditions due to a lack of oxygen in the water, or the presence of certain algae.

7. *Harmful chemicals*. There is a wide range of toxic and harmful organic and inorganic compounds that can occur in water resources. These are absorbed from the soil or occur due to contamination from sewage or industrial wastewaters.

Water treatment and distribution is the process by which water is taken from water resources, made suitable for use and then transported to the consumer. This is the first half of the human or urban water cycle, before water is actually used by the consumer (Figure 1.2). The second half of the cycle is the collection, treatment and disposal of used water (sewage) (Gray, 2004).

The objective of water treatment is to produce an adequate and continuous supply of water that is chemically, bacteriologically and aesthetically pleasing. More specifically, water treatment must produce water that is:

1. *Palatable* – that is, has no unpleasant taste;
2. *Safe* – it should not contain any pathogenic organism or chemical that could be harmful to the consumer;
3. *Clear* – be free from suspended matter and turbidity;
4. *Colourless and odourless* – be aesthetic to drink;
5. *Reasonably soft* – to allow consumers to wash clothes, dishes and themselves without excessive use of detergents or soaps;
6. *Non-corrosive* – water should not be corrosive to pipework or encourage leaching of metals from pipes or tanks;
7. *Low in organic content* – a high organic content will encourage unwanted biological growth in pipes or storage tanks, which can affect the quality of the water supplied.

With the publication of drinking water standards such as the European Union Drinking Water Directive (98/83/EEC) (Appendix 1) and the Safe Drinking Water Act (1974) in the USA, which has given rise to the National Primary and Secondary Drinking Water Standards (Appendix 2), water must conform to the standards laid down for a large number of diverse parameters. In England and Wales, for example, the European Directive is enforced by the Water Supply

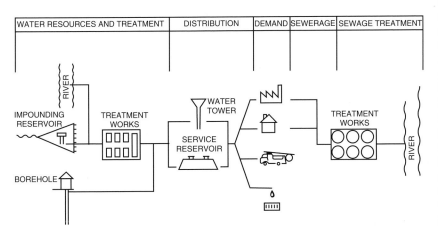

Figure 1.2 Schematic diagram showing the role of the water companies in supplying water to the consumer and subsequently treating it before returning it to the hydrological cycle (Latham, 1990). Demands shown are industrial, domestic, fire-fighting and leakage. Adapted with permission from the Chartered Institution of Water and Environmental Management.

(Water Quality) Regulations (2000), which requires the water supply companies to deliver water to consumers that is wholesome and defines clearly what this term means. Consumers expect clear, wholesome water from their taps 24 hours a day, every day. Although water that is unaesthetic, for example due to colour or turbidity, may be perfectly safe to drink, the consumer will regard it as unpalatable and probably dangerous to health. Problems not only originate from the resources themselves, but during treatment, distribution and within the consumer's home (Chapter 3).

1.4   Water utilities

Water supply has traditionally been a function carried out by state or regional authorities, but throughout Europe, Canada, Australia and the USA this role is increasingly being transferred to the private sector. Regulation and the overall quality control of drinking water remains largely with Governments and their agencies; however, the day-to-day operation is now largely privatized. There is growing concern that there may be a slow globalization of the market with a relatively small number of large companies dominating this vital product; also there is little evidence to support the idea that private companies are any more or less efficient than the public sector in supplying water (Hall and Lobina, 2005). Clearly, it is extremely difficult to generalize, so an example of how drinking water is managed and regulated in a single country, the UK, is given below.

1.4.1   Water undertakers and regulation in the UK

Prior to 1989 a mixture of private companies and public-owned water authorities provided drinking water in England and Wales. Since September of that year all water services have been provided by the private sector. Ten Water Service Companies created by the privatization of the 10 water authorities deliver both water and sewerage services, while 29 water supply companies that had always been in the private sector supplied water only. The number of water supply companies has subsequently been reduced to 16 through a number of amalgamations (Table 1.6).

The situation elsewhere in the British Isles is rather different. In Scotland the three public water authorities, North of Scotland Water, East of Scotland Water and West of Scotland Water were amalgamated to form a single new authority, Scottish Water in April 2002 (www.scottishwater.co.uk). Although answerable to the Scottish Executive it is structured and managed as a private company. Northern Ireland is the only part of the UK where water supply and sewerage provision remains within the public sector. The Water Service is an Executive Agency that was set up after local government reorganization in 1996 within the Department for Regional Development (www.waterni.gov.uk); it became a Government Company in April 2007.

Table 1.6 *List of the 10 water and sewerage companies and 16 water only companies supplying drinking water in England and Wales and their web addresses*

Water and sewerage companies

Anglian Water Services Ltd	www.anglianwater.co.uk
Dwr Cymru Cyfyngedig (Welsh Water)	www.dwrcymru.co.uk
Northumbrian Water Ltd	www.nwl.co.uk
Severn Trent Water Ltd	www.stwater.co.uk
South West Water Ltd	www.southwestwater.co.uk
Southern Water Services Ltd	www.southernwater.co.uk
Thames Water Utilities Ltd	www.thameswater.co.uk
United Utilities Water Plc	www.unitedutilities.com
Wessex Water Services Ltd	www.wessexwater.co.uk
Yorkshire Water Services Ltd	www.yorkshirewater.com

Water only companies

Albion Water Ltd	www.albionwater.co.uk
Bournemouth & West Hampshire Water Plc	www.bwhwater.co.uk
Bristol Water Plc	www.bristolwater.co.uk
Cambridge Water Company Plc	www.cambridge-water.co.uk
Cholderton and District Water Company Ltd	www.water-guide.org.uk/cholderton-water.html
Dee Valley Water Plc	www.deevalleygroup.com/DVW/DVW.htm
Essex & Suffolk Water (Now part of Northumbria Water Ltd)	www.eswater.co.uk
Folkestone & Dover Water Services Ltd	www.fdws.co.uk
Hartlepool Water Plc (Now part of Anglian Water Services Ltd)	www.hartlepoolwater.co.uk
Mid Kent Water Plc	www.midkentwater.co.uk
Portsmouth Water Plc	www.portsmouthwater.co.uk
South East Water Plc	www.southeastwater.co.uk
South Staffordshire Water Plc	www.south-staffs-water.co.uk
Sutton and East Surrey Water Plc	www.waterplc.com
Tendring Hundred Water Services Ltd	www.thws.co.uk
Three Valleys Water Plc	www.3valleys.co.uk

There are three forms of regulation on water utilities, economic, quality and environmental. Economic and financial regulation depends on whether companies are privately or publicly owned, whereas quality and environmental regulation are imposed on all water utilities by the implementation of EC Directives or other legislation, although the rate of implementation and the exact nature of their implementation may differ slightly.

So while largely privatized, UK water utilities are regulated by the government through legislation and standards, and in three key areas by government-appointed organizations. In England and Wales the Water Act 1989 enabled privatization of the water industry, and through its numerous amendments makes the principal regulators the Secretary of State, the Water Services Regulation Authority and the Environment Agency.

The water utilities in England and Wales are different to other business sectors in that they do not have to compete for domestic customers and only compete in a very limited way for industrial customers. For that reason the price they can charge for water is regulated to protect customers from being exploited. This function is carried out by the Water Services Regulation Authority in England and Wales who is charged with protecting customers' interests while ensuring that the privately owned water companies carry out and finance their functions properly. The Water Services Regulation Authority replaced the Office of Water Services (Ofwat) as the Government's statutory watchdog (www. ofwat.gov.uk) in April 2006 in order to bring it in line with the other economic regulators, although the acronym Ofwat has been retained. It is to this authority that consumers ultimately take their complaints and problems relating to pricing and standards of service when they have failed to obtain satisfaction from either the water company itself or through the Consumer Council for Water, which is an independent organization that represents customers' interests. The cost of this is borne by the licence fee paid by the water service companies and statutory water companies (known as appointees) in England and Wales.

More specifically, the Water Services Regulation Authority, in consultation with the Secretary of State for Environment, Food and Rural Affairs, the Welsh Assembly and other interested groups, has the primary duty to (1) ensure that water and sewerage functions are properly carried out in England and Wales and (2) ensure that the water undertakers are able to finance the proper operation of these functions by securing reasonable returns on their capital. Subject to these primary duties the Authority is also responsible to: (i) protect the interests of customers and potential customers in respect of charges (having particular regard to the interests of customers in rural areas and to ensure that in fixing charges there is no undue preference towards, or undue discrimination against, customers or potential customers); (ii) protect the interests of customers and potential customers in respect of other terms of supply, the quality of services (taking into account in particular those who are disabled or of pensionable age) and the benefits that could be secured from the proceeds of the disposal of certain land transferred to companies when the former Regional Water Authorities were privatized; (iii) promote economy and efficiency on the part of utilities in the carrying out of the water and sewerage functions in a sustainable manner; and (iv) facilitate effective competition between persons holding or seeking appointments under the Water Act as water or sewerage utilities.

To act as a buffer between consumers with complaints and Ofwat itself, the Consumer Council for Water (www.ccwater.org.uk) is an independent group that deals with consumer complaints at a personal level when they have not received satisfaction from the Company complaints procedure. Ofwat deals with complaints of a more general nature in relation to pricing, competition and levels of service provided by companies.

Since 1 September 1989 all the water utilities in England and Wales have operated under the terms of an individual appointment, which is in essence a kind of operator's licence. The terms of the appointment set out the maximum charge increase *(K)* for all regulated services, mainly water supply and sewerage charges. Also defined are the circumstances under which unforeseen or previously unquantifiable obligations placed on the companies may be eligible for costs to be passed onto customers, a process known as *cost pass through*. The Water Act enables Ofwat to monitor standards of customer service, which includes the assessment of the state of underground assets such as mains and sewers to ensure that these are protected against progressive deterioration. Among the major financial problems facing the water service companies is the rehabilitation of thousands of kilometres of old and leaking water mains and sewers, and the replacement of lead service pipes.

The Water Act requires each water company to develop and maintain an efficient and economic system of water supply within its area. It must also ensure that all such arrangements have been made for providing supplies of water to premises in its area, making such supplies available to people who demand them and for maintaining, imposing and extending its distribution system as necessary, to enable it to meet its water supply obligation.

In Scotland the Water Industry Commissioner has the same function as Ofwat in relation to Scottish Water and oversees charging policy and service standards. The Commissioner is supported by five regional panels that represent the views of customers. Water services in Northern Ireland are controlled centrally by government who also set charges.

The Drinking Water Inspectorate (DWI) (www.dwi.gov.uk) was formed at the beginning of January 1990 and has nine main tasks: (1) *To carry out technical audits of water companies*. This is a system used by the DWI to check that water supply companies are complying with their statutory obligations and whether they are following good practice. There are three elements to this technical audit. Firstly an annual assessment based on information provided by the companies of the quality of water in each supply zone, water treatment works and service reservoirs, compliance with sampling and other requirements and the progress made on improvement programmes. The second element is the inspection of individual companies covering all the above points at the time of the inspection, but also an assessment of the quality and accuracy of the information collected by the company. The third and final element is interim checks, which are made based on the information provided by the companies. (2) *To instigate action as necessary to secure or facilitate compliance with legal requirements.* (3) *To investigate incidents that adversely affect water quality.* (4) *To advise the Secretary of State in the prosecution of water companies* who have supplied water found to be unfit for human consumption. (5) *To provide technical and scientific advice to Ministers and Officials of the Department for Environment, Food and Rural Affairs (Defra) and Welsh Office on drinking water issues.* (6) *To assess and respond to consumer complaints when local procedures have been exhausted.* Almost all of these problems are resolved by referring the complaint back to the water supply company concerned, requesting them to investigate the matter and report back to the DWI. The DWI may also ask the local environmental health officer to investigate the matter and to report back. The DWI also works closely with Ofwat and will investigate and liaise with them on complaints. Ofwat receives complaints from Consumer Council for Water and these may be passed on to the DWI. (7) *To identify and assess new issues or hazards relating to drinking water quality and initiate research as required.* (8) *To assess chemicals and materials used in connection with water supplies.* The DWI operates a statutory scheme that assesses and approves (if appropriate) the use of chemicals in treating drinking water. This scheme also covers the construction materials used to build water treatment plants and distribution systems. This scheme is to protect the public by ensuring that all chemicals added to water are safe, and that the chemicals which leach from construction materials are also safe. It is also important that such chemicals do not encourage microbial growth in distribution systems that would affect the taste or odour of the water. (9) *To provide authoritative guidance on analytical methods used in the monitoring of drinking water.*

All the water supply companies are inspected annually, although the level of inspection will vary from year to year. In the report that follows the inspection, areas of compliance and non-compliance are identified and recommendations made to ensure full compliance with statutory requirements, which includes prosecution where necessary. The DWI checks the sampling procedures and the location of sampling points to ensure that they are representative of water quality within each water supply zone. About 90 different laboratories analyze water supplies for the water companies. During inspection the procedures used and the training and competence of laboratory staff are examined. The actual results from sampling are also scrutinized, as is the data-handling system used to analyze them, which is done to ensure that the integrity of the data is maintained. The DWI concentrates on those results that have not complied with the National Standards, and ensures that the correct follow-up action has been taken. Every day procedures are checked and all the data recorded on public registers.

The DWI only covers England and Wales so neither Scotland nor Northern Ireland have a specific agency concerned with drinking water quality, which remains the responsibility of government in the form of the Scottish Executive and the Northern Ireland Water Service respectively.

Environmental quality regulation in England and Wales is carried out by the Environment Agency (www.environment-agency.gov.uk), which replaced the National Rivers Authority in April 1996. The local authorities have retained their public health responsibilities as to the wholesomeness of drinking water, and have special powers to deal with private supplies. The regulations referring to water abstractions, impoundments and discharges of wastewaters is the responsibility of the Environment Agency. Their main regulatory functions cover (1) water resources (mostly abstractions, for which they issue licences as well as independently monitoring river quality); (2) pollution control through the issuing of discharge licences, also known as consents, to both industries and sewage treatment works operated by the water service companies; (3) fisheries; (4) land drainage and flood protection; and finally (5) conservation, amenity, recreation and navigation. While Defra also has regulatory functions, it is the activities of the DWI that are of most importance with respect to drinking water.

The Scottish Environment Protection Agency (SEPA) (www.sepa.org.uk) carries out a similar function to the Environment Agency and has the duty to control discharges to rivers and seas, conserve water resources, prevent pollution and promote conservation throughout Scotland. In Northern Ireland, the Water Service is monitored by the Environment and Heritage Service (www. ehsni.gov.uk).

1.4.2   Charges

One of Ofwat's most important tasks is reviewing increases in charges made by the water companies. Under a complex pricing formula the water companies in England and Wales can impose price increases in line with the current rate of inflation plus an individual sum known as the K factor. This extra charge K takes account of the need to finance the major improvements programme and the amount is decided by Ofwat each year. The producer prices construction output index is used when considering cost pass through. For capital expenditure pricing, an index specifically applied to the water industry, the public works non-roads index (PWNRI) is used. The representative weighting in the retail price index (RPI) for gas, telecoms and water are 2.1%, 1.6% and 0.7% respectively. The basic regulation of charge increases (%) is by the formula $RPI + K$, so most charges rise each year by K percentage points more than the annual rise of inflation. Those charges covered by this formula are the basket

Table 1.7 *Price limits, including K, used in the calculation of water charges by water service companies in England and Wales during 2006/7. (Adapted from Ofwat (2006) with permission from Ofwat.) The retail price index (RPI) for this period is 2.43. Unused K is carried forward (U)*

Water Company	K for 2006/7 (%)	Price limit (K+RPI+U) (%)	Actual increase in average charge (%)	U carried forward (%)	U from 2005/6 (%)
Anglian	0.0	2.43	2.43	–	–
Dwr Cymru	3.6	11.09	6.41	4.68	5.06
Northumbrian	3.7	6.16	6.16	–	0.03
Severn Trent	4.8	7.23	6.58	–	–
South West	9.8	12.23	12.23	–	–
Southern	3.9	6.33	6.33	–	–
Thames	2.1	4.53	4.53	–	–
United Utilities	6.4	8.83	8.82	0.01	–
Wessex	4.9	7.33	7.33	–	–
Yorkshire	4.9	7.33	7.33	–	–
Weighted average	4.0				

items, which are unmeasured water supply and sewerage services, measured (metered) water supply and sewerage services, and also trade effluent, which covers most water company charges. The value of K was set for the first time in January 1991 when Ofwat calculated the RPI as 9.7%, compared to 2.43% in 2005, and allowed varying K values for each water undertaker. In recent years the formula $K + \text{RPI} + U$ has been used where U is any unused K the company wishes to carry forward for use in future years (Table 1.7) (Ofwat, 2006). Pricing is now done using a financial model called Aquarius 3 which can be examined on Ofwat's website.

Charging household customers for water and sewerage is based on three different systems: (1) unmetered charges that are either fixed or based on the rateable value (RV) of the customer's property; (2) metered charges that are based on the amount of water recorded by the customer's meter; or (3) assessed charges that are based on the amount of water a customer would be likely to use if he/she had a meter. In England and Wales unmetered consumers have the right to request a water meter, which is fitted free of charge. On the other hand companies have the right to compulsorily install meters where there is a change of occupier or where a customer (1) waters his/her garden using a non-hand-held appliance; (2) has a swimming pool or pond with a capacity $>10\,000$ litres that is replenished automatically; (3) has a bath, spa or Jacuzzi with a capacity >230 litres; (4) has certain types of shower, including power showers; (5) has a water treatment unit that incorporates reverse osmosis; or (6) lives in an area of water scarcity as determined by the Secretary of State.

Table 1.8 *The standing and volumetric charges applied to metered households in England and Wales by the water service companies in 2006/7 and the percentage of households currently metered. Adapted from Ofwat (2006) with permission from Ofwat*

Water Company	Standing charge (£ per year)	Volumetric charge (pence per m^3)	Households metered (%)
Anglian	24.00	106.46	57.4
Anglian SoLow[a]	0.00	138.46	–
Dwr Cymru	27.00	113.93	15.2
Northumbrian	26.40	78.95	39.6
Severn Trent	19.80	108.77	27.5
South West	23.76	124.97	55.3
Southern	24.57	76.00	33.4
Thames	23.00	95.10	22.7
United Utilities	25.00	112.60	21.4
Wessex	17.00	128.76	37.4
Yorkshire	24.05	102.00	30.5
Mean	23.46	104.75[b]	30.3

[a] Anglian SoLow is a metered tariff system for low-volume users.
[b] Excluding Anglian SoLow.

Metered charges comprise of two elements, a fixed standing charge and a volumetric charge based on the amount of water used (Table 1.8). Only two companies have introduced metered tariffs to reward low-volume users. These are Anglian Water (which now owns Hartlepool Water Company) and Mid Kent Water Company. The tariff includes no standing charge but consumers pay a higher volumetric charge. So in order to break even they must use $<75\,\text{m}^3$ per annum (i.e. the volume above which the tariff is no longer beneficial). There are special tariffs for vulnerable groups within the community.

The current mean level of metering is 30.3% (2006/7). Level of metering varies between companies from 15.2% (Dwr Cymru) to 57.4% (Anglian) for the water service companies (Table 1.8) and from 7.8% (Portsmouth) to 64.9% (Tendring Hundred) for the water only companies. The level of metering generally reflects the degree of water scarcity in supply areas.

Ofwat now undertakes a five-yearly price review to determine customer prices. The most recent review was completed in December 2004 to set prices for 2005 to 2010. As only a quarter to a third of domestic customers are currently metered, most pay a flat rate charge. Non-domestic customers are normally metered and like domestic customers pay according to the amount they use plus a standing charge. The average household bill in 2006/7 was £142 for water and £152 for sewerage. The average metered cost was £127 compared to £149 for unmetered supplies. The average unmetered water supply bill varied for this period from £112 (Southern Water) to £169 (South West Water) for the service companies and from £80 (Portsmouth Water) to £167 (Tendring Hundred Water) for the water only companies. So a litre of tap water costs on average about £0.10 or £0.19 including the full cost of sewage treatment.

Together the UK water companies supply drinking water to over 20 million properties and operate 1000 reservoirs and over 2500 water treatment works. Together with 9000 sewage treatment works and in excess of 700 000 kilometres of mains and sewers they have an annual turnover of more than £7 billion making water a very big business indeed. Each year more than £3 billion are invested back into the industry in England and Wales alone.

1.5   Water conservation

1.5.1   Water demand management

Water demand management (WDM), arose from the key principles of the Dublin Statement (ICWE, 1992), which were restated at the Second World Water Forum held at The Hague in 2000. Water demand management, sometimes referred to as demand-side management, uses a range of tools such as conservation, pricing, water-efficient technologies and public education in conjunction with existing water supply infrastructure (i.e. supply-side management) to address the problems of dwindling water supplies and escalating water demand. The shortfall between supply and demand is a worldwide problem that is being exacerbated in many areas by climate change. Although applicable to all water users including industry and the biggest user agriculture, the section below deals only with drinking water supplies. Water demand management has been pioneered in Canada through the POLIS water sustainability project (POLIS, 2005) and in Australia, where the first water use efficiency labelling scheme has recently been introduced (Australian Government, 2005). Specific actions may include the integrated use of conservation measures, metering, charging, building regulations incorporating water use minimization, and the increased water use efficiency of appliances and fixtures. To be successful, WDM also requires the development of new management techniques and structures, the use of decentralized technologies, and a change in user attitudes and behaviour.

Water demand management has identified the need to move away from expensive, unrestrained and ecologically damaging infrastructural development associated with continually increasing water production. Rather it aims to replace traditional engineered solutions with a more sustainable approach where existing resources are used more effectively so there is no longer a need to exploit new surface or ground water resources, or damage existing resources further by increasing abstraction or by the construction of impounding reservoirs. The POLIS project has proposed ten key actions to achieve a more sustainable use of water resources (Table 1.9) (Brandes, 2006). All had to

Table 1.9 *The ten key steps identified by the POLIS project to achieve water sustainability. Adapted from Brandes (2006) with permission from POLIS project, University of Victoria*

Priority area	Actions

Leakage control	Locate and repair leaks:
	    Distribution network
	    Supply pipes
	    Household leaks

Water-efficient appliances and fixtures	Replace the following with water-efficient models:
	    Toilet
	    Showerhead
	    Taps/faucets
	    Washing-machine (laundry)
	    Dishwasher

Implementation of water demand management (WDM)	Creation of permanent WDM staff Integration with existing supply-side management Sufficient financial support Long-term commitment to WDM
Linking water conservation and development	Make water infrastructural funding dependent on WDM Capping local water use so that further development is dependent on offsetting new demand through conservation
Conservation-orientated pricing	Universal metering Volume-based pricing
Planning sustainably	Long-term strategic planning Soft-path approach to planning
Rainwater harvesting	Promote decentralized infrastructure New buildings to rely on rainwater as primary water source Develop new gardening methods
Water reuse	Promote decentralized infrastructure Develop high-profile demonstration projects to build community support
Water-sensitive urban design	Integrate land use decisions and planning with catchment management and water conservation
Education	Development of a water ethic Identify and target high-water-use groups Promote community involvement Promote practical advice and solutions

meet the basic criteria of being technically feasible, broadly applicable, socially acceptable and cost effective compared to normal infrastructural development. These are considered below in detail.

1.5.2   Leak reduction in water mains

All water distribution systems suffer from leaks, with the degree of loss related to the age of the pipework. So as our systems age more water is lost requiring ever increasing investment to mend leaks. In Canada 13% of the total volume of treated water is lost via leaks, while in the UK it is 24% equivalent to 3608 Ml d^{-1}. It is relatively easy to save between 5% and 10% by having a dedicated leak detection and repair service. However, in the UK, the national leakage rate has remained at approximately the same level for a decade even though increasing effort is spent in leak detection and repair each year (Figure 1.1).

There are a number of ways in which leaks can be detected in water networks: water audits, sonic leak detection and passive detection. Water audits compare the amount supplied to the amount consumed using water meters at the supply and householders ends of the distribution network (i.e. integrated metering). The International Water Association (IWA) and the American Water Works Association (AWWA) have jointly developed a sophisticated water audit system that takes into account system-specific features such as pressure and length of pipework. This model allows whole sections of the water supply network to be checked at the same time allowing a water balance to be drawn up so that all water can be accounted for and losses can be identified as either apparent or real (Table 1.10, Fig 1.3). The model adopts a theoretical reference value, the unavoidable annual real loss (UARL), that represents the lowest level of leakage that can be realistically achieved if all of today's best technology could be successfully applied.

$$\mathrm{UARL} = (5.41\,Lm + 0.15\,Nc + 7.5\,Lp) \times P \quad \text{(gallons per day)}$$

where Lm is the length of water mains in miles, Nc the number of service connections, Lp the total length of private (i.e. supply) pipe in miles calculated as $Nc \times$ the average distance from curb stop to customer meter, and P the average pressure in the system in psi. The UARL gives the level of leakage control that utilities should strive to reach. The software can be downloaded free from the AWWA website (www.awwa.org/WaterWiser/waterloss/Docs/WaterAuditSoftware.cfm).

Alternatively the distribution pipework can be tested manually from the surface using a sonic leak detector. A number of water utilities, such as the Las Vegas Valley Water District, use fixed underground noise detection systems that allows subsurface leaks to be rapidly identified and located. This is a proactive approach where leaks can be detected early before the losses become too severe.

Table 1.10 *Components and definitions of the water balance used in the IWA/
AWWA leakage model. Extrapolated from the IWA/AWWA leakage model at
www.awwa.org/WaterWiser/waterloss/Docs/WaterAuditSoftware.cfm with permission from the American Water Works Association*

Water balance component	Definition
System Input Volume	The annual volume input to the water supply system
Authorized Consumption	The annual volume of metered and/or unmetered water taken by registered customers, the water supplier and others who are authorized to do so
Water Losses	The difference between System Input Volume and Authorized Consumption, consisting of Apparent Losses plus Real Losses
Apparent Losses	Unauthorized Consumption, all types of metering inaccuracies and data-handling errors
Real Losses	The annual volumes lost through all types of leaks, breaks and overflows on mains, service reservoirs and service connections, up to the point of customer metering
Revenue Water	Those components of System Input Volume that are billed and produce revenue
Non-Revenue Water (NRW)	The difference between System Input Volume and billed Authorized Consumption (i.e. Revenue Water)

It also allows the integrity of the pipes to be quantified so that replacement of
distribution mains can be prioritized more effectively. However, where water
meters are not installed then leaks may only be detected when water is seen on
the surface or enters the basement of buildings. This reactive approach is known
as passive detection and results in greater water loss and is more expensive as
reactive repairs cannot be managed or anticipated to any great extent.

Leakage detection and repairs to the distribution system do not involve
customers and so are relatively straightforward; leaks in the customer's supply
pipe linking the water main, after the meter, to the house are the responsibility of
the householder. For example in the UK $1024\,\mathrm{Ml\,d^{-1}}$ is lost from leaking supply
pipes after the company meter, and so are not picked up by normal water audits.

Figure 1.3 The IWA/
AWWA Water Balance
model used to determine
leaks and loss of water
from the distribution
system. Volume data
measured normally for a
reference period of 12
months. Extrapolated
from the IWA/AWWA
Lakage model at www.
awwa.org/WaterWiser/
Waterloss/Docs/
WaterAuditSoftware.cfm
with permission from the
American Water Works
Association.

System Input Volume (corrected for known errors)	Authorized Consumption	Billed Authorized Consumption	Billed Metered Consumption (including water exported)	Revenue Water
			Billed Unmetered Consumption	
		Unbilled Authorized Consumption	Unbilled Metered Consumption	Non-Revenue Water (NRW)
			Unbilled Unmetered Consumption	
	Water Losses	Apparent Losses	Unauthorized Consumption	
			Customer Metering Inaccuracies	
			Data-Handling Errors	
		Real Losses	Leakage on Transmission and Distribution Mains	
			Leakage and Overflows at Utility's Storage Tanks	
			Leakage on Service Connections up to point of Customer metering	

Detection of these leaks requires the involvement of the customer with household and business audits saving customers 5% on average on their bills.

The POLIS project recommends that utilities adopt a comprehensive leak detection and system maintenance programme; and adopt integrated metering. This will need a large financial investment by companies and may need to be included into existing regulations or legislation (Brandes, 2006).

Leaks also occur within households through poor maintenance or damage to household plumbing systems. While only metered customers will be paying for this wasted water, it is creating unnecessary demand. Leaking taps and cisterns are not always obvious but can waste significant volumes of water, while it will be even less likely that the householder will realize whether the supply pipe connecting the house to the mains is leaking. If a water meter is installed and accessible then the simplest way to check for leaks is to ensure that all the taps are turned off and that no water-using appliances are running. The meter is then read and again after an hour. If the reading has increased then there is a leaking supply pipe, dripping tap or faulty toilet cistern. If the household is not metered the only alternative is to inspect all the taps and appliances for leaks or check for the sound of water movement in the pipework when all appliances are turned off and taps are fully closed. Modern toilet cisterns rarely have external overflows with excess water discharged into the bowl, so householders are rarely aware if the cistern is overflowing. Toilets can be checked for leaks by putting a little food colouring into the cistern. If the colouring begins to appear in the bowl without flushing, then the ballcock in the cistern needs to be either adjusted or replaced. A leaking toilet can waste more than 60 000 litres of water per year while each dripping tap can waste between 30 and 200 litres of water each day.

1.5.3 Efficient water-using appliances and fixtures

Residential water use has been rising steadily and has been associated with greater ownership and use of certain household appliances. However, mainly as a response to the need to reduce energy usage, these appliances have become increasingly water efficient. So a key area where water demand could be reduced without compromising current standards of living could be through the adoption of water-efficient appliances and fixtures.

With low rainfall and high evaporation rates, Australia is one of the driest continents in the world. With a rapidly expanding population and increasing standard of living, a serious shortfall between demand and supply of water is developing. The Australian Government has responded by the development of a range of water conservation measures to stabilize and potentially reduce demand, including the introduction in July 2006 of a new water efficiency labelling scheme for consumers and manufacturers. The mandatory Water Efficiency Labelling and Standards Scheme (WELS) applies to seven product types that must now be labelled for water efficiency. Using a six star rating system, the labelling scheme is very similar to that currently employed in many parts of the world for energy efficiency of white electrical goods. The new labels give a star rating of between one to six, with six being the best, as well as an actual water consumption figure. The WELS label is very distinctive and must be fixed to the product, and if packaged then it must also appear on the outer packaging. The actual label or the details contained on the label must also appear in any product specifications such as brochures, magazines, advertizements or website promoting a registered product.

The products covered by WELS are washing-machines, dishwashers, showers, toilets, urinals, all taps, except those used over the bath, and flow controllers. All these products must now be tested and rated according to new water efficiency standards, and registered with the National Regulator. Compliance with WELS is monitored by a permanent team of Inspectors who can impose a range of fines and penalties. Under the WELS legislation (i.e. the *Water Efficiency Labelling and Standards Act 2005*) the regulator can compel products to be withdrawn from the market, deregister products and also issue fines for inaccurate advertizing claims.

The majority of water savings from the scheme is expected to be from replacement of inefficient washing-machines and dishwashers. A water-efficient washing-machine only uses 35–45% the water of an older model and it is anticipated that by 2016 WELS-approved washing machines will be saving about 25 600 Ml of water per year. The most efficient dishwashers currently use half the water of average models and it is expected that the WELS scheme will save a further 1200 Ml per annum from improved water use by these machines.

Toilet flushing contributes between 25% and 30% of normal household water usage with between 6 and 12 litres of water used per flush, depending on design and age of the cistern. Dual flush cisterns are increasingly common with normally 6 litres used for full flush and 3 litres for a half flush. Compared to the more traditional single flush action cistern, this represents an average water usage of just 3.8 litres per flush or a 67% overall reduction in water used for toilet flushing by a family of four. The WELS sets performance requirements for toilets including minimum water efficiency targets. The average water consumption for new toilets must not exceed 5.5 litres per flush. In the regulations, the average water consumption of a dual flush cistern is taken to be the average of one full flush and four half flushes. This means dual flush cisterns of 9 litre full flush and 4.5 litre half flush are the least efficient products that can now be sold. Most countries, including the UK, have now adopted a maximum flush volume for toilets of 6 litres. Compost toilets and separation toilets also save significant amounts of water as well as recycling the organic and liquid content of the waste. Urinals are also covered by the scheme, and generally use about 2.2 litres per flush. The most efficient urinals available through the scheme can reduce flush volumes by 30–35% but when used with 'smart control' to reduce unnecessary flushing, then savings in water use can be as high as 50%.

Flow restriction devices on taps and showers are widely employed where the water pressure is variable or where pressure becomes erratic when other taps within the home are turned on. They are ideal in the bathroom where they prevent splashing and also prevent pipe vibration or hammering due to a shock wave of high pressure through the pipework when the tap is suddenly turned off. Flow can be reduced by up to 50% saving as much as $7–10\,l\,min^{-1}$ per tap while in use. There are four options: (1) pressure-limiting valves fitted to the supply pipe that can be set at a maximum value; (2) flow regulators can be fitted to individual taps and are sensitive to varying water pressure giving a flow from the tap at equalized pressure; (3) flow restrictors are cheaper than regulators as they constrain the flow but are not sensitive to pressure variation; (4) aerators are fitted to tap nozzles and reduce the flow and introduce air into the water at the same time. Aerators also make the water more pleasant to use during personal washing. A typical Australian tap discharges water at a rate of between

15 and 18 l min^{-1}. Maximum recommended flow rates are 12 l min^{-1} for laundry and bathroom taps, 9 l min^{-1} for kitchen taps and 6 l min^{-1} for hand basin taps. Taps now have to comply with new WELS specifications to reduce flow rates. Low-flow taps and those fitted with either an aerator or flow restrictor are estimated to reduce flows to less than a third of standard taps, with low-flow and aerating models using as little as 2 l min^{-1}.

Showers are very popular in Australia but have been identified as a major consumer of water, with a standard showerhead using up to 25 l min^{-1}. On average each shower session uses between 120 and 150 litres of water, which could be reduced by 40% if a water-efficient showerhead that uses as little as 6 or 7 l min^{-1} was installed. Replacing old showerheads with a water-efficient model is estimated to save 14 500 litres per household each year in Australia. This could also reduce energy usage due to showering by 47%.

The new labelling system will allow consumers to easily compare the water efficiency of products before they purchase, and individual manufacturers and specific products can be compared on-line at the WELS website (www. waterrating.gov.au). The on-line facility contains exhaustive information about an enormous range of products, both Australian and imported, and allows potential customers to calculate water consumption for each product using their own usage data over variable set periods. It is confidently predicted that labelling will be an incentive for manufacturers to continue to improve the water efficiency of their products.

With the per capita water usage in Australia amongst the highest in the world at 350 l d^{-1}, the Government is hoping to reduce the average domestic water demand in the 7.4 million households throughout the country by 5% or 87 200 Ml each year over the next decade. It is estimated that half the water savings will come from more efficient washing-machines, about 25% from the installation of improved showerheads and 22% from optimum dual flush toilets. The greater efficiency of products will not only save energy used in treatment and supply of drinking water, but also through more efficient usage of hot water in the home.

Apart from water-efficient equipment, the better design of plumbing systems can also contribute to water conservation. For example, using small-bore pipework and reducing the distance between the hot-water cylinder and the most frequently used taps, usually the kitchen, reduces the volume of cold water that must be drawn off each time the hot-water tap is used.

Conservation measures can be classified as either structural, which involves an investment in water-efficient technology, or behavioural, which involves a change in daily habits. Almost all water utilities offer conservation advice to consumers, and although there are local and regional variations in the advice offered there are a number of key actions that are universally recommended. Whatever conservation techniques are adopted, hygiene and public health must not be compromised. What is important, however, is for families to develop a philosophy about the importance of water conservation and to maintain good habits not just at home but also at work, school and on holiday.

The POLIS project has recommended the development and implementation of cost-effective replacement programmes for fixtures, including incentives for the replacement of toilets with dual flush systems. They also recommend the use of legal tools such as water byelaws and building regulations to ensure water-efficient products are installed in new houses. An interesting idea is the requirement of home water audits and retrofits when houses are resold.

The potential of reducing demand by introducing water-efficient appliances and fixtures is enormous, and at the consumer level will reduce both water and energy bills. For example a 120-room hotel that replaced toilets, showers, taps and urinals with new water-efficient systems during its refurbishment achieved a 47% reduction in its water use. What was also very interesting was the high level of customer satisfaction with this sustainable approach (Brandes, 2006).

1.5.4   Adoption of water demand management

The importance of including WDM as part of the overall water supply management programme is crucial. In the past, water engineers have been reluctant to incorporate WDM solutions into long-term planning, preferring to rely on traditional infrastructural solutions, but this is slowly changing. There are three important aspects to the successful implementation of water conservation programmes: appropriate staffing, sufficient financial resources and sufficient time to achieve results (Maddaus and Maddaus, 2006). It is important that suitably trained permanent staff are appointed, and this will include those with an understanding of economics, psychology and education. Success of WDM depends on bringing all stakeholders on board and in particular gaining the support of consumers. Apart from the financial cost of specialist staff, it will also be necessary to fund incentive programmes and to offer financial assistance where required. Timescale is also important and water conservation programmes normally take up to ten years before significant results are achieved, but far less time than required bringing a new reservoir on stream from scratch.

The POLIS recommendations are for utilities to establish a permanent WDM team and fully integrate them with all aspects of water supply management including operations, finance, planning and all strategic decision making (Brandes, 2006).

1.5.5   Linking water conservation to development

The POLIS project has identified that the current system of funding urban water infrastructure does not promote either conservation or innovation. So they have proposed linking water conservation to development by making funding for water infrastructure and development permits contingent on WDM planning. The concept of stabilizing the water footprint, in a similar way to carbon footprints, on a community basis by acknowledging that there is a limit to local growth in terms of available resources is a contentious idea. However, it does not necessarily prevent further development, it merely caps the amount of resources, in this case water, that is available to the community as a whole. This places the burden on developers to offset the water demand resulting from new development by reducing water in existing homes and businesses through water efficiency.

1.5.6 Water charging and levies

Flat-rate charges provide no incentive to conserve water and indeed encourage a wasteful approach to use. In contrast metering can reduce water consumption by between 20% and 45% overall and is seen by the Organisation for Economic Co-operation and Development (OECD) as the single most effective measure to encourage efficiency and reduce water use (OECD, 1987). The full price for water is paid either directly through an annual charge levied by the company, or indirectly through taxation. Indirect funding does nothing to instil either the importance or the value of water and so does not promote its conservation. It is important that the full cost of water is levied directly to customers with incentives that will encourage them to reduce their usage and thereby save money. Pricing must achieve two things, reflect the full cost of supplying water and penalize excessive consumption. To achieve this, all supplies must be metered.

A range of pricing options can be adopted and these are summarized in Table 1.11, although a tiered charging system based on actual volume used appears to be the most effective in reducing water use. In order for access to water to be equitable and to ensure that all basic human water needs are met, a fixed but sufficient volume of water for the size of the household should preferably be provided at either low or no cost to the customer, and subsequent water usage charged at increasingly higher rates as various volume barriers are past. However, installing meters is relatively expensive and requires more staff to administer the new system of charging. Also, unlike flat-rate charges, revenue from meters is less predictable making financial management more problematic. However, putting consumers in control of their own water usage through volume-based charging will drive demand for more efficient appliances and encourage conservation-based behaviour (Want et al., 2005). Not surprisingly, the POLIS project has recommended the introduction of universal metering of supplies and appropriate but equitable volume-based pricing. Metering supplies in the UK has been considered in detail in Section 1.4.

Table 1.11 *The various options for pricing metered water. Adapted from Brandes (2006) with permission from POLIS project, University of Victoria*

Charging system	How it works	Expected impact on demand
Uniform rates	Price per unit volume is constant	Reduces average demand
Increasing block rates	Price per block (i.e. set volume) increases as consumption increases	Reduces both average and peak demand, by providing increasing incentives to reduce waste
Seasonal (drought) rates	Prices during peak periods (e.g. summer) are higher	Sends a stronger signal during periods of high demand or water scarcity
Excess-use rates	Prices significantly higher for above-average use	Targets excessive users thus reducing peak demands
Indoor/outdoor rates	Prices for indoor uses are lower than prices for outdoor uses	Reduces seasonal peak demand, which is mainly from outdoor use, and is considered more responsive to price changes
Feebacks	High water users pay a premium that is distributed to those who use less	Promotes revenue neutrality and provides incentives by penalizing heavy users and rewarding low users

1.5.7   Planning sustainably

In the past, water conservation has been seen as a short-term mechanism to overcome temporary periods of water scarcity or to allow enough time for further expansion of water supplies to meet the deficit between supply and demand. So conservation programmes are normally designed with a terminal life-expectancy of 2–5 years. In contrast, WDM requires a strategic planning approach for the water supply chain that looks at least 20–30 years ahead or more preferably 50 years into the future. Conventional water planning has largely isolated the engineer and planner from the other stakeholders, including the consumer, creating a spiral of increasing demand. The objective is to avoid further infrastructural expansion by making demand-side management as important as supply-side management. Sustainable planning is designed to ensure utilities are able to meet the future demands for water as well as mitigate the effects of climate change and so protect ecological health of resources through the adoption of a more integrated approach using the tools of WDM. This can only be achieved by creating community-based partnerships with utilities and increased stakeholder involvement. An effective water conservation programme should result in water savings as high as 50% of current usage and in theory could even be higher.

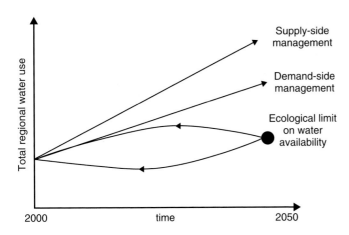

Figure 1.4 Comparison of traditional forecasting using both supply- and demand-side management approaches with soft-path backcasting. Adapted from Brandes, and Brooks (2005) with permission from the POLIS project, University of Victoria.

A key tool in sustainable planning is soft-path planning. Whereas conventional planning treats water solely as an end-product, soft path identifies water as a means to accomplish certain tasks, thus focussing on demand rather than supplying water simply to satisfy demand. So conventional planning uses traditional forecasting to extrapolate future demand from past use, linked to factors such as population trends and economic growth (Section 4.5). Incorporating standard conservation measures simply reduces demand by a fixed percentage and so reduces the slope of the predicted demand on the forecast chart (Figure 1.4). However, in both cases demand continues to rise steadily. In contrast, soft path relies on backcasting, the starting point of which is the desired future end-point in terms of human need and ecological limits. So a future limit is placed on water use based on the potential resources available and on a rate of withdraws that are both ecologically and socially acceptable. The planners then work backwards to find feasible paths to meet long-term social and economic needs (Figure 1.4). The core principles in soft-path water planning are: containing water demand within local eco-hydrological limits; providing services rather than water per se; maximizing productivity of water withdraws; matching quality of water supplied to quality required by end user; open, democratic and participatory planning; and finally, planning backwards to connect a desired future state to present conditions (i.e. backcasting) (Gleick, 2003; Brooks, 2005).

1.5.8   Rainwater harvesting

One of the most effective water conservation measures that can be undertaken is the collection and storage of rainwater for use inside the home for toilet flushing and washing clothes and for outside use such as watering the garden (Section 29.4). Estimates in both the UK and Canada have shown that 45–50% of water used in the home could potentially be replaced by rainwater. This is not a new idea, as many Islands such as Bermuda have relied on rainwater collection systems as the primary source of water in the home for centuries. It is also feasible that rainwater could be used for drinking purposes, but in Europe this would normally require a point-of-use treatment system in order to meet drinking water standards and general quality expectations (Section 29.4).

The key benefits of using rainwater include enhanced local water security; reduced environmental impacts due to reduced demands on central water resources; improved urban stormwater control; and reduced centralized infrastructure needs for water supply, wastewater and stormwater treatment (Brandes, 2006).

1.5.9 Water reuse

A major flaw in our water supply chain is that all water that is supplied must be of drinking water standard, yet less than a third needs to be of this quality. Up to 30% is used to flush toilets and in the summer months large quantities of high-quality water are used to water gardens with only about 5% used for drinking and cooking. The concept of dual water systems, one high quality and one only partially treated has been muted for decades, and in the 1970s many new buildings were constructed with dual plumbing systems to facilitate this futuristic idea. However, the excessive cost of providing two distribution networks and concerns associated with the misuse of microbially unsafe water led to this early initiative being abandoned. The reality is now, as we approach almost universal water scarcity in our cities, that reusing or recycling water for toilet flushing and outdoor irrigation could save up to 50% of supplies during periods of high demand in the summer. The idea of double use is an attractive one as it does not require a second set of water mains supplying partially treated water as was proposed in the 1970s and also prevents open access to the used water. In several southern states in the USA reclaimed water is supplied for outdoor use on a local basis via a separate distribution system, the so-called purple mains (http://srcsd.com/purplepipes/index.htm).

There are three options in relation to 'double use' of drinking water. Reclamation is the direct use of treated wastewater effluents. The use of these effluents is dependent on the degree of treatment given, and while they can be used in theory to flush toilets, they are mostly used for irrigation of municipal parks and golf courses where the added nutrients are considered beneficial. The use of stormwater and surface runoff after simple filtration is also increasingly common. Reuse is the recovery of water within the home that is then reused without treatment. A typical example is the collection of bath, shower and laundry water for toilet flushing. Recycling is where water is used again for the same purpose and is normally associated with industrial processes. So in domestic situations drinking water can only be reused. Domestic wastewater can be separated into a number of different waste streams that offer a range of reuse opportunities. These are summarized in Table 1.12.

Table 1.12 *Different components of domestic wastewater that can be separated and the potential for reuse*

Type	Content	Potential use
Black water	Urine and faeces	None
Brown water	Faeces only	None unless dry composted over several years
Yellow water	Urine only	Can be used as fertilizer in garden
Grey water	Washing water	Flushing toilets
White water	Runoff	Unfiltered: flushing toilets Filtered: laundry, hot water

Reusing used water reduces the amount of water withdrawn by the household from the mains and also the volume of wastewater generated and subsequently needing treatment. Public concerns over health are currently a major limitation on the widescale use of used water by householders, although it is becoming increasingly common practice in industry and new commercial buildings. The low cost of treated water is also a constraint on its use, being reflected in new homes rarely incorporating the necessary infrastructure, this having to be retrofitted at many times the cost.

Garden use currently represents 6% of average water use in the UK. This has increased tenfold in just 30 years and is predicted to double by 2025. In actual per capita terms it does not appear to be significant; however, use in the garden is restricted to periods of greatest overall demand when supplies are at their lowest. As the summers become warmer, then traditional gardens need more frequent watering. This can be offset by reusing grey water from the house or collecting rainwater. Alternatively gardens could be designed to better match the climate, by using drought-resistant plants, creating more shade and replacing lawns with paved areas.

1.5.10   Water-sensitive urban design

The introduction of the Water Framework Directive in Europe was a measure to integrate all activities within catchments to protect the ecological quality of all water resources. This approach will need the support of planners if water conservation is to be fully implemented in the design of houses, businesses and communities. Key issues include high water demand associated with excessive outdoor water use, lack of opportunity to reuse grey and white water, or harvest rainwater due to poor building design, reduced groundwater recharge by stormwater unable to percolate back into the ground due to widespread impermeable surfaces, and finally erosion and flooding due to inadequate stormwater management. Water sustainability is only possible by ensuring that

Table 1.13 *Breakdown of water usage for an average US family. Reproduced with permission from the American Water Works Association*

Use	Gallons per capita	Litres per capita	Per cent of total use
Showers	11.6	43.9	16.8
Clothes Washers	15.0	56.8	21.7
Dishwashers	1.0	3.8	1.4
Toilets	18.5	70.0	26.7
Baths	1.2	4.5	1.7
Leaks	9.5	36.0	13.7
Faucets	10.9	41.3	15.7
Other Domestic Uses	1.6	6.1	2.2

all land use and planning decisions are assessed for impacts on the catchment and water supplies (USEPA, 2006).

1.5.11 Education

The success of water conservation relies very heavily on the public understanding the problems, accepting the need for conservation and actively participating. So education must not only inform it must also inspire and achieve a permanent change in behaviour and a willingness to invest in structural solutions. Education must be supported by realistic technology and a reward system. The POLIS project lists the main benefits of a good education programme as (1) instilling conservation habits; (2) increasing public awareness to the point where other measures such as volume-based pricing are accepted, (3) changing personal attitudes towards water use, creating a lifelong water ethic; and (4) creating a proactive attitude towards water conservation leading to making water sustainability a political issue (Brandes, 2006).

1.5.12 Potential for water conservation in the home

It is generally accepted that only 50–80 litres per capita of high-purity water is required each day to maintain the current standard of living. Through simple conservation measures it is possible to reduce this even further. For example, the indoor water usage in the USA for a typical single family home is 69.3 gallons (262 litres) per capita (Table 1.13). The percentage breakdown is similar to that for the UK, except the volume used by the US family is almost double. Vickers (2001) has studied water use in the USA and estimates that households could reduce their daily per capita water usage by about 35% to just 45.2 gallons (171 litres) per day by installing more efficient water fixtures and regularly checking for leaks. The average US household uses 350 gallons (1325 litres) per day, which is equivalent to 127 400 gallons (482 260 litres) per year. The adoption of water conservation measures up to 1998 saved 44 million gallons (166.6 Ml) of water each day. It is estimated that if all US households installed water-saving features, water use would decrease by 30% overall, saving an estimated 5.4 billion gallons (20 400 Ml) per day. According to Vickers (2001) this would result in dollar-volume savings of $11.3 million per day or more than $4 billion per year!

1.6 Conclusions

Throughout the developed world pressure on water supplies is expected to increase due to rising population, new housing development and reducing household size. Increasing migration within countries, for example increasing population trends in south-east England and Ireland, is leading to serious demand–supply shortfalls. The effects of climate change are predicted to produce more-extreme weather patterns. In the south-east of England, for example, a reduction in summer rainfall of between 30% and 40% is expected, resulting in more frequent exceptionally dry summers. So the long-term forecast is for a reduction in water resources and an increase in demand, a situation that is clearly not sustainable.

Considerable efforts have been made to educate consumers in their water use, and there is now a plethora of good advice widely available to consumers worldwide. Some countries have made significant efforts to reduce water usage, such as the introduction of a mandatory water labelling scheme in Australia for a range of appliances, or setting maximum flush volumes for toilets and the use of banning orders for the use of hosepipes in gardens during droughts in the UK. Metering has been shown to significantly reduce water usage, while the incorporation of new water efficiency targets for water appliances in National Building Regulations (i.e. low and dual flush toilets, water-efficient heating and plumbing systems) in new houses is also making a significant impact on demand. While simple conservation measures can reduce household water usage by up to 40%, water recycling and rainwater harvesting have huge potentials for saving water (Table 1.14). Although extremely costly to retrofit in existing homes, they are easily incorporated into new houses. For most homes in the UK rainwater harvesting could supply enough water for toilet flushing, washing clothes and watering the garden, which represents currently half the water used by most households. Just reusing grey water, collected from the shower, bath and washbasins, to flush the toilet in the home could save up to 18 000 litres per household each year, equivalent to a third of the household water demand (Table 1.14).

For all water utilities leakage control is an on-going and costly battle. Although many have already reached their calculated economic level of leakage (i.e. the level of leakage where it is cheaper to develop new sources of water than to

Table 1.14 *Potential water and energy savings by employing different levels of water conservation action*

	Mains water supplied (l ca^{-1} d^{-1})	Per cent water saving	Power consumed (kWh ca^{-1} yr^{-1})	Per cent energy saving
Standard household	170	–	55	–
Water conservation measures	102	40.0	32	41.8
Water conservation measures plus rainwater harvesting	55	67.7	29	47.3
Water conservation measures plus rainwater harvesting plus grey water recycling	21	87.7	22	60.0

reduce leakage), it is clear that much more could be done to reduce leakage. For example, currently the life-expectancy of distribution pipes is approximately 100 years; however, accelerating the replacement programme would significantly reduce the occurrence of leaks (Environment Agency, 2006).

The urban water cycle (Figure 1.2) of abstraction, storage, treatment and distribution followed by the subsequent treatment and disposal of the wastewater generated all requires energy. It is estimated that the power requirement for water supply and wastewater treatment is equivalent to CO_2 emissions of between 0.3 and 0.8 g equivalent per litre depending on the efficiency of the supplier, although this dramatically increases where desalination or transportation is used. So simple water conservation, and particularly how we use water within the home can significantly help to reduce an individual's carbon footprint (Table 1.14).

Increasing demand for water is seriously affecting water resources, with over-abstraction of groundwater resulting in rivers drying up. Water scarcity is bringing with it a rapid infrastructural expansion for water supply including dam and reservoir construction, river diversion and new irrigation projects. The mismanagement and excessive use of water is destroying river and lake habitats throughout the world, with fish species particularly at risk. To ensure long-term water security in terms of adequate supplies, a community must live within its ecological water budget. This is the only way to ensure that future generations will have the same access to adequate drinking water (Postel, 2000). There exists within Europe, the USA, Canada and Australia strong enabling legislation and water supply infrastructure. What is needed is the widespread adoption of the principles underlying WDM to achieve an ecologically sustainable water supply for future generations.

References

Archibald, G. (1986). Demand forecasting in the water industry. In *Water Demand Forecasting*, ed. V. Gardiner and P. Herrington. Proceedings of a workshop sponsored by the Economic and Social Research Council, University of Leicester.

Australian Government (2005). *Water Efficiency Labelling and Standards Act 2005*. Canberra: Attorney-General's Department.

Bailey, J., Jolly, P. K. and Lacey, R. F. (1986) *Domestic Water Use Patterns*. Technical Report: 225, Water Research Centre, Medmenham.

Brandes, O. M. (2006). *Thinking Beyond Pipes and Pumps*. POLIS Project of Ecological Governance, University of Victoria.

Brandes, O. M. and Brooks, D. B. (2005). *The Soft Path in a Nutshell*. POLIS Project of Ecological Governance, University of Victoria.

Brooks, D. B. (2005). Beyond greater efficiency: the concept of water soft paths. *Canadian Water Resources Journal*, **30**(1), 83–92.

Defra (2006). *Digest of Environmental Statistics*. London: Department for Environment, Food and Rural Affairs. (www.defra.gov.uk/environment/statistics/index.htm).

Environment Agency (2006). *Do We Need Large-scale Water Transfers for South-east England?* Bristol: Environment Agency.

Gleick, P. H. (2003). Global, freshwater resources: soft path solutions for the 21st century. *Science*, **302**, 524–8.

Gray, N. F. (2004) *Biology of Wastewater Treatment*. London: Imperial College Press.

Hall, D. and Lobina, E. (2005). *The Relative Efficiency of Public and Private Sector Water*. PSIRU Report, Public Services International Research Unit, University of Grenwich, London.

ICWE (1992). *The Dublin Statement*. International Conference on Water and the Environment, Dublin.

Latham, B. (1990). *Water Distribution*. London: Chartered Institution of Water and Environmental Management.

Maddaus, W. O. and Maddaus, L. A. (2006). *Water Conservation Programs: a Planning Manual*. AWWA Manual of Water Supply Practice: M52. Denver, CO: American Water Works Association.

National Water Council (1982). *Components of Household Water Demand*. Occasional Technical Paper 6. London: National Water Council.

OECD (1987). *Pricing of Water Services*. Paris: Organisation for Economic Co-operation and Development.

Ofwat (2006). *Water and Sewerage Charges, 2006–7 Report*. Birmingham: Office of Water Services.

POLIS (2005). *At a Watershed*. POLIS Project of Ecological Governance, University of Victoria.

Postel, S. (2000). Entering an era of water scarcity: the challenges ahead. *Ecological Applications*, **10**(4), 941–8.

Smith, A. and Ali, M. (2006). Understanding the impact of cultural and religious water use. *Water and Environment Journal*, **20**, 203–9.

USEPA (2006). *Growing Toward More Efficient Water Use: Linking Development, Infrastruture, and Drinking Water Policies*. Washington DC: US Environmental Protection Agency.

Vickers, A. (2001). *Handbook of Water Use and Conservation*. Amherst, MA: Waterplow Press.

Want, Y. D., Smith, W. J. and Byrne, J. (2005). *Water Conservation-Orientated Rates-Strategies to Extend Supply, Promote Equity, and Meet Minimum Flow Levels*. Denver, CO: American Water Works Association.

Chapter 2
Drinking water standards and risk

2.1 Introduction

The realization of high drinking water quality requires integrated control measures at all points along the supply chain starting with catchment management and the protection of water resources, throughout treatment, storage and distribution, as well as the home plumbing system. Thus, maintaining high-quality drinking water is extremely expensive, and may at times be unnecessary where no threat to human health has been identified. Therefore, drinking water standards must be a compromise between cost and risk to both consumers and the environment. However, water scarcity and sustaining the increasingly high levels of demand may compromise standards, which must be realistic and achievable under local operating conditions.

The World Health Organization (WHO) has proposed a preventative management framework to ensure safe drinking water (Figure 2.1). This comprises five components: (1) health-based targets; (2) assessment of the supply system to ensure that targets can be met on a continuous basis; (3) operational monitoring; (4) assessment and monitoring procedures within a management plan that also incorporates operational and emergency procedures; and finally (5) independent surveillance of the entire system, which feeds back to all the other components of the framework. Also included within this management framework is the constant revision of the published health-based literature in relation to drinking water quality and the effects of individual substances and pathogens found in water. This water safety management framework is universally being adopted both by rich and poor countries alike (Section 2.4).

2.2 Development of quality standards

Ideally water quality standards are based on health-based targets, which in turn are based on a review of the current epidemiological and medical research. For that reason, drinking water quality criteria, from which standards are derived, are constantly being reviewed with standards equally likely to be tightened or relaxed depending on the most reliable information. Health-based targets offer a

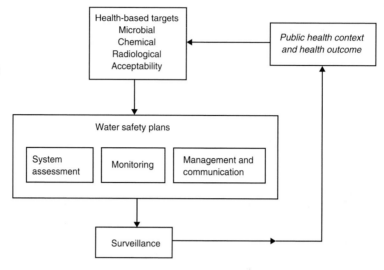

Figure 2.1 The WHO framework for safe drinking water. Adapted from WHO (2004) with permission from the World Health Organization.

wide range of benefits during their formulation, implementation and subsequent evaluation (WHO, 2004). These are outlined below:

Formulation

- Provides insight into the overall health of the community
- Reveals gaps in our current knowledge
- Supports priority setting for contaminants
- Promotes consistency between national health programmes
- Provides insight and transparency to health policy
- Stimulates debate

Implementation

- Inspires and motivates collaborating authorities to take meaningful action
- Improves commitment
- Fosters accountability
- Guides the rational allocation of resources

Evaluation

- Supplies established milestones for incremental improvements
- Provides opportunity to take action to correct deficiencies and/or deviations
- Identifies data needs and discrepancies

Enteric pathogens and hazardous chemicals are the contaminants most associated with drinking water. However, in practice water is rarely the only source, with food, air and person-to-person contact also important sources of contaminants. Therefore, if drinking water is only supplying a small portion of a specific contaminant to the consumer, then introducing unnecessarily strict standards may not only be ineffective in protecting the public but also divert money and effort from other more important contaminants or other areas of health protection.

Health-based targets are difficult to develop due to the variable nature, wide range and differing concentrations of contaminants and pathogens found in drinking water. There are four main types of health-based targets employed by utilities. These are: (1) health outcome targets, (2) water quality targets, (3) performance targets, and finally (4) specified technology targets (Table 2.1). Each differs in respect to the amount of resources required to develop them, and in the benefits that arise from their implementation. Targets based on (1) and (2) require significant scientific and technical support and are more precisely linked to a specific level of health protection than targets based on (3) and (4). These latter targets require little interpretation and are simple to implement, but are based on numerous assumptions derived from data provided from health outcome and water quality targets.

Health outcome targets are based on quantitative risk assessment using exposure and dose–response relationships. This approach has developed from reducing the exposure of the population to a waterborne disease with a resultant measurable reduction of the disease. Now strongly epidemiological based, health outcome targets form the starting point for the development of national drinking water standards. The approach has a limited application used primarily for microbial pathogens and hazardous chemical contaminants that are clearly associated with drinking water such as fluoride. *Water quality targets* are guideline values that have been developed for all drinking water contaminants that normally represent a health risk after long-term exposure. These contaminants do not fluctuate in concentration to any appreciable amount in the short term. *Performance targets* are employed where short-term exposure, or large fluctuations, in usually a microbial pathogen can result in a public health risk. Performance targets set reduction goals in water treatment and distribution to reduce the presence of pathogens to a tolerable risk level. *Specified technology targets* are precise actions that may be specified including operational procedures, protection of wellheads, treatment processes, pipework and chemicals. They are generally employed where supplies serve small communities or even individual homes. These different types of health-based targets are compared in Table 2.1.

2.3 Risk assessment in the development of health-based targets

The contaminants and pathogens associated with drinking water are very diverse in terms of their health effects, the time it takes for symptoms to develop and those who are most at risk. For example, the effects range in severity from minor conditions such as dental fluorosis to very severe life-threatening conditions that include birth defects and cancer. Many effects are rapid such as

Table 2.1 Development of health-based targets. Reproduced from WHO (2004) with permission from the World Health Organization

Type of target	Nature of target	Typical applications	Assessment
Health outcome			
• epidemiology based	Reduction in detected disease incidence or prevalence	Microbial or chemical hazards with high measurable disease burden largely water associated	Public health surveillance and analytical epidemiology
• risk assessment based	Tolerable level of risk from contaminants in drinking water, absolute or as a fraction of the total burden by all exposures	Microbial or chemical hazards in situations where disease burden is low or cannot be measured directly	Quantitative risk assessment
Water quality	Guideline values applied to water quality	Chemical constituents found in source waters	Periodic measurement of key chemical constituents to assess compliance with relevant guideline values
	Guideline values applied in testing procedures for materials and chemicals	Chemical additives and by-products	Testing procedures applied to the materials and chemicals to assess their contribution to drinking water exposure taking account of variations over time
Performance	Generic performance target for removal of groups of microbes	Microbial contaminants	Compliance assessment through system assessment and operational monitoring
	Customized performance targets for removal of groups of microbes	Microbial contaminants	Individually reviewed by public health authority; assessment would then proceed as above
	Guideline values applied to water quality	Threshold chemicals with effects on health that vary widely (e.g., nitrate and cyanobacterial toxins)	Compliance assessment through system assessment and operational monitoring
Specified technology	National authorities specify specific processes to adequately address constituents with health effects (e.g., generic WSPs for an unprotected catchment)	Constituents with health effect in small municipalities and community supplies	Compliance assessment through system assessment and operational monitoring

Note: Each target type is based on those above it in this table, and assumptions with default values are introduced in moving down between target types. These assumptions simplify the application of the target and reduce potential inconsistencies.

diarrhoea or methaemoglobinaemia, while some may take weeks (e.g. infectious hepatitis) or even years (e.g. cancer) to develop. Adverse effects can be caused by a single exposure, as with pathogens, or only after prolonged exposure, which is the case with most chemicals. Young children and the elderly are most at risk from waterborne diseases and chemical contaminants in drinking water. However, cryptosporidiosis, normally a mild and self-limiting infection, causes a high mortality for those with HIV, while the hepatitis E virus has a high mortality amongst pregnant women. While there is a need to quantify the risks associated with all water-related diseases and contaminants, it is extremely difficult to compare such diverse hazards with very diverse health outcomes.

Risk is expressed in terms of specific health outcomes such as a maximum frequency of infection or cancer incidence. In the development of the WHO drinking water guidelines a *reference level of risk* is used to enable comparison between a range of different water-related diseases and contaminants to ensure a consistent approach. The level of risk for each contaminant is calculated using the DALY, which stands for disability-adjusted life-year. Each health effect is weighted in terms of severity from normal good health (0) to death (1). Next the weighting is multiplied by the duration of the disease, this can be either life-expectancy or, more commonly, the time the disease is apparent. Finally, this is multiplied by the number of people affected by a particular outcome (i.e. death or disability):

$$DALY = YLL + YLD$$

where YLL is the number of years of life lost by premature mortality and YLD is years of life lost in a state below normal expected health. Both are standardized by means of the severity weighting. Using this approach the net effects can be compared of various treatments against the original contaminant. The reference level of risk is set at 10^{-6} DALYs per person per year. This is equivalent to a lifetime excess cancer risk of 10^{-5}. In practice this is equivalent to one excess case of cancer per 100 000 of the population drinking water containing the contaminant at the guideline value over a lifetime. As there is a theoretical risk at any level of exposure to a contaminant, then this is deemed as a tolerable level of risk. While risk factors vary locally, in practice health-based targets have largely been derived from epidemiological evidence and to some extent historical precedent. As can be seen in Section 2.5, most countries have adopted health-based targets from international practice and guidance, especially those published by the WHO.

Exposure to contaminants occurs not only through drinking water but also from food, skin contact and inhalation. This varies between countries and is also affected by culture and diet. So in assessing risk it is necessary to make an allocation of the tolerable (or acceptable) daily intake (TDI) of a contaminant specifically through drinking water. For example, if 80% of a contaminant is associated with the diet, then it would normally be pointless to set an excessively strict drinking water standard. Many volatile substances in water will be released into the atmosphere during use, especially showering, while some may be absorbed during washing or bathing. The bacterium *Legionella pneumophila* and asbestos fibres both attack the lungs and so require to be inhaled rather than ingested, which most likely occurs during showering with contaminated water. Such factors should also be taken into consideration. Risk assessment comprises an important step in the development of water safety plans, and is considered further in Section 2.4.

2.4   Water safety plans

Water safety plans (WSPs) have been used for many years to improve water quality control strategies, in conjunction with excreta disposal and personal hygiene, to deliver sustainable health gains within the population (WHO, 2004). These plans use a combination of risk assessment and risk management techniques such as a multi-barrier approach to pathogen control and hazard analysis critical control point (HACCP) principles that are employed primarily by the food industry (Rasco and Bledsoe, 2005). While WSPs are principally used to achieve health-based targets in developing countries, they equally apply to good water management practice and quality assurance systems (e.g. ISO 9001:2000) used in developed countries. The underlying principles used in the development of WSPs are very similar to those employed in water security plans (Chapter 30).

2.4.1   Development of WSPs

Water safety plans protect public health by ensuring safe palatable water through good management practice. This includes the minimization of contamination of water resources, the removal or reduction of contaminants by appropriate treatment, and subsequent prevention of contamination within the distribution mains and the household plumbing system of the consumer. Water supply systems vary significantly both in size and complexity, but each requires a unique WSP.

While the water supplier or utility is the major stakeholder and is responsible for the development of a WSP, other stakeholders, including consumers, must also be involved in its development and subsequent implementation. The WSP provides a framework that allows hazards to be identified, their risk to be assessed and then for a risk management protocol to be developed that includes control measures, the development of monitoring, incident and emergency plans, and the collation of the necessary information about the operation and management of the water supply chain. As outlined in Figure 2.1, a WSP has three key components that are directed by health-based targets and overseen by

Table 2.2 *Summary of the key objectives of a water safety plan (WSP)*

- Complete understanding of each step of the water supply chain from source to tap and its capability to supply water to meet the health-based targets set to protect consumers
- Identifying within each step of the water supply chain potential sources of contaminants and other potential hazards
- Carrying out a risk assessment for each contaminant or hazard present at each step of the supply chain
- Selection of appropriate and effective control measures for each identified risk and the validation of these measures
- Implementation of a system of routine monitoring for each control measure adopted, including the setting of operational targets, which trigger specific remedial actions when exceeded
- Development of remedial actions each with a specific assessment produce to ensure actions have been successful in restoring compliance with health-based target
- Validation monitoring to ensure the water supply system is performing as assumed in the system assessment
- Independent surveillance to ensure WSP is being implemented correctly and that water supplies conform fully with National health-based targets and are safe to drink

---

a surveillance programme (WHO, 2004). These are system assessment, operational monitoring, and management and communication. *System assessment*: it is critical at the outset to know whether the existing water supply chain is capable of satisfying water demands and at the required quality (i.e. meets the health-based targets) at the point of consumption. The assessment must identify all the potential hazards at each step within the supply chain, assess the level of risk each hazard identified represents, and finally identify the appropriate control measures required to reduce or remove this hazard to ensure that the health-based targets are met in full. New design criteria must also be assessed before adoption. *Operational monitoring*: once risks have been assessed and control measures adopted to ensure drinking water meets the health-based targets, then monitoring is required. This must be of an appropriate nature and frequency to be able to rapidly detect any deviation from the required targets. Operational monitoring must be in place throughout the supply chain and clearly defined for each control measure in place. *Management and communication*: this documents the management arrangements, which includes details of system assessment and operational monitoring, as well as collating details of the physical process system and staff responsibilities. It should include detailed procedures for normal operational management and also in the event of an incident or emergency that leads to a risk of non-compliance with the health-based targets. Details of how investigations are to be carried out, how remedial actions are identified and implemented, reporting procedures and communication

Figure 2.2 The key steps in the development of a water safety plan. Reproduced from DWI (2005) with permission from the Drinking Water Inspectorate.

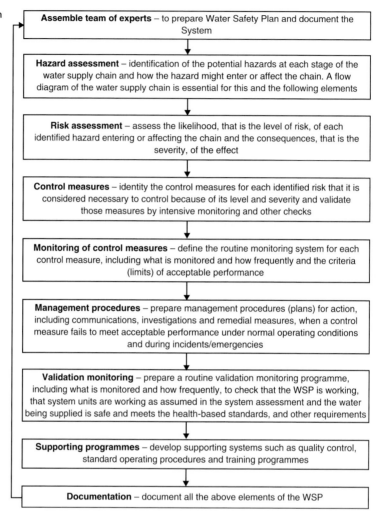

within the utility and with other stakeholders including consumers must also be included (Table 2.2).

The first task in the preparation of a WSP is to appoint a team of experts with a full understanding of each step of the supply chain including catchment and raw water sources, water treatment, distribution network and operations management, physicochemical and microbial aspects of drinking water quality, and domestic plumbing systems. The team will include members of the water utility as well as outside experts and consumer representatives. Next a flow diagram is prepared to describe the key steps in the plan's development (Figure 2.2), which are then carried out by the relevant members of the assembled team. The results of this exercise are the formulation of a WSP, which must be implemented as quickly as possible, then validated. Such plans must be reviewed at regular intervals and also when significant changes are made to any part of the supply system or there has been a system failure that has affected water quality.

2.4.2   Information required for WSP development

The first key step in the development of a WSP is to document the water supply system. In order to carry out the required risk assessment a significant amount of information is required, examples of which are given below.

The starting point of the water supply chain is the catchment. Water quality is largely determined by catchment geology and soils, but contamination arises primarily from land use and other human activities. In order to assess the nature and quality of water resources significant information is required, much of which has been collected in the formulation of River Basin Management Plans required under the Water Framework Directive (2000/60/EC) (Section 2.5). The following information should be taken into consideration when making an assessment of the catchment, surface and ground waters:

Catchment

- Geology and soils
- Hydrology, meteorology and weather patterns
- Nature and intensity of land use including natural land and its wildlife, urbanization, industrial development, intensive and traditional farming, mining and other activities that could release contaminants into the catchment
- Catchment protection and control areas
- Competing water uses including irrigation and river compensation
- Planned future activities that will affect source water

Groundwater

- Unconfined or confined aquifer, the hydrology and recharge area
- Dilution characteristics, flow rate and direction of flow
- Rate of response to surface activities and events
- Detail of wellhead protection including depth of casing
- Rate of abstraction, number and location of boreholes
- Inventory of potentially polluting activities within the recharge area

Surface water

- Description of water body (e.g. river, impounding or storage reservoir, lake, canal)
- Physical characteristics of water body (e.g. size, depth, altitude, thermal stratification)
- Flow and reliability of source including retention times where appropriate
- Water quality and variability with weather patterns and season
- Inventory of point discharges (e.g. sewage and industrial effluents, mining discharge)
- Inventory of potentially polluting diffuse impacts (e.g. agriculture, forestry)
- Recreational and other human activities that could affect quality
- Existing source protection schemes

The most important step in the supply chain is treatment. A systems flow diagram of the treatment system is required initially, which forms the basis of the assessment. Amongst the information required is:

- Details of each treatment unit process, including design and operational management, and what contaminants they are designed to remove
- Details of the monitoring of unit processes, whether discrete, continuous, automatic or manual
- Criteria for deciding that each process is working efficiently
- Water treatment chemicals used including storage and handling
- Disinfection contact time and potential for residuals to be formed
- Monitoring of finished water to test efficiency of treatment, especially in terms of pathogens and toxicity
- Any hazards identified during the assessment of the catchment that cannot be controlled either in the catchment nor removed by existing treatment processes

The distribution system can have significant impacts on water quality and so must be fully evaluated for inclusion in the WSP. It is necessary to include information on service reservoirs and maps of the distribution network including location of all hydrants, valves and connections. This is best achieved using a Geographic Information System. Specific information to be collated includes:

- Service reservoir design, including capacity, compartments, position of inlet/outlet, pumps, construction materials, retention time, protection devices to prevent contamination by unauthorized humans or animals
- Pipework materials used in network including liners
- Hydraulic conditions in network including pressure range, flow rates, retention times.
- Normal methods of operation and permitted variations in operating conditions to satisfy demand
- Condition of network including structural integrity, frequency of bursts, leakage rates, water quality, corrosion, deposition and biofilm development

At the household level, contamination of drinking water has to be considered on a case-by-case basis due to wide variations in the age of the plumbing, the materials used, installation and the varying corrosiveness of the supplied water. Water utilities do not take responsibility for household plumbing and water storage, the proper installation of which is covered in the UK by the Water Byelaws, and elsewhere generally by National Building Regulations (Chapter 26). The information that is relevant to the development of the WSP is:

- Number of connections by category (e.g. domestic, commercial, industrial) and location.
- Details of major users in terms of demand, also water distribution and storage on site.
- Details of connections where water requires pumping to central supply tank (e.g. high-rise buildings, both office and residential).

Table 2.3 *Typical definitions of likelihood and severity categories used in risk scoring*

Category	Definition
a. Likelihood categories	
Almost certain	Once per day
Likely	Once per week
Moderately likely	Once per month
Unlikely	Once per year
Rare	Once every five years
b. Severity categories	
Catastrophic	Potentially lethal to large population
Major	Potentially lethal to small population
Moderate	Potentially harmful to large population
Minor	Potentially harmful to small population
Insignificant	Impact not detectable if any

• Detail of proportion of plumbing materials used in housing areas normally assumed by age of property. Number of homes with lead connection pipes, use of lead or galvanized storage tanks, use of covers for storage tanks.

2.4.3   Hazard identification and risk assessment

The next step in the preparation of a WSP is hazard assessment, which identifies potential hazards and how they enter the supply chain. These hazards must then be prioritized in terms of risk to consumers' health, as not every hazard will require the same degree of attention.

The object of risk assessment is to distinguish between important and less important contaminants or hazardous incidents. It does this by asking two specific questions. What is the chance of the hazardous event occurring? How severe will the consequences be? The likelihood of the health-based targets being exceeded depends on what existing catchment controls are in place and the type of treatment processes that are employed. It must then be quantified using a scoring method (e.g. certain, likely, rare) (Table 2.3, section a). Likewise the severity of the consequences of any potential hazardous incident is also scored (e.g. insignificant, major, catastrophic) (Table 2.3, section b). Priorities are then set for the control of individual contaminants and hazards using Table 2.4. Scoring can be quite subjective and so it is important that it is done by a panel of experts using the most recent scientific and technical information available.

The characteristics of raw water can vary quite significantly due to both natural and human factors within the catchment. This variation in turn affects treatment

Table 2.4 *Simple risk assessment procedure for contaminants and associated hazards in drinking water based on a simple scoring system (Table 2.3). Total score indicates priority for control action. Scores <4 are unlikely to require attention, as the risk is so small*

Likelihood of occurrence	Severity of consequences				
	Insignificant	Minor	Moderate	Major	Catastrophic
Almost certain	5	20	30	40	50
Likely	4	16	24	32	40
Moderately likely	3	12	18	24	30
Unlikely	2	8	12	16	20
Rare	1	4	6	8	10

efficiency and the risk to consumers from drinking the finished water. Examples of typical hazards associated with the catchment and raw water source are (1) rapid variations in raw water quality due to changing meteorological conditions within the catchment; (2) point sources of pollution such as sewage and industrial effluents, septic tank discharges, stormwater overflows, mining activity and landfill leachate; (3) diffuse pollution from land use including agricultural use of chemicals and fertilizers, runoff from tillage, waste from animal rearing and forestry; (4) pollution from recreational activity and wildlife; (5) inadequate protection of surface water resources (e.g. buffer zones, sediment traps) and groundwater supplies (e.g. wellhead protection, inadequately cased boreholes); and (6) short-circuiting, stratification and cyanobacterial blooms in raw water reservoirs. Problems associated with the catchment and raw water sources are dealt with in detail in Part II of this text.

The use of chemicals during treatment will mean that even during optimal operation some traces of these chemicals will remain. Also as the nature of the raw water varies (e.g. colour, acidity) treatment efficiency will also vary. For example, increased turbidity in raw water is common, especially during or after heavy rainfall. This can result in solids not being fully removed during coagulation and filtration and pathogens not being destroyed by the disinfection processes, allowing unsafe water to enter the distribution system. Typical hazards at the treatment stage are: (1) flow and quality variations outside the design limits of the treatment plant; (2) inappropriate or insufficient treatment, including inadequate back-up equipment or staff; (3) failure or poor reliability of process control and monitoring equipment, including alarms; (4) contamination arising from chemical-dosing problems, including inadequate mixing, formation of disinfection by-products; (5) natural disasters, power failures, deliberate acts of vandalism or pollution; and (6) internal cross-contamination between process water and wastewater streams. The problems associated with the water treatment are dealt with in detail in Part III of this text.

The distribution system is made up of service reservoirs and also many kilometres of pipework with numerous connections. This makes the network very vulnerable to vandalism, poor connections and structural damage that can lead to serious microbial or chemical contamination (Part IV). Any contamination at this stage will inevitably reach the consumer. Contamination can arise due to (1) contaminated surface or subsurface water, especially near damaged sewers, entering distribution system due to loss of internal pressure or through the effect of a pressure wave (infiltration/ingress); (2) ingress of contaminated water or other material when the service reservoir or network is opened for repairs; (3) contamination via birds or other medium at open service reservoirs or viaducts, or due to breech of security; (4) back flow of contaminated water from consumer's premises during periods of low pressure or interrupted flow; (5) leaching of contaminants from inappropriate pipework materials or liners; (6) unauthorized tampering of the network and illegal connections, including to hydrants; (7) corrosion of pipework and the build-up of deposits and microbial growths due to inadequate treatment or poor operation; and (8) diffusion of solvents and hydrocarbons through plastic pipes.

Those problems associated with those consumers who rely on private boreholes or who are using rainwater are discussed in Chapter 29.

2.4.4 Control measures

The development of a WSP requires the selection of appropriate control measures to reduce or eliminate the risks identified. The importance of catchment management has long been recognized in preventing contamination of surface and ground water resources that may be used for supply. In Europe this has been recognized by the introduction of the Water Framework Directive. Each control measure must be subject to initial validation as well as subsequent monitoring to ensure effectiveness. Effective resource and source protection is achieved by: (1) the development and implementation of a catchment management plan that includes measures to protect both surface and ground waters; (2) development and implementation of planning regulations that include the protection of water resources against future development; (3) protection of abstraction points; and (4) appropriate management of individual resources.

Water treatment has to be designed specifically for each raw water supply and the potential risks identified. Some contaminants, such as endocrine-disrupting chemicals and the pathogen *Cryptosporidium*, are not easily controlled at the catchment level and are both difficult to remove and monitor at the treatment stage. The control measures required can be summarized as the correct selection and optimal operation of the unit processes required to deal with the risks and contaminants identified, the use of approved water treatment chemicals and materials, and the use of reliable monitoring and back-up systems including alarms.

Water entering the distribution system must be microbially safe, have a low concentration of dissolved organic matter and be non-corrosive. The distribution system must be operated to minimize sudden changes in flow to prevent scouring of biofilm and associated animals, and the resuspension of solids, which can be controlled by routine flushing. Adequate positive pressures must be maintained at all times as pressure failure is the single most serious potential threat to water safety. A disinfectant residual must also be maintained throughout the network, where practicable, to control microbial regrowth in the water and on pipe surfaces. Protocols are required for carrying out repairs and dealing with incidents and emergencies. Where the network has been opened then appropriate disinfection is required before reconnection to supply. The status of the pressure and flow, as well as whether valves are open should be continuously monitored to identify unauthorized connection or use. Unauthorized materials should not be used in the network and replaced if possible when found. A risk assessment must be carried out before significant actions or alterations to the network are authorized.

To prevent leaching of metals from household plumbing, water should be treated with appropriate chemicals to prevent corrosion. Where possible lead connection pipes should be replaced with plastic. Consumers should also be educated to the potential risks and how to minimize them (Part V).

2.4.5   Operational monitoring and management

Once the WSP has been prepared then the whole supply chain must be monitored to ensure that health-based targets are being met and all the control measures are functioning correctly. The concept of operational monitoring is to select the most appropriate parameters to assess the performance of each control measure, and at appropriate intervals that may range from continuous on-line measurement of the chlorine residual to annual inspections of the integrity of the treatment structures (Table 2.5).

The most critical monitoring point should be the consumer's tap, which should be selected to reflect the varying nature of the distribution system and the nature of the homes and premises connected. Key parameters normally included in routine monitoring at consumers' taps are: (1) disinfection residuals, coliforms, *Escherichia coli*, and colony counts; (2) lead, copper, nickel, arsenic, chromium; and (3) taste and odour. Where it can be shown that neither the distribution system nor the consumer's plumbing is affecting quality then routine monitoring can be concentrated at the outlet point of the treatment plant or other more appropriate locations within the supply chain. In practice the major parameters, the location of sampling points and the frequency of sampling, are normally prescribed in national drinking water legislation.

Table 2.5 *Examples of operational monitoring parameters used in monitoring control measures. Reproduced from WHO (2004) with permission from the World Health Organization*

Operational parameter	Raw water	Coagulation	Sedimentation	Filtration	Disinfection	Distribution system
pH		✓	✓		✓	✓
Turbidity (or particle count)	✓	✓	✓	✓	✓	✓
Dissolved oxygen	✓					
Stream/river flow	✓					
Rainfall	✓					
Colour	✓					
Conductivity (total dissolved solids, or TDS)	✓					
Organic carbon	✓		✓			
Algae, algal toxins and metabolites	✓					
Chemical dosage		✓			✓	
Flow rate		✓	✓	✓	✓	
Net charge		✓				
Streaming current value		✓				
Headloss				✓		
Ct^a					✓	
Disinfectant residual					✓	✓
DBPs[b]					✓	✓
Hydraulic pressure						✓

[a] Ct = Disinfectant concentration × contact time.
[b] DBPs = Disinfection by-products.

However, the WSP may identify local and regional variations in water quality, treatment techniques or the nature of the distribution system that may necessitate the selection of additional parameters requiring monitoring. Conversely, the WSP may demonstrate a minimal risk of a particular contaminant exceeding the health-based target; but if it is included in the national standards then it will have to be monitored even though risk assessment has shown this to be unnecessary. Details of national and international standards and sampling protocols are discussed below in Section 2.5.

In conclusion, the WSP must be fully documented to produce a working management plan. The plan must be regularly reviewed and be subject to revision whenever significant material changes are made to the supply chain. It needs to be independently verified through independent surveillance, which in the case of the UK water utilities is done by the Drinking Water Inspectorate.

2.5 Drinking water standards

2.5.1 WHO guidelines

Health-based targets are used to develop international and national drinking water standards including the drinking water guidelines produced by the WHO (2004), which have been widely adopted and provide the health-based targets that form the basis for both the US and EU drinking water standards. Drinking water that complies with these guidelines poses no significant risk to health over a lifetime's consumption, which includes those periods of an individual's lifetime when sensitivities to certain contaminants may be higher (e.g. infancy, pregnancy, old age). In order to take account of the most recent toxicological and scientific evidence these guidelines are constantly being reviewed with new guidelines published approximately every ten years (WHO, 1993, 2004). The current guidelines were published on 21 September 2004 and include microbiological (Table 19.4) and physicochemical parameters (Appendix 3). The revised guidelines also give health-based targets for almost 200 individual radionuclides. However, due to the complexity and expense of identifying and quantifying individual radioactive species, the WHO have recommended screening drinking water for gross alpha and beta activity and taking further action if their guideline values are exceeded (Section 12.4). Health-based guidelines have not been set for a number of chemicals that are no longer considered to pose a significant health risk at the concentrations normally found in drinking water (Table 2.6). Some of these compounds can however cause consumer complaints on aesthetic grounds (i.e. colour, taste, odour, staining).

2.5.2 US National Primary and Secondary Drinking Water Standards

In the USA the Safe Drinking Water Act of 1974, and its subsequent amendments, protects both drinking water and water resources. The US Environmental Protection Agency (USEPA) was set up to oversee this legislation and to develop the necessary standards and regulations. It introduced two types of regulations for drinking waters. Primary Drinking Water Standards are mandatory and cover those contaminants considered to be potentially harmful to health; and are broken down into disinfectants, disinfectant by-products, inorganic chemicals, organic chemicals, micro-organisms and radionuclides (Appendix 2). Secondary Drinking Water Standards are non-mandatory and cover parameters that are not considered to pose a significant

Table 2.6 *Health-related guide values have not been set by the WHO for a number of chemicals that are not considered hazardous at concentrations normally found in drinking water. Some of these compounds may lead to consumer complaints on aesthetic grounds*

Chemicals excluded from guideline value derivation

Chemical	Reason for exclusion
Amitraz	Degrades rapidly in the environment and is not expected to occur at measurable concentrations in drinking water supplies
Beryllium	Unlikely to occur in drinking water
Chlorobenzilate	Unlikely to occur in drinking water
Chlorothalonil	Unlikely to occur in drinking water
Cypermethrin	Unlikely to occur in drinking water
Diazinon	Unlikely to occur in drinking water
Dinoseb	Unlikely to occur in drinking water
Ethylene thiourea	Unlikely to occur in drinking water
Fenamiphos	Unlikely to occur in drinking water
Formothion	Unlikely to occur in drinking water
Hexachlorocyclohexanes (mixed isomers)	Unlikely to occur in drinking water
MCPB	Unlikely to occur in drinking water
Methamidophos	Unlikely to occur in drinking water
Methomyl	Unlikely to occur in drinking water
Mirex	Unlikely to occur in drinking water
Monocrotophos	Has been withdrawn from use in many countries and is unlikely to occur in drinking water
Oxamyl	Unlikely to occur in drinking water
Phorate	Unlikely to occur in drinking water
Propoxur	Unlikely to occur in drinking water
Pyridate	Not persistent and only rarely found in drinking water
Quintozene	Unlikely to occur in drinking water
Toxaphene	Unlikely to occur in drinking water
Triazophos	Unlikely to occur in drinking water
Tributyltin oxide	Unlikely to occur in drinking water
Trichlorfon	Unlikely to occur in drinking water

Chemicals for which guideline values have not been established

Chemical	Reason for not establishing a guideline value
Aluminium	Owing to limitations in the animal data as a model for humans and the uncertainty surrounding the human data, a health-based guideline value cannot be derived; however, practicable levels based on optimization of the coagulation process in drinking water plants using aluminium-based coagulants are derived: $0.1\,mg\,l^{-1}$ or less in large water treatment facilities, and $0.2\,mg\,l^{-1}$ or less in small facilities
Ammonia	Occurs in drinking water at concentrations well below those at which toxic effects may occur

Table 2.6 (cont.)

Chemicals for which guideline values have not been established

Chemical	Reason for exclusion
Asbestos	No consistent evidence that ingested asbestos is hazardous to health
Bentazone	Occurs in drinking water at concentrations well below those at which toxic effects may occur
Bromochloroacetate	Available data inadequate to permit derivation of health-based guideline value
Bromochloroacetonitrile	Available data inadequate to permit derivation of health-based guideline value
Chloride	Not of health concern at levels found in drinking water[a]
Chlorine dioxide	Guideline value not established because of the rapid breakdown of chlorine dioxide and because the chlorite provisional guideline value is adequately protective for potential toxicity from chlorine dioxide
Chloroacetones	Available data inadequate to permit derivation of health-based guideline values for any of the chloroacetones
Chlorophenol, 2-	Available data inadequate to permit derivation of health-based guideline value
Chloropicrin	Available data inadequate to permit derivation of health-based guideline value
Dialkyltins	Available data inadequate to permit derivation of health-based guideline values for any of the dialkyltins
Dibromoacetate	Available data inadequate to permit derivation of health-based guideline value
Dichloramine	Available data inadequate to permit derivation of health-based guideline value
Dichlorobenzene, 1,3-	Toxicological data are insufficient to permit derivation of health-based guideline value
Dichloroethane, 1,1-	Very limited database on toxicity and carcinogenicity
Dichlorophenol, 2,4-	Available data inadequate to permit derivation of health-based guideline value
Dichloropropane, 1,3-	Data insufficient to permit derivation of health-based guideline value
Di(2-ethylhexyl) adipate	Occurs in drinking water at concentrations well below those at which toxic effects may occur
Diquat	Rarely found in drinking water, but may be used as an aquatic herbicide for the control of free-floating and submerged aquatic weeds in ponds, lakes and irrigation ditches
Endosulfan	Occurs in drinking water at concentrations well below those at which toxic effects may occur
Fenitrothion	Occurs in drinking water at concentrations well below those at which toxic effects may occur

Table 2.6 (cont.)

Chemicals for which guideline values have not been established

Chemical	Reason for exclusion
Fluoranthene	Occurs in drinking water at concentrations well below those at which toxic effects may occur
Glyphosate and AMPA	Occurs in drinking water at concentrations well below those at which toxic effects may occur
Hardness	Not of health concern at levels found in drinking water[a]
Heptachlor and heptachlor epoxide	Occurs in drinking water at concentrations well below those at which toxic effects may occur
Hexachlorobenzene	Occurs in drinking water at concentrations well below those at which toxic effects may occur
Hydrogen sulphide	Not of health concern at levels found in drinking water[a]
Inorganic tin	Occurs in drinking water at concentrations well below those at which toxic effects may occur
Iodine	Available data inadequate to permit derivation of health-based guideline value, and lifetime exposure to iodine through water disinfection is unlikely
Iron	Not of health concern at concentrations normally observed in drinking water, and taste and appearance of water are affected below the health-based value
Malathion	Occurs in drinking water at concentrations well below those at which toxic effects may occur
Methyl parathion	Occurs in drinking water at concentrations well below those at which toxic effects may occur
Monobromoacetate	Available data inadequate to permit derivation of health-based guideline value
Monochlorobenzene	Occurs in drinking water at concentrations well below those at which toxic effects may occur, and health-based value would far exceed lowest reported taste and odour threshold
MX	Occurs in drinking water at concentrations well below those at which toxic effects may occur
Parathion	Occurs in drinking water at concentrations well below those at which toxic effects may occur
Permethrin	Occurs in drinking water at concentrations well below those at which toxic effects may occur
pH	Not of health concern at levels found in drinking water[b]
Phenylphenol, 2- and its sodium salt	Occurs in drinking water at concentrations well below those at which toxic effects may occur

Table 2.6 (cont.)

Chemicals for which guideline values have not been established

Chemical	Reason for exclusion
Propanil	Readily transformed into metabolites that are more toxic; a guideline value for the parent compound is considered inappropriate, and there are inadequate data to enable the derivation of guideline values for the metabolites
Silver	Available data inadequate to permit derivation of health-based guideline value
Sodium	Not of health concern at levels found in drinking water[a]
Sulphate	Not of health concern at levels found in drinking water[a]
TDS	Not of health concern at levels found in drinking water[a]
Trichloramine	Available data inadequate to permit derivation of health-based guideline value
Trichloroacetonitrile	Available data inadequate to permit derivation of health-based guideline value
Trichlorobenzenes (total)	Occurs in drinking water at concentrations well below those at which toxic effects may occur, and health-based value would exceed lowest reported odour threshold
Trichloroethane, 1,1,1-	Occurs in drinking water at concentrations well below those at which toxic effects may occur
Zinc	Not of health concern at concentrations normally observed in drinking water[a]

[a] May affect acceptability of drinking water.
[b] An important operational water quality parameter.

risk to health but could affect the aesthetic quality of drinking water or cause minor cosmetic effects (e.g. aluminium, chloride, colour, copper, corrosivity, fluoride, foaming agents, iron, manganese, odour, pH, silver, sulphate, total dissolved solids (TDS) and zinc) (Appendix 2).

The Office of Drinking Water at the USEPA have set two standards for parameters covered in the Primary Drinking Water Standards. Maximum contaminant level goals (MCLGs) are set at the concentration of contaminants in drinking water below which there is no known or expected risk to health. They include a margin of safety and are non-mandatory. All known carcinogens have a MCLG of zero based on the possibility that all such compounds pose a risk of cancer, regardless of exposure level. Maximum contaminant levels (MCLs) are the highest concentration of contaminants allowed and are set as close to the MCLG as feasible using the best treatment technology and taking cost into consideration. These are mandatory standards and outlined in Appendix 2. The

Primary Drinking Water Standards must be complied with by all States, although in practice they may set more stringent standards for certain parameters. The standards cover only public supplies that are defined as those serving 25 or more people or that have at least 15 service connections (AWWA, 1990). While the USEPA has set drinking water standards for more than 90 contaminants, the Safe Drinking Water Act requires them to identify and list unregulated contaminants that may require a regulation in the future. The Contaminant Candidate List (CCL) is constantly being reviewed and the USEPA carries out research on these contaminants to determine whether they should be regulated (www.epa.gov/safewater/ccl/). Selection of contaminants to become regulated is determined by (1) the projected adverse health effects from the contaminant, (2) the occurrence of the contaminant in drinking water and the normal concentrations expected, and (3) whether regulation of the contaminant would realistically reduce any associated health risk. The Act requires that at least five contaminants from each prepared CCL are considered for inclusion into the Primary Drinking Water Standards, a process known as regulatory determination. Using this process nine contaminants from the first CCL were considered. The USEPA concluded that sufficient data and information was available to make the determination not to regulate *Acanthamoeba*, aldrin, dieldrin, hexachlorobutadiene, manganese, metribuzin, naphthalene, sodium and sulfate. The most recent CCL was published in 2005 and contains 51 contaminants including 9 micro-organisms (Appendix 7).

2.5.3 European standards

Most European countries also introduced drinking water standards at the same time as the USA, but these were harmonized by the European Union in 1980 through the introduction of the EC Directive relating to the quality of water intended for human consumption, more widely known as the Drinking Water Directive (80/778/EEC). This Directive also covered water used in the production of food or in the processing or marketing of products intended for human consumption. This first Directive was originally drafted in its final form in 1975 and so when the revised WHO guidelines were published in 1993 the Drinking Water Directive was seen to be both scientifically and technically out of date and was accordingly revised in 1998 (98/83/EEC). The number of listed parameters in the new Drinking Water Directive was reduced from 66 to 48, or 50 for bottled waters, and of these 15 were new parameters. The key changes in water quality standards introduced by the revised Drinking Water Directive are: (1) faecal coliforms are replaced by *E. coli*, and *Pseudomonas aeruginosa* is now measured in bottled water; (2) antimony is reduced from 10 to 5 μg l^{-1}; (3) lead is reduced from 50 to 10 μg l^{-1}. A 15-year transition period is allowed for the replacment of lead distribution pipes; (4) nickel (a precursor for eczema) is reduced from 50 to 20 μg l^{-1}; (5) disinfection by-products and certain flocculants are also included (e.g. trihalomethanes, trichloroethene and tetracholoroethene, bromate, acrylamide etc.); (6) copper is reduced from 3 to $2\,\text{mg l}^{-1}$; (7) maximum permissible concentrations for individual and for total pesticides are retained at $0.1\,\mu\text{g l}^{-1}$ and $0.5\,\mu\text{g l}^{-1}$ respectively with more stringent standards introduced for certain pesticides (i.e. $0.03\,\mu\text{g l}^{-1}$). The new parameters are listed under three Parts: Part A. Microbiological (Table 19.2); Part B. Chemical and Part C. Indicator parameters (Appendix 1).

On 25 December 1998 the new EC Drinking Water Directive came into force with all member states required to transpose the Directive into national legislation within 24 months. Full compliance had to be achieved by 25 December 2003 although exceptions were made for bromate, trihalomethanes and lead for which full compliance is required by 25 December 2008 for the first two parameters and by 25 December 2013 for lead. Interim values of 25, 150 and $25\,\mu\text{g l}^{-1}$ respectively had to be achieved by 25 December 2003 for the three parameters. Some individual member states have opted for stricter standards for certain listed parameters. How this affects the drinking water standards in the UK is summarized in Table 2.7.

Sampling EU drinking water is now done exclusively at consumers' taps, which *must be free from any micro-organisms, parasite or substance which, in numbers or concentrations, constitute a potential danger to human health, as well as meeting the minimum quality standards listed in the Directive.* The Directive introduces some new management functions to ensure better quality drinking water is supplied. Regular check monitoring, using the parameters listed in Part C, ensures that the basic organoleptic and microbial quality is maintained, while audit monitoring is done less frequently to ensure compliance with all the listed quality parameters in the Directive. Exact details of the frequency of monitoring, which depends on the parameter, treatment employed, population served or volume of water supplied by the water supply zone and previous result, and the location of sampling points are all specified in the Directive and National legislation. For example in England and Wales this is fully explained in the Water Supply (Water Quality) Regulations 2001 (National Assembly for Wales, 2001).

All parametric values in the Directive are to be regularly reviewed and where necessary adjusted in accordance with the latest available scientific knowledge. Two principal sources of information will be the WHO Water Quality Guidelines and the Scientific Committee on Toxicology and Ecotoxicology. The Directive also seeks to increase transparency and safety for consumers. This is achieved by (1) ensuring that compliance is at the point of use, (2) obligation on suppliers to report on quality, and (3) an obligation on suppliers to inform the consumer on drinking water quality and measures that they can take to comply with the requirements of the Directive when the non-compliance is because of the domestic distribution system (e.g. internal pipes, plumbing etc.).

The Drinking Water Directive only covers public water supplies while private supplies are excluded (for example springs, wells, etc.). The UK

Table 2.7 *New or tighter drinking water mandatory standards introduced in England and Wales in response to the revised EC Drinking Water Directive (98/83/EEC)*

Parameter	Previous Directive regulatory standard	New Directive mandatory standard	Unit	Comment
Faecal coliforms now *E. coli*	0	0	No.100 ml^{-1}	Parameter name change
Faecal streptococci now *Enterococci*	0	0	No.100 ml^{-1}	Parameter name change
Acrylamide		0.10	μg l^{-1}	Control by product specification
Antimony	10	5.0	μg l^{-1}	
Arsenic	50	10	μg l^{-1}	
Benzene		1	μg l^{-1}	
Benzo 3,4 pyrene	0.01[a]	0.01	μg l^{-1}	
Boron	2.0[a]	1.0	mg l^{-1}	
Bromate		25	μg l^{-1}	By end 2003
		10	μg l^{-1}	By end 2008
Copper	3.0	2.0	mg l^{-1}	
1,2 dichloroethane		3.0	μg l^{-1}	
Epichlorohydrin		0.10	μg l^{-1}	Control by product specification
Lead	50	25[b]	μg l^{-1}	By end 2003
		10[b]	μg l^{-1}	By end 2013
Nickel	50	20	μg l^{-1}	
Nitrate/nitrite		Formula[c]		
Tetrachloroethene	10	Sum of both not	μg l^{-1}	Sum of two
Trichloroethene	30	to exceed 10	μg l^{-1}	substances
Trihalomethanes	100[d]	150	μg l^{-1}	By end 2003
(sum of 4 THMs)		100	μg l^{-1}	By end 2008
Vinyl chloride		0.5	μg l^{-1}	Control by product specification

[a] annual average; [b] weekly average; [c] the formula is [nitrate]/50 + [nitrite]/3 $\leq$ 1; [d] three monthly average.

Water Industry Act 1991 defines a private water supply as any supply of water not provided by a statutorily appointed (licensed) water company. In England and Wales there are over 50 000 private supplies supplying about 350 000 people for domestic purposes. Of these, approximately 30 000 supplies serve a single dwelling. The actual number of people consuming private water supplies at some time is very much greater due to their use at hotels, schools and other isolated locations. The Private Water Supplies Regulations 1991 require local authorities to monitor such supplies to protect public health and, where necessary, demand improvements to be made. However, in practice, private supplies within the European Union fall outside the regulatory system.

Natural mineral and medicinal waters are also excluded from the Drinking Water Directive and are covered by a separate Directive (i.e. Natural Mineral Water Directive 80/777/EEC). However, bottled waters not legally classified as natural mineral water are required to conform to the Drinking Water Directive (Section 29.2).

Water Safety Plans now deal with the entire supply chain, including the catchment and raw water supplies. All European surface and ground waters, which are used for water abstraction, are now covered by Article 7 of the Water Framework Directive (2000/60/EC). This will eventually replace the existing Surface Water Directive (75/440/EEC). Under the Water Framework Directive, water bodies used for the abstraction of water intended for human consumption, and either providing $> 10 \, \text{m}^3 \, \text{d}^{-1}$ or serving more than 50 people, are designated under Article 6 as protected areas. Likewise, those water bodies intended for abstraction in the future are also classified as protected areas. This requires member states to protect such waters to avoid any deterioration in their quality in order to minimize the degree of treatment required. This includes establishing, where necessary, protection zones around water bodies to safeguard quality. The Directive also requires that these waters receive sufficient treatment to ensure that the water supplied conforms to the Drinking Water Directive.

At present, all European surface waters abstracted for human consumption must be classified under the Surface Water Directive. This classifies surface waters into three categories (A1, A2 or A3) based on raw water quality using 46 parameters (Table 2.8). For each category there is a mandatory minimum degree of treatment required (Table 2.9) (Section 14.2).

The methods and sampling frequency for this Directive were not published until 1979 (79/869/EEC), with sampling taking place at designated sites where water is abstracted. Surface water that falls outside the mandatory limits for A_3 waters is normally excluded for use, although it can be blended with better quality water prior to treatment. The Surface Water Directive will continue in force until 2012 when all measures in the first River Basin Management Plans (RBMP) become fully operational.

Table 2.8 *Standards for potable water abstractions set by the EC Surface Water Directive (75/440/EEC). Reproduced with permission from the European Commission*

Parameter (mgl⁻¹ except where noted)	A1 Guide limit	A1 Mandatory limit	A2 Guide limit	A2 Mandatory limit	A3 Guide limit	A3 Mandatory limit
Treatment type	A1		A2		A3	
pH units	6.5–8.5		5.5–9.0		5.5–9.0	
Colour units	10	20	50	100	50	200
Suspended solids	25					
Temperature (°C)	22	25	22	25	22	25
Conductivity (µS cm⁻¹)	1000		1000		1000	
Odour (DN[a])	3		10		20	
Nitrate (as NO₃)	25	50		50		50
Fluoride	0.7–1.0	1.5	0.7–1.7		0.7–1.7	
Iron (soluble)	0.1	0.3	1.0	2.0	1.0	
Manganese	0.05		0.1		1.0	
Copper	0.02	0.05	0.05		1.0	
Zinc	0.5	3.0	1.0	5.0	1.0	5.0
Boron	1.0		1.0		1.0	
Arsenic	0.01	0.05		0.05	0.05	0.1
Cadmium	0.001	0.005	0.001	0.005	0.001	0.005
Chromium (total)		0.05		0.05		0.05
Lead		0.05		0.05		0.05
Selenium		0.01		0.01		0.01
Mercury	0.0005	0.001	0.0005	0.001	0.0005	0.001
Barium		0.1		1.0		1.0
Cyanide		0.05		0.05		0.05
Sulphate	150	250	150	250	150	250
Chloride	200		200		200	
MBAS	0.2		0.2		0.5	
Phosphate (as P₂O₅)	0.4		0.7		0.7	

Table 2.8 (cont.)

Treatment type	A1		A2		A3	
Parameter (mgl^{-1} except where noted)	Guide limit	Mandatory limit	Guide limit	Mandatory limit	Guide limit	Mandatory limit
Phenol		0.001	0.001	0.005	0.01	0.1
Hydrocarbons (ether soluble)		0.05		0.2	0.5	1.0
PAH[b]		0.0002		0.0002		0.001
Pesticides		0.001		0.0025		0.005
COD[c]					30	
BOD[d] (with ATU[e])	<3		<5		<7	
DO[f] per cent saturation	>70		>50		>30	
Nitrogen (Kjeldahl)	1		2		3	
Ammonia (as NH$_4$)	0.05		1	1.5	2	4
Total coliforms/100 ml	50		5000		50 000	
Faecal coliforms/100 ml	20		2000		20 000	
Faecal strepto-cocci/100 ml	20		1000		10 000	
Salmonella	absent in 5 l		absent in 1 l			

Mandatory levels 95% compliance, 5% not complying should not exceed 150% of mandatory level.
[a] DN, dilution number; [b] PAH, polycyclic aromatic hydrocarbons; [c] COD, chemical oxygen demand; [d] BOD, biochemical oxygen demand; [e] ATU, allylthiourea; [f] DO, dissolved oxygen.

Table 2.9 *Categories of surface waters intended for supply and the required level of treatment as specified in the Surface Water Directive (75/440/EEC). Reproduced with permission from the European Commission*

A1	simple physical treatment and disinfection (e.g. rapid filtration and chlorination);
A2	normal physical treatment, chemical treatment and disinfection (e.g. pre-chlorination, coagulation, flocculation, decantation, filtration and final chlorination);
A3	intensive physical and chemical treatment, extended treatment and disinfection (e.g. chlorination to break point, coagulation, flocculation, decantation, filtration, adsorption (activated carbon) and disinfection (ozone or final chlorination).

The national drinking water guidelines for Australia and New Zealand have been prepared by a joint committee of the National Health and Medical Research Council (NHMRC) and the Agriculture and Resource Management Council of Australia and New Zealand (ARMCANZ). The Australian drinking water guidelines and details of the management of water resources and water supplies can be downloaded at: www.waterquality.crc.org.au/AboutDW_ADWG.htm. Details of the regulation and management of drinking water quality in Canada, including drinking water standards can be accessed at: www.hc-sc.gc.ca/ewh-semt/pubs/water-eau/doc_sup-appui/sum_guide-res_recom/index_e.html.

2.6 Conclusions

A guideline value is normally defined as the concentration of a substance at which a tolerable risk to the health of a consumer over a lifetime's consumption is not exceeded. These guidelines apply to all consumers, even though they have been set to protect vulnerable groups within the population. Examples include nitrate to protect bottle-fed infants from methaemoglobinaemia and lead to protect fetuses and infants from birth defects and learning impairment respectively. Many of the values apply to transient problems that may be seasonal such as toxins produced in surface waters by toxic cyanobacteria. How an acceptable or tolerable risk is defined will vary from country to country and will depend on a wide range of socio-economic as well as local factors. Local factors will include environmental, social, cultural and dietary aspects that affect overall exposure to contaminants. For this reason health-based targets are typically national in character, although strongly influenced by the current WHO guidelines. Standards based on modest but realistic goals, where a reduced number of priority contaminants are regulated, are more likely to succeed in providing a reasonable degree of protection to consumers.

Water safety plans are pivotal in protecting water resources and ensure safe and continuous supplies of drinking water to consumers. However, natural water quality can vary quickly and no system is exempt from occasional failure. One of the most serious problems is the increased level of microbial contamination of water sources after rainfall. In many countries the more intense rainfall events that have been occurring due to climate change has led to an increase in the incidence of waterborne disease outbreaks following heavy rain. The very success of WSPs often leads to complacency within utilities and with consumers alike, leading to possible disastrous decisions in relation to reducing costs that can weaken the plan leaving supplies at increased risk of contamination. It is only human nature that benefits from adopting new and often costly treatment technologies are only appreciated after emergencies lead to breaches of quality standards with subsequent health impacts within the community. For that reason water should be considered in the same way as any other food product, with the same strict controls applied.

Drinking water safety should also be viewed in the wider context of better waste management, improved personal hygiene, effective sanitation, better pollution control and safer use of hazardous chemicals. In Europe this will largely be achieved through the New Water Framework Directive and the development of River Basin Management Plans. There are many countries in the world today where water scarcity, rather than quality, is the major issue in relation to health (Section 31.7).

References

AWWA (1990). *Water Quality and Treatment: A Handbook of Community Water Supplies*, 4th edn. New York: McGraw-Hill.

DWI (2005). *A Brief Guide to Drinking Water Safety Plans*. London: Drinking Water Inspectorate.

National Assembly for Wales (2001). *Water, England and Wales: The Water Supply (Water Quality) Regulations 2001*. Statutory Instrument No. 2001:3911. Cardiff: National Assembly of Wales.

Rasco, B. A. and Bledsoe, G. E. (2005). *Bioterrorism and Food Safety*. Boca Raton, FL: CRC Press.

WHO (1993). *Guidelines for Drinking Water Quality*, Vol. 1. *Recommendations*, 2nd edn. Geneva: World Health Organization.

WHO (2004). *Guidelines for Drinking Water Quality*, Vol. 1. *Recommendations*, 3rd edn. Geneva: World Health Organization.

Chapter 3
A quick guide to drinking water problems

3.1 Introduction

The development of water treatment towards the end of the nineteenth and beginning of the twentieth century arose from the need to control the primary pathogens that caused infectious diseases such as cholera, typhoid and paratyphoid. The control of waterborne and water-related diseases remains the primary objective of water treatment today. However, in the 1950s and 1960s the need to control chemical contaminants in drinking water was also realized. This arose from two factors, the greater understanding of the causative toxic effects of some metals such as lead, and the increasing development of organic chemicals and their widespread use in industry and agriculture. The severe impact of some of these organic chemicals on wildlife observed during the 1960s quickly raised fears about their potential toxicity and carcinogenic properties in relation to humans. While pathogens cause a rapid response in consumers (24–48 hours), chemical contaminants in drinking water are more likely to have medium- (1–5 years) to long-term (>10 years) toxic effects, including cancer. This has led to the widespread development and adoption of drinking water quality standards that include microbial, chemical and physical parameters, rather than previously employed microbial-based hygiene standards (Section 2.5). Another milestone in drinking water quality has been its use to promote public health through the addition of fluoride to water to prevent dental caries within the community; although the use of drinking water for community mass-medication has given rise to considerable controversy.

In this book, contamination of drinking water is classified by the point at which individual contaminants enter the supply chain, with each dealt with in depth. Problems arise from the resource due to leaching from natural rocks and soils or from chemicals used by man that subsequently enter the hydrological cycle (Part II), from the treatment of surface and ground waters (Part III), as the water travels from the treatment plant to the consumer via the distribution network (Part IV) or within the household plumbing system (Part V) (Table 3.1).

A quick reference guide to problems from each stage in the supply chain is given in Table 3.2.

Table 3.1 *Main sources of drinking water contamination*

Resource	Natural geology
	Land use
	Pollution
Treatment	Unit process efficiency
	Chemicals added to clarify water
	Chemicals added for consumer protection
Distribution	Material of pipework, coating
	Organisms
	Contamination
Home plumbing	Materials of pipework or tank
	Contamination
	Poor installation

Table 3.2 *Sources of principal drinking water problems. Relevant chapters or sections are given in parenthesis*

Resource based	Arising from water treatment	Arising from distribution system	Arising from home plumbing system
Nitrate (5)	Aluminium (15)	Sediment (21.3)	Lead (27.3)
Pesticides (6.2)	Discolouration (14.4)	Discolouration (21.2)	Copper (27.4)
Industrial solvents (6.3)	Chlorine (14.2)	Asbestos (22)	Zinc (27.5)
Odour and taste (8)	Odour and taste (16)	Odour and taste (21.1)	Odour and taste (28.3)
Iron (9.1)	Iron (14.2)	Iron (21.2)	Fibres (28.2)
Manganese (9.2)	Trihalomethanes (18)	PAHs (23)	Corrosion (27.2)
Pathogens (13)	Pathogens (19)	Pathogens (25)	Pathogens (28.1)
Hardness (10)	Fluoride (17)	Animals/biofilm (24)	
Algal toxins (11)	Nitrite (5)		
Radon/radionuclides (12)	Acrylamide (15)		
Arsenic (9.3)			
PPCP[a] (7.2)			
EDCs[b] (7.3)			

[a] Pharmaceutical and personal care products; [b] Endocrine-disrupting compounds.

3.2 The problems

What are the critical health issues relating to our drinking water? These can be broken down into infectious diseases, cancer, endocrine-disrupting compounds and fertility, mineral content, metals and organic compounds. While there are also aesthetic quality problems such as taste, odour and staining, which can be very problematic for consumers, it is the problems that result in a risk to health that must take priority (Section 2.4).

3.2.1 Infectious diseases

Micro-organisms that cause disease via drinking water are generally known as pathogens, and can be categorized in diminishing size as helminths ($>100\,\mu m$), protozoa ($5-100\,\mu m$), bacteria ($0.5-1.0\,\mu m$), and viruses ($0.01-0.1\,\mu m$) (Table 13.1). They originate from either human or animal faeces and if they are not removed by water treatment and disinfection and so reach the consumer's tap, then they may cause outbreaks of the disease within the community. Pathogens from human faeces are the most serious if they contaminate drinking water supplies (e.g. cholera, typhoid, viral gastroenteritis), however pathogens from animals, including birds, which are known as zoonotic diseases or zoonoses, are also a potentially serious threat to health (e.g. salmonella). Zoonoses can be spread from pets such as cats and dogs, farm animals and even wildlife, so all surface water resources have the potential to be microbially contaminated and so require treatment before supply. Microbial contamination, however, can arise at every stage of the water supply cycle (Table 3.2).

Helminths are rarely found in treated water and are primarily associated with untreated surface waters in developing countries. There are, however, a wide range of protozoans connected with water supplies, the most important being *Cryptosporidium*, *Giardia*, *Cyclospora*, *Naegleria*, *Entamoeba* and *Acanthamoeba*. Bacterial pathogens of faecal origin found in water supplies include *Campylobacter*, *Salmonella*, *Shigella*, *Vibrio* and *Yersinia*; while those of less risk to human health of an environmental origin include *Aeromonas*, *Legionella*, *Mycobacterium* and *Pseudomonas aeruginosa*. Viruses of most concern are adenovirus, enterovirus, hepatitis viruses, Norwalk viruses and rotavirus. These are all considered in detail in Chapter 13.

In poorer countries pathogen transfer can occur in water via a number of different routes (Table 3.3), with the amount of illness within a community affected by the quantity of water available as well as its microbial quality. Water scarcity results in poor personal hygiene, inability to adequately wash cooking utensils or clothes, factors that all lead to high rates of gastroenteritis. So increasing the volume of water in these circumstances will result in health benefits, even if the water is of poor quality (Chapter 30).

3.2.2 Cancer

In 1965 the World Health Organization (WHO) set up the International Association of Research on Cancer (IARC) to co-ordinate and conduct research into the causes of human cancer. Based at Lyon in France it classifies potential

Table 3.3 *Water-related diseases can be classified into four broad groups according to their mode of transmission. Only the waterborne diseases are of importance in terms of drinking water quality*

Type of disease	Mode of transmission	Notes
Waterborne diseases	Person contracts the disease by drinking water contaminated with the disease-causing organism	Only spread by the faecal-oral route
Water-washed diseases	Lack of personal hygiene caused by water scarcity	Transmitted by the faecal-oral route Infections of skin and mucous membranes are non-faecal in origin (e.g. scabies, bacterial and fungal skin infections and trachoma) Other health problems, such as lice and ticks, are also reduced by better hygiene
Water-based diseases	Pathogen spends part of its life cycle in a water snail or other aquatic animal	Diseases are caused by parasitic worms, and include schistosomiasis and dracunculiasis (Guinea-worm)
Water-related insect-borne diseases	Carried by blood-sucking insects that breed in water or by insects that bite near water	Diseases include malaria, yellow fever and dengue fever that may be carried by mosquitoes, and trypanosomiasis (sleeping sickness) carried by tsetse flies

Table 3.4 *Potential carcinogens are classified into five groups by IARC*

Group	Description of carcinogenicity of agent (mixture)
1	Carcinogenic to humans
2A	Probably carcinogenic to humans
2B	Possibly carcinogenic to humans
3	Not classified as to its carcinogenicity to humans
4	Probably not carcinogenic to humans

carcinogens using the available human and animal data into one of five groups (Table 3.4). A list of compounds found in drinking water thought to be carcinogenic are given in Table 3.5, along with their IARC groups; full reports on each compound on which these grouping are based are available on the Internet (www.IARC.fr). The IARC groupings are also given for individual

Table 3.5 *The IARC grouping (Table 3.4) for drinking water contaminants thought to be carcinogenic. The USEPA health-based contaminant level goals (MCLG) are also given. Likely sources are given in Appendix 2 along with the MCLs for each contaminant. Contaminants rated NR indicates that no rating is currently available from the IARC*

Contaminants	IARC Group	MCLG (mg l^{-1})
Inorganic agents		
Antimony	2B	0
Asbestos	1	$-^a$
Beryllium	1	0.004
Cadmium	1	0.005
Chromium	1	0.1
Nitrate	NR	10
Nitrite	NR	1
Organic agents		
Acrylamide	2A	0
Alachlor	NR	0
Chlordane	2B	0
Dibromochloropropane	2B	0
Dichloromethane	2B	0
Dioxin	1	0
Epichlorohydrin	2A	0
Ethylene dibromide	2A	0
Heptachlor	2B	0
Heptachlor epoxide	NR	0
Hexchlorobenzene	2B	0
PAHs (benzo[a]-pyrene)	2A	0
PCBs[b]	2A	0
Phthalate (di[2-ethylhexyl])	2B	0
Simazine	NR	0.004
Tetrachloroethylene	2A	0
Toxaphene	2B	0
Volatile organic agents		
Benzene	1	0
Carbon tetrachloride	2B	0
p-Dichlorobenzene	2B	0.075
1,2-Dichloroethane	2B	0
1,1-Dichloroethylene	NR	0.007
Trichloroethylene	2A	0
Vinyl chloride	1	0
Other interim standards		
Alpha-emitters	NR	0^c

Table 3.5 (cont.)

Contaminants	IARC Group	MCLG (mg l^{-1})
Arsenic	1	0.05
Beta-photon-emitters	NR	0^d
Combined radium 226/228	NR	0^c
Total THMs	NR	0

^{a} 7×10^6 fibres l^{-1}; ^{b} PCBs: polychlorinated biphenyls; ^{c} picocuries l^{-1}; ^{d} millirems yr^{-1}.

compounds and ions in the WHO Drinking Water Guidelines (WHO, 2004). The key drinking water contaminants associated with cancer are arsenic, asbestos, nitrates, organic chemicals and radionuclides (resource based); trihalomethanes (THMs), haloacetates and acrylamides (derived from water treatment); and polycyclic aromatic hydrocarbons (PAHs) and asbestos (associated with the distribution of water) (Table 3.2).

3.2.3   Endocrine-disrupting compounds and fertility

There are now known to be a large number of natural and synthetic compounds that can interfere with the normal functioning of the endocrine system in animals and humans, resulting in a range of developmental, neurobehavioural and reproductive problems (Table 7.1). While it is accepted that these compounds cause reproductive malformation and sex changes in fish and other aquatic organisms, the potential of these compounds, when present in drinking water, to have a similar effect on the human population is still under investigation. While there are numerous industrial oestrogen (estrogen)-mimicking compounds (e.g. alkylphenols, alkylphenol ethoxylates, Bisphenol A, pesticides, dioxins and organotin compounds), it is the presence in drinking water of the natural female steroid hormones oestrogen and 17-oestradiol, and its synthetic analogue ethinyl oestradiol, that are thought to be primarily associated with human fertility problems.

3.2.4   Inorganic compounds

Inorganic compounds and elements arise in drinking water from a variety of sources. As rainfall flows over and through soils and through permeable rocks it dissolves traces of anything with which it comes into contact. So within individual catchments, both surface and ground waters develop a unique chemistry. These anions and cations are measured as total dissolved solids with a generally acceptable upper limit for drinking water of 500 mg l^{-1}. Divalent cations, in particular calcium and magnesium, cause hardness in water with excessive amounts ($>300\,$mg CaCO$_3$ l^{-1}) causing significant aesthetic problems. Similarly, high concentrations of iron and manganese significantly affect taste and stain laundry. Other essential elements are also found naturally in water resources, especially groundwaters, including arsenic, selenium, chromium, copper, molybdenum, nickel, sodium and zinc. Antimony, arsenic, beryllium, cadmium and chromium are all known carcinogens while many other elements are toxic at elevated concentrations (Appendix 2). Both arsenic and fluoride are causing widespread health problems in certain semi-arid countries where some groundwaters have naturally high concentrations of these elements, and where there are no alternative uncontaminated supplies available. Nitrogen, used as an agricultural fertilizer, can also build up in groundwaters and is also to be found in elevated concentrations in surface water during periods of high rainfall. Contamination of drinking water by either nitrate or nitrite poses significant health risks due to infantile methaemoglobinaemia or cancer from *N*-nitroso compounds. Some flocculants such as iron and aluminium salts or polyacrylamides used in water treatment can also contaminate the finished water. Fluoride is added deliberately to finished water to prevent dental caries and so is a very commonly found element in drinking water. Iron can also be leached from ductile iron distribution pipes, but once water enters the household plumbing system then metals such as lead, a serious toxin, copper and zinc can all be leached from pipework and other fittings by corrosion to contaminate drinking water (Table 3.2).

3.2.5   Organic compounds

Drinking water often contains trace amounts of hundreds of organic compounds, both natural and man-made. They derive from a number of different sources and vary significantly in their toxicity. The breakdown products of naturally occurring organic material are usually quite benign, although they can result in aesthetic problems such as colour, odour and taste. In contrast, toxins produced by some blue-green algae (cyanobacteria) are complex organic compounds that are highly toxic and potentially lethal. Natural organic compounds can be quite difficult to remove during treatment requiring extra chemical coagulant to be used. Dissolved organic matter that is not removed at the treatment stage can cause microbial growth in the distribution system. Most of the manufactured organic chemicals arise from their use within catchments resulting in contamination of water resources. These include pesticides, industrial solvents, pharmaceutical and personal care products (PPCPs). Organic contaminants can also be formed during water treatment. For example, natural organic compounds react with chlorine to form disinfection by-products such as trihalomethanes, haloacetic acids, halonitriles, haloaldehydes and chlorophenols, while acryl-amides are used as coagulants during treatment. Treated water can become contaminated by PAHs leached from distribution pipes that have been lined with

Table 3.6 *Most commonly employed point-of-use system for the removal of specific contaminants in drinking water. See Table 27.5 for more details*

Contaminant	Point-of-use systems
Arsenic	Carbon filtration, reverse osmosis
Bacteria	Filtration plus UV disinfection
Chlorine	Carbon filtration
Cryptosporidium	Ultrafiltration, reverse osmosis
Fluoride	Reverse osmosis
Giardia	Ultrafiltration, reverse osmosis
Iron	Carbon filtration
Lead	Carbon filtration
Nitrate	Reverse osmosis
Odour and taste	Carbon filtration
Organics	Carbon filtration, reverse osmosis, ultrafiltration
Pesticides	Carbon filtration, reverse osmosis
Radon	Carbon filtration
Turbidity	Filtration
Volatile organics	Reverse osmosis

coal tar or other bitumen-based materials. While many of these compounds are classified as either carcinogenic or toxic, little is known about the long-term exposure to cocktails of these chemicals at trace levels that are common in many finished drinking waters.

3.3   Conclusions

The concept of safe drinking water on tap is a luxury not shared by the majority of the world's population and taken for granted by the majority of those who have it. More than a billion people have no access to safe drinking water, and over the past 20 years over 2 million people, mainly children, have died unnecessarily each year from water-related diarrhoea. In the developing world it is estimated that 45% of all deaths are due to contaminated drinking water. In these affected countries chemical quality is unimportant compared to the need for pathogen-free water to drink. Safety in this context is relative, and the success in preventing waterborne diseases in the developed world has focussed attention on other contaminants. Yet the risk from microbial pathogens remains ever present in the developed world and a daily challenge for the water treatment engineer and scientist.

The objective of water treatment is to produce water that is wholesome and fit for the purpose it is used for, in this case consumption. The majority of trace elements naturally occurring in water are essential to health as well as giving water a satisfactory flavour, so it is generally undesirable and unnecessary to remove them. Conventional, and where necessary advanced, treatment can produce water of the highest quality and so in the vast majority of cases further treatment is unnecessary. However, some consumers find the taste of chlorine or the idea of fluoride and trace amounts of coagulants added to the water in order to conform to treatment standards, unacceptable. Also contamination during distribution or from household plumbing may also raise concerns or actual problems. In those circumstances, and for those on private supplies, point-of-use treatment systems may be appropriate (Table 3.6)

As monitoring and analytical techniques have improved allowing increasingly lower concentrations to be accurately measured and new compounds to be detected, fresh concerns have arisen about the contaminants in our drinking water causing problems such as infertility, cancer, dementia and heart disease. Also our acceptance of risk based on a lifetime exposure to single compounds rather than the cocktails of hundreds of chemicals that are found in our drinking water has changed, with the result that consumers are demanding ever cleaner and safer drinking water. This greater awareness of consumers has led to escalating sales of bottled water and point-of-use treatment systems, even though drinking water quality has never been so high for the vast number of consumers served by mains supplies.

References

WHO (2004). *Guidelines for Drinking Water Quality*, Vol. 1. *Recommendations*, 3rd edn. Geneva: World Health Organization.

PART II
PROBLEMS WITH THE RESOURCE

Chapter 4
Sources of water

4.1 Introduction

Unlike distilled or deionized water all drinking water supplies will contain a range of dissolved chemical compounds. Even rainwater, which is the purest water found naturally, contains a wide range of ions and cations. In the early days of chemistry, water was known as the universal solvent as a result of its ability to slowly dissolve into solution anything it comes into contact with, from gases to rocks. So as rain falls through the atmosphere, flows over and through the Earth's surface, it is constantly dissolving material, forming a chemical record of its passage from the clouds. Therefore, water supplies have a natural variety in quality, which depends largely on the source of the supply. All our water comes from the water cycle and it is this process that controls our water resources.

4.2 The water cycle

Water is constantly being recycled in a system known as either the water or more correctly the hydrological cycle. Hydrologists are the people who study the chemical and physical nature of water and its movement on and below the ground. While the total volume of water in the world remains constant its quality and availability varies significantly. In terms of total volume, 97.5% of the world's water is saline with 99.99% of this found in the oceans, the remainder making up the salt lakes. This means that only 2.5% of the volume of water in the world is actually non-saline. However, not all of this fresh water is readily available for use by humans. About 75% of this fresh water is currently locked up as ice caps and glaciers, with a further 24% located underground as groundwater, which means that less than 1% of the total fresh water is found in lakes, rivers and the soil. Therefore, only 0.01% of the world's water budget is present in lakes and rivers, with another 0.01% present as soil moisture but unavailable to humans for supply. So although there appears to be a lot of water about, there is in reality very little which is readily available for human use (Table 4.1). Within the cycle, water is constantly moving, driven by solar energy. The sun causes evaporation from the oceans, which forms clouds and precipitation (rainfall). Evaporation also occurs from lakes, rivers and the soil,

Table 4.1 *Total volume of water in the global water cycle*

Type of water	Area (10^3 km^2)	Volume (10^3 km^3)	Percentage of total water
Atmospheric vapour	510 000	13	0.0001
(water equivalent)	(at sea level)		
World ocean	362 033	1 350 400	97.6
Water in land areas	148 067	—	—
Rivers (average channel storage)	—	1.7	0.0001
Fresh water lakes	825	125	0.0094
Saline lakes; inland seas	700	105	0.0076
Soil moisture; vadose water	131 000	150	0.0108
Biological water	131 000	(Negligible)	—
Groundwater	131 000	7 000	0.5060
Ice caps and glaciers	17 000	26 000	1.9250
Total in land areas (rounded)		33 900	2.4590
Total water, all realms (rounded)		1 384 000	100
Cyclic water			
Annual evaporation			
From world ocean		445	0.0320
From land areas		71	0.0050
Total		516	0.0370
Annual precipitation			
On world ocean		412	0.0291
On land areas		104	0.0075
Total		516	0.0370
Annual outflow from land to sea			
River outflow		29.5	0.0021
Calving, melting and deflation from		2.5	0.0002
ice caps			
Groundwater outflow		1.5	0.0001
Total		33.5	0.0024

with plants contributing significant amounts of water by evapotranspiration. Although about 80% of precipitation falls back into the oceans, the remainder falls onto land. It is this water that replenishes the soil and groundwater, feeds the streams and lakes, and provides all the water needed by plants, animals and, of course, humans (Figure 4.1). The cycle is continuous and so water is a renewable resource (Franks, 1987; Cech, 2005). In essence, the more it rains the greater the flow in the rivers and the higher the water table rises as the underground storage areas (i.e. the aquifers) fill with water as it percolates downwards into the earth. Water supplies depend on the rainfall so when the amount of rain decreases then the volume of water available for supply will also decrease, and in cases of severe

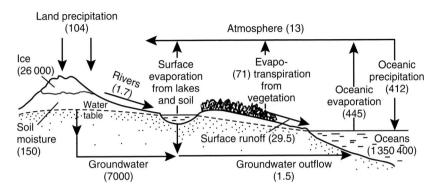

Figure 4.1 Hydrological cycle showing the volume of water stored and the amount cycled annually. Volumes expressed in 10^3 km^3.

drought it may fall to nothing. To provide sufficient water for supply all year round careful management of resources is required.

Nearly all our supplies of fresh water come from the precipitation that falls within a catchment area. Also known as a watershed or river basin, the catchment is the area of land, often bounded by mountains, from which any water that falls into it will drain into a particular river system. A major river catchment will be made up of many smaller subcatchments, each draining into a tributary of the major river. Each subcatchment will be comprised of different rock and soil types, and each will have different land-use activities, which also affect water quality. Therefore the water draining from each subcatchment will be different in terms of chemical quality. As the tributaries enter the main river they mix with water from other subcatchments upstream, constantly altering the chemical composition of the water. Thus water from different areas of the country will be chemically unique.

When precipitation falls into a catchment one of three major fates befalls it. (1) It may remain on the ground as surface moisture and eventually be returned to the atmosphere by evaporation. Alternatively, it may be stored as snow on the surface until the temperature rises sufficiently to melt it. Storage as snow is an important source of drinking water in some regions. For example, throughout Scandinavia lagoons are constructed to collect the runoff from snow as it melts, and this provides the bulk of their annual drinking water supply. (2) Precipitation flows over the surface into small channels to become surface runoff entering streams and lakes. This is the basis of all surface water supplies and will eventually evaporate into the atmosphere, percolate into the soil to become groundwater, or continue as surface flow in rivers back to the sea. (3) The third route is for precipitation to infiltrate the soil and slowly percolate into the ground to become groundwater, which is stored in porous sediments and rocks (Section 4.4). Groundwater may remain in these porous layers for periods ranging from just a few days to possibly millions of years. Eventually groundwater is removed by natural upward capillary movement to the soil surface, plant uptake, groundwater seepage into surface rivers, lakes or directly to the sea, or artificially by pumping from wells and boreholes.

Water supplies therefore come from two principal resources within the water cycle: surface water and groundwater. Each of these resources is interrelated and each has its own advantages and disadvantages as a source of drinking water. Clearly, as water moves through this system of surface and underground pathways its quality is altered, often dramatically, so that the quality of water leaving the catchment will be different from the water that entered it as precipitation.

Our almost exclusive dependence on rainfall to provide drinking water requires careful and long-term management. Although in theory the amount of rain falling on most countries is currently more than adequate to meet all foreseeable needs, there are two practical problems. The first is that more water than is required for immediate needs must be collected and stored during periods of heavy rainfall, usually during the winter, so that this excess can be used to supplement supplies during periods of low rainfall. Secondly, the areas where rainfall is highest are generally the areas of least population, with most of the population, as is the case in the UK, centred in the areas of lowest rainfall. This means transferring water from areas of high rainfall to areas where the demand is greatest, and finding and exploiting as many alternative supplies as possible.

Water for drinking is abstracted from rivers, reservoirs, lakes or underground aquifers (groundwater). In England and Wales, the Environment Agency licenses all abstractions of water greater than $20\,m^3\,d^{-1}$ by the water supply companies, industry, agriculture and private supplies (www.environment-agency.gov.uk/). Some industries require large volumes of water regardless of quality, whereas others treat their own water to the standard required by their process, which is often much higher than drinking water standards. It is therefore common for industry to abstract water directly from a water source. Domestic supplies are treated and are increasingly metered. Currently over 99% of the population in England and Wales are connected to a public water supply. This is comparable with the best in Europe, with the exception of the Netherlands, where nearly 100% of the population is connected. In contrast, nearly all industrial and many commercial supplies are metered. There are currently about 100 000 private supplies in England and Wales, of which only 200 or so supply more than 500 people.

All the major water resources are considered below, with some newer resources and management techniques considered in Section 4.5. The effect of climate change and details of water resources in the USA have been reviewed by Waggoner (1990).

4.3 Surface waters: lakes, reservoirs and rivers

4.3.1 Surface water

Surface water is a general term describing any water body that is found flowing or standing on the surface, such as streams, rivers, ponds, lakes and reservoirs.

Surface waters originate from a combination of sources: (1) *surface runoff*: rainfall that has fallen onto the surrounding land and that flows directly over the surface into the water body; (2) *direct precipitation*: rainfall that falls directly into the water body; (3) *interflow*: excess soil moisture that is constantly draining into the water body; and (4) *water table discharge*: where there is an aquifer below the water body and the water table is high enough, the water will discharge directly from the aquifer into the water body (Bowen, 1982).

The quality and quantity of surface water depends on a combination of climatic and geological factors. The recent pattern of rainfall, for example, is less important in enclosed water bodies such as lakes and reservoirs where water is collected over a long period and stored, whereas in rivers and streams where the water is in a dynamic state of constant movement, then the volume of water is dependent on the preceding weather conditions.

In rivers the discharge rate is generally greater in winter than in summer due to a greater amount and longer duration of rainfall. Short fluctuations in discharge rate, however, are more dependent on the geology of the catchment. Some catchments yield much higher percentages of the rainfall as stream flow than others. Known as the runoff ratio, the rivers of Wales and Scotland can achieve values of up to 80% compared with only 30% in lowland areas in southern England. So although the Thames, for example, has a vast catchment area of $9869 \, \text{km}^2$, it has only half the annual discharge of a river such as the Tay, which has a catchment of only $4584 \, \text{km}^2$. Of course, in Scotland there is a higher rainfall than in south-east England, and also lower evaporation rates due to lower temperatures.

Even a small reduction in the average rainfall in a catchment area (e.g. 20%), may halve the annual discharge from a river. This is why when conditions are only marginally drier than normal a drought situation can readily develop. Severe droughts are quite rare in England and Wales although the recent drought in south-east England of 2005–6 was similar in severity to the celebrated droughts of 1975–6 and 1933–4.

As we have seen in drought areas of the UK, it is not always the case that the more it rains the more water there will be in the rivers. Groundwater is also an important factor in droughts (Beran and Rodier, 1985). In some areas during the dry summer of 1975, despite the rainfall figures, which were far below average, the stream flow in rivers that received a significant groundwater input was higher than normal due to the excessive storage built up in the aquifer over the previous wet winter. The drought beginning in 1989, with three successive dry winters, resulted in a significant reduction in the amount of water stored in aquifers with a subsequent fall in the height of the water table (Section 4.4). This resulted in some of the lowest flows on record in a number of south-eastern rivers in England, with sections completely dried up for the first time in living memory. Before these rivers return to normal discharge levels, the aquifers that feed them must be fully replenished and this may take several years.

Groundwater contributes substantially to the base flow of many lowland rivers, so any steps taken to protect the quality of groundwaters will also indirectly protect surface waters.

Precipitation carries appreciable amounts of solid material to earth such as dust, pollen, ash from volcanoes, bacteria, fungal spores and even, on occasions, larger organisms. The sea is the major source of many salts found dissolved in rain such as chloride, sodium, sulphate, magnesium, calcium and potassium ions. Atmospheric discharges from the home and industry also contribute materials to clouds that are then brought back to earth in precipitation. These include a wide range of chemicals such as organic solvents and the oxides of nitrogen and sulphur that cause acid rain. The amount and type of impurities in precipitation varys with the location and time of year, and can affect both lakes and rivers. Land use, including urbanization and industrialization, significantly affects water quality, with agriculture having the most profound effect on supplies due to its dispersed and extensive nature (Eriksson, 1985).

The quality and quantity of water in surface waters are also dependent on the geology of the catchment. In general, chalk and limestone catchments result in clear, hard waters, whereas impervious rocks such as granite result in turbid, soft waters. Turbidity is caused by fine particles, both inorganic and organic in origin, which are too small to readily settle out of suspension and so the water appears cloudy. The reasons for these differences is that rivers in chalk and limestone areas rise as springs or are fed from aquifers through the river-bed. Because appreciable amounts of the water come from groundwater resources, the river retains a constant clarity, constant flow and indeed a constant temperature throughout the year, except after periods of the most prolonged rainfall. The chemical nature of these rivers is also very stable and rarely alters from year to year. The water has spent a very long time in the aquifer before entering the river and during this time dissolves the calcium and magnesium salts comprising the rock, resulting in hard water with a neutral to alkaline pH. In comparison, soft water rivers usually rise as runoff from mountains, so the flow is linked closely to rainfall. Such rivers suffer from wide fluctuations in flow rate, with sudden floods and droughts. Chemically, these rivers are turbid due to the silt washed into the river with the surface runoff, and because there is little contact with the bedrock they contain low concentrations of cations such as calcium and magnesium, which makes the water soft with a neutral to acidic pH. Such rivers often drain upland peaty soils and the water therefore contains a high concentration of dissolved and colloidal organic matter, giving the water a clear brown-yellow colour, similar to beer in appearance.

As large cities expanded during the nineteenth century they relied on local water resources, but as demand grew they were forced to invest in reservoir schemes, often remote from the point of use. Examples include reservoirs built in Wales, the Pennines and the Lake District to supply major cities such as Birmingham, Manchester and Liverpool, with water in some instances being pumped over 80 km to consumers. Most are storage reservoirs where all the water collected is used for supply purposes. Such reservoirs are sited in upland areas at the headwaters (source) of rivers. Suitable valleys are flooded by damming the main streams. They can take many years to fill and once brought into use for supply purposes must be carefully managed. A balance must be maintained between the water removed for supply and that being replaced by surface runoff. Usually the surface runoff during the winter far exceeds demand for supply so that the excess water can be stored and used to supplement periods when surface runoff is less than the demand from consumers. There is, of course, a finite amount of water in a reservoir and water rationing is often required to prevent storage reservoirs drying up altogether during dry summers. A major problem is when there is a dry winter and so the expected excess of water does not occur, resulting in the reservoirs not being adequately filled at the beginning of the summer. Under these circumstances water shortages may occur even though the summer is not excessively dry.

Most lakes have an input and an output, and so in some ways they can be considered as slowly flowing rivers. The long period of time that water remains in the lake or reservoir ensures that the water becomes cleaner due to bacterial activity removing any organic matter present, and physical flocculation and settlement processes, which remove small particulate material. Storage of water therefore improves the quality, which then reduces the treatment before supply to a minimum (Section 4.2). However, this situation is complicated by two factors. Firstly, in standing waters much larger populations of algae can be supported than in rivers and, secondly, deep lakes and reservoirs may become thermally stratified, particularly during the summer months. These two factors can seriously affect water quality.

Thermal stratification is caused by variations in the density of water in lakes and reservoirs. Water is at its densest at 4 °C, when it weighs exactly 1000 kg m^3. However, either side of this temperature water is less dense (999.87 kg m^3 at 0 °C and 999.73 kg m^3 at 10 °C). During the summer the sun heats the surface of the water reducing its density, so that the colder denser water remains at the bottom of the lake. As the water continues to heat up, then two distinct layers develop. The top layer, or epilimnion, is much warmer than the lower layer, the hypolimnion. Owing to the differences in density, the two layers, separated by a static boundary layer known as the thermocline, do not mix but remain separate (Figure 4.2).

The epilimnion of lakes and reservoirs is constantly being mixed by the wind and so the whole layer is a uniform temperature. As this water is both warm and exposed to sunlight it provides a favourable environment for algae. Usually the various nutrients required by algae for growth, in particular phosphorus and nitrogen, are not present in large amounts (i.e. limiting concentrations). When excess nutrients are present, due to agricultural runoff, for example, then massive algal development may occur in the water (i.e. algal blooms), a

Figure 4.2 Thermal
stratification in deep lakes
and reservoirs.

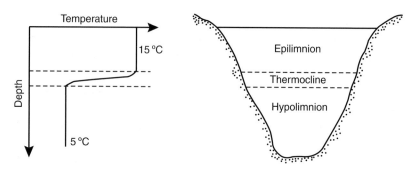

phenomenon known as eutrophication. The algae are completely mixed
throughout the depth of the epilimnion and in severe cases the water can
become highly coloured. This top layer of water is usually clear and full of
oxygen, but if eutrophication occurs then the algae must be removed by
treatment. The algae can result in unpleasant tastes in the water even after
treatment (Chapter 8), and some species can also release toxins (Chapter 11).
Like all plants, algae release oxygen during the day by photosynthesis, but at
night they remove oxygen from the water through respiration. When
eutrophication occurs the high numbers of algae will severely deplete the
oxygen concentration in the water during the hours of darkness, possibly
resulting in fish kills, and certainly causing further problems at the water
treatment plant. In contrast, there is little mixing or movement in the
hypolimnion, which rapidly becomes deoxygenated and stagnant. Dead algae
and organic matter settling from the upper layers are degraded in this lower layer
of the lake. As the hypolimnion has no source of oxygen to replace that already
used, its water may become completely devoid of oxygen. Under anaerobic
conditions iron, manganese, ammonia, sulphides, phosphates and silica are all
released from sediments in the reservoir into the water, whereas nitrate is reduced
to nitrogen gas. This makes the water unfit for supply purposes. For example, iron
and manganese will result in complaints about discoloured water and bad taste.
Ammonia interferes with chlorination, depletes oxygen faster and acts as a
nutrient to encourage eutrophication (as do phosphorus and silica). Sulphides also
deplete oxygen and interfere with chlorination; they also have an awful smell and
impart an obnoxious taste to the water.

The thermocline, the zone separating the two layers, has a tendency to move
slowly to lower depths as the summer progresses. This summer stratification is
usually broken up in autumn or early winter as the air temperature falls and the
temperature of the epilimnion decreases. This increases the density of the water
in the epilimnion to the same density of the water comprising the hypolimnion,
making stratification unstable. Stratification is eventually broken up by wind
action at the surface causing the whole water body to turn over and the layers to
become mixed. Throughout the rest of the year the whole lake remains

completely mixed, resulting in a significant improvement in water quality. Limited stratification can also develop during the winter as surface water temperatures approach $0\,^{\circ}$C while the temperature of the lower waters remains at $4\,^{\circ}$C. This winter stratification is broken up in the spring as the temperature increases and the high winds return. Stratification is mainly a phenomenon of deep lakes and reservoirs.

During stratification each layer or zone has its own characteristic water quality and this can pose serious operational problems to the water supply company. Some are able to abstract water at various depths, thus ensuring that the best quality water or combination of waters is always used. Others try to prevent nutrients entering reservoirs by controlling agricultural activity and other possible sources of nutrients within the catchment area. Occasionally algal growth is controlled by chemical addition. Another method widely used is to pump colder water from the base of the lake to the surface, thus ensuring that the lake does not stratify. The management of reservoirs and lakes for supply is a complex business, dependent on a number of external factors such as air temperature, amount of sunshine, nutrient inputs (both natural and from human activity) and many others. The operation and management of both reservoirs and lakes for abstraction has been reviewed by Cooke *et al.* (2004).

The water supply company usually owns most of the land comprising the catchment area around reservoirs. They impose strict restrictions on farming practice and general land use to ensure that the quality of the water is not threatened by indirect pollution. Restricted access to catchment areas and reservoirs has been relaxed over the past few years, although it is still strictly controlled. This controlled access and restrictions on land use have caused much resentment, especially in Wales where as much as 70% of all the water in upland reservoirs is stored for use in the English Midlands. The problem is not only found in Wales, but in England as well. For example, as much as 30% of the Peak District is occupied by reservoir catchment areas. Water supply companies want to ensure that the water is kept as clean as possible, because water collected in upland reservoirs is of a very high quality. Storage also significantly improves its purity further. The cleaner the raw water, the cheaper it is to treat. Restricted access to catchment areas therefore means that this quality is less likely to be reduced. Conflict is inevitable between those who want access to the land for recreation or other purposes and the water companies, who want to supply water to their consumers at the lowest price possible. There is also an increasing realization of the vulnerability of water resources to deliberate contamination (Chapter 30).

4.3.2   Water abstraction

As a result of the very large capital investment in land and building reservoirs, and the enormous opposition from the public to such schemes on environmental, social and aesthetic grounds, there has been a swing away from the construction of storage reservoirs in recent decades. During this period river abstraction has been exploited along with groundwater resources, although the increasing problem of water scarcity has resulted in renewed interest in reservoir construction.

Water is abstracted from rivers by constructing weirs to ensure a minimum depth of water behind the weir, or by using floating pontoons. Abstraction must not interfere with other river uses such as navigation, but must ensure that water can be removed at all times of the year. In Europe porous banks are used along a short section of the river and the water allowed to filter passively into lagoons where the water is stored before treatment. The amount of water that can be abstracted is limited by the minimum discharge rate required to: (1) protect the biological quality of the river including fisheries; (2) dilute industrial and domestic wastes, as rivers are vital for removing wastes and to a certain extent treating wastes through natural self-purification processes; (3) to ensure other river uses are not affected by abstraction (such as sufficient depth for navigation); and (4) to allow adequate discharge rate to prevent the tide encroaching further upstream and turning freshwater sections brackish. To maintain the integrity of rivers the minimum flow under dry weather conditions must be calculated and maintained at all times. Once calculated, any water in excess of this minimum dry weather flow can in theory be used for abstraction. Conversely, if the discharge rate in the river falls close to or even below this minimum value, then abstraction must be reduced or stopped altogether in order to protect the ecology of the river below the abstraction point.

The quality of river water is also an important factor. River water requires complex and expensive treatment before being supplied to the consumer. The complexity and cost of treatment increases as the quality of the raw water deteriorates. Also, as rivers drain large areas of land, pollution is inevitable. All the waste disposed of, or chemicals used, within a catchment area will eventually find their way into the hydrological system, and so extreme care must be taken to ensure that the water quality is protected and monitored continuously. Most intakes have a storage capability so that raw water can be stored for up to seven days before being treated and supplied. This has a dual function. Firstly, it protects the consumer from the effects of pollution in the river, or accidental spills of toxic materials, allowing sufficient time for the pollution to disperse in the river before abstraction is resumed without cutting off supplies to consumers. Secondly, storing water in this way improves the quality of the water before treatment (Section 14.2).

After supply the water is returned to the river as treated sewage effluent and may well be abstracted again further downstream. This is certainly the case in many major lowland rivers such as the Thames, Severn and Trent (Section 4.5). However, as demands have continued to grow the natural flows of many rivers have proved inadequate to meet the volumes currently needed for abstraction. Also, the quality of many rivers has deteriorated through our exploitation of rivers as receiving waters for effluents.

To maximize water availability for supply, hydrologists examine the hydrological cycle within the catchment, measuring rainfall, discharge rate and surface runoff, and where applicable groundwater supplies. Often they can supplement water abstracted from rivers at periods of very low discharge by taking water from other resources such as groundwater or small storage reservoirs, using these limited resources to top up the primary source of supply at the most critical times. More common is the construction of reservoirs at the headwaters, which can then be used to control the discharge rate in the river itself, a process known as compensation. Compensation reservoirs are designed as an integral part of the river system. Water is collected as surface runoff from upland areas and stored during wet periods. The water is then released when needed to ensure that the minimum dry weather flow is maintained downstream to allow abstraction to continue. In winter, when most precipitation falls and high discharge rates are generated in the river, all this excess water is lost. Storing the excess water by constructing a reservoir and using it to regulate the flow in the discharge rate maximizes the output from the catchment area. A bonus is that such reservoirs can also play an important function in flood prevention. The natural river channel itself is used as the distribution system for the water, unlike supply reservoirs where expensive pipelines or aqueducts are required to transport the water to the point of use. River management is also easier because most of the water abstracted is returned to the same river. Among the more important UK rivers that are compensated are the Dee, Severn and Tees.

Reservoirs are not a new idea and were widely built to control the depth in canals and navigable rivers. Smaller reservoirs, often called header pools, were built to feed mill races to drive water wheels. Without a reservoir, there are times after dry spells where there is negligible discharge from the soil so that the only flow in the river will be that coming from groundwater seeping out of the underlying aquifer. Some rivers, rising in areas of permeable rock, may even dry up completely in severe droughts. In many rivers the natural minimum rate of discharge is about 10% of the average stream discharge rate. Where river regulation is used the minimum dry weather flow is often doubled, and although this could in theory be increased even further, it would require an enormous reservoir capacity.

Reservoirs are very expensive to construct. However, there are significant advantages to regulating rivers using compensation reservoirs rather than supplying water directly from a reservoir via an aqueduct or pipeline. With river regulation much more water is available to meet different demands than from the stored volume only, as the reservoir is only fed by the upland section of the river contained by the dam. Downstream all the water draining into the system is also available. Compensation reservoirs are also generally much smaller in size and so cheaper to construct.

Reservoirs can only yield a limited supply of water and so the management of the river system to ensure adequate supplies every year is difficult. Owing to the expense of construction, reservoirs are designed to provide adequate supplies for most dry summers. However, it is not cost effective to build a reservoir large enough to cope with the severest droughts that may only occur once or twice a century. Water is released from the reservoir to ensure the predicted minimum dry weather flow. Where more than one compensation reservoir is available within the catchment area, water will first be released from those that refill quickly. Water regulation is a difficult task requiring operators to make intuitive guesses as to what the weather may do over the next few months, even with all the new software and predictive forecasting models now at their disposal. For example, many water companies were severely criticized for maintaining water restrictions throughout the winter of 1990–1 in order to replenish reservoirs that failed to completely fill during the previous dry winter. Imposing bans may ensure sufficient supplies for essential uses throughout a dry summer and autumn; however, if it turns out to be a wet summer after all, such restrictions will be deemed by the consumers to have been quite unnecessary. Although water planners have complex computer models to help them predict patterns in water use and so to plan the best use of available resources, it is all too often impossible to match supplies with demand. This has mainly been a problem in south-east England, where the demand is greatest due to a high population density and also a high industrial and agricultural demand, but where the least rainfall is recorded.

Dry winters are often more of a problem than dry summers as reservoirs are not fully replenished. This is, of course, a major problem with storage reservoirs, where most of the water for supply in the summer and autumn is collected during the winter. If the reservoir is not full by the beginning of spring, restrictions in supply are almost inevitable. Similar problems occur with compensation reservoirs, for if the winter is dry the reservoir will be required to augment low flows early in the summer and so will be dangerously depleted if required to continue augmenting the flow throughout the rest of the summer. Where reservoirs are used to prevent flooding there needs to be room to store the winter flood water. This may mean allowing the level of the reservoir to fall deliberately in the autumn and early winter to allow sufficient capacity to contain any flood water. This is the practice at the Clywedog reservoir on the upper Severn. However, if the winter happens to be drier than expected, then the reservoir may be only partially full at the beginning of the summer. Operating reservoirs and regulating rivers is a delicate art, and as the weather is so unpredictable decisions made on the best available information many months previously may prove to have been incorrect (Parr *et al.*, 1992). Further information about water resources can be found in Cech (2005).

4.4 Groundwater sources

4.4.1 Groundwater supplies

About a quarter of potable water supplies in Britain come from groundwater resources, although in other countries the dependence on groundwater is much greater (Tables 4.2 and 4.3). Economically groundwater is much cheaper than surface water, as it is available at the point of demand at relatively little cost and it does not require the construction of reservoirs or long pipelines. It is usually of good quality, usually free from suspended solids and, except in limited areas where it has been affected by pollution, free from bacteria and other pathogens. Therefore it does not require extensive treatment before use (Section 14.2).

British groundwater is held in three major aquifer systems, with most of the important aquifers lying south-east of a line joining Newcastle upon Tyne and Torquay (Figure 4.3). An aquifer is an underground water-bearing layer of porous rock through which water can flow after it has passed downwards (infiltration) through the upper layers of soil. On average 7000 Ml of water are abstracted from these aquifers each day. Approximately 50% of this vast amount of water comes from Cretaceous chalk aquifers, 35% from Triassic sandstones and the remainder from smaller aquifers, the most important of these being Jurassic limestones. Of course, groundwater is not only abstracted directly for supply purposes; it often makes a significant contribution to rivers also used for supply by discharging into the river as either base flow or springs (Figure 4.4). The discharge of groundwater into rivers may be permanent or seasonal, depending on the height of the water table within the aquifer. The water table separates the unsaturated zone of the porous rock comprising the aquifer from the saturated zone; in essence it is the height of the water in the aquifer (Figure 4.5). The water table is measured by determining the level of the water in boreholes and wells. If numerous measurements are taken from wells over a wide area, then the water table can be seen to fluctuate in height depending on the topography and climatic conditions. Rainfall replenishes or recharges the water lost or taken from the aquifer and so raises the level of the water table. If the level falls during periods of drought or due to over-abstraction for water supply, then this source of water feeding the river may cease. In periods of severe drought groundwater may be the only source of water feeding some rivers and so if the water table falls below the critical level, the river itself could dry up completely (Owen, 1993).

Groundwater abstraction in the southern regions of England has increased substantially over the past 30 years. In some areas water levels in wells have been reported to be falling year by year. Many southern chalk streams are drying up due to over-abstraction, not only for drinking water supply, but for crop irrigation and industrial use. This has led to an outcry by fishermen and conservationists alike. In the past water companies have struggled through years of drought to maintain an ever-increasing demand for water by abstracting more and more

Table 4.2 *Percentage of drinking water derived from groundwater resources in various European countries*

Country	Percentage from groundwater	Country	Percentage from groundwater
Denmark	98	Luxemburg	66
Austria	96	Finland	49
Portugal	94	Sweden	49
Italy	91	Greece	40
Germany (former West)	89	UK	25
Switzerland	75	Ireland	25
France	70	Spain	20
Belgium	67	Hungary	10

Table 4.3 *Approximate amount of drinking water supplied from surface and groundwater in the UK*

Country	Amount (Ml d^{-1})	Source (%)	
		Surface water	Groundwater
England and Wales	19 500	72	28
Scotland	2210	97	3
Northern Ireland	680	92	8
Total	22 390	75	25

groundwater. Aquifers, especially chalk aquifers, refill with water (recharge) very slowly. However, the Water Act 2003 has given new powers to the Environment Agency to manage abstractions from all water resources more sustainably.

As drinking water can come either directly or indirectly from groundwater sources, its quality is important to many more than simply those people who receive supplies directly via boreholes and wells. The principal aquifers lie beneath extensive areas of farmland in eastern, central and southern England, and in many of these areas groundwater may contribute more than 70% of the drinking water supplies (Table 4.4).

4.4.2 Aquifer classification

Aquifers are classified as either confined or unconfined. An unconfined aquifer is one that is recharged where the porous rock is not covered by an impervious

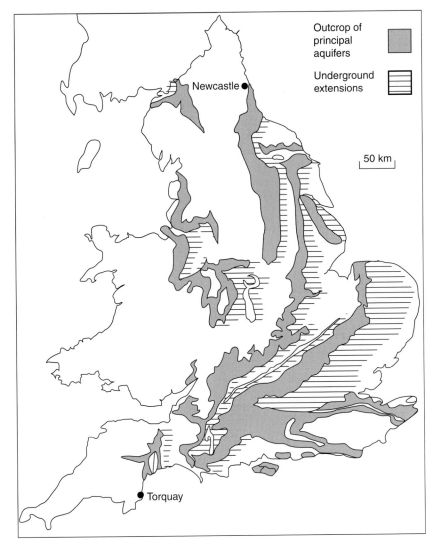

Figure 4.3 Location of principal aquifers in England and Wales. Reproduced from Open University (1974) with permission from the Open University.

Outcrop of
principal
aquifers

Underground
extensions

50 km

Newcastle

Torquay

layer of soil or other rock. The unsaturated layer of porous rock is separated from the saturated water-bearing layer by an interface known as the water table. The unsaturated layer is rich in oxygen. Where an impermeable layer overlies the aquifer (i.e. a confined aquifer), no water can penetrate into the porous rock from the surface; instead, water slowly migrates laterally from unconfined areas. There is no unsaturated zone because all the porous rock is saturated with water as it is below the water table level, and of course there is no oxygen (Figure 4.4) (Brown *et al.*, 1983). Because confined aquifers are sandwiched between two impermeable layers the water is usually under considerable hydraulic pressure, so that the water will rise to the surface under its own pressure via boreholes and wells, which are known as artesian wells. Artesian wells are well known in parts

Figure 4.4 Schematic diagram of groundwater systems. Reproduced from Lawrence and Foster (1987) with permission from the British Geological Survey.

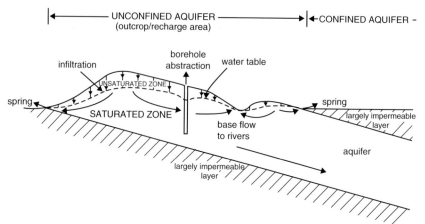

Figure 4.5 Cross-section through soil and aquifer showing various zones in the soil and rock layers and their water-bearing capacities. Reproduced from Open University (1974) with permission from the Open University.

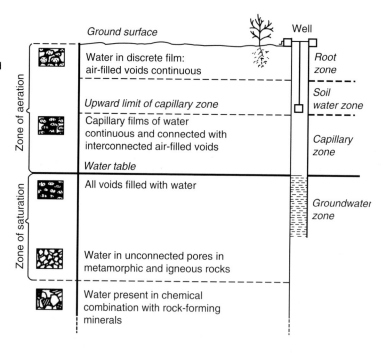

of Africa and Australia, but are also found on a smaller scale in the British Isles. The most well known artesian basin is that on which London stands. This is a chalk aquifer that is fed by unconfined aquifers to the north (Chiltern Hills) and south (North Downs) (Figure 4.6). In the late nineteenth century the pressure in the aquifer was such that the fountains in Trafalgar Square were fed by natural artesian flow. However, if the artesian pressure is to be maintained then the water lost by abstraction must be replaced by recharge, in the case of London by infiltration at the exposed edges of the aquifer basin. Continued abstraction in excess of natural recharge has lowered the piezometric surface (i.e. the level that

Table 4.4 *Proportion of regional supplies abstracted from groundwater in England and Wales. Dependence on groundwater may be very much higher locally within these regions*

Water supply region	Percentage of supplies from groundwater
Southern	73
Wessex	50
Anglian	44
Thames	42
Severn Trent	37
North West	13
Yorkshire	13
South West	10
Northumbria	9
Welsh	4

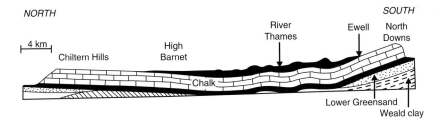

Figure 4.6 Cross-section through the London artesian basin. Reproduced from Open University (1974) with permission from the Open University.

the water in an artesian well will naturally rise to) by about 140 m below its original level in the London basin. A detailed example of an aquifer is given later in this section.

It is from unconfined aquifers that the bulk of the groundwater supplies are abstracted. It is also from these aquifers, in the form of base flow or springs, that a major portion of the flow of some lowland rivers in eastern England arises. These rivers are widely used for supply purposes and so this source of drinking water is largely dependent on their aquifers; thus good management is vital.

Groundwater in unconfined aquifers originates mainly as rainfall and so is particularly vulnerable to diffuse sources of pollution, especially agricultural practices and the fallout of atmospheric pollution arising mainly from industry. Increased chemical and bacterial concentrations in excess of the EU limits set out in the Drinking Water Directive have been recorded in isolated wells for many years. This was a local phenomenon with the source of the pollution usually easily identified to a local point source such as a septic tank, a leaking sewer or farmyard drainage. In the 1970s there was concern over increasing nitrate levels in particular, which often exceeded the EU limit for drinking water of $50\,\mu g\,l^{-1}$. This was not an isolated phenomenon, but increased levels were found throughout all the principal unconfined aquifers in the UK. The areas affected were so large that clearly only a diffuse source could be to blame, rather than point sources (Royal Society, 1984). However, it was not until the mid 1970s that the widespread increase in nitrate concentrations in groundwater was linked to the major changes in agricultural practices that had occurred in Britain since the Second World War. The major practice implicated at that time was regular cereal cropping, which has to be sustained by the increasing use of inorganic fertilizers (Section 5.4). Trace amounts of organic compounds are another major pollutant in groundwaters. As a result of the small volumes of contaminant involved, once dispersed within the aquifer it may persist for decades. Many of these compounds originate from spillages or leaking storage tanks. A leak of l,l,l-trichloroethane (also used as a thinner for typing correction fluids) from an underground storage tank at a factory manufacturing microchips for computers in San Jose, California in the USA, caused extensive groundwater contamination resulting in serious birth defects, including miscarriages and stillbirths, in the community receiving the contaminated drinking water (Section 6.3). Agriculture is also a major source of organic chemicals. Unlike spillages or leaks, which are point sources, agricultural-based contamination is a dispersed source. In the UK, pesticides that were banned in the early 1970s such as DDT (dichloro-diphenyl-trichloroethane) only started to appear in many groundwater supplies during the early 1990s. Many different pesticides are also being reported in drinking water and this is discussed in Section 6.2. Other important sources of pollution are landfill and industrial waste disposal sites, impoundment lagoons including slurry pits, the disposal of sewage sludge onto land, runoff from roads and mining (OECD, 1989).

Unlike confined aquifers, unconfined aquifers have both an unsaturated and a saturated zone. The unsaturated zone is situated between the land surface and the water table of the aquifer. Although it can eliminate some pollutants, the unsaturated zone has the major effect of retarding the movement of most pollutants, thereby concealing their occurrence in groundwater supplies for long periods. This is particularly important with major pollution incidents, where it may be many years before the effects of a spill or leakage from a storage tank, for example, will be detected in the groundwater due to this prolonged migration period (Eriksson, 1985).

Most of the aquifers in Britain have relatively thick unsaturated zones. In chalk they vary from 10 to 50 m in thickness, which means that surface-derived pollutants can remain in this zone for decades. Another problem is that the soils generally found above aquifer outcrops are thin and highly permeable, and so allow the rapid infiltration of water to the unsaturated zone taking the pollutants with them. Soil bacteria and other soil processes therefore have little opportunity to utilize or remove pollutants. This unsaturated zone is not dry; it does in fact hold large volumes of water under tension in a process matrix, along with varying proportions of air. However, below the root zone, the movement of this water is predominantly downwards, albeit extremely slowly (Figure 4.5).

It is in the saturated zone of unconfined aquifers that the water available for abstraction is stored. The volume of water in unconfined aquifers is many times the annual recharge from rainfall. It varies according to rock type and depth, but, for example, in a thin Jurassic limestone the ratio may be up to 3, whereas in a thick porous Triassic sandstone it may exceed 100. The saturated zone also contains a large volume of water that is immobile, locked up in the micro-porous matrices of the rock, especially in chalk and Jurassic limestone aquifers.

Where aquifers have become fissured (cracked), movement is much more rapid. However, the movement of pollutants through unfissured rock, by diffusion through the largely immobile water that fills the pores, will take considerably longer. This, combined with the time lag in the unsaturated zone, results in only a small percentage of the pollutants in natural circulation within the water-bearing rock being discharged within a few years of their originally infiltrating through agricultural soils. Typical residence times vary but will generally exceed 10–20 years. The deeper the aquifer, the longer this period will be. Other factors, such as enhanced dilution effects where large volumes of water are stored, and the nature of the porous rock all affect the retention time of pollutants. An excellent guide to aquifers and groundwater has been prepared by Todd and Mays (2005). The management and assessment of the risk of contamination in groundwaters is reviewed by Reichard et al. (1990) and also Appelo and Postma (2005).

4.4.3 Quality

The quality of groundwater depends on a number of factors: (1) the nature of the rainwater, which can vary considerably especially in terms of acidity due to pollution and the effects of wind-blown spray from the sea, which affects coastal areas in particular; (2) the nature of the existing groundwater, which may be tens of thousands of years old; (3) the nature of the soil through which water must percolate; and (4) the nature of the rock comprising the aquifer.

In general terms groundwater consists of a number of major ions that form compounds. These are calcium, magnesium, sodium, potassium and to a lesser extent iron and manganese. These are all cations (they have positive charges) which are found in water combined with an anion (which have negative charges) to form compounds referred to as salts. The major anions are carbonate, hydrogencarbonate, sulphate and chloride. Most aquifers in the UK have hard water. Total hardness is made up of carbonate (or temporary) hardness caused by the presence of calcium hydrogencarbonate ($CaHCO_3$) and magnesium hydrogencarbonate ($MgHCO_3$), whereas non-carbonate (or permanent) hardness is caused by other salts of calcium and magnesium (Chapter 10). It is difficult to generalize, but limestone and chalk aquifers

Table 4.5 *Mineralization of groundwater can be characterized by conductivity*

Conductivity in $\mu S\ cm^{-1}$, at 20 °C	Mineralization of water
<100	Very weak (granitic terrains)
100–200	Weak
200–400	Slight
400–600	Moderate (limestone terrains)
600–1000	High
>1000	Excessive

contain high concentrations of calcium hydrogencarbonate, whereas dolomite aquifers contain magnesium hydrogencarbonate. Sandstone aquifers are often rich in sodium chloride (NaCl), whereas granite aquifers have elevated iron concentrations. The total concentration of ions in groundwater, the total dissolved solids (TDS), is often an order of magnitude or more higher than in surface waters. The Department of the Environment (1988) has published an excellent review of groundwater quality for England and Wales.

The total amount of anions and cations present also increases with depth due to less fresh recharge to dilute existing groundwater and the longer period for ions to be dissolved into the groundwater. In very old, deep waters the concentrations are so high that they are extremely salty (i.e. mineralized). Such high concentrations of salts may result in problems due to over-abstraction or in drought conditions when old saline groundwaters may enter boreholes through upward replacement, or due to saline intrusion into the aquifer from the sea. In Europe conductivity is used as a replacement for TDS and is employed to measure the degree of mineralization of groundwaters (Table 4.5).

In terms of the volume of potable water supplied, confined aquifers are a less important source of groundwater than unconfined aquifers. However, they do contribute substantial volumes of water for supply purposes and can locally be the major source of drinking water. Groundwater in confined aquifers is much older than in unconfined aquifers and so is characterized by a low level of pollutants, especially nitrates and micro-organic pollutants, including pesticides. This source is currently of great interest for use in diluting water from sources with high pollutant concentrations, a process known as blending (Section 5.6).

Confined aquifers are generally not used if there is an alternative source of water because of low yields from boreholes and quality problems, especially high salinity in some deep aquifers, excessive iron and/or manganese, problem gases such as hydrogen sulphide and carbon dioxide and the absence of dissolved oxygen. These problems can be overcome by water treatment, although this, combined with the costs of pumping, makes water from confined aquifers comparatively expensive.

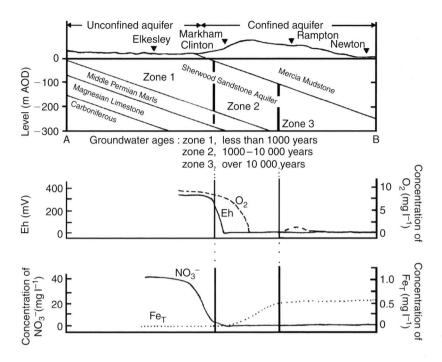

Figure 4.7 Hydrogeochemistry of the Sherwood sandstone aquifer. Reproduced from Kendrick *et al.* (1985) with permission from WRc Plc.

An example of a major British aquifer is the Sherwood Sandstone aquifer in the East Midlands, which is composed of thick red sandstone. It is exposed in the western part of Nottinghamshire, which forms the unconfined aquifer, and dips uniformly to the east at a slope of approximately 1 in 50. It is overlain in the east by Mercia Mudstone to form a confined aquifer (Figure 4.7). Three groundwater zones have been identified in this aquifer, each of increasing age. In zone 1 the groundwater is predominantly modern and does not exceed a few tens or hundreds of years in age, whereas in zone 2 the groundwater ranges from 1000 to 10 000 years old. In zone 3 the groundwater was recharged 10 000–30 000 years ago. In this aquifer the average groundwater velocity is very slow, just $0.7 \, \text{m yr}^{-1}$ (Edmunds *et al.*, 1982). The variations in chemical quality across these age zones is typical of unconfined and confined aquifers. This is seen in the other major aquifers, especially the Lincolnshire limestone aquifer in eastern England where the high nitrate concentrations in the youngest groundwater zone are due to the use of artificial fertilizers in this intensively cultivated area. There are large groundwater resources below some major cities; in the UK these include London, Liverpool, Manchester, Birmingham and Coventry. However, these resources are particularly at risk from point sources of pollution from industry; especially solvents and other organic chemicals (Section 6.3), as well as more general contaminants from damaged sewers and urban runoff. Such resources are potentially extremely important, although due to contamination they are often under-exploited (Lerner and Tellam, 1993).

4.4.4   Demand and groundwater

Water demand is continuing to increase rapidly in southern England. However, there is already a danger that current abstraction rates for some major lowland rivers are approaching the total usable discharge rate under dry weather conditions. In some localized areas this has already been exceeded with serious environmental consequences. There is now considerable opposition to flooding valleys for new reservoirs, mainly on landscape quality and aesthetic grounds, so the development of groundwaters has been proposed as the obvious way forward. Supporters cite a number of facts to support this approach, namely: (1) groundwater abstraction does not affect the amenity value or current land use and appears to have a low environmental impact; (2) the major aquifers in Britain coincide with centres of maximum water demand, where alternative supplies are scarce and where land prices are highest; (3) reservoir construction and operation is far more expensive than groundwater abstraction; and (4) the large losses of water from reservoirs due to evaporation in the summer do not occur with groundwaters, and so it is more efficient in terms of resource utilization. However, in practice it seems unlikely that existing groundwater resources can be exploited much further, with chalk aquifers in particular, such as the London Basin, becoming seriously depleted and many others becoming increasingly contaminated, often through artificial recharge. Future increases in supplies must come from the conservation of existing supplies, reuse and the more prudent management of water supplies generally (Section 1.5).

Throughout Europe groundwater supplies are being taken out of service as the levels of contamination exceed legal limits or become uneconomic to treat using conventional water treatment methods (OECD, 1989). Treating pollution within aquifers is currently not possible without first abstracting the water, treating it and then returning it to the aquifer by injection. However, there are some interesting developments. For example, nitrate concentrations tend to decrease with depth due to dilution. Deepening wells, or encasing the upper sections of boreholes to prevent contamination from the upper sections of the aquifer ensures that more of the supply water is abstracted from the less contaminated water lower in the aquifer, thus ensuring a reduction in nitrate concentrations. This is only a short-term measure. However, the use of shallow scavenging wells located around a main supply borehole can intercept and draw off the high-nitrate water preventing it from contaminating the rest of the supply. The high-nitrate water is then used for irrigation where the nitrate is used by plants. This can reduce nitrate levels in water supplies in the long term, especially where linked with improvements in agricultural practice (Section 5.6). Other ideas include seeding aquifers with bacteria and organic matter to encourage anaerobic denitrification (where the nitrate is converted to gaseous nitrogen). This process occurs naturally in some parts of confined aquifers due to the absence of oxygen, so the principle is to speed up this natural activity. There are a number of other options still very much in the experimental phase which may offer some hope for the future, especially on a local scale. Hydrology relating to drinking water is considered further by Wanielista (1990).

4.5 Other sources of water

Although groundwater and surface water currently supply most of our water, increasing demand and increasing pollution of existing resources in some areas are intensifying the search for new resources. In many rivers abstraction already exceeds the total usable dry-weather flow, the danger being that rivers could simply be sucked dry, destroying their flora and fauna as well as eliminating their potential for other uses. In the late 1970s and 1980s there was a general move away from reservoir construction, not only on a cost basis, but also due to intense opposition from the public on environmental and social grounds, as explained in the earlier sections. However, by the end of the 1980s it was clear that groundwater sources in southern England especially, but also throughout the UK, could not be significantly exploited any further to meet increased demands. Indeed, there is considerable evidence that many aquifers have been overexploited and are seriously depleted.

Perhaps the major reason for moving away from constructing new reservoirs is the difficulty of accurately predicting or forecasting water demand. Water usage soared from the end of the Second World War up to the end of the 1960s, by as much as 35% each decade. Water planners began to extrapolate from this rapid increase in water usage that demand for water would double again by the end of the century. New schemes were examined such as barrages and even a national water grid, and worries about massive water shortages were mooted. However, demand for water slackened and during the 1970s it increased by just 12%, and by only 10% in the 1980s. Demand was predicted to rise by only 6% in the 1990s, but in reality it stabilized and has slightly fallen since 1990 due in part to a decrease in industry (Table 1.1).

One of the most serious manifestations of the incorrect forecasting of the late 1960s was the construction of Kielder Water to satisfy totally erroneous water demands. Kielder Water was built to supply water to the great companies in the north-east such as British Steel. It was approved in October 1973 and its construction started in 1976, spurred on by the drought. However, by the time it was constructed the demand for water in the area had decreased dramatically due to a reduction in heavy industry and a fall off in the rate of increase in domestic usage. The type of forecast on which the decision to construct Kielder Water was based used a linear extrapolation of demand over the previous 20–30 years (Figure 1.4) (Section 1.5). This type of forecasting did not take into account other important factors, such as a decline in manufacturing industry, so that overall demand actually slowed (Figure 4.8). One of the largest reservoirs in

Figure 4.8 Type of forecasting used to predict future water consumption in the early 1970s compared with actual demand in England and Wales over the same period 1971–88. Reproduced from Clayton and Hall (1990) with permission from the Foundation for Water Research.

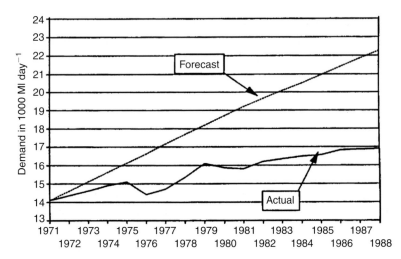

Europe, Kielder Water has a storage capacity of 200 000 Ml, more than enough to satisfy the current shortfall between supply and demand in the south-east England in full, if only it could be transported. Kielder Water, which required one and a half million trees to be felled during its construction, is just about as far north as you can get in England, just 5 km from the Scottish border. It captures water from the River Tyne and doubled the availability of water in the north-east. Since its completion in 1980 (it took two years to fill), demand has risen by just 8% and so far neither the reservoir nor the expensive tunnels and pipes built to supply the cities on the lower Tyne and Tees have ever supplied customers. The mistake of the construction of Kielder Water is perhaps a major contributing factor to the decrease in reservoir construction since that time. However, although demand has decreased in the north, it has increased in the south and south-east. To meet this demand the water supply companies will most likely resort to constructing new reservoirs. In September 2006, Thames Water announced plans to construct a new reservoir with a capacity of 150 billion litres of water near Abingdon in Oxfordshire. Thames Water currently only obtains 17% of its daily 2 822 Ml of water it supplies to customers from groundwater resources. Opponents to the scheme have suggested that the 895 Ml of treated water it loses each day through leakage should be tackled first. However, with a predicted population increase for London of 800 000 over the next decade, the new reservoir is key to meeting the projected increased demand.

With existing resources largely fully exploited, alternative sources of water will have to be developed by using poorer quality resources or reusing and recycling water. Aquifers are now managed, often with groundwater levels maintained by artificial recharge. Where alternative supplies are not available attention has focussed on conservation, with metering becoming increasingly common (Section 1.5). Transferring water from one catchment where water is plentiful to another where supplies are low has led to the idea of a national grid for water, the water being transferred from the wet north-west to the dry south-east. Other newer techniques are also being examined such as desalination, ion-exchange and even importing water. All these are considered below with further details on resource management given by Cech (2005).

4.5.1   Use of poor quality resources

Where new surface or groundwater supplies cannot be exploited, the only option may often be the utilization of water resources previously rejected for supply purposes due to poor quality or pollution. New advanced methods of water treatment such as granular activated carbon or ion-exchange can in theory produce pure water from raw water of any original quality, no matter how poor. Although technically feasible, the quality of raw water permitted to be abstracted from surface waters, including lakes and reservoirs, is controlled by legislation. The quality of surface water abstracted for public supply must conform to EC Surface Water Directive (75/440/EEC). This lays down the required quality of water that can be abstracted for public supply and classifies all suitable waters into three broad categories: A1, A2 and A3 (Tables 2.8 and 2.9). It also specifies the level of water treatment necessary to transform each category of water to drinking water quality (Section 2.5).

Clearly, the better the quality of the surface water (A1), the less treatment is required and the cost is kept to a minimum. Conversely, the poorest quality waters may often be prohibitively expensive to treat. Most surface waters currently used in the UK fall into either the A1 or A2 category. However, many of the lowland rivers that flow through major conurbations where water demand is greatest, so that many will have to be used in the future as sources of drinking water, fall into the poorest category A3. Where the quality of surface waters falls short of the mandatory limits for category A3, then they may not be used for abstraction. However, exceptions can be made when special treatment must be used, such as blending, to bring the supply up to acceptable drinking water standards. The European Commission itself must be notified of the grounds for such exemption on the basis of a water resources management plan for the area concerned. This ensures that other more suitable resources are not being ignored and that there is no alternative to using the poorer quality water. In the list of exceptional circumstances are: (1) a reduction in quality due to floods or other natural disasters; (2) increased levels of specified parameters such as nitrate, temperature or colour, copper (for A1 waters), sulphate (for A2 and A3 waters) and ammonia (A3 waters) due to exceptional meteorological or geographical conditions; (3) where surface waters undergo natural enrichment (eutrophication); and (4) nitrate, dissolved iron, manganese, phosphate, chemical and biochemical oxygen demand and dissolved oxygen concentrations in waters abstracted from lakes and stagnant waters which do not receive any discharge of

waste, which are less than 20 m in depth, and which have a very slow exchange
of water (less than once each year). Exemptions cannot be made where public
health is threatened, and of course the EC Drinking Water Directive is
unaffected, so the water still has to be treated to an acceptable standard. The EC
Surface Water Directive only covers surface waters abstracted for human
consumption, it does not, for example, cover the quality of water used to
recharge aquifers used for public supply. Many member states considered the
Surface Water Directive to be obsolete once the Drinking Water Directive had
been adopted in 1980. It will eventually be replaced in 2012 when the EC Water
Framework Directive comes fully into place (Sections 2.5 and 14.5).

4.5.2 Reuse and recycling

In areas where water is scarce, treated sewage effluent may be reused after
sufficient disinfection for uses not associated with human or animal
consumption. In arid areas it is common to use sterilized treated effluents for
non-consumable activities such as washing cars, flushing toilets and even
washing clothes (Fewkes and Ferris, 1982). In the UK approximately 35% of
the raw water used for public supply is obtained from recycled effluent. This is a
mean value for the whole of England and Wales, and in areas where supplies of
upland water are very restricted, such as the south-east of England, this figure
may be as high as 70% at times. Treated effluent is discharged from one
consumer area into a lowland river and abstracted for reuse at the next urban
area downstream. It is incredible, but true, that the rivers Thames and Lee
consist of 95% treated effluent during dry summers. With the major areas of
population centred in the Midlands and south-east England, it is unlikely that
new upland supplies will be made available in the future and any subsequent
increase in demand will have to be largely met by using groundwater or
reclaimed water. The Thames River basin in England is an example of open
cycle reuse, where the sewage from one community is converted to drinking
water in another. Overall, the population of London is in excess of 10 million,
with water being supplied from boreholes, the River Thames and its tributaries,
including 1 200 000 m^3 of sewage. The recycle rate is, on average, 13%, but
during the 1975–6 drought it exceeded 100%. The sewage receives full
biological treatment followed by nitrification and denitrification when required,
so that there is no ammonia or nitrate in the water. The water abstracted for
supply from the Thames is stored for seven days before treatment, which
normally involves slow sand filtration followed by chlorination. The reuse of the
River Thames water is shown schematically in Figure 4.9. It is interesting that
one community, Walton Bridge, actually discharges its effluent upstream of its
own intake.

All municipal wastewaters discharging to rivers used for public supply are
fully treated biologically and the abstracted water is then subjected to full water

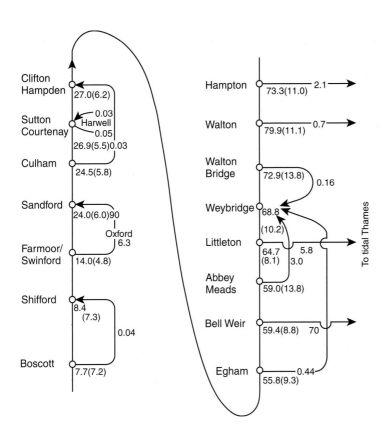

Figure 4.9 Example of the reuse of surface water for drinking water supply. The River Thames is shown schematically with the average river flows given in m³ s⁻¹ and the percentage of this flow that consists of sewage effluent given in parentheses. Reproduced from Dean and Lund (1981) with permission from Elsevier Ltd.

treatment (which may also include treatment with activated carbon, membrane filtration or ion-exchange as necessary) (Section 14.2). However, when water is recycled many times dissolved salts will accumulate in it, particularly the end-products from biological sewage treatment. These include nitrate, sulphate, phosphate and chloride, all of which can cause unpleasant tastes and a decrease in quality, corrosion and scaling in pipes, and even toxicity. If there is no alternative source of supply then these inorganic salts may eventually have to be removed by advanced water treatment methods such as ultrafiltration, reverse osmosis or ion-exchange, which makes the water expensive. Some pollutants, such as organic compounds including pesticides, pharmaceuticals and industrial compounds, are not easily removed and are very persistent. Although they are only normally present in trace amounts they are still toxic or in some instances carcinogenic. Numerous surveys have indicated that there may be long-term effects in reusing water after it has been consumed, with a higher incidence of cancer among people drinking river water that has already been used for supply purposes compared with those drinking groundwater. However, a study carried out in London over the period 1968–74 showed that there was no significant difference between the mortality rate of those drinking reused River Thames water and groundwater (Beresford, 1981). This work did, however, provide epidemiological evidence consistent with the earlier studies that a small risk to health exists from the reuse of drinking water. In this study, the percentage of domestic sewage effluent in the water was positively associated with the incidence of stomach and bladder cancers in women (Beresford *et al.*, 1984). The health aspects of the reuse of wastewater for human consumption is considered further in Part III of the text.

4.5.3   Artificial recharge and aquifer management

Water management schemes often involve two or more sources of supply. Usually one is used in normal circumstances, whereas the other is used only at times of peak demand or when there is a shortage of water from the normal source. Such schemes usually include a surface water supply, most probably a river, and a groundwater resource. The river is the primary source of water and the aquifer is used at times of low river discharge rates.

Artificial recharge is now widely practised in the USA and the rest of Europe, although it is still uncommon in the British Isles (David and Pyne, 1995). Unaided, aquifers recharge comparatively slowly, which means that the rate of water abstraction for supply is limited by the rate of natural recharge. Artificial recharge rapidly feeds water back into the aquifer during periods when it is plentiful on the surface. This ensures maximum utilization of the available storage capacity underground, thereby increasing the amount of water available to be abstracted during periods of shortages that would otherwise require reservoir construction for surface storage. There are a number of methods of recharging aquifers. One of the simplest is to excavate a shallow lagoon and allow the water to percolate rapidly through the bottom into the aquifer. More commonly, recharge boreholes are used and the water is pumped directly into the aquifer. Its use is limited due to geological conditions in the UK, unlike the Netherlands where artificial recharge is widely practised. However, there are major concerns about this technique. River water is mainly used for artificial recharge and so the quality of this water will be very different from the groundwater and may alter the overall chemical nature of the water within the aquifer. More importantly surface water is of a lower quality and is also very susceptible to pollution. Therefore there is a very real risk of polluting all the groundwater within the aquifer by recharging with contaminated surface water. The significant amount of organic material and suspended solids in river waters in particular will increase biological activity within the aquifer and may even cause physical clogging. This is more of a problem where direct pumping is used via recharge boreholes, rather than infiltration lagoons where the water must percolate through the porous sediment in the unsaturated zone, which improves the overall water quality. The high rate of infiltration can lead to clogging of the underlying sediment of lagoons if the water is too turbid, which seriously reduces the infiltration rate.

There has been much discussion of the possibility of using treated sewage effluent to recharge aquifers, thus conserving what water there is in areas of low rainfall. This is becoming an increasingly important issue as the winters have been comparatively dry, resulting in incomplete natural recharge of aquifers in some areas, with consequently less water for supply during the summer. In rural areas where the population is dispersed the effluent has traditionally been disposed of onto land either by surface irrigation or via percolation areas. In some areas, such as the chalk aquifer of southern England, there are so few rivers that it is sometimes difficult to find surface waters to discharge effluents into, so disposal to land in some instances has been going on for a long time. Water companies argue that in some areas the dependence on groundwater resources is so great (e.g. 70% in some parts of the Southern Water area, 50% in areas serviced by Thames Water and Anglian Water), that artificial recharge using treated effluent is the only practical option available if abstraction from aquifers is to be maintained at the same rate (Baxter and Clark, 1984). Although it appears to be an efficient method of sewage disposal, it does cause concern about the long-term quality of groundwater resources. Although there is some removal of contaminants, including pathogenic micro-organisms, within the soil and the unsaturated zone, there is a tendency for most of these contaminants to slowly migrate downwards and eventually find their way into the saturated zone of the aquifer itself. Dilution is currently reducing the concentrations of contaminants to well below that required by the EC Drinking Water Directive (Appendix 1), but with time concentrations will increase until an equilibrium is reached. There is a possibility of increased concentrations of organic chemicals and metals, in particular pesticides, pharmaceutical and personal care products, endocrine-disrupting compounds and the so-called sewage metals zinc, copper, chromium and lead, which are all found in slightly increased concentrations in treated sewage effluents. So far the most obvious increases have been in chloride and nitrate concentrations, but only time will tell exactly how safe effluent recharge is. Work carried out in other countries has shown that there may be a time lag of several years between the start of recharge and subsequent contamination of observation or supply boreholes, although sometimes rapid migration is seen, especially in fissured chalk and limestone aquifers. If problems are detected it may require a much longer period for the full recovery of water quality once recharge has stopped. As recharge begins a mixture of recharge effluent and natural groundwater will eventually be pumped from the supply borehole, so regular monitoring of water quality is essential to ensure potable standards are maintained. Owing to the potential long-term risks to important groundwater resources, where infiltration lagoons are used for recharge, it is advisable to use only wastewater treated to as near drinking water quality as possible and to regard the lagoon system solely as a means of recharging the aquifer, much in the same way that injection wells are used. Any benefits from the potential treatment capacity of the soil and the unsaturated zone as the water percolates through to the aquifer should be regarded as incidental. The recharge method requires large areas of land, up to $7\,m^2$ per person, and so could never be feasible for large communities. The bulk of the water for artificial recharge will therefore most probably come from rivers, although there is increasing interest in the use of stormwater.

However, it is possible to use shallow aquifers primarily as water treatment plants. For example, water from the River Rhine, which is of extremely poor quality, is pumped into settling basins near the town of Essen (Germany). After settlement for 12 hours the water flows into an enormous recharge lagoon ($200\,000\,m^2$ in area), which has been excavated in highly porous river gravels and sands. Under the sediments through which the water percolates perforated pipes have been laid about 10 m deep via which the water is continuously abstracted. The water is then chlorinated and supplied directly to the 850 000 inhabitants of Essen at a very low unit cost.

The potential for the development of aquifers for water storage, treatment and supply is enormous, although at present many are being over-abstracted. To achieve the maximum from these valuable resources water companies are appointing aquifer managers to control and develop specific groundwater resources. In certain areas of Britain, such as the south-east, aquifers are so important to the community in terms of household, industrial and agricultural water supplies that every effort must be made to preserve them and manage them wisely. A failure to do so could lead to severe social and economic hardship for many areas, simply because there is not enough water to go around.

4.5.4   Water transfer

Rivers and long aqueducts are used to convey water collected in upland reservoirs to principal centres of demand. However, it has been suggested that larger regional transfer systems to transport water from areas of high water availability in the north of England, and especially Scotland, to the high demand areas where supplies are currently short in the south-east should be developed. Such transfer schemes could be co-ordinated to form a national network or grid, rather like the electricity power grid, allowing water to be transferred to wherever it was needed (Figure 4.10). The Environment Agency have carried out a number of studies on the provision of a National Grid for the UK and have concluded that the expected population growth in the south-east of two million by 2030 will result in a water demand increase of 8%, and that there are cheaper, more secure local solutions available. A major factor, apart from construction costs that in itself probably makes the idea unsustainable, is the energy required to operate a national water grid (Environment Agency, 2006). Similar schemes have been suggested throughout the world, especially in North America. There is also considerable interest in the idea of importing water from France for use in East Anglia, via a water main laid in the Channel Tunnel. The construction of

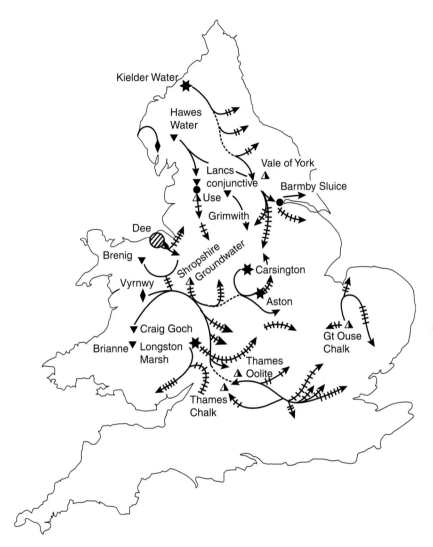

Figure 4.10 The idea of a water distribution network in England and Wales was first proposed in the early 1970s although little progress has been made. Reproduced from Kirby (1979) with permission from Dr Celia Kirby.

barrages across the mouths of major estuaries also offers a potential solution to the provision of more surface water supplies. The estuarine barrage would form a huge fresh water lake behind it, providing vast amounts of surface water that would become available for use without the loss of further agricultural land for reservoir construction. It also offers the potential to provide water supplies to areas where the populations are greatest, releasing existing supplies for other users. Of course, there are disadvantages: the quality of such water is usually fairly poor, often contaminated by heavy industry, and there are many ecological, physical and social disadvantages.

4.5.5 Desalination

The process of evaporating fresh water from seawater using the Sun's energy is old technology. However, modern distillation systems are able to produce continuous supplies of high-purity water and are widely used on board ships at sea. To do this on a large enough scale to supply drinking water to even a small town requires enormous amounts of energy to remove the salt, so water produced in this way is very expensive and has a heavy carbon footprint. Currently the cost of desalination using the state of the art technology is between three and four times the cost of A2 water, although it is very difficult to put a realistic cost on conventional treatment using existing water resources. Such comparisons may be meaningless in the long term. It is more realistic to compare desalination to new schemes such as water transfer, new reservoir construction, aquifer recharge and the use of poor quality resources. However, as poorer quality water is used and more advanced treatment technologies are employed to achieve the EC Drinking Water Directive required standards, especially for trace amounts of organic compounds and nitrates, then the cost of desalination becomes less prohibitive. Linked with the cost, indeed the feasibility, of providing new resources in certain areas such as the south-east of England, then desalination appears to be a strong possibility, especially during drought periods. The lack of water and the unpredictability of supplies in some areas is so restrictive, especially for industry and agriculture, that users in these areas may be prepared to pay significantly more for the water they use. In many arid countries where there is usually no surface water at all, and where groundwater supplies are inadequate, then desalination is the only method of water supply. It is these countries, in particular the Middle East, from which most of our current expertise in desalination technology derives. Desalination is also popular in arid countries such as Spain and Australia, and of course highly populated islands such as the Virgin Islands. Desalination is also used on Jersey to supplement the limited amount of surface water available. Worldwide there are about 7600 plants producing 14×10^9 m^3 of water annually. Sixty per cent of the capacity is in the Middle East, 13% in North America, 10% in Europe (including eastern Europe) and 7% in Africa.

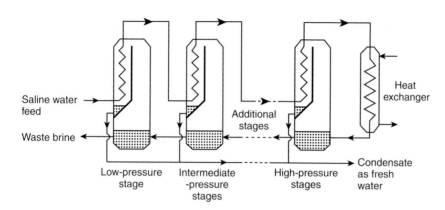

Figure 4.11 Principles of the multi-stage flash distillation process.

Modern distillation plants such as that in Jersey are based on the multi-stage flash process (Figure 4.11). Cold salt water is fed into the system through a long pipe that passes through each chamber where it acts as a condenser. The salt water absorbs heat during condensation and is then heated in a steam-fed heat exchanger. The hot salt water enters the first chamber where part of the salt-free water flashes to salt-free vapour that condenses on the cold feed-pipe as fresh water. The salt water passes to additional chambers, which are each operated at slightly higher pressures than the preceding chamber, ensuring that further flashing occurs. The fresh water is collected for use as drinking water whereas the concentrated waste brine solution has to be carefully disposed of back to the sea (Wood, 1987). Currently, the high energy demand of distillation and the concerns over carbon emissions makes the process one of last resort.

Other processes of desalination, apart from distillation or evaporation, include freeze distillation and reverse osmosis. Freezing a salt solution makes crystals of fresh water form and grow, leaving a concentrated brine solution behind. Although the technology exists, no large-scale facility has so far been developed. However, as freezing only needs 10–15% of the energy required by evaporation, it is attractive both economically and environmentally (Coughlan, 1991).

Osmosis is the movement of water (or any solvent) from a weak solution to a strong solution through a semi-permeable membrane. Therefore if the membrane is placed between fresh water and salt water the solvent (i.e. pure water) will move through the membrane until the salt concentration on either side is equal. Only water can pass through the membrane so the salts are retained The movement of water across the membrane is caused by a difference in pressure and continues until the pressure in both solutions is equal, limiting further passage. The pressure difference, which causes osmosis to occur, is known as the osmotic pressure. Reverse osmosis uses this principle to make the solvent (pure water) move from the concentrated solution (salt water) to the weak solution (fresh water) by exerting a pressure higher than the osmotic pressure on the concentrated solution, thus reversing the direction of flow across the membrane (Figure 4.12).

Sources of water

Figure 4.12 Principles of the reverse osmosis process.

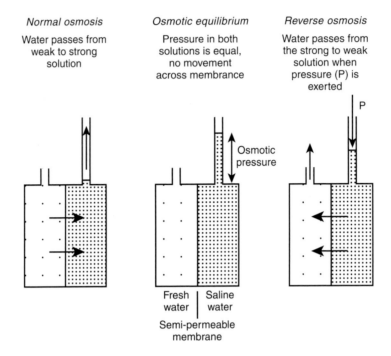

Normal osmosis

Water passes from weak to strong solution

Osmotic equilibrium

Pressure in both solutions is equal, no movement across membrance

Reverse osmosis

Water passes from the strong to weak solution when pressure (P) is exerted

By subjecting saline water to pressures greater than the osmotic pressure, pure water passes through the semi-permeable membrane and can be collected for use as drinking water. Reverse osmosis is not membrane filtration, i.e. purification is not the result of passing the water through defined pores or holes. It is a far more complex action, where one molecule at a time diffuses through vacancies in the molecular structure of the membrane material. Filtration physically removes particles, although these particles can be very small. Commercial units are available, with membranes made from cellulose acetate, triacetate and polyamide polymers. Multi-stage units are generally used with the best results from the use of brackish water, rather than seawater. Reverse osmosis is also used for treating polluted river water and removing common contaminants such as nitrate (Parekh, 1988). Although expensive, reverse osmosis can also be used to soften water (Chapter 29).

4.5.6 Other options

Transferring water from one region to another has already been examined. However, there are other transfer options that are currently under serious consideration. The importation of water by bulk tank to areas with water shortages is common during times of drought or when water mains burst. The importation of water by bulk sea tankers is common in many arid countries including the Greek Islands. Modern transport of water by sea now employs large inflatable cigar-shaped polyurethane bags, each between 750 and 2000 m^3 in size, which are towed in large groups behind any sea-going vessel, so that a specialized tanker is no longer required. As fresh water is less dense than seawater the bags float, although only about 5% of the bag is visible above water. As the ship is not required to be pumped out it can immediately continue on its way after delivery, making delivery far cheaper. The drinking water is then conveniently stored in the floating bags until required for use or treatment, with the emptied bags returned for refilling by air. This has not been widely adopted in the UK but has been seriously considered by some water companies during drought periods in coastal areas severely hit by water shortages. For example, during the drought of 1989 and 1990, Northumbria Water considered the idea of importing water from Gibraltar using ocean-going tankers. The disadvantage of importing water from another country is that in the long term you become very vulnerable to economic and political forces. Everyone wants to be independent in terms of water supply.

Rain-making has become fairly successful, especially in south-western Australia. Once the province of the witch doctor or medicine man, it is now a high-tech scientific art. Silver iodide smoke is released from aircraft onto the tops of clouds where the temperature is less than $-10\,^\circ$C. It is estimated that 1 g of silver iodide will release as much as 250 000 m^3 of rainwater, which makes it comparatively cheap. However, large-scale interference in natural rainfall patterns could prove ecologically and politically very serious. It may also result in localized flooding. For example, there have been suggestions that experimental rain seeding experiments carried out in England during 1952 may have contributed to the Lynmouth flood disaster in Devon, although the evidence is circumstantial.

The most widely studied transfer method in recent years is the use of natural ice, i.e. icebergs, as a source of drinking water as well as a cheap form of energy collected as they slowly heat up. It has been suggested that the recent increase in the breaking up of the ice sheets due to the greenhouse effect driving global warming will make this an easier and cheaper operation, although this remains to be seen. Towing icebergs the size of small islands is technically very difficult and extremely hazardous, but for arid countries, icebergs seem to be an ideal although bizarre source of pure drinking water. With increasing global warming, attention is also focussing on the increase in glacial runoff as a potential source of high-quality drinking water.

More recently in arid coastal areas in Africa existing water supplies have been successfully supplemented using fog water. Water is collected from fog by erecting mesh or sheet traps, rather like long curtains, onto which water from the fog condensates and is then collected. The volume of water collected is very variable, but Olivier (2004) using experimental fog traps 70 m^2 in area on the west coast of South Africa reported average yields of 4.6 litres of water per m^2 of collection surface per day. The water is of excellent quality and could be an extremely important alternative water supply for small rural communities.

4.6 Conclusions

Our management of water resources is becoming more difficult due to increasing demands in areas where there is less rainfall and also rapid urbanization. Due to the hydrological cycle being so closely interconnected with temperature and precipitation, climate change will continue to exacerbate this already difficult situation. As temperatures rise evaporation increases leading to more rainfall; and as glaciers increasingly melt then it would seem that the overall amount of available fresh water is in fact increasing. However, climate change will alter local and regional climates resulting in more frequent drought and floods. This will require new management techniques to capture and store precipitation during periods of less frequent but more intense rainfall. For the UK climate change will have only a minor effect on water resources over the next 30 years, although after this time there is expected to be increasingly wetter winters and drier summers.

How climate change will affect water quality is unclear, but already problems with salinization due to increased evaporation and less rainfall, and also increased algal growth are just two of many possible emerging problems in relation to water resources. In the USA, Europe and Australia water companies are looking primarily at salt or brackish water and stormwater as realistic alternative sources, with growing interest in water conservation including the use of reclaimed water and household rainwater harvesting (Section 1.5). In order to prevent further deterioration of the ecological quality of surface waters, especially during periods of low flow, then it is unlikely that existing resources can be further exploited with any degree of safety. New management approaches, such as water demand management, are required to ensure sufficient water of adequate quality will be available through this new century and beyond (Section 1.5). However, water companies in the UK have decided that the best way to tackle the increasingly serious water shortages in the south-east, where demand is set to increase by 8% over the next 25 years due to population increases, is to build new or extend existing reservoirs supplying an extra 545 Ml d^{-1} (Table 4.6).

Due to the uncertainties surrounding the future of water resources, abstractions are now generally strictly controlled under licence. In England and Wales the publication of the Water Act (2003) made significant changes in the way water resources are operated. Abstraction licences are now issued for a specific time period, normally a maximum of 12 years, and with strict conditions in order to ensure sustainability of supplies and to protect the ecological status of surface waters. Future management of abstractions will also include the concept of water rights trading, where the licensable water rights can be transferred from one party to another for gain. The Environment Agency in England and Wales manages water abstractions through individual catchment abstraction management strategies (CAMS) (www.environment-agency.gov.uk/cams), which are integrated into the River Basin Management Plan which will

Table 4.6 *Proposed new and extended reservoirs in the south-east of England to cope with expected shortfalls until 2030*

Reservoir	Type of work	Water company	Completion date	Expected output (Ml d^{-1})
Bray Reservoir, Berkshire	Enlargement	South East Water	2008	18
Abberton Reservoir, Essex	Raising	Essex and Suffolk Water	2014	50
Bewl Reservoir, Kent	Enlargement	Southern Water; Mid Kent Water	2015	14
Clay Hill, East Sussex	New	South East Water	2015	18
Broad Oak, Kent	New	Mid Kent Water; Southern Water; Folkestone and Dover Water	2019	42
Havant Thicket, Hampshire	New	Portsmouth Water	2020	23
Upper Thames Reservoir, Oxfordshire	New	Thames Water	2020	380
Total				545 Ml d^{-1}

be used in the future to manage catchments under the EC Water Framework Directive and renewed every six years.

The quality of groundwaters is continuing to decline in the UK with almost 50% of all the groundwater abstracted for public supply suffering from contamination. Key problems are nitrates, pesticides, *Cryptosporidium*, arsenic and hydrocarbon solvents. In the past groundwaters have been relatively cheap sources of water as they required minimal treatment; however, declining quality has led to increasingly more sophisticated treatment being required and as the cost of treatment has risen many groundwater sources have become uneconomic and so abandoned for supply purposes. Since 1975 groundwater quality problems has cost the UK water industry £754 000 000 (58% on new treatment, 18% on blending to dilute contaminants and the remainder on replacement sources) (UKWIR, 2004).

Many of the world's larger river systems flow through more than one country, with the countries downstream dependent on the behaviour of those upstream to ensure adequate water volume and quality for their needs. Increasingly, over-abstraction and the construction of dams is dramatically reducing flows in some major international rivers, resulting in severe water shortages for downstream countries. This is leading to increasing hardship and in many cases conflict (Pearce, 1992, 2006; Ward, 2003).

References

Appelo, C. A. J. and Postma, D. (2005). *Geochemistry, Groundwater and Pollution*, 2nd edn. London: Taylor and Francis.

Baxter, K. M. and Clark, L. (1984). *Effluent Recharge*. Technical Report 199. Stevenage: Water Research Centre.

Beran, M. A. and Rodier, J. A. (1985). *Hydrological Aspects of Drought*. Studies and Reports in Hydrology 39. Paris: UNESCO.

Beresford, S. A. (1981). The relationship between water quality and health in the London area. *International Journal of Epidemiology*, **10**, 103–15.

Beresford, S. A., Carpenter, L. M. and Powell, P. (1984). *Epidemiological Studies of Water Reuse and Type of Water Supply*. Technical Report 216. Stevenage: Water Research Centre.

Bowen, R. (1982). *Surface Water*. London: Applied Science.

Brown, R. H., Konoplyantsev, A. A., Ineson, J. and Kovalvsky, V. S. (1983). *Groundwater Studies*. Studies and Reports in Hydrology 7. Paris: UNESCO.

Cech, T. (2005). *Principles of Water Resources: History, Development, Management and Policy*, 2nd edn. New York: John Wiley and Sons.

Clayton, R. C. and Hall, T. (1990). *A Review of the Requirements for Treatment Works Performance Specifications*. Report FR0089. Marlow: Foundation for Water Research.

Cooke, G. D., Welch, E. B., Peterson, S. and Nichols, S. A. (2004). *Restoration and Management of Lakes and Reservoirs*, 3rd edn. Boca Raton, FL: CRC Press.

Coughlan, A. (1991). Fresh water from the sea. *New Scientist*, **131**, 37.

David, R. and Pyne, G. (1995). *Groundwater Recharge and Wells: A Guide to Aquifer Storage Recovery*. Boca Raton, FL: CRC Press.

Dean, R. B. and Lund, E. (1981). *Water Reuse: Problems and Solutions*. London: Academic Press.

Department of the Environment (1988). *Assessment of Groundwater Quality in England and Wales*. London: HMSO.

EC (1975). Council Directive concerning the quality required of surface water intended for the abstraction of drinking water in the member states (75/440/EEC). *Official Journal of the European Community*, **L194**(25.7.75), 26–31.

Edmunds, W. M., Bath, A. H. and Miles, D. L. (1982). Hydrochemical evolution of the East Midlands Triassic sandstone aquifer, England. *Geochimica et Cosmochimica Acta*, **46**, 2069–82.

Environment Agency (2006). *Do We Need Large-Scale Water Transfers for South-East England?* Bristol: Environment Agency.

Eriksson, E. (1985). *Principles and Applications of Hydrochemistry*, London: Chapman and Hall.

Fewkes, A. and Ferris, F. A. (1982). The recycling of domestic wastewater: factors influencing storage capacity. *Building and Environment*, **17**, 209–16.

Franks, F. (1987). The hydrologic cycle: turnover, distribution and utilization of water. In *Handbook of Water Purification*, ed. W. Lorch. Chichester: Ellis Horwood, pp. 30–49.

Kendrick, M. A. P., Clark, L., Baxter, K. M. *et al.* (1985). *Trace Organics in British Aquifers: A Baseline Study*. Technical Report 223. Medmenham: Water Research Centre.

Kirby, C. (1979). *Water in Great Britain*. London: Penguin.

Lawrence, A. R. and Foster, S. S. D. (1987). *The Pollution Threat from Agricultural Pesticides and Industrial Solvents*. Hydrological Report 87/2. Wallingford: British Geological Survey.

Lerner, D. N. and Tellam, J. H. (1993). The protection of urban groundwater from pollution. In *Water and the Environment*, ed. J. C. Currie and A. T. Pepper. Chichester: Ellis Horwood, pp. 322–37.

OECD (1989). *Water Resource Management: Integrated Policies*. Paris: Organisation for Economic Co-operation and Development.

Olivier, J. (2004). Fog harvesting: an alternative source of water supply on the West Coast of South Africa. *GeoJournal*, **61**(2), 203–14.

Open University (1974). *The Earth's Physical Resources* Vol. 5. *Water Resources*, S26. Milton Keynes: Open University Press.

Owen, M. (1993). Groundwater abstraction and river flows. In *Water and the Environment*, ed. J. C. Currie and A. T. Pepper. Chichester: Ellis Horwood, pp. 302–11.

Parekh, B. S. (ed.) (1988). *Reverse Osmosis Technology*. New York: Marcel Dekker.

Parr, N., Charles, A. J. and Walker, S. (ed.) (1992). *Water Resources and Reservoir Engineering*. Proceedings of the seventh conference of the British Dam Society, Telford, London.

Pearce, F. (1992). *The Dammed: Rivers, Dams and the Coming of the World Water Crisis*. London: Bodley Head.

Pearce, F. (2006). *When the Rivers Run Dry: Water the Defining Crisis of the Twenty-First Century*, Boston, MA: Beacon Press.

Reichard, E., Craner, C., Raucher, R. and Zapponi, G. (1990). *Groundwater Contamination and Risk Assessment: a Guide to Understanding and Managing Uncertainties*. IAHS Publication 196. Wallingford, Oxford: International Association of Hydrological Sciences Press.

Royal Society (1984). *The Nitrogen Cycle of the United Kingdom: A Study Group Report*. London: Royal Society.

Todd, D. K. and Mays, L. W. (2005). *Groundwater Hydrology*. New York: John Wiley and Sons.

UKWIR (2004). *Implications of Changing Groundwater Quality for Water Resources in the UK Water Industry*. Report 04/WR/09/8. London: UK Water Industry Research.

Waggoner, P. (ed.) (1990). *Climate Change and US Water Resources*. New York: John Wiley.

Wanielista, M. P. (1990). *Hydrology and Water Quality Control*. New York: John Wiley.

Ward, D. R. (2003). *Water Wars*, New York: Riverhead Books.

Wood, F. C. (1987). Saline distillation. In *Handbook of Water Purification*, ed. W. Lorch. Chichester: Ellis Horwood, pp. 467–87.

Chapter 5
Nitrate and nitrite

5.1 Sources in water

The nitrogen cycle has altered drastically in the past 50 years with nitrate steadily accumulating in both surface and ground waters. Anthropogenic sources of nitrogen, of which nitrogen-based fertilizer is by far the greatest, now exceed the amount of nitrogen fixed by natural processes by 30% (Fields, 2004). Organic and inorganic sources of nitrogen are transformed to nitrate by a number of processes including mineralization, hydrolysis and bacterial nitrification. The resulting nitrate not utilized by plants or denitrified to nitrogen gas under anoxic (reducing) conditions leaches into surface and ground waters. However, recent evidence has shown that the rapid increase in nitrates in water sources is due not only to increasing fertilizer usage and the spreading of livestock wastes but also to the release of nitrogen oxides (NO_x) from burning fossil fuels. This increase in fixing nitrogen from the air to produce ammonia fertilizers while at the same time releasing nitrogen from the use of fossil fuels has led to the creation of a man-induced nitrogen cycle that dominates the original natural cycle. The nitrogen cascade, as it is widely known, is resulting in acid rain formation, global warming, ground level ozone and smog formation, and increasing runoff and leaching of nitrogen into water resources (Figure 5.1) (Hooper, 2006).

Nitrate fertilizer is the single most important and widely used chemical in farming. Its use on farms throughout Europe has rapidly increased over the past 30 years in particular, reaching the current phenomenal levels. In the UK the use of nitrate fertilizer increased from just 6000 tonnes per year in the late 1930s to 190 000 tonnes by the mid 1940s. This rise was due to the need to grow more food during the Second World War. However, the reasons for the increase in its annual use to a staggering 2 000 000 tonnes by 1995 are less clear. Similar trends have been seen in the USA rising from <1 to 11 million tons (1 ton is equivalent to 0.907 tonnes) of nitrogen equivalent between 1945 and 1992. This equates to a worldwide increase from 4 to 81 million tonnes between 1950 and 1990 (Cantor, 1997).

Nitrogen is an essential plant nutrient that is usually absorbed as nitrate or ammonium from the soil. It is used to form plant proteins, which in turn are used

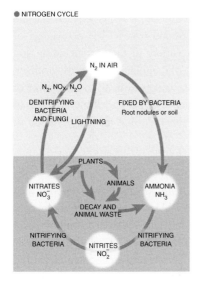

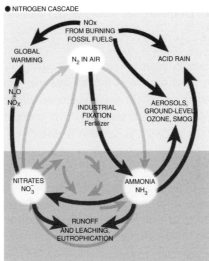

Figure 5.1 Comparison of the pre-industrial nitrogen cycle with the current nitrogen cycle, the nitrogen cascade, that now overshadows the natural cycle. Reproduced from Hooper (2006) with permission from *New Scientist*.

as major dietary sources of amino acids by humans and animals. The nitrogen absorbed from the soil must be replaced to maintain the fertility of the soil and therefore its long-term productivity. Farmers replace the nitrogen by either spreading manure or applying artificial fertilizers. Most of the nitrogen in the soil is in an organic form that is bound up either in plant material or organic matter and humus. As plants can only use inorganic (mineral) nitrogen this natural reserve of organic nitrogen is largely unavailable to plants for growth, unless it is broken down by microbial action from its organic form to nitrate. This slow process continuously releases mineral nitrogen in low concentrations, but not always when and in the amounts required by crops.

There are two main sources of nitrate contamination of water resources. (1) Nitrate is released when organic matter is broken down by bacteria in the soil. However, if the crops are not actively growing then the nitrate produced by microbial activity is not used by the plants and so is carried through the soil by rainwater into the aquifer to contaminate groundwater. (2) Inorganic nitrogen is added directly to fields by the farmer as artificial fertilizer. Where application exceeds plant needs or the ability of the plant to use the nitrate, then the excess will either be bound up in the soil or, more likely, washed out of the soil by rain into either surface or ground water. Both of these key sources are diffuse, potentially affecting vast areas of land. Other diffuse sources of nitrogen are the spreading of manure, slurry and sewage sludge to agricultural land. Point sources may also cause significant contamination of groundwaters used for supply. These include manure heaps, leaky slurry tanks, waste storage lagoons, silage clamps, farmyard runoff, soakaways and septic tanks.

Under ideal conditions 50–70% of the nitrogen applied to land as artificial fertilizer is taken up by plants, 2–20% is lost by volatilization, 15–25% is bound up with the organic matter or clay particles in the soil, leaving between 2% and 10% to be leached directly into surface or ground waters. The percentage of nitrate leaching from the soil depends on a variety of factors, including soil structure, plant activity, temperature, rainfall, the application rate of fertilizer, the water content of the soil and many more, so it is difficult to generalize or predict accurately the amount of nitrate that will be lost from an area of agricultural land. However, is it correct to conclude that the main source of nitrate pollution is artificial fertilizer?

Published work on nitrate pollution of water resources clearly concludes that the leaching of agricultural fertilizers is a major source. The Department of the Environment (1988) review on the subject, *The Nitrate Issue*, supports this view, suggesting that a cut of just 20% in the 190 kg of nitrogen now typically applied to each hectare of winter wheat would reduce nitrate leaching by up to 42%. Although there appears to have been some success in controlling nitrate by drastically reducing the application rates of nitrate fertilizers, research carried out at the Rothamsted Experimental Station indicates that the interaction between fertilizer application and nitrate leaching is far more complex (Addiscott, 1988; MacDonald *et al.*, 1989). All soils contain large amounts of organically bound nitrogen. Arable soils contain between 3000 and 8000 kg N ha^{-1}, whereas grassland can contain up to 15 000 kg N ha^{-1}, of which 90–95% is bound up with organic matter, mainly humus. Depending on factors such as the soil type, weather conditions and the previous crop grown, up to 3% of this organically bound nitrogen can be mineralized by soil bacteria into ammonia. This is a slow process, but subsequent nitrification of ammonia to nitrate occurs very rapidly (Figure 5.2). The nitrate is then available for a number of soil processes: for example, to be taken up by the crop, fixed biologically in the soil as microbial biomass, denitrified by micro-organisms to nitrogen gas, which is lost to the atmosphere, or leached from the soil into water resources. It is difficult to predict whether the nitrate will be used in the soil or whether it will be lost via leaching, but tillage certainly significantly increases leaching. Where grassland is intensively fertilized by manure, due to a high stocking density of animals or the excessive disposal of animal manure or sewage sludge, then there will be an excess of nitrogen that will be readily leached as nitrate. On certain soil types, such as sandy soils, the rate of leaching may approach those levels observed for arable land. Whenever grassland is

Figure 5.2 The breakdown of urea to nitrogen gas requires a number of microbial processes, both aerobic and anaerobic.

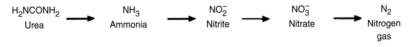

H_2NCONH_2		NH_3		NO_2^-		NO_3^-		N_2
Urea		Ammonia		Nitrite		Nitrate		Nitrogen gas

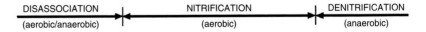

DISASSOCIATION	NITRIFICATION	DENITRIFICATION
(aerobic/anaerobic)	(aerobic)	(anaerobic)

ploughed up there is a large release of nitrogen into the soil that can be leached into water resources. The actual amount depends on the age of the grass, but $280 \, \text{kg N ha}^{-1}$ ploughed would be an average figure.

Even when crops have no fertilizer added about $20 \, \text{kg N ha}^{-1}$ leaches into the groundwater. Studies have shown that most of the nitrogen applied as fertilizer is used by the plant, whereas the excess nitrate, which is the most likely source of leachate, comes from the soil's own vast reserve of organically bound nitrogen, on average about $5000 \, \text{kg ha}^{-1}$, by the action of the microbes living in the soil. The problem is that the microbes are most active when conditions are ideal for them, not when the crop needs the nitrogen. For example, it is in the autumn when the soil is warm and its moisture content is increasing that the microbes are stimulated into producing most nitrate. This is also the time that the rainfall is beginning to exceed evaporation so that water flows downwards into the groundwater, taking any soluble nitrate with it. More land in Britain is put over to growing winter cereals than any other crop and so the nitrogen applied as fertilizer is more prone to being washed out of the soil by heavy rain. When spring barley is grown, however, the soil is left bare during the winter allowing more nitrate to be leached out of the soil than when winter wheat is grown. In Europe, nitrate will be mainly leached from the soil and into water resources in late autumn, winter and early spring. Water, not used by the plants nor lost by evaporation, percolates through the soil and eventually reaches the water-bearing aquifer, or alternatively surface waters. Forms of inorganic nitrogen other than nitrate found in fertilizers are ammonium and urea, both of which are rapidly transformed by soil micro-organisms into nitrate. The problem is therefore farming practice in general rather than specifically the overuse of fertilizer.

Inorganic fertilizers also contain smaller amounts of phosphorus and potassium. These nutrients do not generally cause a problem as they are effectively bound up and held by the soil particles, and unlike nitrate are not readily leached out of the soil by rainfall. However, although phosphorus and potassium do not affect drinking water quality directly, phosphorus, along with nitrogen, can cause eutrophication in surface waters with all the associated problems of excessive plant growth (Chapter 11).

5.2   Water quality standards

The units used to express nitrate concentrations have caused much confusion. They are expressed as either milligrams of nitrate per litre $(\text{mg NO}_3^- \, \text{l}^{-1})$ or as milligrams of nitrogen present as nitrate per litre $(\text{mg NO}_3^- - \text{N l}^{-1})$. There is considerable difference between the two. For example, $50 \, \text{mg NO}_3^- \, \text{l}^{-1}$ is equivalent to $11.3 \, \text{mg NO}_3^- - \text{N l}^{-1}$. In this book the actual nitrate concentration is used, i.e. $\text{mg NO}_3^- \, \text{l}^{-1}$, unless specified when the corrected value is given in parentheses. Care is needed when examining reports or data to ensure which units are being used. To correct mg $NO_3^- \, l^{-1}$ to mg $NO_3^- - N \, l^{-1}$ multiply by 0.226; conversely, to correct mg $NO_3^- - N \, l^{-1}$ to mg $NO_3^- \, l^{-1}$ multiply by 4.429.

In the EC Directive on Drinking Water (98/83/EEC), nitrate is included in Part B dealing with the chemical parameters with a maximum value of 50 mg $NO_3^- \, l^{-1}$. This is in line with the recommended guideline concentration specified by the World Health Organization (WHO, 2004) to protect against methaemoglobinaemia in bottle-fed infants (Section 5.3). Because of the close relationship between nitrate and nitrite, standards have also been set for nitrite (NO_2^-). The WHO has set two guideline nitrite values, a maximum value of 3 mg $NO_2^- \, l^{-1}$ for short-term exposure and a provisional long-term exposure value of 0.2 mg $NO_2^- \, l^{-1}$. In the 2004 revised guidelines for drinking water, the WHO also proposed that the sum of the concentrations of each of its guideline values should not exceed one. Many counties have adopted the nitrate–nitrite formula:

$$[NO_3^-/50] + [NO_2^-/3] = \leq 1$$

where NO_3^- and NO_2^- are the concentration of nitrate (mg $NO_3^- \, l^{-1}$) and nitrite (mg $NO_2^- \, l^{-1}$) respectively in the sample of water being evaluated.

The EC has set a maximum standard of 0.5 mg l^{-1} for nitrite. A lower prescribed value of 0.1 mg l^{-1} for nitrite has been set in the UK where supplies are disinfected using chloramination, that is where chlorine and ammonia are used together. Both nitrate and nitrite are included in the US Primary Drinking Water Regulations with maximum contaminant levels (MCLs) of 10 mg $NO_3^- - N \, l^{-1}$ and 1 mg $NO_2^- - N \, l^{-1}$ respectively, which are similar to the WHO nitrate and short-term nitrite exposure guideline values. The inclusion in the Primary Regulations is to protect against methaemoglobinaemia in infants under the age of 6 months.

5.3   Effect on consumers

Nitrate is a common component of food, with vegetables usually being the principal source in the daily diet. Intake of nitrate varies according to dietary habits, with the intake by vegetarians much higher than non-vegetarians. Some vegetables have low nitrate concentrations; for example, peas, mushrooms and potatoes all contain less than 200 mg kg^{-1}, whereas others such as beetroot, celery, lettuce and spinach are all very rich in nitrate, in excess of 2500 mg kg^{-1}. Vegetables grown out of season, or forced, generally contain higher than normal concentrations of nitrate (National Academy of Sciences, 1981). Therefore because our diets are so variable, estimates of nitrate intake from food vary over a wide range, from 30 to 300 mg $NO_3^- \, d^{-1}$. Table 5.1 shows that water can contribute significantly to the intake of nitrate; for example, when water contains the maximum limit of 50 mg l^{-1} it is contributing about half of the daily nitrate intake of an average consumer (Chilvers et al., 1984).

Table 5.1 *Contribution of nitrate in drinking water to the daily intake of nitrate in the diet*

Concentration of nitrate in water (mg l^{-1})	Daily nitrate intake (mg)		Percentage derived from drinking water
	Water	Food	
10	14	57	20
50	71	57	55
75	107	57	55
100	143	57	71
150	214	57	79

Breast-fed infants have a low nitrate intake whereas those fed on infant formula feeds receive nitrate from the water used in their preparation. Formula milk powder used for infant feeds also contains some nitrate in its own right, equivalent to about 5 mg l^{-1}. Thus the main source of nitrate for bottle-fed infants is the water; for example, if drinking water contains 25 mg l^{-1} nitrate, then the infant's feed will contain about 30 mg l^{-1} of nitrate. There is a significant risk of elevated nitrate exposure to bottle-fed infants when bottled water is used to make up feeds, so care is needed in the selection of suitable brands for this purpose (Section 29.2).

There is little nitrite, as opposed to nitrate, in drinking water, although some foods, especially cooked and cured meats, can contain high levels of sodium nitrite, which is a food preservative. Nitrite can often occur in distribution systems in warmer months due to nitrification, which can add an additional 0.2–1.0 mg l^{-1} of nitrite, or be formed after chloramination. Nitrate itself appears to be harmless at the concentrations found in water and most foodstuffs. In the body it is rapidly assimilated in the small intestine and taken up in the blood. Once absorbed, nitrate is excreted mostly unchanged in urine, but some will be reduced by bacterial action to nitrite. Part of the ingested nitrate, about 25%, is recirculated by excretion in saliva. It is in saliva that most of the nitrate is reduced to nitrite by bacteria. A similar conversion to nitrite occurs in the stomach (Tannenbaum *et al.*, 1976; Eisenbrand *et al.*, 1980). All the health considerations relating to nitrate are related to its conversion to nitrite, which is a reactive molecule associated with a number of problems, most commonly conversion to *N*-nitroso compounds, the formation of methaemoglobin and cancer (Section 5.7).

5.3.1 Infantile methaemoglobinaemia

The main concern associated with high nitrate concentrations in drinking water is the development of methaemoglobinaemia in infants. To cause enhanced methaemoglobin levels in blood, nitrate must first be reduced to nitrite, as nitrate itself does not cause the disorder. The nitrite combines with haemoglobin in red blood cells to form methaemoglobin, which is unable to carry oxygen and so reduces oxygen uptake in the lungs. Normal methaemoglobin levels in blood are between 0.5% and 2.0%. Research has shown that the use of water containing levels of nitrate up to double the EC maximum concentration, 50–100 mg l^{-1}, for infant feed preparation results in increased levels of methaemoglobin in the blood, but still within the normal physiological range. As methaemoglobin does not carry oxygen, excess levels lead to tissue anoxia (i.e. oxygen deprivation). It is only when the methaemoglobin concentration in the blood exceeds 10% that the skin takes on a blue tinge in infants, the disorder known as methaemoglobinaemia or blue-baby syndrome. The progressive symptoms resulting from oxygen deprivation are stupor, coma and eventual death. Death ensues when 45–65% of the haemoglobin has been converted. However, the disorder can be readily treated using an intravenous injection of methylene blue, which results in a rapid recovery.

Infants aged less than three months have a different respiratory pigment that combines with nitrite more readily than haemoglobin, making them especially susceptible to the syndrome. Their nitrate intake at this age is also high relative to their body weight compared with older children and they have an increased capacity to convert nitrate to nitrite. Children up to 12 months of age may have an incompletely developed system for methaemoglobin reduction, being naturally deficient in two specific enzymes that convert methaemoglobin back to haemoglobin (e.g. cytochrome b5 reductase), and so are also at risk. Nitrate is reduced to nitrite by gastric bacteria, and as infants have lower gastric acidity than older children, they are also potentially more at risk from this conversion process. Although methaemoglobinaemia is not usually a problem in adults, pregnant women are thought to be at risk, although the reasons are unclear (WHO, 1984).

It appears unlikely that infantile methaemoglobinaemia is caused by bacteriologically pure water supplies containing nitrate concentrations up to 100 mg l^{-1}. With 98% of Europe's population using piped mains, with the water treated to remove bacteriological contamination, the incidence of infantile methaemoglobinaemia has been significantly reduced. The disorder has not been reported in children drinking mains piped water, except when it has been contaminated by bacteria. The remaining 2% of the population is supplied with well or spring water of variable quality and acute infantile methaemoglobinaemia has been largely associated with bottle-fed infants using high-nitrate (>100 mg l^{-1}) well water. This is commonly referred to as well-water methaemoglobinaemia. Where lower concentrations of nitrate were found, then the bacterial status of the water was poor and/or the infants had gastroenteritis. Acute infantile methaemoglobinaemia can be a rare complication in gastroenteritis irrespective of nitrate intake. It is thought that this is due to the enhanced conversion of nitrate to nitrite from increased bacterial activity associated with enteritis. Like all cases of the disorder, well-water methaemoglobinaemia has become extremely rare over the past 40 years in the European Union with the reduction in incidence linked with the development of public supplies to rural areas (ECETOC, 1988). This is discussed further in Section 5.7. Avery (1999) has written an interesting and comprehensive review on the role of nitrates in drinking water and infantile methaemoglobinaemia.

5.3.2 *N*-nitroso compounds and cancer

Nitrate can be reduced to nitrite by the acidic conditions found in the stomach or by commensal bacteria in the saliva, small intestine and the colon, which then reacts with certain compounds in food under acidic conditions to produce *N*-nitroso compounds with amines and amides, *N*-nitrosamines and *N*-nitrosamides (Tricker, 1997), which are among the most serious carcinogens known (IARC, 1978) (Section 3.2). Indeed there is a direct relationship between nitrate intake and the formation of *N*-nitroso compounds (Moller *et al.*, 1989). Ward *et al.* (2005) describes this process, known as endogenous nitrosation, in detail. While *N*-nitroso compounds are extremely potent animal carcinogens, associated with cancer of the oesophagus, stomach, colon, bladder, lymphatics and haematopoietic system, there is little epidemiological evidence to link nitrate directly with cancer in humans (Cantor, 1997; Ward *et al.*, 2005). This is considered further in Section 5.7.

5.4 Nitrate in groundwaters

Nitrate leaches into the water table throughout the year, although the rate of leaching depends on factors such as geology, soil type, rainfall pattern, crop utilization rate of the nitrogen, the microbial conversion rate of nitrate and the fertilizer application pattern. Greatest leaching occurs in the autumn and winter. Nitrate is very soluble and is dissolved by rainwater and percolates deeper into the soil where it either enters the groundwater by direct percolation or, if it meets an impermeable layer such as clay, by sideways migration through the soil until it finds a way into the groundwater. This makes nitrate the commonest chemical contaminant of groundwater (Spalding and Exner, 1993). In some areas over 80% of the public water supply may come from groundwater sources, whereas small community and private supplies are almost exclusively from a groundwater source via boreholes, shallow wells or springs. In the USA an estimated 42% of the population relies on groundwater for their drinking water (Hutson *et al.*, 2004). This means that a considerable number of people are receiving groundwater that contains increased nitrate concentrations. Where there is no intervening impermeable layer to prevent nitrate percolating into the aquifer, then in areas of intensive arable production high nitrate concentrations are inevitable. In fact the difference in groundwater nitrate concentrations between agricultural and undisturbed areas can be as high as 60-fold (Hallberg and Keeney, 1993). Owing to the variable migration period it is difficult to predict the rate of movement of nitrate through the aquifer or the resulting concentration in the water supply. Depending on the thickness of the overlying rock, and whether or not the rock is fissured, this migration of nitrate can take up to 40 years. The extent of nitrate accumulation in groundwater depends on the climate and geology, but over the past half century the concentration of nitrate in rivers and waters from deep boreholes has increased steadily. The increase in some groundwaters observed in the 1980s, especially some chalk aquifers where percolation may be $<1.0\,\mathrm{m\,yr^{-1}}$, may reflect the intensification of agriculture during and after the Second World War when 8×10^6 acres of grassland were ploughed up and brought into arable production.

The 1970s and 1980s have been periods of massive nitrate use, equivalent to an eightfold increase in fertilizer use in Britain alone. This is reflected by many aquifers showing a steady increase in nitrate concentrations. However, when and at what concentration nitrate will stabilize in groundwater supplies is unknown. Predictions for the most severely affected areas in the UK suggest that the maximum concentration in drinking water supplies will level out at $150\,\mathrm{mg\,l^{-1}}$, assuming a continuation of current agricultural practice. In 1986 the UK Nitrate Co-ordination Group admitted that most groundwater sources were showing an overall increase in nitrate concentrations. They concluded that there would be a continuing and slow increase in groundwater nitrate concentration in most unconfined aquifers. This has indeed been so, although the rate of increase has generally been much faster than expected. The fact is that due to the time delay between the application of fertilizer and its appearance in drinking water, in many areas the concentration of nitrate in aquifers will continue to increase even if nitrates are banned altogether or land is taken out of production. Where the migration times are much less, however, remedial measures will have a more rapid effect on stabilizing and even reducing nitrate concentrations. As a general guide, the time lag between the reduction in nitrate leaching from the soil and a measurable reduction in groundwater nitrate concentrations will vary from a few

Figure 5.3 Trends in groundwater nitrate concentration in three representative aquifers in England. Reproduced with permission from the Environment Agency, Bristol. www. environment-agency.gov. uk/yourenv/eff/1190084/ water/2105666/? version=1&lang=_e.

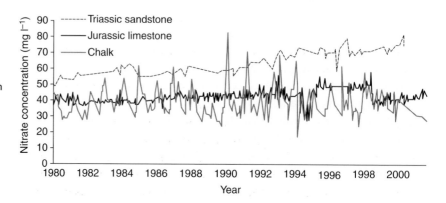

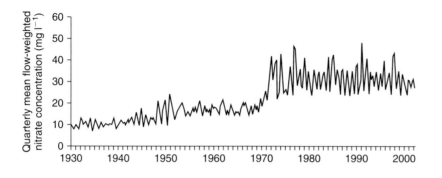

Figure 5.4 Nitrate concentrations recorded over 70 years in the River Thames at Walton, Surrey. In recent decades the nitrate level has stabilized with the expectation that the new measures adopted to reduce nitrogen leaching from agricultural land and nitrogen in discharges from treatment plants will see a gradual fall in the nutrient in this and other European rivers. Reproduced with permission from the Environment Agency, Bristol. www.environment-agency.gov.uk/yourenv/eff/1190084/water/210666/?version=1&lang=_e.

years for limestone aquifers to 10–20 years for sandstone aquifers, and up to 40 years in chalk aquifers. General trends in groundwater nitrate concentrations are shown in Figure 5.3.

5.5   Nitrate in surface waters

Those people obtaining their drinking water from surface waters are also at risk from nitrates. Nitrates enter rivers and lakes as surface runoff or as interflow, which is water moving sideways through the soil layer into a watercourse. Some rivers are also fed by groundwater via springs either above or below the surface of the stream. The nitrates are mainly added as runoff from agricultural land, with the highest nitrate concentrations clearly associated with areas of intensive farming. According to the Royal Society (1984) the concentrations of nitrates in British rivers have been increasing dramatically since the early 1960s, by about 400% during 1960–80 alone. However, nitrate concentrations in rivers of England and Wales have stabilized over the past decade with 28% of rivers in 2005 still having concentrations >30 mg l^{-1} (Figure 5.4). Areas of highest concentration are associated with intensive agricultural production.

Rivers have a seasonal pattern of nitrate concentration, with levels at their highest in autumn and winter when drainage from the land is greatest. This situation is exacerbated by climate change, which is making summers drier so that nitrogen builds up in the soil leading to more leaching when significant rainfall does occur. Many rivers in the UK have winter peaks in excess of 100 mg l^{-1}. This seasonal variation is clearly seen in the River Stour, for example, with nitrate peaks reaching way beyond the 50 mg l^{-1} EC limit (Figure 5.5). Nitrogen is added from other sources apart from agriculture, such as sewage and industrial discharges, which can at times be significant. Private water sources are particularly at risk as by their very nature these tend to be in rural areas and are susceptible to surface runoff and the rapid migration of pollution. Nitrate contamination originates not only from fertilizer and manure application, ploughing up grassland, tillage and other agricultural practices, but also from silage pits, slurry-holding tanks and septic tanks. In some areas land

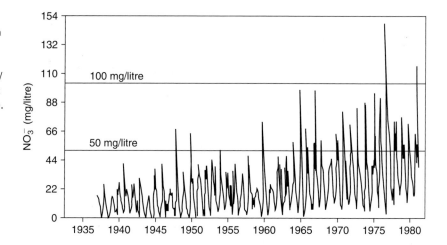

Figure 5.5 Increase in the concentration of nitrate in the River Stour at Langham over the period 1937–82 showing autumn/winter peaks (Department of the Environment, 1988). The period 1965 to 1975 showed the most rapid increase in nitrate concentration in British rivers due to the intensification of agriculture. Reproduced with permission from Defra.

drainage has significantly increased nitrate leaching, whereas sewage treatment plant effluents, silage liquor and slurry disposal are all important sources. Dutch farmers are currently producing 94 million tonnes of animal slurry each year, although their land can only absorb about 50 million tonnes safely. This has resulted in serious contamination of drinking waters in some provinces. Water blending is now common in the Netherlands and France to reduce nitrate levels; in the UK this practice is also becoming increasingly widespread. In the long term, a reduction in animal stocking density appears to be the only answer in the Netherlands. Treated sewage effluents also contain nitrate and these can contribute a significant proportion of the nitrate in rivers used for supply purposes. This is especially so in rivers such as the Thames and Trent where the water is reused a number of times (Section 4.5) (Figure 4.9). In these instances denitrification of effluents before they leave the treatment plant has become necessary, not only to reduce the nitrate concentration in the water being abstracted for supply, but to control eutrophication. Nutrient removal is required under the EC Urban Waste Water Treatment Directive (91/271/EEC) for receiving waters classed as sensitive (i.e. susceptible to eutrophication) (Gray, 2005).

Increased nitrate leaching is clearly associated with the intensification of agriculture. The problem in terms of control is that nitrate from agriculture comes from diffuse sources, that is from large undefined areas of land adjacent to rivers. In contrast, sewage and industrial nitrogen usually enter the river at point sources, that is a single point along the river, usually a clearly defined outlet pipe, which can be more easily controlled and treated.

5.6 Control of high-nitrate water

There are two approaches to reducing the nitrate concentrations in drinking water supplies. These are either to improve agricultural practice (prevention) or subsequently to reduce their concentration in water supplies (cure). The approach selected is dictated purely by economics and curiously enough prevention of contamination may be more expensive than removing the nitrate at the water treatment plant. Water utilities, however, take the view that those responsible for polluting water resources with contaminants should pay for their removal.

5.6.1 Treatment options

The methods employed to remove nitrates from water supplies include: (1) *Replacement.* In theory the easiest option is to replace high-nitrate supplies with low-nitrate supplies. However, this can be very expensive as new water mains and a suitable alternative supply are required. So, in practice, this option is usually restricted to small and isolated contaminated groundwater resources. (2) *Blending.* This is the controlled reduction of nitrate to an acceptable concentration by diluting nitrate-rich water with low-nitrate water. This is currently widely practised throughout Europe. Blending becomes increasingly more expensive if the nitrate concentration continues to increase. Also, not only is a suitable alternative supply required, but also facilities to mix water at the correct proportions. Although groundwater nitrate concentrations do not fluctuate widely from month to month, surface water nitrate levels can display large seasonal variations, with particularly high concentrations occurring after heavy rainfall following severe drought periods. Most blending operations involve diluting contaminated groundwater with river water, so such seasonal variations can cause serious problems to water supply companies. (3) *Storage.* Some removal of nitrate can be achieved by storing water for extended periods of time in large reservoirs. The nitrate is reduced to nitrogen gas by denitrifying bacteria under the low oxygen conditions that exist in the sediments of the reservoir. (4) *Treatment.* The nitrate can be removed by ion-exchange or microbial denitrification. Both are expensive and continuous operations. Water purification at this advanced stage is technically complex and difficult to operate continuously at a high efficiency level. Also the malfunction of nitrate reduction treatment plants would cause further water contamination. Ion-exchange systems use a resin that replaces nitrate ions with chloride ions. Although the system is efficient it does produce a concentrated brine effluent that requires safe disposal. Nitrate removal brings about other changes in drinking water composition, and the long-term implications for the health of consumers whose drinking water is treated by ion-exchange are unknown. This is the process in which most water companies within Europe have invested, ensuring few cases of piped drinking water now exceeding the EC limit. Methods of reducing nitrate concentrations by treatment within aquifers and the use of scavenging wells to intercept polluted groundwater in aquifers and prevent it reaching supply wells are being developed. (5) *Selective replacement.* Rather than treating the entire water supply, many water supply companies use a system of identifying those most at risk within the community from high nitrates, and supplying an alternative supply such as bottled water or home treatment units. This is by far the cheapest approach for the water industry but if limits in the supplied water are exceeded, then prosecution will inevitably follow. Bottling plants have been used in some affected areas in Britain and Europe to supply low-nitrate water for infants and pregnant women during emergences (Chapter 29), for example, in the villages around Ripon in the Yorkshire Dales where the local groundwater sources became severely contaminated by nitrates in 1988 bottled water was supplied for infants less than six months of age. Blending eventually reduced the nitrate level. Similar situations are becoming increasingly common throughout Europe, but have received little media attention. Of course, the first response of a member of the public on hearing that his/her drinking water is contaminated is to boil it. However, as with many other inorganic chemical pollutants found in water, boiling will not remove nitrate or render it harmless, but simply increase its concentration in the water due to evaporation. In the home, nitrate can only effectively be removed from drinking water by ion-exchange (Chapter 29). This is considered further by German (1989).

5.6.2   Control of agricultural sources of nitrate

There are a number of options farmers can adopt to reduce nitrate contamination: (1) cutting down on fertilizer use; (2) modifying cropping systems; and (3) light control measures (Department of the Environment, 1988). It is the last category of measures that may have the most significant effect on nitrate leaching. Among these measures are applying nitrogen fertilizer strictly according to professional advice: not applying it in autumn; leaving the soil covered over winter, perhaps by growing a 'catch crop' to mop up the nitrate; sowing winter crops as early as possible; and taking great care with manures. Specifically, good agricultural practice includes: (1) fallow periods should be avoided. The soil should be kept covered for as long as possible, especially during winter, by the early sowing of winter cereals, intercropping or using a straw mulch. (2) Avoid increased sowing of legumes unless the subsequent crop can use the nitrogen released by mineralization of nitrogen-rich residues. (3) Grassland should not be ploughed. (4) Tillage should be minimized and avoided in the autumn. Direct drilling should be used whenever possible. (5) Slopes should, whenever possible, be cultivated transversely to minimize runoff. (6) Manure should not be spread in autumn or winter. It should be spread evenly and the amount should not exceed crop requirements. (7) The amount of nitrogen fertilizer and manure used should be applied at times and in amounts required by plants to ensure maximum uptake. The amount of available nitrogen in the soil should also be considered. Simple computer models are available to help the farmer calculate very accurately the amount of fertilizer required for a particular crop grown on a specific soil type; alternatively, consultants should be used to advise on fertilizer application rates. Farmers can achieve considerable savings in expenditure by optimizing fertilizer use. This is an example of what is good economics is also good for the environment.

Taking land out of arable cropping and putting it down to low-productivity grassland would certainly reduce nitrate leaching. The set-aside policy, where arable land is taken out of production and put back to grassland, or even forestry, will reduce nitrate leaching. Of course, once this land is brought back into arable production, leaching will begin again. In those areas where aquifers are most at risk this may be the only option.

Aquifer protection zones have been suggested throughout Europe. In Ireland much success has been achieved in reducing nitrate contamination of groundwater in aquifer protection zones by improving farming practice, mainly through using less nitrogen fertilizer, especially in the autumn, and by sowing winter wheat earlier. The Wessex Water Company has experimented with restricting the amount of fertilizer used by tenant farmers on its land. The 96 000 people of the city of Bath obtain 80% of their water from springs that feed several supply reservoirs. In 1984 a review of water quality indicated that the nitrate concentration would exceed the EC standard within a few years. The nitrate was being leached from agricultural land and it was found that in several areas rapid infiltration of surface water to shallow springs was the main cause of the problem. The following year, when an opportunity arose to alter farming practice in parts of the catchment, Wessex Water asked tenant farmers to transfer from arable cropping to grass production, and suggested that where cereals were grown they should be autumn-sown varieties. No development of the land was allowed and the amount and timing of fertilizer application was strictly controlled. The shallow springs showed an immediate reduction in nitrate levels, with the downward trend still continuing, although the deeper springs had not shown any reduction in nitrate concentration at the time the report was published. The cost of this operation, in terms of land purchase and compensation to farmers for loss of income due to the restrictions, has been a fraction of the cost of advanced water treatment or replacing the resource (Knight and Tuckwell, 1988). This proved highly successful, with a significant reduction in nitrate concentrations recorded in both groundwater and surface waters.

The EC Nitrates Directive (91/676/EEC) requires all member states to reduce water pollution by nitrates from agricultural sources by two specific actions (EC, 1991): (1) the implementation of codes of good agricultural practice to limit nitrate leaching and (2) the designation of Nitrate Vulnerable Zones (NVZs) and the development of action programmes for their management.

There are a number of specific codes of practice to protect surface and ground waters from agricultural pollution, including nitrates. In the UK these include *The Water Code* (MAFF, 1998) and two excellent guides to best practice published by the Scottish Executive (2005) and the Northern Ireland Department of Agriculture and Rural Development (2005). The latter contains useful appendices giving methods for the calculation of application rates of inorganic fertilizers and manures to crops. Similar codes have been developed throughout the world in an attempt to reduce the effects of nitrate leaching.

While the codes of good agricultural practice are voluntary, they form the basis of the Action Programmes for NVZs, which contain mandatory measures enforced by the Environment Agency. Among the measures that the Action Programme must contain are: (1) The application of inorganic nitrogen fertilizer must be limited to the requirement of the crop, taking into account the residues in the soil and other sources of nitrogen. (2) Organic manure applications must be restricted to 210 kg total nitrogen (N) ha^{-1} yr^{-1} averaged over the entire farm not in grass and reducing to 170 kg N ha^{-1} yr^{-1}after four years. A higher limit of 250 kg N ha^{-1}yr^{-1} is allowed on grassland. (3) No manures should be applied to sandy or shallow soils between 1 September and 1 November (grassland or autumn-sown crop) or 1 August and 1 November (field not in grass without an autumn-sown crop). Farmers must be able to demonstrate that they have sufficient storage capacity for such manures or are able to dispose of them in an alternative environmentally acceptable manner. (4) Adequate records must be kept including livestock numbers, cropping, the use of inorganic fertilizers and organic manures.

Originally 66 NVZs were designated in England covering 600 000 hectares of agricultural land. Areas were selected where the surface or ground waters are at risk of having a high nitrate concentration (> 50 mg l^{-1}). After pressure from the EC, these areas were extended to cover 55% of the country (Figure 5.6). Farmers have had to comply fully with the Action Programmes since 19 December 2002. From 19 December 2006, farms in all NVZs are restricted to a whole-farm limit for organic manure loading of 170 kg N ha^{-1} yr^{-1}.

The EC Water Framework Directive (2000/60/EC) establishes management plans for individual water basins or catchments, including groundwaters and will play an increasingly important role in the control of contaminants such as nitrates in the future. The Groundwater Directive (80/68/EEC) and the Urban Wastewater Treatment Directive (91/271/EEC) also play important roles in controlling nitrates and other dangerous substances discharged into water supplies. The new EC Directive on the protection of groundwater against pollution and deterioration, the so-called Groundwater Daughter Directive (2006/118/EC) will set new threshold values for key water quality parameters and strengthen existing measures to prevent or limit contamination.

Traditionally, sewage sludge has been applied to arable land in the autumn. It is clear that this coincides with the period of greatest potential risk for nitrate leaching. In nitrate-sensitive areas the spreading of slurry or liquid sewage sludge is no longer permitted between 31 August and 1 November to grassland, or between 30 June and 1 November to land under arable cultivation. The

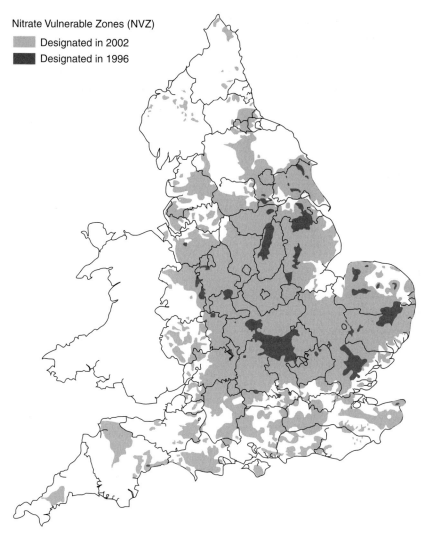

Nitrate Vulnerable Zones (NVZ)

Designated in 2002

Designated in 1996

Figure 5.6 Nitrate vulnerable zones as originally designated in England and Wales in 1996 and subsequently extended in 2002. Reproduced with permission from the Department for Environment, Food and Rural Affairs. www.defra. gov.uk/Environment/ water/quality/nitrate/ maps.htm#summary.

maximum application of slurry or sewage sludge to agricultural land is restricted to 175 kg N ha^{-1} yr^{-1}. The levels of inorganic nitrogen fertilizer must be reduced to below the economic optimum (i.e. 25 kg N ha^{-1} yr^{-1} below the optimum for winter wheat, barley and forage catch crops, and 50 kg N ha^{-1} yr^{-1} below that for oilseed rape). These levels have to be adjusted to take into account any additional nitrogen supplied by the application of organic manures, including sewage sludge.

In some instances a reduction in nitrate fertilizer has a poor effect on reducing nitrate concentrations in water resources for a number of reasons: (1) the organic nitrogen reservoir in the soil is so large that in many instances this will take years to decrease significantly; (2) arable farming, even with reduced fertilizer use and productivity, can still produce high nitrate levels; and finally (3) in some areas nitrate from past leaching has yet to reach the water table. Organic farming can also cause nitrate leaching, especially the overuse of manure on land or the use of legumes in tillage rotation. In the eastern counties of Britain the annual drainage is low, less than 250 mm, whereas farming is predominantly arable. The water draining from the land is therefore rich in nitrate. As reducing the rate of application of fertilizer may only have a marginal effect, the only effective option may be to take several thousand square kilometres of prime arable land in the UK out of production. Conversion to grassland with modest livestock numbers, or even better to forestry, can drastically reduce nitrate leaching. Increasing interest is being shown in alternative crops such as the growth of elephant grass (*Miscanthus*) for biomass. This crop does not require the use of fertilizer thus preventing nitrogen leaching (Clifton-Brown *et al.*, 2004).

5.7 Conclusions

Nitrate concentrations in drinking water are generally below $10 \, \text{mg} \, \text{l}^{-1}$ with better water treatment technology maintaining the WHO guideline of $50 \, \text{mg} \, \text{l}^{-1}$ for the majority of piped supplies. Biological denitrification is generally employed for surface waters and ion-exchange for groundwaters. Both can achieve nitrate concentrations of $<5 \, \text{mg} \, \text{l}^{-1}$. New legislation in Europe is aimed at minimizing the effects of diffuse agricultural sources of nitrate entering water resources, while point sources such as wastewater treatment plant effluents now require nutrient removal where there is a threat of eutrophication in receiving waters. Those supplied by individual wells or small rural schemes are most at risk from elevated nitrate in their drinking water. For example a survey of 3351 private wells in agricultural areas carried out by the US Geological Survey (Mueller *et al.*, 1995) found that approximately 9% exceeded the MCL. Shallow dug wells appear to be far more at risk from contamination than deeper drilled boreholes, with 39% and 22% respectively exceeding the MCL in a study in South Dakota (Johnson and Kross, 1990). In Europe nitrate concentration currently exceeds the WHO guideline in about 35% of groundwater supplies, with some private wells in rural areas exceeding the limit by up to 15 times (European Environment Agency, 2003). However, after treatment very few public supplies exceed maximum nitrate levels. This is not the case with private supplies and in the UK there are some 140 000 private water supplies that utilize mainly groundwater resources. For example, in Aberdeenshire (Scotland) a study carried out between 1992 and 1998 found that 15% of 1100 samples taken from private supplies exceeded the EC limit for nitrate. The average nitrate concentration was $32 \, \text{mg} \, \text{l}^{-1}$ with 50% of samples $>25 \, \text{mg} \, \text{l}^{-1}$. The samples were from predominately agricultural areas with 41% and 30% of the samples also exceeding the EC limits for total and faecal coliforms respectively (Reid *et al.*, 2003). This is a worldwide problem.

While many researchers have suggested that the WHO guideline of $50\,\text{mg}\,l^{-1}$ is overly conservative (Avery, 1999; L'Hirondel and L'Hirondel, 2002), others have identified potential adverse health effects at nitrate concentrations well below this value such as birth defects (Brender *et al.*, 2004), colon and rectal cancer (DeRoos *et al.*, 2003), respiratory infections (Gupta *et al.*, 2000), non-Hodgkin's lymphoma (Ward *et al.*, 1996) and elevated risks of cancer in older women (Weyer *et al.*, 2001).

The WHO guideline for nitrate in drinking water is based on the observation that no cases of infantile methaemoglobinaemia were recorded below this concentration (Walton, 1951). This is supported by very few cases being reported in countries that have adopted this guideline value. So while the guideline value appears to be about right, large numbers of infants on private supplies in agricultural areas are exposed to much higher nitrate concentrations and yet the incidence of infantile methaemoglobinaemia also remains low in these areas. Cofactors such as the presence of microbial contaminants, diarrhoea and respiratory diseases are known to increase methaemoglobin levels, so better awareness of water quality, especially faecal bacterial contamination, better child health and diet, especially the consumption of antioxidants such as vitamin C, all seem to have contributed to the prevention of methaemoglobinaemia. There is also evidence that cytochrome b5 reductase activity is higher in those drinking water with high nitrate concentrations indicating perhaps an adaptation to high-nitrate drinking water (Gupta *et al.*, 1999). In the absence of any strong epidemiological evidence to link drinking water nitrate directly to cancer, then it would appear that the current WHO guideline of $50\,\text{mg}\,l^{-1}$, adopted as the nitrate standard by most countries, is adequate to protect consumers.

Those most at risk from methaemoglobinaemia, those under 12 months of age, should however be protected from high nitrate concentrations, especially when making up dried milk and other liquid feeds. Breast milk contains little nitrate; however, the relatively high intake of water in relation to the body weight of infants puts them at great risk from nitrates in water. With elevated levels of nitrate in some vegetables (Sanchez-Echaniz *et al.*, 2001), it may be prudent to source water and manufactured drinks with as low a nitrate concentration as possible for this age group. Nitrite is ten times more potent than nitrate on a molar basis in the causation of methaemoglobinaemia and so drinking water should contain as little nitrite as possible. The disinfection process chloramination can result in erratic nitrite formation within the distribution system and this requires constant monitoring to prevent elevated nitrite in drinking water.

Clearly a combination of better agricultural practices and some water treatment is the best solution to control nitrates in drinking water. However, there are some very real practical problems in monitoring fertilizer use or good agricultural practice, so that the brunt of the responsibility will inevitably have to be borne by the water supply companies with the high cost being passed on to the consumer (Conrad, 1990).

References

Addiscott, T. M. (1988). Long-term leakage of nitrate from bare unmanured soil. *Soil Use and Management*, **4**, 91–5.

Avery, A. A. (1999). Infantile methemoglobinemia: reexamining the role of drinking water nitrates. *Environmental Health Perspectives*, **107**, 1–8.

Brender, J. D., Olive, J. M., Felkner, M. *et al.* (2004). Dietary nitrates, nitrosatable drugs and neural tube defects. *Epidemiology*, **15**, 330–6.

Cantor, K. P. (1997). Drinking water and cancer. *Cancer Causes and Control*, **8**, 292–308.

Chilvers, C., Inskip, H., Caygill, C. *et al.* (1984). A survey of dietary nitrate in well-water users. *International Journal of Epidemiology*, **13**, 324–31.

Clifton-Brown, J. C., Stampfl, P. F. and Jones, M. B. (2004). *Miscanthus* biomass production for energy in Europe and its potential contribution to decreasing fossil fuel carbon emissions. *Global Change Biology*, **10**, 509–18.

Conrad, J. J. (1990). *Nitrate Pollution and Politics*. Aldershot, UK: Avebury Technical Press.

Department of the Environment (1988). *The Nitrate Issue*. London: HMSO.

DeRoos, A. J., Ward, M. H., Lynch, C. F. and Cantor, K. P. (2003). Nitrate in public water systems and the risk of colon and rectum cancers. *Epidemiology*, **14**, 640–9.

EC (1991). Council Directive concerning the protection of waters against pollution caused by nitrates from agricultural sources (91/676/EEC). *Official Journal of the European Communities*, **L375**, (31.12.91), 1–8.

ECETOC (1988). *Nitrates and Drinking Water*. ECETOC Technical Report 27. Brussels, Belgium: European Chemical Industry Ecology and Toxicology Centre.

Eisenbrand, G., Speigelhalder, B. and Preussmann, R. (1980). Nitrate and nitrite in saliva. *Oncology*, **37**, 227–31.

European Environment Agency (2003). *Europe's Environment: The Third Assessment*. Environmental Assessment Report No.10. Copenhagen: European Environmental Agency.

Fields, S. (2004). Global nitrogen: cycling out of control. *Environmental Health Perspectives*, **112**, A557–63.

German, J. C. (ed.) (1989). *Management Systems to Reduce Impact of Nitrates*. London: Elsevier.

Gray, N. F. (2005). *Water Technology: An Introduction for Environmental Scientists and Engineers*. Oxford: Elsevier.

Gupta, S. K., Gupta, R. C., Gupta, A. B. *et al.* (1999). Adaptation of cytochrome b5 reductase activity and methemoglobinemia in areas with a high nitrate concentration in drinking water. *Bulletin of the World Health Organization*, **77**, 749–53.

Gupta, S. K., Gupta, R. C., Gupta, A. B. *et al.* (2000). Recurrent acute respiratory infections in areas with high nitrate concentrations in drinking water. *Environmental Health Perspectives*, **108**, 363–6.

Hallberg, G. R. and Keeney, D. R. (1993). Nitrate. In *Regional Groundwater Quality*, ed. W. M. Alley. New York, NY: Van Nostrand Reinhold, pp. 297–322.

Hooper, R. (2006). Something in the air. *New Scientist*, **189**(2535) (21 January, 2006), 40–3.

Hutson, S. S., Barber, N. L., Kenny, J. F. *et al.* (2004). *Estimated Use of Water in the United States*, USGS Circular 1268. Denver, CO: US Geological Survey.

IARC (1978). *Some N-nitroso Compounds*. IARC Monograph on the Evaluation of the Carcinogenic Risk of Chemicals to Humans, Vol. 17, International Agency for Research on Cancer, Lyon, France.

Johnson, C. J. and Kross, B. C. (1990). Continuing importance of nitrate contamination of groundwater and wells in rural areas. *American Journal of Industrial Medicine*, **18**, 449–56.

Knight, M. S. and Tuckwell, J. B. (1988). Controlling nitrate leaching in water supply catchments. *Water and Environmental Management*, **2**, 248–52.

L'Hirondel, J. and L'Hirondel, J.-L. (2002). *Nitrate and Man: Toxic, Harmless or Beneficial?* Wallingford: CABI Publishing.

MacDonald, A. J., Powlson, D. S., Poulton, P. R. and Jenkinson, D. S. (1989). Unused fertilizer nitrogen in arable soils-its contribution to nitrate leaching. *Journal of the Science of Food and Agriculture*, **46**, 407–19.

MAFF (1998). *The Water Code: Code of Good Agricultural Practice for the Protection of Water*. London: Ministry of Agriculture Fisheries and Food. www.defra.gov.uk/farm/environment/cogap/index.htm.

Moller, H., Landt, J., Pedersen, E. *et al.* (1989). Endogenous nitrosation in relation to nitrate exposure from drinking water and diet in a Danish rural population. *Cancer Research*, **49**, 3117–21.

Mueller, D. K., Hamilton, P. A., Helsel, D. R., Hitt, K. J. and Ruddy, B. C. (1995). *Nutrients in Groundwater and Surface Water of the United States: An Analysis of Data Through 1992*. Water Resources Investigations Report 95–4031. Denver, CO: US Geological Survey.

National Academy of Sciences (1981). *The Health Effects of Nitrate, Nitrite, and N-nitroso Compounds*. Washington DC: Committee on Nitrate and Alternative Curing Agents in Food, National Academy of Sciences.

Northern Ireland Department of Agriculture and Rural Development (2005). *Code of Good Agricultural Practice for the Prevention of Pollution of Water*. Belfast: Department of Agriculture and Rural Development. http://agrifor.ac.uk/browse/cabi/c1fdca80522f8f9851a2616c44125010.html.

Reid, D. C., Edwards, A. C., Cooper, D., Wilson, E. and Mcgaw, B. A. (2003). The quality of drinking water from private supplies in Aberdeenshire, UK. *Water Research*, **37**, 245–54.

Royal Society (1984). *The Nitrogen Cycle of the United Kingdom. A Study Group Report*. London: The Royal Society.

Sanchez-Echaniz, J., Benito-Fernandez, J. and Mintegui-Raso, S. (2001). Methemoglobinemia and the consumption of vegetables in infants. *Pediatrics*, **107**, 1024–8.

Scottish Executive (2005). *Prevention of Environmental Pollution From Agricultural Activity*. Edinburgh: Scottish Executive. www.scotland.gov.uk/Publications/2005/03/20613/51366.

Spalding, R. F. and Exner, M. E. (1993). Occurrence of nitrate in groundwater-a review. *Journal of Environmental Quality*, **22**, 392–402.

Tannenbaum, S. R., Weisman, M. and Fett, D. (1976). The effect of nitrate intake on nitrite formation in human saliva. *Food and Cosmetics Toxicology*, **14**, 549–52.

Tricker, A. R. (1997). *N*-nitroso compounds and man: sources of exposure, endogenous formation and occurrence in body fluids. *European Journal of Cancer Prevention*, **6**, 226–68.

Walton, G. (1951). Survey of literature relating to infant methemoglobinemia due to nitrate-contaminated water. *American Journal of Public Health*, **41**, 986–96.

Ward, M. H., Mark, S. D., Cantor, K. P. *et al.* (1996). Drinking water nitrate and the risk of non-Hodgkin's lymphoma. *Epidemiology*, **7**, 465–71.

Ward, M. H., deKok, T. M., Levallois, P. *et al.* (2005). Workshop report: drinking-water nitrate and health – recent findings and research needs. *Environmental Health Perspectives*, **113** (11), 1607–14.

Weyer, P. J., Cerhan, J. R., Kross, B. C. *et al.* (2001). Municipal drinking water nitrate levels and cancer risk in older women: the Iowa Women's health study. *Epidemiology*, **12**, 327–38.

WHO (1984). *Guidelines for Drinking Water Quality,* Vol. 2. *Health Criteria and Other Supporting Information*. Geneva: World Health Organization.

WHO (2004). *Guidelines for Drinking Water Quality,* Vol. 1. *Recommendations*, 3rd edn. Geneva: World Health Organization.

Chapter 6
Pesticides and organic micro-pollutants

6.1 Organic micro-pollutants

Under normal circumstances food intake of pesticide residuals is about 10^3 to 10^5 times higher than that induced by drinking water or inhalation (Margni *et al.*, 2002). However, when organic contaminants enter water resources then drinking water becomes an increasingly important factor in human toxicity. Organic compounds found in drinking water fall into three broad categories: (1) naturally occurring organics; (2) synthetic (man-made) organic compounds; and (3) organic compounds synthesized during water treatment (disinfection by-products). The last category is dealt with in Chapter 18, while the first two categories are discussed below.

A large number of organic compounds are found naturally as well as synthesized for industrial uses (as solvents, cleaners, degreasers, petroleum products, plastics manufacture and, of course, their derivatives) and for agricultural use (mainly pesticides). Since the 1940s tens of thousands of new chemicals have been developed and are being used not only by industry and agriculture, but by the service and domestic sectors. In a survey carried out by the Water Research Centre in the UK in the late 1970s, 324 organic compounds were identified in drinking water samples (Fielding *et al.*, 1981).

Organic micro-pollutants, including pesticides, are similar to nitrates in that their toxicity cannot be attributed solely to drinking water and that many of them percolate into the soil and accumulate in aquifers or surface waters. They can contaminate drinking water sources via many routes: agricultural runoff to surface waters or percolation into groundwaters, industrial spillages to surface and ground waters, runoff from roads and paved areas, industrial wastewater effluents, leaching from chemically treated surfaces, domestic sewage effluents, atmospheric fallout, carried in rainfall, and as leachate from industrial and domestic landfill sites.

The Water Research Centre survey showed that a high proportion of the British public is exposed to minute amounts of a wide range of organic chemicals and for most of these there are insufficient toxicological data to make an accurate assessment of their safety. Many of them, however, are known to be toxic and also possibly carcinogenic, even at very low concentrations, leading to concerns about the health effects of long-term exposure to trace concentrations of these organic compounds. Analytical, toxicological and epidemiological studies on organic mirco-pollutants are underway throughout Europe. However, to date no definite health effects have been established for the group at the concentrations usually found in drinking water (Cantor, 1997). Of course, many organic compounds are present at undetectable levels and there is evidence that repeated small doses of some organic chemicals might also lead to chronic diseases. There is a lot of controversy about the presence or absence of a non-effect level, particularly for carcinogens. However, these organic chemicals are not only found in drinking water, they also accumulate in food and the atmosphere. Therefore in calculating the total intake of such chemicals the low concentrations of organic micro-pollutants found in drinking water may indeed be significant in terms of the overall acceptable (or tolerable) daily intake (ADI). Other factors such as possible increased toxicity effects when two or more organic compounds are present together, and the fact that some members of the community, especially the young and old, have a much higher water intake than others, makes the calculation of safe concentrations of these chemicals in water extremely difficult.

At present there is no simple routine technique that can be used to identify and measure all the organic compounds in drinking water. There are a vast number of organic compounds, many of which may form further complex compounds in water. The analytical methods available for organic chemicals are the most sophisticated of all the chemical analytical techniques, namely chromatography and mass spectrometry. Gas chromatography, for example, can identify less than 30% of the compounds present due to volatility limitations, and although mass spectrometry is much more effective it requires a large financial investment. Gas chromatography-mass spectrometry (GC-MS) has been widely used to analyze organic contaminants, although this is being replaced by more efficient methods using liquid chromatography-mass spectrometry (LC-MS) (Petrović et al., 2003). For example, Rodriguez-Mozaz et al. (2004) describe a simultaneous extraction method for key pesticides (e.g. atrazine, simazine desethylatrazine, the main degradation product of atrazine, isoproturon and diuron) and oestrogens (e.g. oestradiol, oestrone, oestriol, oestradiol-17-glucuronide, oestradiol diacetate, oestrone-3-sulphate, ethinyl oestradiol and diethylstilbestrol) using pre-concentration followed by LC-MS. In practice, very little is known about what is in drinking waters in terms of organic chemicals as we are unable to determine a large portion of them and often cannot afford to monitor those we can analyze frequently enough, even though water utilities are generally required to do so. In the UK, for example, the Water Supply (Water Quality) Regulations 2000 requires all companies to monitor pesticides (DETR, 2000). To complicate matters, factors such as temperature, pH and the hardness of the water all affect toxicity. For example, organic micro-pollutants tend to be more toxic in soft waters. There is no doubt that organic micro-pollutants are the most important and potentially the most harmful contaminants found in drinking water.

Micro-pollutants are present in water in infinitesimal amounts. For normal inorganic compounds such as nitrate, concentrations can be adequately expressed in milligrams per litre (mg l^{-1}). However, for micro-pollutants, micrograms per litre (μg l^{-1}) are often used, so 1 μg l^{-1} is the same as 0.001 mg l^{-1} (1000 μg l^{-1} = 1 mg l^{-1}). For some compounds even smaller units are required such as nanograms per litre (ng l^{-1}), so 1 ng l^{-1} is equivalent to 0.000001 mg l^{-1} (1 000 000 ng l^{-1} = 1 mg l^{-1} or 1000 ng l^{-1} = 1 μg l^{-1}).

The legislation governing organic micro-pollutants is quite complex covering concentrations in both finished drinking water as well as water resources. In the EC, groups of organic compounds and specific compounds are covered under Part B of the Drinking Water Directive (98/83/EEC) (EC, 1998). These include pesticides (Section 6.2), industrial solvents (Section 6.3) and polycyclic (polynuclear) aromatic hydrocarbons (Section 6.4) (Appendix 1). The Water Framework Directive (2000/60/EC) now controls the concentrations of these compounds in surface and ground waters and lists 33 priority substances including the organic compounds (Appendix 5), although it has been argued that several of the triazine herbicides should also be included in this list. Similarly the US Environmental Protection Agency (USEPA) have set maximum contaminant levels (MCLs) for pesticides in drinking water and also included many of them in their drinking water contaminant candidate list (Appendix 7). Details of current standards in both Europe and the USA are given below.

6.2 Pesticides

6.2.1 Classification

Pesticides can be classified by use as an insecticide, herbicide, algicide or fungicide. They are used to kill either a broad spectrum of pests or specific groups or species of pest. Occasionally they may be used together, especially for fumigation and sterilization where both a broad-acting insecticide and fungicide are required. Herbicides are often mixed with fertilizers such as lawn improvers, the idea being to kill off weeds but to encourage the growth of the desired crop. Although herbicides are by far the largest group in terms of compounds and weights used, it has been the insecticides that have, in the past, posed the major environmental and health hazards. There are four main groups of insecticides: organochlorines, organophosphorus compounds, carbamates and pyrethroids.

Organochlorines (chlorinated hydrocarbons) were widely manufactured and used in the 1950s and 1960s. They were considered a major breakthrough because they were very persistent, that is, they degraded extremely slowly in the environment and so went on working for very long periods of time. They have largely been banned or their use severely restricted in most countries following the major destruction of natural wildlife, especially birds. The organochlorines are lipophilic, which means they become concentrated in fatty tissues. As predators eat these animals there is a net increase in the concentration of these compounds in animals up the food chain. There are three types of organochlorines.

DDT, and related substances such as DDE were widely used to control malarial mosquitoes and all flying insects. DDT is an acronym for dichlorodiphenyltrichloroethane, although its chemical name is $1,1'$-(2,2, 2-trichloroethylidene)*bis*-(4-chlorobenzene). It is insoluble in water but soluble in organic solvents and is an extremely stable compound that is highly resistant to breakdown in the environment. Many species are able to convert DDT to its metabolite DDE $(1,1'$-(2,2-trichoroethylidene)*bis*-(4-chlorobenzene) (WHO, 1984). Its use became so ubiquitous that many insects began to develop resistance and so it became less effective. It was also found to be accumulating in fatty tissues of all animals, including humans, and was identified in human breast milk. Today its use is severely restricted in developed countries, although unfortunately it is still widely used in some tropical countries. DDT mainly affects the central nervous system and the liver.

Lindane (γ-hexachlorocyclohexane), also written as γ-HCH or γ-BHC, and related chemicals were also widely used at one time as broad-spectrum insecticides for a wide range of applications. Like the others in this group it is only degraded very slowly in soil and has now become ubiquitous in surface waters and more recently has been found in groundwaters as well. This group is thought to cause birth defects, including stillbirths, and has been found to be carcinogenic in laboratory tests on animals.

Aldrin, dieldrin and all the other 'drins' were used mainly for seed dressing. The full chemical names for these two pesticides are warnings in themselves that they are dangerous. Aldrin (1, 2, 3, 4, 10, 10-hexachloro-l, 4, 4a, 5, 8, 8a-hexahydro-*endo*-1, 4-*exo*-5, 8-dimethanoncephthalene) and dieldrin (1, 2, 3, 4, 10, 10-hexachloro-6, 7-epoxy-l, 4, 4a, 5, 6, 7, 8a-octahydro-*endo*-l, 4-*exo*-5, 8-dimethanonaphthalene) are persistent insecticides that accumulate in the food chain and have been found to be responsible for massive deaths of birds and other wildlife. Aldrin is readily converted to dieldrin by chemical oxidation in the soil and by metabolic oxidation in animals and plants. Only dieldrin is therefore normally found in water. Today its use is very severely limited, although it is still used in the control of termites. Owing to its persistence it is still found in surface waters in some areas of the world, although in the UK it has been recorded in some groundwaters. Dieldrin attacks the central nervous system.

Organophosphorus compounds used as pesticides are closely related to nerve gases. Used in a diluted and modified way, they retain the same principle by attacking the central nervous system of insects and animals. Although potentially very dangerous they are not persistent and are rapidly broken down in the environment. Commonly used organophosphorus pesticides include malathion, diazinon and dichlorvos, which is widely used under the trade name of Nuvan as a control for salmon lice in aquaculture.

Carbamates are naturally occurring substances originally used in medicine but also found to be effective insecticides. They also attack the central nervous system. About 20 are widely used, and all are comparatively safe as they are reasonably degradable. Although they are much more expensive than other pesticides, they are increasingly being used (e.g. dimethoroate and aldicarb).

Pyrethroids are naturally occurring compounds although they are now made synthetically. They are relatively non-toxic and widely used by gardeners.

6.2.2 The use of pesticides

The use of pesticides has developed enormously since 1950, with many new and powerful pesticides introduced during the 1960s and 1970s. Pesticides are particularly hazardous as they are chemically developed to be toxic and to some extent persistent in the environment. The most commonly used pesticides in Europe are herbicides of the carboxy acid and phenylurea groups, which are applied to cereal-growing land (Worthing and Walker, 1983). Other pesticides such as fungicides and insecticides are also widely used, and the list of pesticides in use in the British Isles, and their respective target organisms, is vast. Details of the most commonly used pesticides during the late 1980s in the UK are given in Table 6.1.

In 1986, 26 000 000 kg of pure pesticides were used in the UK. Most are so toxic that they have to be diluted hundreds of times before they can be used, so that over a billion gallons of formulated pesticides were sprayed onto crops by British farmers in that year alone. After reaching a peak in the 1980s, pesticide usage by the agricultural and horticultural sectors in England and Wales has remained constant at approximately 19–20 thousand tonnes of active ingredient annually for the past decade (Table 6.2). This does not include all the other pesticides used by local authorities on road verges and in parks, by railway companies along track networks, by industry, forestry, or by individuals in their homes and gardens. Typical pesticide application rates in the British Isles are currently in the range of 1–10 kg of active ingredient per hectare each year. Of this, no more than 10% of the application will reach the target area, whether it is soil insects or plant roots. Further details of pesticide usage can be accessed at http://pusstats.csl.gov.uk/. So where does the remainder go?

Unlike warmer climates, evaporation losses are generally low in the UK and Ireland due to the relatively low air and soil temperatures at the normal application periods (March to May and September to November), so the major proportion of the applied pesticide remains in the soil for some time. In permeable soils they are able to infiltrate to groundwater; in more impermeable

Table 6.1 *Common pesticides found in UK drinking waters*

Pesticide	Main type/use	Pesticide	Main type/use
Aldrin/dieldrin	Insecticide	Isoproturon	Herbicide (cereals)
Atrazine	Herbicide (non-agricultural)	Linuron (mainly in mixtures)	Herbicide (cereals)
Bromoxynil (mainly in mixtures)	Herbicide (cereals)	Malathion	Insecticide
Carbendazim	Fungicide (cereals)	Mancozeb (Maneb plus zinc oxide)	Fungicide (other arable)
Carbetamide	Selective herbicide	Maneb (see Mancozeb)	Fungicide (cereals)
Carbophenothion	Seed treatment (cereals); veterinary use	MCPA	Herbicide (cereals)
Chlordane (total isomers)	Insecticide	MCPB	Herbicide (cereals)
Chloridazon	Herbicide	Mecoprop	Herbicide (cereals)
Chlormequat	Growth regulator (cereals)	Metamitron	Herbicide (other arable)
Chlortoluron	Herbicide (cereals)	Metham sodium	Soil sterilant (other arable)
Clopyralid	Herbicide	Methoxychlor	Insecticide
2,4-D	Herbicide	Methylbromide	Soil sterilant (protected crops)
DDT (total isomers)	Insecticide	Paraquat	Herbicide (cereals)
Dicamba	Herbicide (cereals)	Prometryne	Herbicide
Dichlorprop	Herbicide (cereals)	Propazine	Herbicide
Difenzoquat	Herbicide (cereals)	Propyzamide	Herbicide
Dimethoate	Insecticide (cereals)	Simazine	Herbicide
EPTC	Pre-emergence herbicide	Sodium chlorate	Herbicide (non-agricultural)
γ-HCH	Insecticide	Triadimefon	Fungicide
Glyphosate	Herbicide (cereals)	Triallate	Herbicide (cereals)
Heptachlor/heptachlor epoxide	Insecticide		
Hexachlorobenzene	Seed dressing		
Ioxynil (mainly in mixtures)	Herbicide (cereals)		

Table 6.2 *Summary of the amount of pesticide active ingredient used and total area treated by agriculture and horticulture in England and Wales between 1988 and 2002. Reproduced from Defra (2006) with permission from Defra*

Year	Total agricultural usage 10^3 tonnes	Total area treated 10^6 hectares
1988	22.04	27.04
1989	22.78	27.72
1990	20.93	31.81
1991	21.26	31.83
1992	19.07	32.39
1993	19.94	33.10
1994	18.70	32.51
1995	18.51	32.76
1996	19.93	37.86
1997	19.98	38.08
1998	20.23	41.96
1999	20.60	41.97
2000	19.14	41.35
2001	19.09	41.41
2002	18.69	42.45

soils they find their way into surface watercourses via drainage networks. This is why pesticide levels in drinking waters taken from rivers tend to be seasonal phenomena, although pesticides can be found at relatively high concentrations at any time of the year. In groundwaters they accumulate over the years so that the overall concentration is slowly but constantly increasing, without any discernible large seasonal variation.

Although most pesticides are fairly water-soluble and able theoretically to reach concentrations between 10 and 1000 mg l^{-1}, the potential for these compounds to be leached from the soil depends on a number of factors. Most pesticides in aqueous solution show a strong affinity for soil organic matter, whereas others become concentrated in the fatty tissues of soil organisms and are therefore bound up and unable to be leached into the water. Degradation processes in the soil also substantially reduce the concentration of many of these pesticides before they can find their way into water resources. Degradation (breakdown to harmless end-products) is by a number of routes, but mainly by chemical hydrolysis and bacterial oxidation. However, these processes are generally slow and so where the soil is very permeable some infiltration into the aquifer is probably inevitable. Once past the soil layer and into the unsaturated zone of the aquifer degradation essentially stops. Many pesticides in aquifers have resisted degradation for decades and a particular worrying trend is that many organochlorine pesticides such as DDT and aldrin, which were banned decades ago, have only recently appeared in some water supplies from chalk aquifers. Some intermediate degradation products may be more toxic than the original pesticide and these make complete breakdown more complicated. A summary of the major degradation products of all the widely used European pesticides is given in Appendix 4, along with some idea of their relative toxicity.

In their undiluted form pesticides are all toxic; for example, just a few drops of the herbicide paraquat could be fatal. Spillages of pure pesticide are major pollution incidents. It has been estimated that just a few litres of undiluted pesticide could contaminate tens of millions of litres of groundwater. Yet those who handle, transport and use pesticides rarely appreciate just how potentially dangerous they are, even though there are many licence and training schemes in place in many parts of the world.

6.2.3   Levels in water

Pesticides are defined in the EC Drinking Water Directive (98/83/EEC) as 'organic insecticides, herbicides, fungicides, nematocides, acaricides, algicides, rodenticides, slimicides, related products such as growth regulators, and their relevant metabolites, degradation and reaction products'. Individual pesticides used in significant amounts in catchments and that are most likely to reach water supplies are required to be monitored under the Directive each with a limit value of $0.1 \mu g \, l^{-1}$, except for the banned organochlorine pesticides, aldrin, dieldrin, heptachlor and heptachlor epoxide, that are still found in water supplies, for which a lower limit value of $0.03 \mu g \, l^{-1}$ has been set. Total pesticides means the sum of all individual pesticides detected and quantified in the monitoring procedure excluding the organochlorine pesticides, and is set at $0.5 \mu g \, l^{-1}$. In the UK 44 individual pesticides are monitored under the Directive (Table 6.3). There is considerable variation and uncertainty over the permissible concentrations of pesticides in drinking water between countries. The World Health Organization (WHO) and the USEPA make recommendations for certain individual pesticides (Table 6.4), allowing much higher concentrations than the EC Directive for most of these compounds. In the revised WHO Regulations (2004) guide values range from $0.03 \mu g \, l^{-1}$ for aldrin and dieldrin to $100 \mu g \, l^{-1}$ for dichlorprop while the MCL for pesticides set by the USEPA are generally significantly different (Table 6.4).

Numerous aquifers, especially in eastern England, have been found to exceed the EC limits for total pesticides in drinking water. This is due primarily to the presence of herbicides of the carboxy acid and basic triazine groups. Levels of common agricultural herbicides such as mecoprop (2-(2-methyl-chlorophenoxy) proprionic acid (MCPP)) and 2,4-D (2,4-dichlorophenoxy acetic acid), and two non-agricultural weedkillers atrazine and simazine, have also been detected in some East Anglian chalk aquifers at concentrations of between 0.2 and $2.0 \mu g \, l^{-1}$, well above EC permissible levels (Croll, 1985). Knowledge of the

Table 6.3 *Pesticides monitored in the UK under the EC Drinking Water Directive (98/83/EEC) including limit values*

Aldrin	0.03 μg l⁻¹	Fenitrothion	0.1 μg l⁻¹
Asulam	0.1 μg l⁻¹	Fenpropimorph	0.1 μg l⁻¹
Atrazine	0.1 μg l⁻¹	Flumethrin	0.1 μg l⁻¹
Azinphos methyl	0.1 μg l⁻¹	Flutriafol	0.1 μg l⁻¹
Bentazone	0.1 μg l⁻¹	γ-HCH	0.1 μg l⁻¹
Chlorfenvinphos	0.1 μg l⁻¹	Glyphosate	0.1 μg l⁻¹
Chlorpropham	0.1 μg l⁻¹	Heptachlor	0.03 μg l⁻¹
Chlortoluron	0.1 μg l⁻¹	Heptachlor epoxide	0.03 μg l⁻¹
Clopyralid	0.1 μg l⁻¹	Hexachlorobenzene	0.1 μg l⁻¹
Cypermethrum	0.1 μg l⁻¹	Isoproturon	0.1 μg l⁻¹
Diazinon	0.1 μg l⁻¹	Malathion	0.1 μg l⁻¹
Dicamba	0.1 μg l⁻¹	MCPA	0.1 μg l⁻¹
Dichlorbenil	0.1 μg l⁻¹	Mecoprop	0.1 μg l⁻¹
Dichlorophen	0.1 μg l⁻¹	op-DDT	0.1 μg l⁻¹
Dichlorvos	0.1 μg l⁻¹	pp-DDT	0.1 μg l⁻¹
Dieldrin	0.03 μg l⁻¹	Parathion	0.1 μg l⁻¹
Diquat	0.1 μg l⁻¹	Pentachlorophenol	0.1 μg l⁻¹
Diuron	0.1 μg l⁻¹	Propetamphos	0.1 μg l⁻¹
Endosulphan Total	0.1 μg l⁻¹	Simazine	0.1 μg l⁻¹
Endosulphan-a	0.1 μg l⁻¹	Tecnazene	0.1 μg l⁻¹
Endosulphan-b	0.1 μg l⁻¹	Terbutryne	0.1 μg l⁻¹
Endrin	0.1 μg l⁻¹	Trifluralin	0.1 μg l⁻¹

problem of runoff of pesticides from treated fields into surface water is very poor. It is therefore difficult to predict what will happen due to the large number of factors affecting the degree of leaching. Although leaching is intermittent, research has shown that pesticides can be easily washed out of the soil by rain into nearby watercourses. Williams *et al.* (1991), working in Herefordshire, found that although less than 1% of the total volume of simazine applied to fields actually reached the watercourse, the resultant concentration of the pesticide in the water increased to a maximum concentration of 70 μg l⁻¹, about 700 times higher than the permitted EC limit, after the first fall of rain. Also they found that simazine continued to be leached from the field into the watercourse for months, exceeding 2.0 μg l⁻¹ five months later and still above the EC limit of 0.1 μg l⁻¹ seven months after the last application. The frequency of sampling and the selection of sampling times mean that in reality we still have very little idea about the true level of pesticides in our water.

In the USA the insecticide aldicarb, which is used to control potato cyst eelworm, reached concentrations in excess of 10 μg l⁻¹ in groundwater sources in Long Island, New York, resulting in its ban in the early 1980s. It has also been found in other areas of the USA, most notably in the aquifers of

Table 6.4 *Pesticides listed in the WHO drinking water guidelines (WHO, 2004) with details of WHO guideline values, USEPA maximum contaminant levels, treatment option and achievable effluent standards and health effects. No guidelines have been set by the WHO for those compounds not listed as they do not occur at concentrations in drinking water considered to cause health effects*

Compound	CAS No.	Type	WHO guideline (mg l⁻¹)	USEPA MCL (mg l⁻¹)	Treatment options[a]	Treatment achievable (mg l⁻¹)	Health effects[b]
Alachlor	15972-60-8	Herbicide	0.02	0.002	G	0.001	C, M, T
Aldicarb[c]	116-06-3	Pesticide	0.01	–	G,O	0.001	T
Aldrin	309-00-2	Pesticide	0.00003[d]	–	G,O	0.00002	T
Atrazine	1912-24-9	Herbicide	0.002	0.003	G	0.0001	PC, T
Carbofuran	1563-66-2	Pesticide	0.007	0.04	G	0.001	T
Chlordane	57-47-9	Pesticide	0.0002	0.002	G	0.0001	C
Chlorotoluron	15545-48-9	Herbicide	0.03	–	G	0.0001	PC, T
Chlorpyrifos	2921-88-2	Pesticide	0.03	–	G,O	0.005	T
Cyanazine	21725-46-2	Herbicide	0.0006	–	G	0.0001	PC
2,4-D	94-75-7	Herbicide	0.03	0.07	G	0.001	PC, T
2,4-DB	94-82-6	Herbicide	0.09	–	G	0.0001	PC
DDT[e]	107917-42-0	Pesticide	0.001	–	G	0.0001	PC
DBCP[f]	96-12-8	Pesticide	0.001[g]	0.0002	AS,G	0.001	C
1,2-DCP[h]	78-87-5	Insecticide	0.04[i]	0.005	G	0.001	PM
1,3-Dichloropropene	542-75-6	Insecticide	0.02	–	NI	–	PC
Dieldrin	60-57-1	Pesticide	0.00003[d]	–	G,O	0.00002	T
2,4-DP[j]	120-36-5	Herbicide	0.1	–	NI	–	PC
Dimethoate	60-51-5	Insecticide	0.006	–	G,C	0.001	T
Diquat	2764-72-9	Herbicide	–[k]	0.02	NI	–	T[l]
Endosulfan	115-29-7	Insecticide	–[k]	–	NI	–	T
Endrin	72-20-8	Insecticide	0.0006	0.002	G	0.0002	T
Fenitrothion	122-14-5	Insecticide	–[k]	–	NI	–	T
Fenoprop[m]	93-72-1	Herbicide	0.009	–	G	0.001	PC
Glyphosate	1071-83-6	Herbicide	–[k]	0.7	NI	–	T
Heptachlor	76-44-8	Insecticide	–[k]	0.0004	NI	–	PC, T

Hexachlorobenzene	118-74-1	Seed dressing	–[k]	0.001	NI	0.001	PC, T
Isoproturon	34123-59-6	Herbicide	0.009	–	O	0.0001	T[n]
Lindane	58-89-9	Insecticide	0.002	0.0002	G	0.0001	T
Malathion	121-75-5	Insecticide	–[k]	–	NI	–	T
MCPA[o]	94-74-6	Herbicide	0.002	–	G,O	0.0001	PC
Mecoprop[p]	93-65-2	Herbicide	0.01	–	G,O	0.0001	T, PC
Methoxychlor	72-43-5	Insecticide	0.02	0.04	G	0.0001	T
Methyl parathion	298-00-0	Insecticide	–[k]	–	NI	–	T
Metolachlor	51218-45-2	Herbicide	0.01	–	G	0.0001	T
Molinate	2212-67-1	Herbicide	0.006	–	G	0.001	T
Parathion	56-38-2	Insecticide	–[k]	–	NI	–	T
Pendimethalin	40487-42-1	Herbicide	0.02	–	G	0.001	T
Permethrin	52645-53-1	Insecticide	–[k]	–	NI	–	T
Propanil[q]	709-98-8	Herbicide	–[k]	–	NI	–	T
Pyriproxyfen	95737-68-1	Insecticide	0.3	–	G	0.001	T
Simazine	122-34-9	Herbicide	0.002	0.004	G	0.0001	T
2,4,5-T	93-76-5	Herbicide	0.009	–	G	0.001	T, PC
Terbuthylazine[r]	5915-41-3	Herbicide	0.007	–	G	0.0001	T, PC
Trifluralin	1582-09-8	Herbicide	0.02	–	G	0.001	T, PC[s]

[a] G is granular activated carbon; O is ozonation; AS is air-stripping; NI is no information available; C is chlorination;
[b] C is carcinogenic; M is mutagenic; PM is possible mutagen; T is toxic; and PC is possible carcinogen;
[c] applies also to aldicarb sulphoxide and aldicarb sulphone; [d] aldrin and dieldrin combined; [e] and metabolites;
[f] 1,2-dibromo-3-chloropropane; [g] odour and taste threshold in water of 10 μg l^{-1}; [h] 1,2-dichloropropane; [i] provisional guideline;
[j] dichloroprop; [k] rarely found in concentrations to be a health risk in drinking water; [l] causes cataract formation;
[m] 2,4,5-trichlorophenoxy propionic acid; [n] tumour promoter; [o] 4-(2-methyl-4-chlorophenoxy)acetic acid;
[p] 2-(2-methyl-4-chlorophenoxy)propionic acid; [q] metabolites are more toxic than parent compound; [r] TBA;
[s] impure grades of trifluralin can contain potent carcinogenic compounds

Wisconsin. The situation in groundwater in the USA has been reviewed by Ritter (1990). In Britain aldicarb and other closely related pesticides of the carbamate group are used mainly in the areas of intensive potato and sugar beet production of eastern England, resulting in localized groundwater contamination problems. In the UK the problem areas for pesticides in drinking water are London, the Home Counties, East Anglia and the East Midlands. The two most common pesticides found in drinking water are probably the weedkillers atrazine and simazine. These two herbicides are not used widely in agriculture, and it appears that the widespread contamination of ground and surface waters is the result of non-agricultural uses of these chemicals such as weed control on roadsides and industrial areas. They are, for example, widely used to control plant development along railway tracks, embankments and other areas associated with the railways. Originally British Rail used 2,4-D, notorious as the compound Agent Orange used by the American Air Force to defoliate Vietnam forests during the Vietnam War in the 1960s. Although 2,4-D is fairly stable chemically it is readily broken down in water and soil, and so should only be rarely found in water resources. Where it does occur is usually due to it being directly sprayed onto, or spilt into, surface waters or very permeable soil overlying an aquifer. Under normal circumstances concentrations of 2,4-D in drinking water should therefore be very low. It is excreted almost unchanged in urine and neither stored nor accumulated in the body, and although it is not thought to be carcinogenic, those exposed to it experience fatigue, headache, loss of appetite and a general feeling of being unwell. Studies carried out in Vietnam have indicated birth defects, miscarriage and a host of other effects from exposure to this chemical. Although now banned in the UK, the pesticide 2,4-D is still occasionally found in groundwaters, especially in East Anglia along with the even more potentially hazardous 2,4,5-Trichlorophenoxyacetic acid (2,4,5-T), a synthetic auxin used as a herbicide to defoliate broad-leaf plants.

Blanchoud *et al.* (2004) concluded that the suburban use of herbicides is a significant threat to drinking water resources, especially rivers. In a detailed study of the Morbras and Réveillon catchments, situated close to Paris, the total input of herbicides was estimated at 8 t yr^{-1}, of which 50% came from non-agricultural sources with householders accounting for 30% of the total. With the exception of mecoprop, the herbicides used in urban areas differed to those in agricultural areas, with diuron the most common urban herbicide. They also reported larger application rates in urban areas, especially by householders, and due to their use on impervious surfaces high runoff coefficients for herbicides are also recorded.

In 1989, 550 tonnes of herbicide were used in England and Wales for non-agricultural weed control. Of this active ingredient 25% was atrazine and 14% simazine, with diuron (12%), 2,4-D (9%) and mecoprop (8%) not far behind. Overall the most widely used chemicals are triazines (e.g. atrazine, simazine and terbutryne), representing 39%, and the phenoxy compounds (e.g. 4-(2-methyl-4-chlorophenoxy) acetic acid (MCPA) and mecoprop), representing 21% of total usage.

Table 6.5 *The 12 pesticides most often found in UK drinking waters.*
All are herbicides except for dimethoate, which is an insecticide

Frequently occurring	Commonly occurring
Atrazine	2,4-D
Chlortoluron	Dicamba
Isoproturon	Dichlorprop
MCPA	Dimethoate
Mecoprop	Linuron
Simazine	2,4,5-T

Atrazine, simazine, isoproturon, chlortoluron and mecoprop remain the most widespread contaminants, being widely present in drinking water (Table 6.5). Simazine and atrazine are so widely used that they are found in nearly all water resources in Europe. Simazine is a widely used pre-emergence herbicide. It is only slowly broken down in the soil where it has a low mobility (WHO, 1987) although it is commonly detected in low concentrations in both ground and surface waters. Although its health effects have not been fully evaluated, it does not appear to be carcinogenic, although it does display chronic toxicity in laboratory animals. The WHO (2004) has set a guideline value for simazine in drinking water of $0.002 \, mg \, l^{-1}$. Apart from its use as a non-agricultural weedkiller, it is also widely used in maize-producing areas, especially Italy, where it is often found in water supplies. Atrazine is more mobile in soil than simazine and relatively persistent. It is slowly degraded by photolysis and microbial action in soil although it can remain in water for many years, so it is particularly a problem in groundwaters although concentrations rarely exceed $10 \, \mu g \, l^{-1}$. Surprisingly, little is known about its health effects; although it is thought to be mildly carcinogenic it is a known endocrine-disrupting compound (Section 7.3), so the WHO (2004) has set a guideline value in drinking water of $0.002 \, mg \, l^{-1}$. Atrazine is certainly more dangerous to humans than simazine, but to date no one has been able to determine exactly how toxic either of them actually are. The WHO has recommended that the use of both these chemicals should be carefully controlled, especially in areas where they may eventually contaminate drinking water.

Apart from spraying crops, pesticides are widely used for animal applications, including fish-farming, as insecticides. Pesticides applied to animals will be washed off or excreted, and then may find their way into water resources. The major hazard, however, comes from the disposal of large volumes of chemicals in which the animals have been immersed, or as in the case of fish-farming when the chemical is added directly to the water (Table 6.6).

A major seasonal source of insecticide in drinking water is from sheep-dipping. Organochlorine, organophosphorus compounds, phenolic and synthetic

Table 6.6 *Main chemicals used in aquaculture. Reproduced from Nature Conservancy Council (1989) with permission from English Nature*

Chemical	Use	FW/SW	Method	Remarks
Therapeutants				
Acetic acid	Ectoparasites	FW	D	Use with $CuSO_4$ in hard water areas
Formalin	Ectoparasites	FW/SW	DA	165–250 ppm up to one hour, 20 ppm four hours use in sea cages as bath is common
Malachite Green	Ectoparasites and fungus	FW/SW	DFS, B	Eggs and fish, 100 ppm 30 seconds 4 ppm one hour common in FW, occasional use in cages as a dye marker
Acriflavin (or proflavine hemisulphate)	Ectoparasites, fungus and bacteria	FW	D	Mostly for surface bacteria, fish and eggs, occasional use only
Nuvan (dichlorvos)	Salmon lice	SW	B	1 ppm for one hour, canvas round sea cage
Salt	Ectoparasites	FW	DB	Occasional alternative to formalin
Buffered Iodine	Bacteriocide	FW	B	Use to disinfect eggs 10 minutes 1000 ppm
Oxytetracycline	Bacteriocide	FW/SW	T	Antibiotic widely used for systemic disease

Oxolinic acid	Bacteriocide	FW/SW	T	Antibiotic widely used for systemic disease
Romet 30 (sulphadimethoxine and orthomeprim)	Bacteriocide	FW/SW	T	Antibiotic for systemic disease
Tribrissen (trimethoprim/ sulphadiazine)	Bacteriocide	FW/SW	T	Third most widely used antibiotic
Hayamine 3500	Surfactant/bacteriocide	FW	A	Quaternary ammonium compound used for treating bacterial gill diseases
Benzalkonium	Bacteriocide	FW	A	Surface antibacterial; 'Roccel' (similar to above)
Chloride Chloramine T	Bacteriocide	FW	A	As above, also effective for some protozoa
Vaccines				
Vibrio anguillarum vaccine		SW	B	Not widely used
Enteric Redmouth vaccine		FW	BSI	Widely used in trout culture
Aeromonas salmonicida		SW	I	Not widely used
Anaesthetics				
MS222 (tricaine methane-sulphonate)		FW/SW	B	Widely used approx 1:10 000 dilution
Benzocaine		FW/SW	B	Widely used, requires acetone to dissolve

Table 6.6 (cont.)

Chemical	Use	FW/SW	Method	Remarks
Carbon dioxide		FW/SW	B	Sometimes used at harvest
Disinfectants				
Calcium hypochlorite		FW/SW	S	General disinfectant for tanks etc.
Liquid iodophore e.g. FAM 30		FW/SW	S	For equipment and footbaths
Sodium hydroxide		FW	S	Most commonly used for earth ponds
Water treatment				
Lime		FW	A	Used in earth ponds
Potassium permanganate		FW/SW	BA	Oxidizer and detoxifier
Copper sulphate		FW/SW	A	Algicide and herbicide

B = bath; A = addition to system; F = flush; D = dip; I = injection; S = spray; T = treated food; FW = fresh water; and SW = seawater.

pyrethroid compounds have all been used to control a number of important parasites of sheep. Contamination of water resources by these pesticides generally occurs only at the time sheep-dipping is in progress in the catchment. For example, the organochlorine pesticide γ-BHC, which was found in rivers of the Grampian region of Scotland during July and August, results in concentrations in drinking water occasionally exceeding the EC limit value of $0.1\ \mu g\ l^{-1}$. However, the use of γ-BHC was discontinued in the mid 1980s after this particular insecticide was found in lamb carcasses at levels that prohibited their export to certain countries. Since that date this pesticide has not been found in the surface waters of the area and was widely replaced by organophosphorus insecticides, most notably diazinon, propetamphos, fenchlorphos and chlorfenvinphos and the synthetic pyrethroids cypermethrin and flumethrin. Although less persistent than γ-BHC, these pesticides have also been identified in surface waters. So how does sheep-dip get into water resources?

Apart from trace concentrations from the animals themselves, who are soaked in the chemical, which is slowly washed out of the animal's coat by rain, the main route is through the disposal of the remaining chemical after dipping is complete. The normal disposal method is to either spread it onto land, or to dispose of it to a soakaway. Either way the chemical eventually finds its way into groundwaters or via field drains into the nearest watercourse. Littlejohn and Melvin (1991), who carried out the Scottish study, found that chemicals disposed of to a soakaway designed exactly to Ministry of Agriculture Fisheries and Food (MAFF) guidelines reached a nearby stream within three hours of the sheep-dip being emptied into it, with the concentration increasing over the subsequent 18 hours. Of course, there is still the problem of improper disposal, and farmers occasionally discharge sheep-dip directly into rivers. Sheep-farming is predominantly an upland activity, often the only one, so it is usually small streams that are affected, which often feed supply reservoirs or join large rivers that are used for supplying drinking water to towns downstream. The water from such upland catchments is thought to be very clean, and so both quality surveillance and treatment is often minimal, resulting in the pesticides reaching the consumer's tap.

Although there are alternatives to dipping that minimize the loss of insecticide into the environment such as pour-on or injectable products, these are not as effective as dipping in the complete control of scab, blowfly and external parasites such as lice (Table 6.7). Also, while organophosphorus is potentially more hazardous to drinking water, and so consumers, than the non-organophosphorus alternatives such as synthetic pyrethroid, the latter pose a much greater hazard to aquatic life. Therefore new regulations have been introduced by the Environment Agencies and the Health and Safety Executive in the UK to protect individuals, water resources and wildlife from sheep-dip (HSE, 1999). This includes the prohibition of the use of soakaways for the disposal of the used dip and strict controls on landspreading. The key elements of the new disposal regulations are: (1) No more than 5000 litres of used dip can

Table 6.7 *Major insectides associated in the UK with sheep-dipping and alternative treatments. Reproduced from HSE (1999) with Permission from the Health and Safety Executive*

	Controls		
Dipping treatment	Scab	Blowfly	Lice, tick, ked
Submersion treatment			
Diazinon[a]	Yes	Yes	Yes
Propetamphos[a]	Yes	Yes	Yes
Flumethrin[b]	Yes	No	Yes
Amitraz[b]	No	No	Yes
High-Cis Cypermethrin[b]	Yes	Yes	Yes
Pour-on treatment			
Cyromazine[b]	No	No[c]	No
Deltamethrin[b]	No	Yes[d]	Yes
Cypermethrin[b]	No	Yes[d]	Yes
High-Cis Cypermethrin[b]	No	Yes[e]	Yes
Injectable treatment			
Ivermectin	Yes	No	No
Doramectin	Yes	No	No

[a] Organophosphorus compound; [b] Non-organophosphorus compound;
[c] No treatment but prevention; [d] Treatment but no prevention;
[e] Treatment and prevention.

be spread per hectare, which must be normally diluted with three parts of water or slurry. (2) The land used for spreading used dip must be as flat as possible and not be subject to flooding, be waterlogged or frozen hard, not have effective pipe or mole drains or be where the soil is cracked, severely compacted or where the sub soil is fissured. (3) Spreading is not permitted onto sloping land where the soil is saturated with water, and an untreated strip at least 10 m wide must be left next to all watercourses. (4) Used dip should not be applied to land within 50 m of springs, wells or boreholes. (5) Land that is important for wildlife or that has access for people or animals is not suitable for spreading used dip. Codes of good agricultural practice have been introduced to minimize the risk of pollution from a wide range of agricultural chemicals and include advice on handling, use and disposal of pesticides (MAFF, 1998; Northern Ireland Department of Agriculture and Rural Development, 2005; Scottish Executive, 2005).

6.2.4 Health effects

In general, poisoning from pesticides can occur rapidly. The acute effects are nausea, giddiness, restricted breathing and eventually unconsciousness and death.

More common are the short-term effects of exposure when pesticides can act as irritants, affecting the skin, lungs, eyes and gut. Chronic effects from exposure occur some time, often years later, or as a result of long-term, low-level exposure. These include cancer, tumour formation, birth defects, allergies, psychological disturbance and immunological damage (Table 6.4). It is difficult to be specific, especially as most studies have concentrated on the major organochlorine compounds, many of which are now banned (Cantor, 1997). Little has been done on long-term exposure to very low levels or on the newer pesticides.

6.2.5   Further considerations

There has been an enormous improvement in pesticide concentrations in drinking water in the UK over the past decade with the EC drinking water limits rarely exceeded. However, the effects of mixing two or more pesticides together are unknown, even at concentrations below limit values. There is evidence to suggest that cocktails of pesticides and other organic micro-pollutants could intensify their potential carcinogenic properties by several orders of magnitude (tens or even hundreds of times more). As catchment areas are large, it is inevitable that water will contain chemical residuals of different types, including pesticides. It is not unusual for a single farm to use ten or more different chemicals, so as there are many farms, possibly hundreds or even thousands, within each catchment the complexity and magnitude of this chemical cocktail will be enormous. Of the 324 different trace organic chemicals isolated by the Water Research Centre survey, only 19 were found in all of the samples tested (Fielding *et al.*, 1981). Thus it may be misleading to assess the toxicity of pesticides and other organic chemicals individually. Of course, there are many more organic chemicals that cannot currently be measured in water at the concentrations at which they are suspected to be present, so 324 is probably a conservative estimate, especially as the Drinking Water Inspectorate has indicated that over 450 different pesticides are currently used in the UK. The situation in the rest of Europe is equally, if not more, worrying.

   The presence of any particular contaminant in drinking water, and its possible effect on our health should not be viewed in isolation from other sources of exposure, especially food and the atmosphere. Although it is clearly difficult to assess the exact contribution of drinking water to the exposure of the consumer to pesticides, it is clearly an important route. It has always been assumed that the intake of pesticides was much higher from food than from water. In fact, the WHO set standards for drinking water of 1% of the ADI value. A report by the Foundation for Water Research suggested that this may be an incorrect practice for pesticides and that the intake of such compounds from water may be of the same order and in some instances it may exceed that for food (Fielding and Packham, 1990). Drinking water may therefore be the major source of pesticide exposure for many people.

6.3   Industrial solvents

There is an enormous range of organic solvents used for industrial applications, primarily degreasing operations such as metal cleansing and textile dry-cleaning. In the UK, it was ICI who pioneered the production and use of industrial solvents. Production of trichloroethylene (TCE) began as early as 1910 but it took until 1928 before a suitable container was developed to actually store the solvent! So it was not until the 1930s that the production of TCE and its use began to increase. There are now a large number of solvents, although only six are widely used. They are described here in descending order of importance.

1. *Methylene chloride or dichloromethane (DCM)* has a wide range of applications including paint stripping, metal cleaning, pharmaceuticals, aerosol propellants and in acetate films.
2. *Trichloromethane or chloroform (TCM)* is used in fluorocarbon synthesis and pharmaceuticals.
3. *Methyl chloroform or 1,1,1-trichloroethane (TCA)* is widely used for metal and plastic cleaning and as a general solvent. It was introduced as a substitute for TCE and tetrachloroethene as it is far less toxic (both in terms of acute and chronic toxicity). It is widely used now, not only in industry but also in other sectors. It is used in adhesives, aerosols, inks and is the solvent for a number of correction fluids.
4. *Tetrachloromethane or carbon tetrachloride (CTC)* is also used in fluorocarbon synthesis and in fire extinguishers.
5. *Trichloroethene or trichloroethylene (TCE)* is the most widely used solvent for metal cleaning, although it is still used for dry-cleaning and industrial extractions. Its use has significantly decreased over the past 30 years, being replaced by TCA.
6. *Tetrachloroethene or perchloroethylene (PCE)* is mainly used in dry-cleaning processes, although it is also used for metal cleaning.

The use of industrial solvents reached a peak in the mid 1970s in the UK, with a steady decrease amounting to between 40% and 50% over the past 30 years. This was due not only to the decrease in manufacturing industry, but also to improvements in process design and operation. Only four (TCE, PCE, TCA and DCM) of these chemicals are frequently found in drinking water.

Laboratory studies on mammals have shown that all the chlorinated solvents are carcinogenic and so they must be considered as potentially hazardous chemicals. There are many case studies dealing with the health effects of industrial solvents in drinking water, many as the result of leaching from industrial and chemical toxic waste sites. These have reported elevated cancer rates (Budnick *et al.*, 1984; Najem *et al.*, 1985), bladder cancer (Mallin *et al.*, 1990), leukaemia, non-Hodgkin's lymphoma as well as birth defects (Cohen *et al.*, 1994). For example, after clusters of childhood leukaemia and birth defects had been noted in 1979 drinking water wells in Woburn Massachusetts

Table 6.8 *WHO guideline, EC permissible concentration and USEPA maximum contaminant level for major industrial solvents*

		WHO (mg l^{-1})	EC (mg l^{-1})	USEPA (mg l^{-1})
Methylene chloride or dichloromethane	(DCM)	0.02	$-^{a}$	0.005
Trichloromethane or chloroform	(TCM)	0.2	$-^{a}$	$-^{a}$
Methyl chloroform or 1,1,1-trichloroethane	(TCA)	$-^{b}$	0.01^{c}	0.2
Tetrachloromethane or carbon tetrachloride	(CTC)	0.004	$-^{a}$	0.005
Trichloroethene or trichloroethylene	(TCE)	0.07	$-^{a}$	0.005
Tetrachloroethene or perchloroethylene	(PCE)	0.04	0.01^{c}	0.005

[a] No value set.
[b] Rarely found in concentrations to be a health risk in drinking water.
[c] Tetrachloroethene plus trichloroethane.

were found to be contaminated with TCE ($267\,\mu g\,l^{-1}$), tetrachloroethylene ($21\,\mu g\,l^{-1}$) and trichlorotrifluoroethane ($23\,\mu g\ l^{-1}$) (Costos and Knorr, 1996). Further investigations in test wells in the area later showed that the groundwater was contaminated with 48 USEPA priority pollutants as well as elevated concentrations of 22 metals (Lagakos *et al.*, 1986). Tetrachloroethylene was also associated with increased incidence of leukaemia, and both bladder and kidney cancer when it leached from vinyl-lined distribution pipes contaminating the drinking water (Aschengrau *et al.*, 1993).

The WHO has set permissible concentrations in drinking water for a wide range of solvents (Appendix 3). The EC Drinking Water Directive (98/83/EEC) includes only two of these solvents and sets a joint limit value for PCE plus trichloroethane of $10\,\mu g\ l^{-1}$. These values are compared to those set by the USEPA and the WHO in Table 6.8.

In practice the nature of these solvents reduces the potential for the contamination of surface waters. They are all highly volatile, which means that on exposure to air they rapidly evaporate into the atmosphere. In fact, up to 85% of DCM and 60% of TCE used are lost by evaporation. Once discharged to a watercourse, evaporation continues. Their fate in the atmosphere is less clear. Like other organic molecules these solvents are oxidized by ultraviolet radiation to harmless end-products. In general, this takes less than three months (known as the atmospheric half-life), although TCA has an atmospheric half-life in excess of five years. Due to this moderate persistence, these solvents become concentrated in the atmosphere, often reaching concentrations in excess of $1\,\mu g\ l^{-1}$ above industrial areas. There is evidence to show that they can subsequently be removed by precipitation, with concentrations of $0.1\,\mu g\ l^{-1}$ in rainfall not uncommon. Trace amounts of industrial solvents in all drinking waters are probably inevitable.

The major pollution threat is when solvents are discharged directly into or onto the ground due to illegal disposal or accidental spillage. Since groundwater is not exposed directly to the atmosphere, the solvents are not able to escape by evaporation, so it is groundwaters that are most at risk from these chemicals. Industrial solvents are now stored above ground in specifically designed and lined tanks to contain any spilt chemical. However, it was, and still is, the practice in some countries to store such compounds in underground storage tanks so that if fractures occur in the tank, pipework or pipe couplings the chemical will be lost directly into the ground and pollute the groundwater (Lawrence and Foster, 1987).

Pollution of groundwater by industrial solvents is a very widespread problem (Loch *et al.*, 1989; Rivett *et al.*, 1990). For example, in the Netherlands, of 232 boreholes tested for organic micro-pollutants, 67% contained TCE, 60% TCM, 43% CTC and 19% PCE. In 1982 over 60% of the 621 water supply wells tested in Milan contained a total concentration of chlorinated solvent in excess of $50 \,\mu g \, l^{-1}$, with 10 wells exceeding $500 \,\mu g \, l^{-1}$. In Italy, TCE is the most frequently detected solvent, followed by PCE, TCM and TCA. The problem is especially widespread in the USA. For example, 20% of the 315 wells tested in New Jersey contained solvents, with TCE the most widespread and occasionally reaching in excess of $100 \,\mu g \, l^{-1}$ in some wells. Although data are sparse, with many of the surveys conducted not published, the problem in the UK is equally bad. The Department of the Environment commissioned a survey in the mid 1980s of over 200 boreholes, of which at least three-quarters were used for public supply. The results showed that TCE exceeded $1 \,\mu g \, l^{-1}$ at 35% of the sites with 4% of these above $100 \,\mu g \, l^{-1}$; PCE exceeded the EC guideline value of $1 \,\mu g \, l^{-1}$ at 17% of the sites, of which 1% exceeded $100 \,\mu g \, l^{-1}$.

In essence, aquifers underlying urbanized areas such as Milan, Birmingham, London or New Jersey will contain high concentrations of all solvents. The stability of TCE in particular and the inability of these solvents to readily evaporate mean that such contamination will last for many decades, even if the use of solvents is discontinued. Where solvents are used then spillage to the subsurface is almost inevitable. Careful handling and correct storage is vital if such contamination is to be minimized.

Industrial solvents tend not to be very soluble and when discharged into water remain as an immiscible phase, i.e. they form a separate liquid (solvent) fraction to the aqueous (water) fraction. These compounds are classed as non-aqueous phase liquids (NAPLs), with dense NAPLs (DNAPLs) tending to sink in water while light NAPLs (LNAPLs) tend to float on water. So solvents classed as DNAPLs can quickly penetrate deep into aquifers.

Solvents spilt onto the soil are less likely to contaminate the groundwater than those discharged below the soil layer. This is because the soil can be effective in retaining solvents by sorption processes, allowing longer periods for the soil bacteria to break down the organic molecules, although most polychlorinated organic compounds, such as PCE, are recalcitrant, i.e. they are highly resistant to biodegradation. In the soil layer there may still be some volatility (evaporation) losses. Once below the soil horizon, however, volatility losses will be negligible and the solvents will move much more quickly downwards towards the water table. Bacteriological activity is greatly reduced in the unsaturated zone and this, combined with much shorter contact times between the solvent and micro-organisms, means that losses due to biodegradation will also be extremely low. Once past the soil horizon, or if spills occur below this layer, which is often the case with leaks from buried pipes or tanks, or discharges from lagoons or soakaways, then contamination of the groundwater is almost inevitable.

As mentioned earlier, chlorinated organic solvents have higher densities but lower viscosities than water. In practice this means that they tend to accumulate in the base of aquifers, with their subsequent movement or migration within groundwater dictated largely by the slope and form of the base of the aquifer, rather than the direction of groundwater flow. Solvent therefore tends to accumulate in depressions within the aquifers, forming pools. The immiscible phase is the separated solvent; however, with time this gradually dissolves into the water (miscible phase) forming distinct plumes. The migration of the solvent, once in the miscible phase, is governed largely by the direction and velocity of the groundwater (Figure 6.1). Where the aquifer is highly fissured, as is the case with Jurassic and chalk aquifers, then lateral migration over many kilometres is possible.

It is interesting to compare the behaviour of chlorinated organic solvents with the aromatic hydrocarbons such as benzene, toluene and fuel oils. These have lower densities than water (LNAPLs) and are more viscous, and so behave in the opposite manner to solvents. In the immiscible phase they tend to float on the surface of the aquifer, forming characteristic pancakes on top of the water

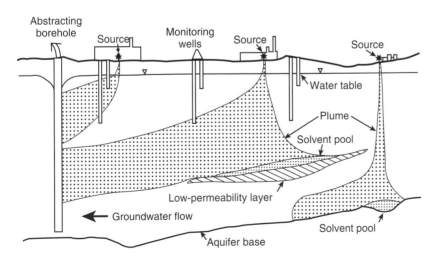

Figure 6.1 Behaviour of chlorinated solvents in groundwaters and the interactions with a pumped extraction borehole and unpumped monitoring wells. Reproduced from Rivett *et al.* (1990) with permission from the Chartered Institution of Water and Environmental Management.

table. Migration follows the groundwater flow, but remains at the surface of the water table. If the water level rises, then the aromatic hydrocarbon also rises, but will be retained within the porous matrix of the unsaturated zone by surface tension once the water table falls again. This process significantly reduces the effects of lateral migration (sideways movement) of the chemical, so that the effects of such spills can be fairly localized compared with chlorinated solvents, which can affect large areas of the aquifer.

6.4   Polycyclic aromatic hydrocarbons

Polycyclic (polynuclear) aromatic hydrocarbons (PAHs) are synthetic compounds that occur in soot, tar, vehicle exhausts and in the combustion products of hydrocarbon fuels such as household oil. While the major source of PAHs is the incomplete combustion of organic matter, some are also formed by bacteria, algae and plants. The PAHs are not very soluble but are strongly adsorbed onto particulate matter, especially clay, resulting in a high concentration where suspended solids are present. Although PAHs are photodegradable, they do require comparatively long exposure to ultraviolet rays in sunlight to do so. Thus in practice, most are either tied up in the subsurface soil where they are eventually degraded by soil micro-organisms, or they find their way into aquifers or surface waters (Clement International Corporation, 1990; Manoll and Samara, 1999). A source of PAH contamination in drinking water resources recently identified has been leaching from reclaimed asphalt pavement. This material is generated during the removal of the road surfaces, which is milled into an aggregate that is often stockpiled at the site for long periods prior to landfilling or recycling. Leaching of PAHs as well as other pollutants including other hydrocarbons and metals rapidly occurs on storage (Brant and De Groot, 2001; Legret et al., 2005). Tea is also a source of PAHs, which are adsorbed onto the leaves from precipitation or dust. Concentrations up to $8800 \, \mu g \, kg^{-1}$ have been recorded with highest values found in black teas. Lin et al. (2005) identified 16 different PAH compounds in various teas tested with between 3.0% and 7.7% of the PAH content released during infusion into the water.

Nearly all PAHs are carcinogenic, although their potency varies; the most hazardous by far is benzo[a]pyrene. The PAHs are readily absorbed in the body and as they are highly lipid-soluble they become localized in fatty tissues. However, they are readily metabolized and do not generally accumulate. Polycyclic aromatic hydrocarbons in drinking water are thought to cause gastrointestinal and oesophageal tumours. The annual adult intake of PAHs is between 1.0 and 10 mg, with benzo[a]pyrene contributing between 0.1 and 1.5 mg. Drinking water contributes only a small proportion of the total PAH intake, less than 0.5% of the total. The PAHs are rarely present in the environment on their own and the carcinogenic nature of individual compounds is thought to increase in the presence of other PAH compounds.

Table 6.9 *The total concentration of PAHs in drinking water is calculated from the concentration of six or four reference compounds*

Compound names used by WHO	Alternative names used by European Commission
Fluoranthene	Fluoranthene
3,4-Benzofluoranthene	Benzo-3,4-fluoranthene[a]
11,12-Benzofluoranthene	Benzo-11,12-fluoranthene[a]
Benzo[a]pyrene	Benzo[a]pyrene
1,12-Benzoperylene	Benzo-1,12-perylene[a]
Indeno-(l,2,3-cd) pyrene	Indeno-(l,2,3-c-d)-pyrene[a]

[a] The reference compounds used in the EC Drinking Water Directive.

While there are over 100 PAH compounds, the WHO has listed just six common reference compounds for the estimation of total PAH in water (Table 6.9). These are all combustion products of hydrocarbons and represent the most widely occurring compounds of this group found in drinking water. Before the recent revision of their drinking water guidelines in 2004, the WHO recommended that total PAH should not exceed $0.2\,\mu g\,l^{-1}$ with an individual guideline of $0.01\,\mu g\,l^{-1}$ for benzo[a]pyrene (WHO, 1984). The latest WHO (2004) guidelines have only specified a value for benzo[a]pyrene, which has been revised upwards to $0.7\mu g\,l^{-1}$, with no total value for PAHs given. Fluoranthene is no longer considered to pose a health risk to consumers at the concentrations normally found in drinking water.

The EC has retained a limit value for total PAH compounds in their revised Drinking Water Directive (98/83/EEC), based now on only four reference compounds (Table 6.9), of $0.1\,\mu g\,l^{-1}$ and a separate limit value of $0.01\,\mu g\,l^{-1}$ for benzo[a]pyrene. The sum of the detected concentrations of the reference substances listed in the EC Drinking Water Directive was originally based on the same six compounds listed by the WHO, although the compound names are slightly different. This list has been modified by the EC with fluoranthene no longer included due to its relatively low carcinogenicity. The USEPA have only set an MCL for benzo[a]pyrene at $0.2\,\mu g\,l^{-1}$.

A UK survey carried out in 1981 of the levels of PAH in water used for abstraction for drinking water found that groundwater sources and upland reservoirs contained low levels of PAHs, less than $50\,ng\,l^{-1}$, whereas lowland rivers used for supply contained between 40 and $300\,ng\,l^{-1}$ during normal flows, increasing in excess of $1000\,ng\,l^{-1}$ at high flows. This appears to be due to the increase in levels of suspended solids in fast rivers onto which the PAHs have adsorbed. Between 65% and 76% of the PAHs in surface waters are bound to particulate matter and so can effectively be removed by physical water treatment processes such as sedimentation, flocculation and filtration. The remainder can be either chemically oxidized or removed by activated carbon. Even when PAH levels in raw water are high, due to localized industrial pollution, water treatment can adequately remove PAH compounds to conform to EC limits.

Fluoranthene, which is the most soluble but least hazardous of the PAH compounds, is occasionally found in high concentrations in drinking water samples collected at consumers' taps. This is due to leaching from the coal-tar and bitumen linings commonly applied to iron water mains (Chapter 23). These linings also contain other PAH compounds that can find their way into the water in the water supply mains as it flows from the treatment plant to individual homes. Conventional water treatment effectively removes the bulk of the PAH compounds adsorbed onto particulate matter, although most of the PAHs remaining is fluoranthene due to its high solubility. Granular activated carbon can remove 99.9% of the fluoranthene and any other PAHs remaining in the water. Owing to its relatively high solubility, fluoranthene is the major PAH to accumulate in groundwater. For the consumer, most of the breaches of the standards for PAHs are due to leaching in the distribution system (Chapter 23).

6.5   Removal of organic contaminants from drinking water

The levels of organic contaminants in water supplies vary with the catchment and are clearly linked with land use and degree of industrialization. In general, groundwaters have fewer organic micro-pollutants than surface waters, although urban aquifers are often locally polluted with industrial solvents. Also lowland surface waters have higher concentrations of such organic compounds compared with upland waters. Careful selection of water resources is therefore vital, and this will include an assessment of the use of organic compounds within the catchment area. Where a problem exists, diluting supplies with uncontaminated water by blending is a widely used option to conform to standards. It may be possible to control the use of key organic micro-pollutants, especially pesticides, within small watersheds, although generally this will not be feasible.

Conventional water treatment is very limited in its ability to remove all trace amounts of organic compounds. However, the development of new advanced water treatment processes now enables water companies to effectively remove organic contamination, although it is expensive. Although membrane processes such as reverse osmosis and nanofiltration can reduce all organic contaminants, they also remove important inorganic compounds including hardness (Baier *et al.*, 1987; Becker *et al.*, 1989; Van der Bruggen and Vandecasteele, 2003). Coagulation followed by sand filtration can remove a significant proportion of colloidal material onto which organics may have complexed, although in practice this only removes a small proportion of the organic contamination (AWWA, 1979). This is followed by membrane filtration process and then activated carbon filtration. Activated carbon is now widely used to remove pesticides, solvents and other organics (Section 14.2.12). Powdered activated carbon (PAC) can be used in conjunction with coagulation and is effective in removing trace concentrations of organic compounds, whereas granular activated carbon (GAC) is used in compact filters (Becker and Wilson, 1978; Croll *et al.*, 1991). The bed-life of a GAC filter can be extended when combined with ozone before filtration to oxidize some of the organic compounds present (Ferguson *et al.*, 1991). Both nitrate and pesticides have been removed simultaneously using biodenitrification. The heterotrophic bacteria responsible for denitrification require a carbon source and as drinking water has a low carbon content pesticides are utilized. However, the pesticides cannot be used as the sole carbon source so a supplement, usually in the form of ethanol, is normally used (Aslan and Turkman, 2006). Removal of organic micro-pollutants from drinking water is considered in detail elsewhere (Kruithof *et al.*, 1989; Akbar and Johari, 1990; Foster *et al.*, 1991; Nyer, 1992; Mather *et al.*, 1998), whereas the removal of disinfection by-products is explored in Chapter 18.

6.6 Conclusions

The toxicological data currently available for pesticides suggest that the current EC limit values for drinking water are adequate for short-term protection. However, a major research initiative is required to determine if constant exposure at these levels is safe in the long term, as we currently just don't know. Those pesticides still registered in the UK of greatest toxicological concern are mainly insecticides. These include chlordane, coumaphos, omethoate, phorate and triazophos. The herbicide amitrole and the fungicide chlorothalonil also pose significant health risks. Cocktails of pesticides and other organic micro-pollutants are inevitable given the diversity of pesticides used. Such mixtures are thought to intensify potential carcinogenic properties of individual pesticides by several orders of magnitude. Efficiency in pesticide usage has increased significantly over the same period with 31% more land treated with the same amount of active ingredient (Table 6.2). Improvements in handling, storage and use of pesticides, as well as the restriction in use of many pesticides, have resulted in a reduction of pesticides in water resources throughout Europe (Table 6.10). However, with more land being treated more resources, especially groundwaters, are at risk from contamination.

With in excess of 40 million sheep in the UK and over 18 000 dipping operations annually, organophosphorus and synthetic pyrethroid insecticides used for the control of sheep parasites will continue to contaminate water resources. While the restriction on the use of soakaways for disposal of used dip and the new codes of practice in relation to land disposal of the waste by spreading will help to protect both ground and surface waters, both remain at risk. In a survey of river monitoring sites in Wales in 1998 by the Environment

Table 6.10 *The frequency of occurrence of pesticides exceeding 0.1 µg l^{-1} in British ground and surface waters in samples collected during 2004. Samples exclude sites where high levels are known to exist. Source: Defra e-digest environmental statistics website: www.defra.gov.uk/environment/statistics/ inlwater/iwpesticide.htm/. Crown copyright material is reproduced with the permission of the Controller of HMSO.*

Pesticides in groundwaters	%	Pesticides in surface waters	%
Atrazine	2.62	Mecoprop	13.8
Bentazone	0.88	MCPA	12.9
Mecoprop	0.74	Isoproturon	10.8
Clopyralid	0.43	Diuron	9.8
Diuron	0.38	2,4-D	6.5
Simazine	0.34	Dichlorprop	5.7
Chlorotoluron	0.26	Simazine	4.1
Pirimicarb	0.25	Chlorotoluron	2.7
Metazachlor	0.25	Bentazone	2.3
Cyanazine	0.21	Atrazine	1.6
Monuron	0.21		

Agency the organophosphorus sheep-dip insecticides diazinan and proptamphos were recorded at 52% and 35% of sites respectively, while the synthetic pyrethroid insecticides cypermethrin and flumethrin were recorded at 34% and 6% of sites. Upland areas are important parts of the catchment that feed water into lakes, reservoirs and rivers, all of which are used for supply, and these very areas are predominantly used for sheep-farming.

Treatment of water to remove pesticides is not considered to be a practical or reliable solution in the short term, although GAC is now widely used in water treatment. An overall reduction in the use of pesticides and/or changes in the formulation of the compounds are thought to be the only long-term solutions. However, in 2003 the water utilities in England and Wales achieved a 99.99% compliance with pesticides in 800 000 samples analyzed, showing widespread conformity with EC standards.

Industrial solvents are generally only problematic in groundwaters from aquifers underlying urbanized or industrial areas, although leaching from toxic landfill sites has caused most serious pollution. Their use in some of the less obviously polluting industries such as computer and microchip manufacture has resulted in unexpected contamination of local wells due to accidental spillage. Better storage, containment and operational practice, linked to a greater awareness of their carcinogenic nature, has resulted in fewer spillages of such chemicals resulting in groundwater contamination than in the past. The USEPA has set very strict MCL values for all the major industrial solvents at 0.005 mg l^{-1} excluding TCA (0.2 mg l^{-1}) and TCM for which no MCL has been set. This contrasts with the EC Drinking Water Directive, which sets a joint permissible concentration of 0.01 mg l^{-1} for TCA + PCE. No limit values have been set for the remaining solvents (Table 6.8). There is considerable evidence supporting a link between industrial solvents in drinking water and birth defects, leukaemia and non-Hodgkin's lymphoma. There are also reported associations with bladder and kidney cancer (Cantor, 1997). With industrial solvents still widely used by the industrial and service industries, strict vigilance and appropriate standards for drinking water are required to adequately protect consumers, especially those on groundwater sources.

With a low solubility and high affinity for particulate matter, PAHs are not found in drinking water at problematic concentrations, being readily removed during the coagulation and filtration stage of treatment. Coal-tar and bitumen linings used to protect distribution mains from corrosion can lead to PAHs being released, although this is primarily fluoranthene which is not considered to be a health risk at the concentrations normally found in drinking water. Most exposure to PAHs comes from exposure to combustion products, including those produced during food preparation by char broiling, grilling, roasting and frying. It is only where coal-tar linings are seriously deteriorated that the intake of PAHs could equal or exceed that from food. Therefore, routine surveillance of such linings for the release PAHs is important (Chapter 23). The maximum permissible values for PAHs in European drinking water does not appear to reflect the potential toxicity of these compounds, especially as even a small spill of just a few litres of solvent could contaminate many millions of litres of water.

References

Akbar, M. and Johari, B. H. (1990). Treatment techniques for the removal of organic pollutants–the SG. Linggi experience. *Water Supply*, **8**(3–4), 664–73.

Aschengrau, A., Ozonoff, D. and Paulu, C. (1993). Cancer risk and tetrachloroethylene (PCE) contaminated drinking water in Massachusetts. *Archives in Environmental Health*, **48**, 284–92.

Aslan, S. and Turkman, A. (2006). Nitrate and pesticide removal from contaminated water using a bionitrification reactor. *Process Biochemistry*, **41**, 882–6.

AWWA (1979). Organic removal by coagulation: a review and research needs. *Journal of the American Water Works Association*, **71**(10), 588–603.

Baier, J. H., Lykins B. W., Fronk, C. A. and Kramer, S. J. (1987). Using reverse osmosis to remove agricultural chemicals from water. *Journal of the American Water Works Association*, **79**(8), 55.

Becker, D. L. and Wilson, S. C. (1978). The use of activated carbon for the treatment of pesticides and pesticide wastes. In *Carbon Adsorption Handbook*, ed. P. Cheremisinoff and F. Ellerbusch. Ann Arbor: Ann Arbor Science, pp. 167–213.

Becker, F. F., Janowsky, U., Overath, H. and Stetter, D. (1989). Die Wirksamkeit von Umehrosmosememmbranen rei der Entfemung von Pestiziden. *Wasser Abwasser*, **130**(9), 425–31.

Blanchoud, H., Farrugia, F. and Mouchel, J. M. (2004). Pesticide uses and transfers in urbanized catchments. *Chemosphere*, **55**, 905–13.

Brandt, H. C. A. and De Groot, P. C. (2001). Aqueous leaching of polycyclic aromatic hydrocarbons from bitumen and asphalt. *Water Research*, **35**, 4200–7.

Budnick, L. D., Sokal, D. C., Falk, H., Logue, J. N. and Fox, J. M. (1984). Cancer and birth defects near Drake superfund site, Pennsylvania. *Archives of Environmental Health*, **39**, 409–13.

Cantor, K. P. (1997). Drinking water and cancer. *Cancer Causes and Control*, **8**, 292–308.

Clement International Corporation (1990). *Toxicological Profile for Polycyclic Aromatic Hydrocarbons*. Atlanta: Agency for Toxic Substances and Disease Registry, Public Health Service, US Department of Health and Human Services.

Cohen, P., Klotz, J., Bove, F., Berkowitz, M. and Fagliano, J. (1994). Drinking water contamination and the incidence of leukaemia and non-Hodgkin's lymphoma. *Environmental Health Perspectives*, **102**, 556–61.

Costos, K. and Knorr, R. (1996). *Woburn Childhood Leukaemia Follow-Up Study, Vol. 1, Analyses*. Draft Final Report. Boston, MA: Massachusetts Department of Public Health.

Croll, B. T. (1985). The effects of the agricultural use of herbicides in freshwater. *Proceedings of the WRC Conference on Effects of Land-use on Fresh Water*, **13**, 201–9.

Croll, B. T., Chadwick, B. and Knight, B. (1991). The removal of atrazine and other herbicides from water using granular activated carbon. *Water Supply*, **10**, 111–20.

Defra (2006). *Digest of Environmental Statistics*. London: Department for Environment, Food and Rural Affairs.

DETR (2000). *The Water Supply (Water Quality) Regulations 2000*, Statutory Instrument No. 3184. London: Department of Environment, Transport and the Regions, Water Supply and Regulation Division, HMSO.

EC (1998). Council Directive 98/83/EEC of 3 November 1998 on the quality of water intended for human consumption. *Official Journal of the European Communities*, **L330** (5.12.98), 32–53.

Ferguson, D. W., Gramith, J. T. and McGuire, M. J. (1991). Applying ozone for organics control and disinfection: a utility perspective. *Journal of the American Water Works Association*, **83** (5), 32–9.

Fielding, M. and Packham, R. F. (1990). *Human Exposure to Water Contaminants*. Report FR 0085. Marlow: Foundation for Water Research Centre.

Fielding, M., Gibson, T. M., James, H. A., McLoughlin, K. and Steel, C. P. (1981). *Organic Micro-Pollutants in Drinking Water*. Technical Report 159. Medmenham: Water Research Centre.

Foster, D. M., Rachwal, A. J. and White, S. J. (1991). New treatment processes for pesticides and chlorinated organics control in drinking water. *Journal of the Institution of Water and Environmental Management*, **5**(4), 466–77.

HSE (1999). *Sheep Dipping*. Publication AS29(rev2) 4/99 C200. London: Health and Safety Executive.

Kruithof, J. C., Puijker, L. M. and Janssen, H. M. (1989). Presence and removal of pesticides. *H₂O*, **22**(17), 526.

Lagakos, S. W., Wessen, B. J. and Zelen, M. (1986). An analysis of contaminated well waters and health effects in Woburn Massachusetts. *Journal of the American Statistical Association*, **81**, 583–96.

Lawrence, A. R. and Foster, S. S. D. (1987). *The Pollution Threat From Agricultural Pesticides and Industrial Solvents*. Hydrological Report 87/2. Wallingford: British Geological Survey.

Legret, M., Odie, L., Demare, D. and Jullien, A. (2005). Leaching of heavy metals and polycyclic aromatic hydrocarbons from reclaimed asphalt pavement. *Water Research*, **39**, 3675–85.

Lin, D., Tu, Y. and Zhu, L. (2005). Concentrations and health risk of polycyclic aromatic hydrocarbons in tea. *Food and Chemical Toxicology*, **43**, 41–8.

Littlejohn, J. W. and Melvin, M. A. L. (1991). Sheep dips as a source of pollution of freshwaters: a study in Grampian Region. *Water and Environmental Management*, **5**, 21–7.

Loch, J.P.G., van Dijk-Looyaard, A. and Zoeteman, B.C.J. (1989). Organics in groundwater. In *Watershed 89, the Future of Water Quality in Europe*, ed. D. Wheeler, M. L. Richardson and J. Bridges. Proceedings of the International Association of Water Pollution Research and Control, 17–20th April 1989. London: Pergamon Press, pp. 39–55.

MAFF (1998). *The Water Code: Code of Good Agricultural Practice for the Protection of Water*. London: Ministry of Agriculture Fisheries and Food. www.defra.gov.uk/farm/environment/cogap/index.htm.

Mallin, K. (1990). Investigations of a bladder cancer cluster in North-western Illinois. *American Journal of Epidemiology*, **132**(Suppl.1), S96–S106.

Manoll, E. and Samara, C. (1999). Polycyclic aromatic hydrocarbons in natural waters: sources, occurrence and analysis. *Trends in Analytical Chemistry*, **18**(6), 417–28.

Margni, M., Rossier, D., Crettaz, P. and Jolliet, O. (2002). Life cycle impact assessment of pesticides on human health and ecosystems. *Agriculture, Ecosystems and Environment*, **93**, 379–92.

Mather, J., Banks, D., Dumpleton, S. and Ferer, M. (ed.) (1998). *Groundwater Contamination and Their Migration*. Geological Society Special Publication No. 128. London: The Geological Society.

Najem, G. R., Louria, D. B., Lavenhar, M. A. and Feuerman, M. (1985). Clusters of cancer mortality in New Jersey municipalities; with special reference to chemical toxic waste disposal sites and per capita income. *International Journal of Epidemiology*, **14**, 528–37.

Nature Conservancy Council (1989). *Fish Farming and the Safeguard of the Natural Marine Environment of Scotland*. Edinburgh: NCC.

Northern Ireland Department of Agriculture and Rural Development (2005). *Code of Good Agricultural Practice for the Prevention of Pollution of Water*. Belfast: Department of Agriculture and Rural Development. http://agrifor.ac.uk/browse/cabi/c1fdca80522f8f9851a2616c44125010.html.

Nyer, E. K. (1992). *Groundwater Treatment Technology*, 2nd edn. New York: Van Nostrand Reinhold.

Petrović, M., Gonzalez, S. and Barceló, D. (2003). Analysis and removal of emerging contaminants in wastewater and drinking water. *Trends in Analytical Chemistry*, **22**(10), 686–96.

Ritter, W. F. (1990). Pesticide contamination of ground water in the United States – a review. *Journal of Environmental Science and Health*, **B25**(1), 1–29.

Rivett, M. O., Lerner, D. N. and Lloyd, J. W. (1990). Chlorinated solvents in UK aquifers. *Water and Environmental Management*, **4**, 242–50.

Rodriguez-Mozaz, S., López de Alda, M. J. and Barceló, D. (2004). Monitoring of estrogens, pesticides and bisphenol A in natural waters and drinking water treatment plants by solid phase extraction-liquid chromatography-mass spectrometry. *Journal of Chromatography A*, **1045**, 85–92.

Scottish Executive (2005). *Prevention of Environmental Pollution From Agricultural Activity*. Edinburgh: Scottish Executive. www.scotland.gov.uk/Publications/2005/03/20613/51366.

Van der Bruggen, B. and Vandecasteele, C. (2003). Removal of pollutants from surface water and groundwater by nanofiltration: overview of possible applications in the drinking water industry. *Environmental Pollution*, **122**, 435–45.

WHO (1984). *Guidelines for Drinking Water Quality*, Vol. 2. *Health Criteria and Other Supporting Information*. Geneva: World Health Organization.

WHO (1987). *Drinking Water Quality: Guidelines for Selected Herbicides*. Copenhagen: World Health Organization.

WHO (2004). *Guidelines for Drinking Water Quality*, Vol. 1. *Recommendations*. 3rd edn. Geneva: World Health Organization.

Williams, R. J., Bird, S. C. and Clare, R. W. (1991). Simazine concentrations in a stream draining an agricultural catchment. *Water and Environmental Management*, **5**, 80–4.

Worthing, C. R. and Walker, S. B. (1983). *Pesticide Manual*, 7th edn. London: British Crop Protection Council.

Chapter 7
Endocrine-disrupting compounds and PPCPs

7.1 Introduction and definitions

The use of cosmetics and prescribed drugs, including antibiotics and synthetic hormones, continues to increase each year. Collectively they are known as pharmaceutical and personal care products (PPCPs), and comprise a wide spectrum of compounds. The PPCPs are either ingested or applied externally and used for both human and veterinary purposes. Antibiotics, growth promoters and other pharmaceuticals are also widely used as food additives for a wide range of livestock. These compounds are excreted in urine and faeces either as the active ingredient or an intermediate metabolite with their fate dependent upon a number of factors (Figure 7.1).

While some PPCPs end up in surface waters due to runoff from land, or directly into groundwater from septic tank leachate, the majority of these chemicals end up in sewage that is normally centrally treated at a wastewater treatment plant. Unfortunately PPCPs are only partially removed by conventional wastewater treatment processes, with adsorption onto suspended solids the main mechanism. This results in approximately 60% of the PPCPs being incorporated into the sludge, with significant quantities of the active ingredient and any metabolic breakdown products lost in the final effluent, which ends up in either surface or ground waters. There is increasing concern that PPCPs are finding their way into drinking water due to increasing surface and ground water pollution, and the reuse of sewage effluents in areas of water scarcity. Among the more commonly recorded PPCPs are antibiotics, painkillers, beta-blockers, lipid-reducing drugs and sex steroids from birth control and hormone replacement therapies (Table 7.1). While pharmaceutical drugs are designed to induce specific biological effects in the target organism for a limited time, most of these compounds retain their active ingredient even after wastewater treatment and subsequent discharge to the environment. Almost nothing is currently known about the risk associated from long-term exposure to these drugs at trace concentrations, or indeed exposure to cocktails of these drugs, on either human or environmental health. It has been suggested that the effects

Table 7.1 *Commonly occurring trace contaminants of a pharmaceutical origin in wastewaters and lowland rivers*

Pharmaceutical	Action
Acetylsalicylic acid	Analgesic/anti-inflammatory
Caffeine	Psychomotor stimulants
Carbamazepine	Analgesic/anti-inflammatory
Carboxyibuprofen	Analgesic/anti-inflammatory
Chloramphenicol	Antibiotic
Ciprofloxacin	Antibiotic
Clofibric acid	Lipid-lowering agent
Diazepam	Psychiatric
Diclofenac	Analgesic/anti-inflammatory
Erythromycin	Antibiotic
17β-oestradiol	Oestrogen
Oestrol	Oestrogen
Oestron	Oestrogen
17α-ethinyl oestradiol	Oestrogen
Hydroxyibuprofen	Analgesic/anti-inflammatory
Ibuprofen	Analgesic/anti-inflammatory
Naproxen	Analgesic/anti-inflammatory
Norflozacin	Antibiotic
Nonylphenol	Oestrogen
Primidone	Antiepileptic
Salicylic acid	Multi-purpose
Sulphadizine	Sulphonamide antibacterial
Sulphomethoxazole	Sulphonamide antibacterial
Sulphonamides	Sulphonamide antibacterial
Trimethoprim	Antibacterial

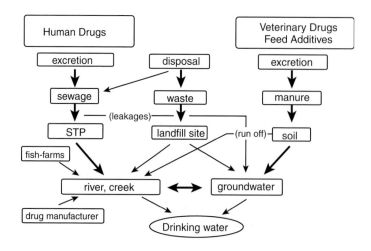

Figure 7.1 Possible pathways of PPCPs into drinking water. STP sewage treatment plant.

from PPCPs may be so subtle that they are not recognizable in real time, and that it is probable they elicit discernible cumulative effects that appear to have no obvious cause.

To date only two classes of PPCPs have been studied in depth in relation to drinking water, these are antibiotics and sex steroids. The misuse and overuse of antibiotics has led to increased resistance to antibiotics by bacterial pathogens, exacerbated by the exposure of micro-organisms to wastewater antibiotics during sewage treatment and while in water resources. On average 12 500 tonnes of antibiotics have been used annually within Europe over the past decade leading to an acceleration in the evolution of antibiotic-resistant bacterial strains in surface waters that can enter the water supply chain. The release of both natural, endogenous steroids, especially the oestrogens (estrogens), into the environment, as well as their synthetic counterparts, such as those used for reproductive control, has resulted in serious disruption of the endocrine system in aquatic animals in contaminated surface waters. Concerns have been expressed over the level of natural and synthetic sex steroids in surface and ground waters used for supply purposes and the potential physiological effects on consumers.

7.2 Pharmaceutical and personal care products (PPCPs)

Clofibric acid, the bioactive metabolite of the lipid-lowering drug clofibrate, was first recorded in a water supply aquifer that had been recharged with treated sewage effluent in the mid 1970s. This led to the realization that pharmaceutical and other drugs such as caffeine and nicotine were also present in treated effluents and so finding their way into water resources (Kümmerer, 2001). It was not until new analytical advances in the 1990s that the extent of the problem was fully realized with treated wastewater effluents found to contain a large range of pharmaceutical drugs and their metabolites, as well as synthetic fragrances (musks) and other components of cosmetics and other personal hygiene products (Daughton and Jones-Lepp, 2001).

Most research into PPCPs has focussed on their concentrations in treated sewage effluents and the resulting concentrations in the environment. However, while there are significant environmental concerns, it is assumed that the concentrations in drinking water recorded at the consumer's tap are normally orders of magnitude lower than that currently found in surface waters. Like other micro-pollutants, measuring such low concentrations (10^{-12} kg l^{-1}) is technically very difficult so little actual data exist on PPCP concentrations in drinking water (Kuch et al., 2001). However, numerous studies have confirmed the presence of PPCPs in drinking water (Heberer et al., 1998; Reddersen et al., 2002; Webb et al., 2003) (Table 7.2).

After being administered, the unused active ingredient in pharmaceutical products and their metabolites are excreted, along with any unwanted drugs that

Table 7.2 *Selected pharmaceuticals in drinking water tested in Germany. Reproduced from Ternes (2001) with permission from the American Chemical Society*

Drugs	LOQ in $\mu g\ l^{-1}$	Number of samples with conc. > LOQ	Number of samples with conc. > 0.010 $\mu g\ l^{-1}$	Median in $\mu g\ l^{-1}$	90-Percentile in $\mu g\ l^{-1}$	Maximum in $\mu g\ l^{-1}$
Antiphlogistics						
Diclofenac	0.001	8 of 30	0	<LOQ	0.002	0.006
Ibuprofen	0.001	3 of 30	0	<LOQ	0.001	0.003
Phenazone	0.010	1 of 12	1	<LOQ	<LOQ	0.050
Lipid regulators						
Clofibric acid	0.001	16 of 30	6	0.001	0.024	0.070
Fenofibric acid	0.005	1 of 30	1	<LOQ	<LOQ	0.042
Bezafibrate	0.025	1 of 30	1	<LOQ	<LOQ	0.027
Contrast media						
Iopamidol	0.010	4 of 10	4	<LOQ	0.070	0.079
Diatrizoat	0.010	5 of 10	5	0.021	0.075	0.085
Iopromid	0.010	1 of 10	1	<LOQ	<LOQ	0.086
Antiepileptics						
Carbamazepine	0.010	1 of 12	1	<LOQ	<LOQ	0.030

LOQ: Limit of quantification

may be disposed of down the toilet, and end up in the wastewater. Pharmaceutical and personal care products pass largely unaltered through conventional wastewater treatment (Ternes, 1998), and although diluted in surface and ground waters, as they are not removed by conventional water treatment processes then there is a high probability that these compounds could end up in drinking water at concentrations close to that found in the water resource from which the supply is taken. Drugs from a wide spectrum of therapeutic groups, and their metabolites, as well as personal care products such as sunscreens and fragrances have been identified in surface and ground waters at very low but measurable concentrations (ng–$\mu g\ l^{-1}$) (Daughton and Ternes, 1999; Debska *et al.*, 2004) (Table 7.3). The most common pharmaceuticals sold are those most frequently recorded in water resources (Table 7.1). Full details of the top 300 prescribed drugs dispensed in the USA during 2006 can be accessed at www.rxlist.com/top200.htm/. In 2005 the top five US drugs dispensed were hydrocodone bitartrate and acetaminophen for pain relief (101.6×10^6 prescriptions dispensed); atorvastatin calcium a lipid-lowering agent (63.2×10^6); amoxicillin an antibiotic (52.1×10^6); lisinopril an ACE (angiotensin-converting enzyme) inhibitor used primarily to treat high blood pressure (47.8×10^6) and hydrochlorothiazide, another drug used for treating high blood pressure (42.7×10^6).

Table 7.3 *Concentrations of PPCPs in groundwater located near contaminated surface waters in Berlin. Reproduced from Herberer et al. (2001) with permission from John Wiley and Sons Ltd*

Drug residues	Concentration range in ng l^{-1}
Clofibric acid	nd[a]–7300
Diclofenac	nd–380
Fenofibrate	nd–45
Gentisic acid	nd–540
Gemfibrozil	nd–340
Ibuprofen	nd–200
Ketoprofen	nd–30
Phenazone	nd–1250
Primidone	nd–690
Propyphenazone	nd–1465
(Salicylic acid)	nd–1225
Clofibric acid derivative	(nd–2900)[b]
N-methylphenacetin	(nd–470)[b]

[a] nd: not detected; [b] concentrations were only estimated, because standards were not commercially available.

Due to the complexity of monitoring these compounds at such low concentrations, analysis has been largely limited to target-based monitoring, thus ignoring all other organic micro-pollutants. In practice monitoring of pharmaceutical PPCPs is limited to those in Table 7.1, although some studies have extended this to in excess of 80 specific compounds of pharmaceutical origin. Thus in practice, our knowledge of what PPCPs are present in our drinking water is far from complete. Risk and hazard analysis on PPCPs in drinking water has been largely based on high doses required for a measurable therapeutic effect, while the concentrations in drinking water will be several orders lower than this resulting in much more subtle non-therapeutic effects (Webb *et al.*, 2003). There are many unanswered questions in relation to risk, which have been discussed by Daughton (2004). However, the key question must be, what are the potential additive or synergistic effects from the simultaneous consumption of trace concentrations of numerous drugs over a lifetime, and the effect on people already taking low therapeutic index medication?

The key to reducing the concentration of pharmaceutical drugs in drinking water is to reduce their use and subsequent release into waste streams. Recommended therapeutic doses are normally much higher than actually required by individuals, which unnecessarily increases the amount of active ingredients going into the waste stream. The concept of individualizing therapy and more accurately prescribing dose rates would significantly reduce PPCPs reaching the environment and then possibly entering the supply chain. Also many countries do not have a mechanism for disposing of expired or unwanted drugs; indeed doctors are often faced with significant problems of disposing of unused samples from suppliers. At worst these should be landfilled, although there is a real risk of the active ingredients being released into the environment via leachate; ideally they should be collected locally and disposed of safely by incineration. The indirect reuse of treated effluents has been identified as a key mechanism for meeting the widening deficit between water supply and demand. However, using resources that have already received treated sewage effluent results in PPCP contamination. This was demonstrated by Heberer *et al.* (1998) who studied the concentration profiles of clofibric acid and diclofenac in the Teltowkanal in Berlin. They observed significant peaks of the drugs after treated sewage effluent entered the river at three separate locations. This problem has been reviewed by Daughton (2004).

The use of septic tanks, which are poor at removing PPCPs, and where concentrations can be relatively high due to low dilution, has been highlighted as a major potential source of such chemicals in groundwaters. This is particularly problematic where people discharging to septic tanks are on long-term medication. Many studies have identified septic tanks as a major source of nitrite and pathogen contamination of groundwaters, and so it is inevitable that when these drugs are used in households employing septic tanks, similar contamination will occur.

7.3   Oestrogen and fertility

There is a surprisingly large number of natural and synthetic substances that are now classified as endocrine-disrupting compounds (EDCs). As the name suggests, these substances interfere with the normal functioning of the endocrine system resulting in a wide range of neurobehavioural, growth, developmental and reproductive problems. Most concern over endocrine disruptors has arisen from their effect on reproductive processes, which have caused a range of worrying effects on wildlife including hermaphrodite fish, reproductive malformation and sex changes in other species. This is most clearly seen in lowland rivers receiving treated and untreated wastewater where the feminization of male fish is now common. There are two key EDCs found in sewage, the natural female steroid hormones oestrogen and 17β-oestradiol, and the synthetic hormone ethinyl oestradiol from the contraceptive pill, both of which arise from the urine of the female population (Environment Agency, 1998b). Oestrogen-mimicking compounds such as alkylphenols (APs), alkylphenol ethoxylates (APEs) and Bisphenol A arise from industrial wastewaters, although a much wider group of compounds are also now classed as oestrogen-mimicking or EDCs including pesticides, dioxins and furans, and

Table 7.4 *Potential endocrine-disrupting compounds and
level of regulatory control in the UK*

Substance	Statutory control
Pesticides	
DDT; 'Drins; Lindane	A,C,D
Dichlorvos; Endosulphan;	B,C,D
Trifluralin; Demeton-*S*-	
Methyl; Dimethoate; Linuron;	
Permethrin	
Herbicides	
Atrazine; Simazine	B,C,D
PCBs	
Polychlorinated biphenyls	C
Dioxins and furans	
Polychlorinated dibenzofuran;	C
Dibenzo-*p*-dioxin congeners	
Antifoul/wood preservative	
Tributyltin	B,C,D
Alkylphenols	
Nonylphenol	None
Nonylphenol ethoxylate	None
Octylphenol	None
Octylphenol ethoxylate	None
Steroids	
Ethinyl osetradiol;	
17β-oestradiol; Oestrone	None

A: EC Dangerous Substances Directive (76/464/EEC) List I
substance; B: List II substance; C: Prescribed substance
under the Integrated Pollution Prevention and Control
Directive (IPPC) (96/61/EEC); D: Statutory environmental
quality standards (EQSs) in place in 2000.

organotin compounds such as tributyltin (TBT). A full list of these compounds
and their effects has been produced by the Environment Agency (1998a) and the
key EDCs are summarized in Table 7.4.

The evidence that the presence of these compounds is having a similar effect
on humans as seen with fish is not well developed and current theories of a link
is based on the work of a Danish paediatric endocrinologist Neils Skakkebaekit.

In the early 1990s he observed that 84% of men tested had sperm quality below the World Health Organization standard although they appeared healthy in every other respect. He found that on average sperm counts had almost halved during the period 1940 to 1990. This has now been confirmed both in the USA and in other European countries with a study in Edinburgh showing males born in the 1940s to have an average sperm count of 128 million compared to only 75 million in males born in the late 1960s. Skakkebaekit has suggested that the steady reduction in sperm counts over the past 50 years, and a similar increase in prostate and testicular cancers, is due to overexposure of the male fetus during gestation to oestrogen-like substances that the mother has obtained from the environment. The exposure of the fetus to excessive oestrogen-like substances results in a delay in the formation of the sex hormone-producing cells in the testicles, which in adulthood results in reduced sperm production. It is also suspected that exposure to excessive oestrogen-like substances during pregnancy can effect the formation of the male fetuses' sex organs. The theory is supported by studies into the use of diethylstilbestrol, an orally active synthetic non-steroidal oestrogen, widely prescribed to prevent miscarriage in the USA during the 1950s and 1960s. These studies showed that high levels of oestrogen and oestrogen-like substances were not beneficial but caused testicular abnormalities and reduced sperm counts in males and anatomic abnormalities in females leading to infertility and in some cases a rare cancer of the vagina. Like many other pharmaceuticals, diethylstilbestrol is now classified as a teratogen. This is seen by many scientists as confirmation that abnormal exposure to oestrogen or oestrogen-like substances during pregnancy could damage the unborn fetus, and that these effects might not be seen for many years. What is clear is that the human body perceives oestrogen-like substances as natural oestrogens, and that even small concentrations of oestrogen can have a physiological effect on human development. Other chemicals can also affect the natural hormonal balance in the body by blocking either oestrogen receptors or the androgen receptors or both, which may make them even more potent than many EDCs. The link, although widely accepted, remains unproven and as such EDCs are not included generically in drinking water standards, although some individual pesticides and industrial organic solvents which are known EDCs are included but due to their carcinogenic or toxic properties. For example atrazine, which has a maximum contaminant level of $3\,\mu g\,l^{-1}$, is a classified EDC that is known to promote the conversion of testosterone to oestrogen. It is one of the most abundant and ubiquitous herbicides found in drinking water, and has even been recorded at concentrations $>40\,\mu g\,l^{-1}$ in precipitation. Yet atrazine can induce hermaphroditism in frogs at concentrations of just $\geq0.1\,\mu g\,l^{-1}$ with tadpoles developing extra testes and even ovaries (Hayes *et al.*, 2002). Atrazine is now accepted to be a potent endocrine disrupter, with the induction of aromatase, the enzyme that converts androgens to oestrogens, the mechanism that causes feminization in amphibians; a mechanism that also occurs in all mammals including humans (Hayes, 2005).

The number of organic compounds that are known to be endocrine disrupting is growing each year, and with the widespread use of these chemicals it is inevitable that the majority of these will find their way into water resources and probably to consumers' taps. Many of these chemicals are persistent, fat-soluble and bioaccumulate in the food chain, resulting in exposure not only from drinking water but also from food. Our chance of avoiding these chemicals in food is slim, although with proper treatment they can be excluded from drinking water. In an experiment that examined the degree of human exposure to various pharmaceutical compounds in drinking water, Webb *et al.* (2003) found the average daily intake of oestrogen from drinking water to be negligible. They concluded that as humans naturally produce and intake various forms of oestrogen at levels up to two orders of magnitude greater than that found in drinking water, that current increased levels of oestrogen in the environment should not cause harmful effects to humans. However, oestrogens represent only a small fraction of the total EDCs found in drinking water.

7.4 Conclusions

Water scarcity is an increasingly common problem in developed as well as developing countries that is leading to a greater dependence on water reuse and exploitation of poorer resources to meet water demands. With less dilution and a greater reuse of sewage effluents, then PPCPs will become more common in drinking water and at higher concentrations than seen at present. Our current knowledge of the toxicology of low-level cocktails of these drugs and their metabolites is far from complete, and currently we have no idea of how they will affect the population if it is exposed to them on a continuous basis in drinking water.

Long-term exposure to trace quantities of EDCs is resulting in reduced fertility in males and even feminization in some aquatic species. So it is very possible that similar exposure to other pharmaceuticals and their metabolites may also be having serious effects.

Preventing contamination of drinking water by PPCPs requires a multiple barrier approach. Firstly, the use of PPCPs must be better regulated, especially in relation to the use of prescribed drugs such as antibiotics and hormones. Secondly, unwanted drugs must be prevented from entering the water cycle through inappropriate disposal methods. Thirdly, better removal of PPCPs is required at the wastewater treatment stage. Removal is primarily by adsorption on to the solids present for hydrophobic compounds and those with positively charged functional groups such as amines. The longer the sludge residence time then the greater the possibility that compounds may be degraded biologically, both during aerobic and anaerobic treatment, while oestrogens have been reported as being removed where nitrification–denitrification steps are employed. However, effective removal of PPCPs can only be achieved by treating the final effluent using advanced technologies such as activated carbon, nanofiltration or ozonation. Ozone doses of between 1 and 15 mg l^{-1} have been shown to destroy PPCPs in a treated sewage effluent at Braunschweig in Germany (Ternes, 2004). The effluent contained 34 detectable PPCPs including antibiotic, beta-blockers and oestrogen residues, all of which were effectively removed by ozonation. Source control may be one way of preventing groundwater contamination from septic tank leachate. As 70–80% of pharmaceutical compounds in wastewater enter via urine, urine separation may be a particularly effective method of preventing such compounds from escaping into water resources where advanced treatment cannot be given. Pharmaceutical and personal care products are not effectively removed by conventional water treatment with chlorination ineffective in oxidizing such compounds so effective removal is only achieved by advanced unit processes. Where these advanced technologies are in place then contamination of drinking water will be very unlikely, although due to cost this will only be likely at larger plants using surface waters that are receiving sewage effluents upstream. Therefore it is the smaller treatment plants, those on group or private supplies, that will be most at risk from pharmaceutical residuals in their drinking water. The treatment of oestrogens and oestrogen-mimicking (EDCs) compounds has been reviewed by Walker (2000). Daughton (2003a,b) has proposed a stewardship programme for PPCPs in which he examines the actions that could be taken to minimize the release of these chemicals into the environment.

A major obstacle in effectively dealing with this emerging problem is the lack of suitably sensitive analytical techniques for measuring the concentrations of the organic compounds that are already suspected of being present in drinking waters, which may be as low as 0.000001 µg l^{-1} which is 1 picogram per litre. This will require some form of enrichment and pre-concentration techniques using a new solid-phase extraction method yet to be devised.

There is a high probability of measurable quantities of PPCPs occurring in drinking water where the supply has been subject to contamination by treated or untreated domestic or livestock wastewaters. Controlling the use of such products and their disposal, better wastewater treatment and the use of advanced water treatment technologies together can minimize any risk to consumers via drinking water. So the indirect reuse of treated municipal wastewaters is feasible but expensive. Where individuals and group schemes rely on groundwaters that are subject to contamination by septic tanks then a suitable point-of-use treatment system, such as activated carbon filtration, will be required if PPCPs are to be removed.

Pharmaceutical and personal care products are increasingly being released on a continuous basis into water resources with little regulation in their use or disposal. Although the US Environmental Protection Agency have over 170 drinking water standards none of them deal with pharmaceuticals although some specific EDCs are included such as pesticides. The quantity of these chemicals currently being discharged is quite staggering and numerous studies have indicated that PPCPs can have potential adverse human and environmental effects from indirect exposure (Halling-Soresen *et al.*, 1998; Cleuvers, 2003; Harder, 2003). Yet the research has been hampered by a lack of analytical capability to accurately measure these compounds at the very low concentrations at which they are found in drinking water. There is enough evidence to evoke the precautionary principle for both PPCPs and EDCs in drinking water, and to review methods of reducing their release into the environment.

References

Cleuvers, M. (2003). Mixture toxicity of the anti-inflammatory drugs diclofenac, ibuprofen, naproxen, and acetylsalicylic acid. *Ecotoxicology and Environmental Saftey*, **59**(3), 309–15.

Daughton, C. G. (2003a). Cradle-to-cradle stewardship of drugs for minimizing their environmental disposition while promoting human health. 1. Rationale for the avenues towards a green pharmacy. *Environmental Health Perspectives*, **111**(5), 757–74.

Daughton, C. G. (2003b). Cradle-to-cradle stewardship of drugs for minimizing their environmental disposition while promoting human health. 2. Drug disposal, waste reduction and future directions. *Environmental Health Perspectives*, **111**(5), 775–85.

Daughton, C. G. (2004). Groundwater recharge and chemical contaminants: challenges in communicating the connections and collisions of two disparate worlds. *Ground Water Monitoring and Remediation*, **24**(2), 127–38.

Daughton, C. D. and Jones-Lepp, T. L. (eds.) (2001). *Pharmaceuticals and Personal Care Products in the Environment: Scientific and Regulatory Issues*. ACS Symposium Series 791. Washington, DC: American Chemical Society.

Daughton, C. G. and Ternes, T. A. (1999). Pharmaceuticals and personal care products in the environment: agents of subtle change? *Environmental Health Perspectives*, **107**(6), 907–38.

Debska, J., Kot-Wasik, A. and Namiesnik, J. (2004). Fate and analysis of pharmaceutical residues in the aquatic environment. *Critical Reviews in Analytical Chemistry*, **34**, 51–67.

Environment Agency (1998a). *Endocrine Disrupting Substances in the Environment: What Should Be Done?* Consultative Report. Bristol: Environment Agency.

Environment Agency (1998b). *The State of the Environment in England and Wales: Freshwaters*. London: The Stationery Office.

Halling-Sorensen, B., Nors Nielsen, S. Lansky, P. F. *et al.* (1998). Occurrence, fate, and effects of pharmaceutical substances in the environment – a review. *Chemosphere*, **36**(2): 357–93.

Harder, B. (2003). Extracting estrogens. *Science News*, **164**(5), 67–8.

Hayes, T. B. (2005). Welcome to the revolution: integrative biology and assessing the impact of endocrine disruptors on environmental and public health. *Integrative and Comparative Biology*, **45**, 321–9.

Hayes, T. B., Collins, A., Lee, M. *et al.* (2002). Hermaphrodotic, demasculinized frogs after exposure to the herbicide atrazine at low ecologically relevant doses. *Proceedings of the National Academy of Science (USA)*, **99**, 5476–80.

Heberer, Th., Schmidt-Bäumler, K. and Stan, H.-J. (1998). Occurrence and distribution of organic contaminants in the aquatic system in Berlin. Part I: Drug residues and other polar contaminants in Berlin surface and ground water. *Acta Hydrochimica et Hydrobiologica*, **26** (5), 272–8.

Heberer, Th., Fuhrmann, B., Schmidt-Baumler, K. *et al.* (2001). Occurrence of pharmaceutical residues in sewage, river and drinking water in Greece and Berlin (Germany). In *Pharmaceuticals and Personal Care Products in the Environment: Scientific and Regulatory Issues.* ed. C. D. Daughton and T. L. Jones-Lepp. ACS Symposium Series 791. Washington, DC: American Chemical Society, pp. 70–83.

Kuch, H. M. and Ballschmiter, K. (2001). Determination of endocrine-disrupting phenolic compounds and estrogens in surface and drinking water by HRGC-(NCI)-MS in the picogram per liter range. *Environmental Science and Technology*, **35**, 3201–6.

Kümmerer, K. (ed.) (2001). *Pharmaceuticals in the Environment: Sources, Fate, Effects and Risks.* Heidelberg, Germany: Springer-Verlag.

Reddersen, K., Heberer, T. and Dunnbier, U. (2002). Identification and significance of phenazone drugs and their metabolites in ground and drinking water. *Chemosphere* **149**, 539–44.

Ternes, T. A. (1998). Occurrence of drugs in German sewage treatment plants and rivers. *Water Research*, **32**, 3245–60.

Ternes, T. A. (2001). Pharmaceuticals and metabolites as contaminants of the aquatic environment. In *Pharmaceuticals and Personal Care Products in the Environment: Scientific and Regulatory Issues.* ed. C. D. Daughton and T. L. Jones-Lepp. ACS Symposium Series 791. Washington, DC: American Chemical Society, pp. 39–54.

Ternes, T. A. (2004). *Assessment of Technologies for the Removal of Pharmaceuticals and Personal Care Products in Sewage and Drinking Water Facilities to Improve the Indirect Potable Water Reuse. POSEIDON*, EU Contract EVK1-CT-1000-00047. Brussels: European Union.

Walker, D. (2000). Oestrogenicity and wastewater recycling: experience from Essex and Suffolk Water. *Journal of the Chartered Institution of Water and Environmental Management*, **14**, 427–31.

Webb, S., Ternes, T., Gibert, M. and Olejniczak, K. (2003). Indirect human exposure to pharmaceuticals via drinking water. *Toxicology Letters.*, **142**(3), 157–67.

Chapter 8
Odour and taste

8.1 Introduction

In its purest state water is both odourless and tasteless; however, as inorganic and organic substances dissolve in the water, it begins to take on a characteristic taste and sometimes odour. Generally, the inorganic salts at the concentrations found naturally in drinking water do not adversely affect the taste; in fact, many of the bottled mineral waters are purchased because of their characteristic salty or sulphurous taste. Both tastes and odours are caused by the interaction of the many substances present. These may include soil particles, decaying vegetation, organisms (plankton, bacteria, fungi), various inorganic salts (for example, chlorides and sulphides of sodium, calcium, iron and manganese), organic compounds and gases (Cohen *et al.*, 1960; Baker, 1963). Water should be palatable rather than free from taste and odour, but with people having very different abilities to detect tastes and odours at low concentrations this is often difficult to achieve. Offensive odours and tastes account for most consumer complaints about water quality.

In 1970 it was estimated that 90% of water supplied in the UK occasionally suffered from odour and taste problems (Bays *et al.*, 1970). This is similar to the figure for the USA and Canada where 70% of water supplies are affected. The situation has improved enormously over the past 36 years due largely to the introduction of the EC Drinking Water Directive. According to the annual reports of the Drinking Water Inspectorate (www.dwi.gov.uk) the problem is now much smaller. However, odour and taste problems can be very transitory phenomena, so it is unlikely that the standardized monitoring carried out by some water supply companies involving taking just four or five samples a year will detect all but the most permanent odour and taste problems.

8.2 Standards and assessment

There are two widely used measures of taste and odour. Both are based on how much a sample must be diluted with odour- and taste-free water to give the least perceptible concentration. The threshold number (TN) is the most widely used parameter:

$$TN = (A + B)/A$$

where A is the volume of the original sample and B the volume of dilutent. The value $A + B$ is the total volume of sample after dilution to achieve no perceptible odour. Therefore a sample with a TN of unity has no odour or taste. The analysis is usually carried out at either 40 or 60 °C as the higher temperatures allows the panellists to be more sensitive; also many complaints are related to hot rather than cold water when any odour-producing compounds are more volatile (APHA, 1998).

Dilution number (DN) was originally used in the first EC Drinking Water Directive (80/778/EEC):

$$DN = B/A.$$

The two are related as $DN = TN - 1$, so a sample with a DN of three has a TN of four. The revised Drinking Water Directive (98/83/EEC) requires taste to be acceptable with no DN set.

Standards are generally the same. In the USA the Environmental Protection Agency (EPA) includes odour in the non-mandatory National Secondary Drinking Water Standards with a maximum contaminant level (MCL) for odour of a TN of three (no temperature specified), whereas the World Health Organization only requires that the taste and odour are acceptable. Originally the EC set a guide DN value of zero and two mandatory values. These were 2 at 12 °C and 3 at 25 °C. The problem is that the Directive did not specify how to interpret the standard and no methodology was specified. However in the new EC Drinking Water Directive (98/83/EEC) taste is listed in Part C as an Indicator Parameter although no maximum permissible value has been set, rather it states that the taste of drinking water must be acceptable to consumers and show no abnormal change. In the UK, the Water Supply (Water Quality) Regulations 2000 set out both European and National requirements. Here taste is included under Part B Chemical Parameters and not under Part C as in the Directive, although it is still included in check monitoring carried out by the water utilities. A National requirement is given for taste of a DN of 3 at 25 °C as measured at the consumer's tap.

It is extremely difficult to chemically analyze tastes and odours, mainly because of the very low concentrations at which humans can detect many of these chemicals. Also, offensive odours can be caused by mixtures of several different chemicals (ASTM, 1968). Odours of substances taken into the mouth can often be detected when they are not detectable by sniffing (Hudson, 2000). This is due to the enclosed space of the mouth and the higher temperature, increasing the concentrations of volatile materials present. When water or food is taken into the mouth and swallowed, air is exhaled through the nose. So as the nose and sinus cavities are far more sensitive to many chemicals than the tongue, odour appears to be the primary sensation in relation to water quality. The ability to sense tastes and odours varies considerably between individuals. Anosmia is a condition where the sense of smell is reduced or entirely lost, and is caused by many different factors the most common being traumatic head injury, colds or infections, and Alzheimer's disease. In fact everyone's ability to detect smells deteriorates with age (Keneda *et al.*, 2000). The physiological and biochemical basis of sense of smell is extremely complex and is considered further by Schneider *et al.* (1966). Odours alter with concentration as well as temperature; for example, nonadienal has a cucumber odour at low concentrations ($2-5\,ng\,l^{-1}$) intensifying to a grassy odour at higher concentrations ($10-20\,ng\,l^{-1}$).

Calculating odour and taste threshold levels determine acceptable levels of odour and taste in drinking water. The measurement is semi-quantitative and is carried out subjectively by trained technicians or by a panel of testers who taste or smell samples randomly. The sample is tested as near to body heat as possible. Testing is carried out by decanting the sample into a conical flask where it is shaken vigorously; the vapour is then sniffed by the panel. If an odour is detected then the panel members are asked to describe it in terms such as fishy, musty, earthy, chlorinous, oily or medicinal. A majority verdict is taken and comparisons may be made to reference samples containing known odorous compounds. If an estimate of intensity is required then the sample is diluted with odour-free water until no odour is detectable. These results can then be compared with standard samples containing odorous compounds of known concentration to give some idea of the concentration of the odour-forming compound in the water.

Assessment by tasting is rarely carried out due to the potential health hazards to the tester. However, panels of testers are used to calculate at what concentrations specific compounds are no longer detectable by taste, which gives the treatment plant operator an idea of how much of a particular compound needs to be removed to maintain its wholesomeness in terms of taste and odour. There are a number of standard methods for assessing taste and odour published in the UK (Environment Agency, 1998, 2004a, b), and in the USA (ASTM, 1984; Clesceri *et al.*, 1989a, b; APHA, 1998). The threshold odor test (Standard Method 2150B) and flavour profile analysis (Standard Method 2170B) are currently the most widely used drinking water odour sensory methods (APHA, 1998). The threshold odor number (TN) measures the overall odour intensity, while flavour profile analysis compares the sample to accepted taste reference standards, i.e. sour, salty, bitter and sweet, using a 7 point scale (0–12 using even numbers only) for each attribute (APHA, 1998). Sampling odours is fully reviewed by Bartels *et al.* (1986).

8.3 Classification

Odour problems are categorized according to the origin of the substance causing the problem. Substances can be present in the raw water, they can be added or

Figure 8.1 Schematic diagram of step approach of investigating an odour problem. Reproduced from Ainsworth *et al.* (1981) with permission from WRc Plc.

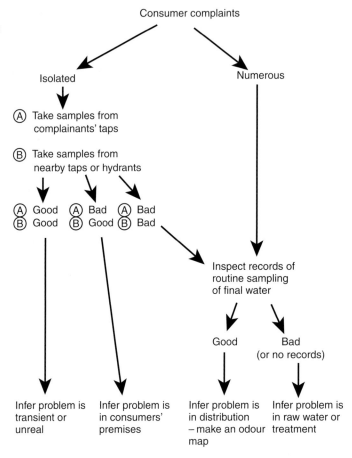

created during water treatment (Chapter 16), or they can arise within the distribution (Chapter 21) or domestic plumbing systems (Chapter 28). Of course, the quality of the raw water will often contribute to the production of odours during treatment and distribution.

Many problems are related to transient problems with the raw water, although it is often the case that all the offending water will have passed through the treatment plant before the consumer identifies the problem. For this reason finished water is often routinely monitored for odour at the treatment plant. Figure 8.1 outlines a scheme used by water supply companies for investigating the source of odour problems. The problems of odour and taste in raw water are due to either substances of natural origin or man-made pollutants.

It is not always straightforward to identify the source of an odour or taste in drinking water (Table 8.1). There are seven common causes:

1. *Decaying vegetation.* Algae produce fishy, grassy and musty odours as they decay, and certain species can cause serious organoleptic problems when alive.

Table 8.1 *Chemicals causing specific groups of odours and their possible sources*

Odour	Odour-producing compounds	Possible source
Earthy/musty	Geosmin, 2-methylisoborneol, 2-isopropyl-3-methoxy pyrazine mucidine, 2-isobutyl-3-methoxy pyrazine, 2,3,6-trichloroanisole	Actinomycetes, blue-green algae
Medicinal or chlorophenolic	2-Chlorophenol, 2,4-dichlorophenol, 2,6-dichlorophenol	Chlorination products of phenols
Oily	Naphthalene, toluene	Hydrocarbons from road runoff, bituminous linings in water mains
Fishy, cooked vegetables or rotten cabbage	Dimethyltrisulphide, dimethyl disulphide, methyl mercaptan	Breakdown of algae and other vegetation
Fruity and fragrant	Aldehydes	Ozonation by-products

2. *Moulds and actinomycetes*. These organisms produce musty, earthy or mouldy odours and tastes. They tend to be found where water is left standing in pipework and also when the water is warm. Usually found in the plumbing systems of large buildings such as offices and flats. They are also associated with waterlogged soil and unlined boreholes.

3. *Iron and sulphur bacteria*. Both bacteria produce deposits, which release offensive odours as they decompose.

4. *Iron, manganese, copper and zinc*. The products of metallic corrosion all impart a rather bitter taste to the water.

5. *Sodium chloride*. Excessive amounts of sodium chloride will make the water taste initially flat or dull, then progressively salty or brackish.

6. *Industrial wastes*. Many wastes and by-products produced by industry can impart a strong medicinal or chemical taste or odour to the water. Phenolic compounds that form chlorophenols on chlorination are a particular problem.

7. *Chlorination*. Chlorine by itself does not produce a pronounced odour or taste unless the water is overdosed during disinfection. Chlorine will react with a wide variety of compounds to produce chlorinated products, many of which impart a chlorinous taste to the water.

8.4   Odour-causing substances of natural origin

Most problems of odours in raw water treatment supplies involve organic substances of natural origin, mainly from algae and/or decaying vegetation. Large crops of algae can be present in reservoirs without causing any problems. However, there are a few uncommon algae that produce organic by-products such as essential oils, which can produce offensive odours even when the algae are present in small numbers. The odour is formed when these materials are released into the water, usually when the algae die.

There are four main groups of algae that are able to cause odour problems:

1. Blue-green algae (*Cyanophyceae*). The important genera (groups) are *Anabaena, Anacystis, Aphanizomenon, Gomphosphaeria* and *Oscillatoria*. These generally give rise to grassy odours intensifying into piggy, almost septic, odours as the cells disintegrate. Some algae can also produce toxic substances (Chapter 11).
2. Diatoms (*Bacillariophyceae*). Important genera are *Asterionella, Fragillaria, Melosira, Tabellaria* and *Synedra*. They secrete oils that impart a fishy or aromatic odour. When moderate densities are present then geranium (*Asterionella* and *Tabellaria*) odours are produced.
3. Green algae (*Chlorophyceae*). Can impart a fishy or grassy odour. Important genera include *Volvox* and *Staurastrum*.
4. Yellow-brown algae (*Chrysophyceae*). Can produce pungent odours. *Uroglenopsis* can produce a very strong fishy odour, whereas *Synura* produces a cucumber odour at low numbers, a fishy odour at moderate numbers and a piggy odour at high numbers.

It is not only algae that cause odours; rooted aquatic plants (macrophytes) can also be the source of taste and odour problems. This only occurs in shallow lakes and reservoirs when the vegetation dies back and begins to decompose. Eutrophication in rivers can cause similar problems. Odour and taste problems due to algae are generally seasonal.

The most common odour complaint in drinking water is of musty or earthy odours. They are generally caused by actinomycetes and to a lesser extent by blue-green algae and fungi. They are associated with supplies taken from lowland rivers, which is probably due to the fact that such rivers are generally rich in suitable organic material on which these micro-organisms thrive. There is evidence to support the idea that some algae, which do not normally produce odorous compounds, do so when actinomycetes are also present, especially blue-green algae. The main organic compounds produced that cause musty/earthy odours are geosmin and 2-methylisoborneol, and consumer complaints will follow if concentrations of either of these exceed 8–10 ng l^{-1} (Tables 8.1 and 8.2). Dimethyltrisulphide is also produced by blue-green algae and although it has a higher odour threshold than geosmin, it produces a grassy odour that intensifies into a very unpleasant septic/piggy odour with increasing concentrations of algal biomass.

Phenols can arise in water from the natural decay of organic matter, although more commonly they are associated with industrial pollution. When these are present strong medicinal (or chlorophenolic) odours are produced when the water is chlorinated.

As mentioned previously, inorganic substances rarely cause odour or taste problems in the UK, although a saline-tasting water can be unpleasant if used to

make tea or coffee in particular. Water will taste salty if chlorides are present at 500 mg l^{-1} or more. This can occur in groundwaters due to over-pumping boreholes near the coast, leading to saline intrusion. Other commonly found salts in groundwater can also impair this taste, such as magnesium sulphate and sodium sulphate. Water that has lost its oxygen (e.g. groundwaters or the hypolimnion of lakes) (Chapter 4) allows sulphide to be formed, which has a most unpleasant odour and taste, even at concentrations below 0.5 mg l^{-1}. Iron and manganese have a characteristic bitter taste and can occur in appreciable amounts in certain underground waters and springs (Sections 9.1 and 9.2 respectively).

Figure 8.2 Drinking water taste and odour wheel used. BHT = butylated hydroxytoluene; MTBE = methyl tertiary-butyl ether; *commonly recorded in water. Reproduced from Suffet *et al.* (1999) with permission from IWA Publishing Ltd.

Table 8.2 *Comparative odour thresholds for some common odour-causing compounds*

Odour-causing compound	Aqueous odour threshold concentration (μg l^{-1})
1-butanol	0.002–0.005
Free chlorine	280–360 pH dependent
Geosmin	0.006–0.01
n-hexanal	0.06–1.9
Isobutanal	0.15–2.0
2-methylisoborneol	0.002–0.02
trans-2 *cis*-6-nonadienal	0.002–0.013

Suffet *et al.* (1999) has produced a unique system for identifying the chemicals in drinking water causing specific tastes and odours (Figure 8.2).

8.5   Man-made odour-causing substances

There are an enormous number of industrial organic pollutants that find their way into surface and ground waters that are offensive in an organoleptic sense (e.g. alkyl benzenes, chlorobenzenes, alkanes, benzaldehydes and benzothiazole) (Lillard and Powers, 1975). The degradation products of these compounds are often even more offensive. Pesticides are highly odorous, as are many of the solvents used in their formulation. However, the odour threshold concentration is above the level at which most pesticides are considered to be safe, so toxicity considerations are more important than organoleptic ones. Chlorophenolic compounds may also be discharged to water resources from industry, although phenolic compounds discharged mainly from petroleum and wood-processing industries (many agricultural chemicals are also phenolic) will be modified by chlorination at the treatment plant to form these odorous compounds (Chapter 14).

8.6   Conclusions

Odours and tastes in drinking water are very subjective and transitory problems, making them notoriously difficult to identify or quantify, and subsequently to track down their source. Yet they represent the single most common complaint concerning drinking water quality. Many engineers and scientists do not take taste monitoring or complaints seriously, seemingly due to the subjective manner of assessment. However, with the new laboratory techniques now available such as gas chromatography-mass spectrometry, it is possible to screen for individual odour-causing compounds and identify when they are exceeding taste thresholds (Table 8.2).

Taste and odour are amongst the most challenging parameters for water utilities as consumers are constantly evaluating the aesthetic quality of their water every time they use it. So problems relating to taste and odour will consistently generate more consumer complaints and adverse attention in the media than more serious quality problems (Hack, 2000).

The removal of offensive odours and tastes is primarily by advanced water treatment using activated carbon and ozone (Section 14.2), whereas the problems of taste are examined further in Chapters 16, 21 and 28. Amoore (1986) has written a most interesting article on odours and our sensitivity to them. He suggests the main reason why humans have such a highly developed sensitivity to the metabolites produced by these naturally occurring micro-organisms, and which gives rise to so many complaints, is because they are biological indicators of the presence of water. Primitive tribes still display a remarkable ability to find water, even in the most arid regions, using such indicators.

References

Ainsworth, R. G., Calcutt, T., Elvidge, A. F. *et al.* (1981). *A Guide to Solving Water Quality Problems in Distribution Systems*. Technical Report 167. Medmenham: Water Research Centre.

Amoore, J. E. (1986). The chemistry and physiology of odour sensitivity. *Journal of the American Water Works Association*, **78**(3), 70–6.

APHA (1998). *Standard Methods for the Examination of Water and Wastewater*, 20th edn. Washington, DC: American Public Health Association.

ASTM (1968). *Manual on Sensory Testing Methods*. ASTM STP 434. Washington, DC: American Society for Testing and Materials.

ASTM (1984). Standard test method for odor in water. D 1292–80. In *Annual Book of ASTM Standards, Section 11*: Water and Environmental Technology, 11.01, 218. Washington, DC: Society for Testing and Materials.

Baker, R. A. (1963). Threshold odors of organic chemicals. *Journal of the American Water Works Association*, **55**(7), 913–16.

Bartels, J. H.M., Burlingame, G. A. and Suffet, I. H. (1986). Flavour profile analysis: taste and odour control of the future. *Journal of the American Water Works Association*, **78**(3), 50–5.

Bays, L. R., Burman, J. P. and Lewis, W. M. (1970). Taste and odour in water supplies in Great Britain: a survey of the present position and problems for the future. *Water Treatment and Examination*, **19**, 136–60.

Clesceri, L. S., Greenberg, A. E., Trussell, R. H. and Franson, R. R. (1989a). Taste 2160. In *Standard Methods for the Examination of Water and Wastewater*, 17th edn. Washington, DC: American Public Health Association, 2.23.

Clesceri, L. S., Greenberg, A. E., Trussell, R. H. and Franson, R. R. (1989b). Odor 2150. In *Standard Methods for the Examination of Water and Wastewater*, 17th edn. Washington, DC: American Public Health Association, 2.16.

Cohen, J. M., Kamphake, L. J., Harris, E. K. and Woodward, R. L. (1960). Taste threshold concentrations of metals in drinking water. *Journal of the American Water Works Association*, **52**(5), 660–70.

Environment Agency (1998). *The Assessment of Taste, Odour and Related Aesthetic Problems in Drinking Waters 1998*. Blue Book No. 171. London: Standing Committee of Analysts, Environment Agency.

Environment Agency (2004a). *The Microbiology of Drinking Water 2004*. Part 11. *Taste, Odour and Related Aesthetic Problems*. Blue Book No. 197. London: Standing Committee of Analysts, Environment Agency.

Environment Agency (2004b). *The Microbiology of Drinking Water 2004*. Part 12. *Methods for the Isolation and Enumeration of Micro-Organisms Associated with Taste, Odour and Related Aesthetic Problems*. Blue Book No. 197. London: Standing Committee of Analysts, Environment Agency.

Hack, D. J. (2000). Common consumer complaint: my hot water stinks. *Opflow*, (June) 11–12.

Hudson, R. (2000). Odor and odorant. A terminological classification. *Chemical Senses*, **25**(6), 693.

Keneda, H., Maeshima, K., Goto, N. *et al.* (2000). Decline in taste and odour discrimination abilities with age, and relationship between gustation and olfaction. *Chemical Senses*, **25**, 331–7.

Lillard, D. A. and Powers, J. J. (1975). *Aqueous Odour Thresholds of Organic Pollutants in Industrial Effluents*. EP A-660/4-75-OO2, Environmental Monitoring Series. Washington, DC: US Environmental Protection Agency.

Schneider, R. A., Schmidt, C. E. and Costiloe, J. P. (1966). Relation of odour flow rate and duration to stimulus intensity needed for perception. *Journal of Applied Physiology*, **21**(1–3), 10–14.

Suffet, I. H., Khiari, D. and Bruchet, A. (1999). The drinking water taste and odour wheel for the millennium: beyond geosmin and 2-methylisoborneol. *Water Science and Technology*, **40**(6), 1–13.

Chapter 9
Metals

9.1 Iron

Iron is an extremely common metal and is found in large amounts in soil and rocks, although normally in an insoluble form. However, due to a number of complex reactions that occur naturally in the ground, soluble forms of iron can be formed, which can then contaminate any water passing through. Therefore excess iron is a common phenomenon of groundwaters, especially those found in soft groundwater areas.

Iron is an essential element and is very unlikely to cause a threat to health at the concentrations occasionally recorded in water supplies. It is undesirable in excessive amounts and can cause a number of problems. Iron is soluble in the ferrous state (Fe^{2+}) and is oxidized in the presence of air to the insoluble ferric form (Fe^{3+}), so when groundwaters are anaerobic, or have low dissolved oxygen concentrations, all the iron will be in a soluble form. Oxygen can enter aquifers via boreholes. This, combined with the removal of carbon dioxide from solution, raises the pH and causes iron in the groundwater to precipitate out as ferric hydroxide (solubility threshold pH 4.2). Even small changes in water chemistry can affect iron solubility, and the conditions in many groundwaters are on the boundary between soluble and insoluble iron (Figure 9.1). At the treatment plant most of the iron is removed by aerating the water or by using coagulants, with the particles of insoluble iron removed by filtration. In private supplies the water will only start to become aerated as it enters the storage tank. Here the ferric iron particles will settle to the bottom of the tank and form an orange sediment, which can then be resuspended when large volumes of water are being drawn from the tank. Alternatively, as the water leaves the tap it is very effectively aerated and it is here the water may become discoloured and turbid. Sediment will also form in the pipework and this will contribute to the discolouration of the water. Some stored waters can also contain iron in its soluble form, especially in deeper parts of reservoirs where the dissolved oxygen concentration will be low. In river waters, which are well aerated, iron, if present, will be in its insoluble particulate form. Iron also encourages the development of a microbial slime comprised of iron bacteria on piped surfaces

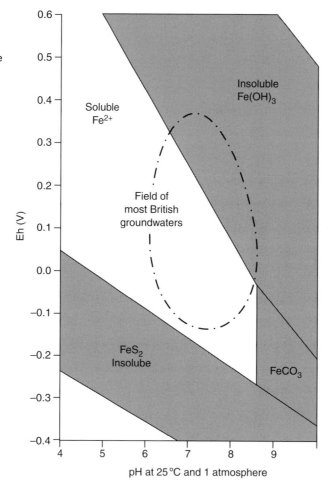

Figure 9.1 Phase-stability diagram for the solubility of iron in groundwaters. The dotted line shows the range for most UK groundwaters. Reproduced from Clarke (1988) with permission from John Wiley and Sons Ltd.

that can affect flow and cause consumer complaints (Chapter 24). Iron bacteria may also grow on the wall of boreholes and even block intake screens and damage submersible pumps.

Although discolouration is inconvenient, it is only an aesthetic problem. It becomes more irritating when the iron causes staining of laundry and discolouration of vegetables such as potatoes and parsnips during cooking. More importantly, iron has a fairly low taste threshold for such a common element, giving the water a strong unpleasant bitter taste that ruins most beverages made with the tap water. High concentrations of iron in water can react with the tannin in tea causing it to turn odd colours, usually inky black.

Iron is so ubiquitous that some will find its way into nearly all supplies. The taste threshold is about $0.3 \, \text{mg} \, l^{-1}$ ($300 \, \mu\text{g} \, l^{-1}$), although it varies considerably between individuals. The EC Drinking Water Directive (98/83/EEC) lists iron as an indicator parameter in Part C at a maximum permissible value of $0.2 \, \text{mg} \, l^{-1}$

(Appendix 1). In the former Directive (80/778/EEC) a guide value, the level considered to be ideal, had been set at 0.05 mg l^{-1}. The World Health Organization (WHO) originally set a guideline at 0.3 mg l^{-1} based on taste thresholds, although in the latest revision a guideline value has been removed as the taste and appearance of water is severely affected below the health-based value (WHO, 2004). The US Environmental Protection Agency (USEPA) includes iron in the National Secondary Drinking Water Standards, which are not mandatory, and has adopted the former WHO guideline of 0.3 mg l^{-1}. The taste threshold for iron is discussed further in Section 21.1.

Iron is an essential element with the recommended minimum daily requirement ranging between 10 and 50 mg d^{-1}, depending on age, sex and physiological status. Lethal doses are generally between 200 and 250 mg kg^{-1} of body weight, although death has been reported at 40 mg kg^{-1}. There has been renewed interest in drinking water as a source of iron in Ireland due to the high incidence of haemochromatosis in that country. It is a genetic disorder that causes increased iron absorption resulting in iron being stored in the liver causing irreversible damage. However, as many foods are rich in iron the contribution from drinking water, even at the highest concentrations tolerable due to taste, would not contribute significantly to the daily intake (WHO, 2003a).

9.2 Manganese

Manganese is also widely found in ores and rocks, and like iron turns up in groundwaters due to reducing conditions in soils and rocks bringing it into a soluble form. Once exposed to air by aeration, soluble manganese (Mn^{2+}) compounds are oxidized into an insoluble black precipitate that encrusts the pipework and becomes deposited in storage tanks. Even at water concentrations of 0.02 mg l^{-1} encrustation of pipes can occur, with the black precipitated particles being sloughed off. It is similar to iron causing all the same problems of taste, discolouration and staining. Staining is far more severe, however, with manganese than iron, as is the unacceptable taste it imparts to the water. The mean taste threshold for manganese is 0.1 mg l^{-1}. Like iron, the EC Drinking Water Directive (98/83/EEC) lists manganese as an indicator parameter in Part C but much stricter limits have been imposed at 0.05 mg l^{-1}; the same value as set by the USEPA in the National Secondary Drinking Water Standards. The EC guide level in the former Directive (80/778/EEC) was set at just 0.02 mg l^{-1}, the concentration at which laundry and sanitary ware become discoloured grey-black. The World Health Organization (WHO, 2004) has a health-related guide value of 0.4 mg l^{-1}, but as staining occurs at much lower concentrations, a guide value of 0.1 mg l^{-1} is recommended to prevent consumer complaints.

Although the major source of manganese in the diet is food, all the manganese in water is readily bioavailable. However, unless the concentration in drinking water is excessively high, it is unlikely that it will significantly contribute to the daily intake of the metal. Tea is a rich source of the metal with an average cup of tea containing between 0.4 and 1.3 mg Mn l^{-1}. Although manganese is found in both surface and ground waters, problematic concentrations are associated in low oxygen resources, in particular groundwaters. The reducing conditions found in groundwaters and in deep lakes and reservoirs results in high solubilities of manganese with concentrations as high as 1.3 mg l^{-1} and 9.6 mg l^{-1} reported in neutral and acidic groundwaters respectively (ATSDR, 2000a).

Toxicity is not a factor with manganese, as consumers will reject it due to taste long before any threshold concentration is reached. Excessive manganese intake does cause adverse physiological effects, particularly neurological, but not at the concentrations normally associated with water in Europe or the USA. Most consumer complaints in relation to manganese are due to taste, although the black manganese precipitate often builds up within the distribution system, causing discolouration of the water at the tap. The toxicity of manganese is reviewed elsewhere (Criddle, 1992; WHO 2003b).

9.3 Arsenic

Arsenic is a widely distributed element commonly found naturally as arsenic sulphide although other metal arsenates and arsenides do occur. The heavy metal is associated with a wide range of minerals and ores, especially those of lead and copper. The major inorganic forms of arsenic found in drinking water are the trivalent (arsenite; As^{3+}) and pentavalent (arsenate; As^{5+}) compounds. Surface waters contain predominately As^{5+} while anoxic groundwaters contain As^{3+}, which is the more reactive and toxic form. Organic forms of arsenic are less toxic and include monomethylarsonic acid (MMA), dimethylarsinic acid (DMA) and trimethylarsine oxide (TMAO). Arsenic is used in the manufacture of lead-free batteries, glass and alloys for electronic equipment such as transistors, lasers and semi-conductors; and was, until recently, widely used as a pesticide in orchards and cotton fields. Now only the organic arsenical pesticides are allowed, generally for use on cotton. The primary use of arsenic (90%) is now in the manufacture of wood preservative. Chromated copper arsenate (CCA) is widely used in the pressure treatment of timber. It is water-soluble and is leached out from the preserved timber resulting in localized soil contamination (Stilwell and Gorny, 1997) with the possibility of localized surface and ground water contamination. The increase in the use of decking and other preserved wood within the garden may represent a real risk of arsenic contamination. Under no circumstances should wood preserved with CCA be burnt in the home due to arsenic contamination via the smoke or ash, the latter containing between 100 and >1000 ppm of arsenic. The widespread use of CCA is a serious point and diffuse source of arsenic contamination of water resources.

Although it is found in windblown dust and soils, especially around mining and smelting sites and agricultural areas where arsenic pesticides have been used, exposure to arsenic is generally via drinking water or food irrigated with contaminated water (Chou and De Rosa, 2003). It has become a major problem in a number of semi-arid areas where there is a high dependency on groundwater that has become contaminated by dissolution of naturally occurring arsenic minerals and ores. Arsenic is generally higher in groundwaters than surface waters, the latter being contaminated primarily from industrial release or agricultural runoff.

Arsenic in groundwater can be due to a number of factors. Soils that naturally contain high levels of arsenic cause the metal to be leached into the groundwater. For example, in Bangladesh many of the shallow aquifers are composed of arsenic-rich sediment. The arsenic is bound to iron and manganese oxyhydroxides but when the water becomes anaerobic the arsenic is released into solution. Arsenic is generally tasteless and odourless making it difficult to detect its presence (ATSDR, 2000b).

Arsenic is an infamous poison, but large oral doses are required to be fatal. It is absorbed into the bloodstream from the gastrointestinal tract and attacks vital organs. Health effects are stomach ache, nausea, vomiting, diarrhoea, fatigue, abnormal heart rhythm, pins and needles in the hands and feet, and characteristic bruising caused by damage to blood vessels. Long-term exposure to inorganic arsenic causes skin lesions, with hyperpigmentation and hypopigmentation occurring with the formation of small corns or warts on the palms, soles and torso. These corns can become cancerous (ATSDR, 2000b). Arsenic is a well-documented carcinogen (Basu et al., 2001) primarily causing cancer of the skin, but significantly increasing the risk of bladder, lung, kidney, liver, colon and prostate cancers (Chen et al., 1999; Afzal, 2006). It is a major cause of cancer on a regional basis being directly related to contaminated drinking water supplies. Research also supports associations between the metal and cardiac and cerebro vascular disease, diabetes mellitus, and late fetal and infant mortality (Lewis et al., 1999; Hopenhayn-Rich et al., 2000; Tseng et al., 2000).

There are many reported cases of chronic arsenic poisoning from drinking contaminated water. It is estimated that more than 100 million people are affected, the worst areas being in Bangladesh, India, China and Taiwan. Tondel et al. (1999) studied drinking water from four villages in Bangladesh and found levels of arsenic between 10 and 2040 µg l^{-1}. In a study of 20 000 tube wells in West Bengal (India), serving over 1 million people, 62% of the wells had arsenic levels >10 µg l^{-1}, with a maximum concentration of 3700 µg l^{-1} (Bagla and Kaiser, 1996). In just 6 affected districts over 175 000 people showed arsenical skin lesions, which are the late stages of arsenic poisoning (Das et al., 1995). Similar problems have been reported from elsewhere with maximum contaminant levels (MCLs) of 4400 µg l^{-1} in Shanxi province in China (Sun et al., 2001), 1820 µg l^{-1} in south-west Taiwan (Tseng, 1968) and 3590 µg l^{-1} in north-east Taiwan (Chiou et al., 1997). Minor cases of chronic arsenic toxicity associated with drinking water have occurred in the USA (Minnesota and California), Canada (Ontario), Poland, Hungary and Japan.

The WHO has set a guideline value for arsenic since the first standards were published in 1958, which has steadily been revised downwards over the years. The current drinking water guideline for arsenic was set in 1993 at $10 \mu g\,l^{-1}$, however given the uncertainty of long-term exposure to low concentrations this value is designated as provisional (WHO, 2004). The USEPA revised its $50 \mu g\,l^{-1}$ MCL to $10 \mu g\,l^{-1}$ in January, 2006, and the EC Drinking Water Directive sets a maximum value of $10 \mu g\,l^{-1}$.

9.4   Other metals

9.4.1   Antimony

Concentrations in natural waters are generally very low ($0.1–0.2 \mu g\,l^{-1}$) although in drinking water this may rise to $5 \mu g\,l^{-1}$ due to dissolution of the metal from plumbing and associated fittings. There are a number of different forms of the metal in water, but antimony (V) oxo-anion, which is associated with pipes and fittings, is the least toxic form. While there is evidence that some antimony compounds are carcinogenic by inhalation, no such evidence exists for antimony in water (Table 3.5). Known health risks by the oral route include an increase in blood cholesterol and a decrease in blood sugar. The WHO (2004) has set a guideline value of $20 \mu g\,l^{-1}$, while stricter limits have been set by the EC ($5 \mu g\,l^{-1}$) and USEPA (MCL $6 \mu g\,l^{-1}$).

9.4.2   Cadmium

Cadmium is a common contaminant primarily associated with wastewaters although it is also a diffuse pollutant resulting from the use of fertilizers. It is widely used in batteries and is also released when galvanized pipes corrode. Generally levels in finished drinking water are $<1 \mu g\,l^{-1}$. Intake from drinking water represents about 10% of the average daily oral intake of between 10 and $35 \mu g\,d^{-1}$, although additional exposure to cadmium via smoking is significant. While cadmium is carcinogenic if inhaled, there is little evidence to support its carcinogenicity by the oral route (Table 3.5). Cadmium toxicity is focussed on the kidneys, where it slowly accumulates. Strict guidelines have been set for this toxic metal by the EC at $5 \mu g\,l^{-1}$ and the USEPA who have set an MCL of $5 \mu g\,l^{-1}$. This compares to the lower guideline of $3 \mu g\,l^{-1}$ set by WHO (2004).

9.4.3   Chromium

Chromium is a widely occurring metal that exists in various oxidation states. The form depends on a number of factors including the redox potential, pH and the presence of oxidizing or reducing compounds. However, the ratio of chromium (III) to chromium (VI) varies widely between water resources. Chromium occurs naturally at low levels, mainly in the trivalent (Cr^{3+}) form, but can become elevated in surface and ground waters close to where it is mined. The metal is widely used in industry where it can contaminate groundwaters. The hexavalent (Cr^{6+}) is more soluble than the trivalent form making it very mobile. Concentrations in drinking water depend on the degree of industrialization. In the USA 18% of the population drink water containing between 2 and 60 µg l^{-1} and <0.1% drink water with levels between 60 and 120 µg l^{-1} (WHO, 2003c). This compares to 76% <1 µg l^{-1} and 98% <2 µg l^{-1} in the Netherlands (Fonds *et al.*, 1987). Chromium (VI) is carcinogenic by inhalation causing lung cancer although it is not known if it causes cancer via the oral route (Table 3.5). It is genotoxic, however, and is suspected of causing a variety of serious conditions if present at high concentrations in drinking water. Due to the problem of analyzing the more toxic form of chromium (Cr^{6+}), total chromium is normally measured. The WHO guideline for total chromium is set at 50 µg l^{-1} as is the EC limit value. In the USA, the MCL for total chromium is 100 µg l^{-1}.

9.4.4 Mercury (inorganic)

Mercury concentrations in water resources are generally very low, between 0.005 and 0.1 µg l^{-1}, with an average concentration in drinking water of 0.025 µg l^{-1}. Occasionally concentrations >2 µg l^{-1} are reported in shallow wells (Ware, 1989). The WHO health-related guideline is 1 µg l^{-1}, which has been adopted by the EC in the Drinking Water Directive (98/83/EEC), while the USEPA MCL has been set at 2 µg l^{-1}. Mercury is now a rare contaminant with dental amalgam currently the most important source of exposure to the metal (IPCS, 2003). Inorganic mercury is predominately the form found in water. The main effects of inorganic mercuric poisoning are liver and renal damage leading to death. Organic methylmercury affects the central nervous system.

9.4.5 Molybdenum

Although found naturally in soils the concentration of molybdenum in finished drinking water is generally low (<0.01 mg l^{-1}), although close to mining areas concentrations can reach 0.2 mg l^{-1} (Greathouse and Osborne, 1980). While molybdenum is an essential element, toxicity does occur at doses >100 mg kg^{-1} of body weight resulting in diarrhoea, anaemia and elevated uric acid in the blood. There is some evidence that gout, caused by excess uric acid, is caused by the stimulation of xanthine oxidase through a high molybdenum intake. There is no toxicological evidence to show that it causes health problems at concentrations normally found in drinking water. The WHO has set a health-related guideline value of $0.07 \text{ mg } l^{-1}$, however, neither the EC or the USEPA have set standards for this metal.

9.4.6 Nickel

This metal is not found in elevated concentrations in natural waters with drinking water a minor source of nickel in the diet ($<10\%$). Drinking water rarely contains more than $20 \mu\text{g } l^{-1}$ of nickel, although elevated levels are possible in ground waters when contaminated by natural nickel deposits, or both surface and ground waters contaminated by industrial sources. Nickel is widely associated with wastewater effluents and sewage sludge, and discharges from stainless steel and nickel alloy manufacture. However, within the home nickel can be leached from nickel- or chromium-plated taps or from alloys used in kettles or borehole equipment. It is from these domestic sources that nickel contamination is most likely to be significant with concentrations at the consumer's tap reaching $1 \text{ mg } l^{-1}$. While inhaled nickel compounds are carcinogenic, there is little evidence to show a similar risk from oral intake of the metal. Nickel and its salts are potent skin sensitizers. Perinatal mortality is considered the major risk from nickel in drinking water with a WHO guideline set at $20 \mu\text{g } l^{-1}$. The EC has also adopted this limit in the Drinking Water Directive (98/83/EEC) but no MCL has been set by the USEPA.

9.4.7 Selenium

Although not strictly a metal, selenium is a trace element found mainly in food, especially fish. The concentration in drinking water depends on the geology but normally is $<10 \mu\text{g } l^{-1}$ at which the WHO guideline is set. Selenium produces low-level toxicity with long-term exposure resulting in hair and fingernail loss, numbness in fingers and toes, liver and circulatory problems. The EC Drinking Water Directive has set a maximum value of $10 \mu\text{g } l^{-1}$ although the USEPA MCL is $50 \mu\text{g } l^{-1}$.

9.5 Removal by treatment

9.5.1 Iron and manganese

Iron can be precipitated out of solution by a combination of aerating the water and increasing the pH before filtration. Manganese is more difficult to remove from solution as it requires a pH of between 8.5 and 9.0, with aeration playing a minor role. The problem of using such high pH values in the removal of manganese is that any insoluble aluminium coagulant left in the water after sedimentation will be dissolved. To overcome this two filtration stages are required, the first to remove turbidity and coagulant and the second for manganese removal at the higher pH. This higher pH also means that disinfection using chlorine will require about three times the chlorine residual at pH 9 than at pH 8. Groundwaters are occasionally treated by the addition of sodium silicate to complex out the manganese and iron before aeration. Certainly for the water treatment operator, the presence of manganese in particular can produce fairly awkward operational difficulties. Iron can also be added to the water during coagulation as iron salts (Section 14.2), or can occur due to corrosion in the distribution system (Section 21.2).

9.5.2 Other metals

Conventional surface water treatment comprising chemical coagulation, sedimentation and filtration can achieve up to 80% removal of most metals (e.g. arsenic, cadmium, chromium, inorganic mercury, nickel), especially where the water contains a high concentration of suspended solids. At low suspended solids concentrations removal can be enhanced by adding powdered activated carbon at the coagulation stage (Welté, 2002). Granular activated carbon or ion-exchange can be used to remove any remaining metal (Stetter *et al.*, 2002). Ferric sulphate is a particularly effective coagulant for the removal of mercury. Activated carbon and ion-exchange can be used to ensure concentrations in finished water of $<1\,\mu g\,l^{-1}$ (Chiarle *et al.*, 2000). Of the two forms of selenium in water, only selenium (IV) is removed by coagulation, with selenium (VI) unaffected by conventional water treatment. Molybdenum is not effectively removed from drinking water during conventional treatment and reverse osmosis is required. Conventional treatment is also ineffective in removing antimony at the concentrations normally found in raw waters; but as the metal generally originates from household plumbing, then only the careful selection of fittings can prevent contamination. Avoiding nickel alloys as well as chrome- and nickel-plated fitting can also prevent nickel contamination. Lead is fully discussed in Section 27.3.

9.6 Conclusions

Both iron and manganese are essential elements for humans and animals and while they can exhibit mild toxicity at high concentrations, consumers normally reject water on aesthetic grounds before the metals approach toxic concentrations. While they are both removed during conventional treatment, private groundwater resources can be high in iron and manganese, which may form an insoluble precipitate on being aerated. This results in staining and taste problems. In contrast, arsenic is a very dangerous metal and both the WHO and USEPA indicate that there is no safe limit in drinking water. This is reflected by the constant revision downwards of guideline and permissible values. This heavy metal has brought indescribable misery to tens of millions of people in some of the poorest regions of the world, especially in terms of skin cancer. Arsenic poisoning due to natural leaching into groundwater is endemic in large areas of Bangladesh, India, China and Taiwan. As the metal is tasteless and odourless in water it is difficult to identify contaminated wells. Monitoring is expensive and sporadic resulting in wide exposure to the contaminant. Also, as contamination levels are generally related to the depth and the draw-down characteristics of the well, adjacent sources often have different arsenic levels, so there is frequently no way of knowing which well is safe or contaminated. Treatment is technically difficult and usually impractical for small or poor communities. Arsenic is a worldwide problem. For example, in the USA some groundwater systems in California, Nevada and Texas have levels in excess of 0.05 mg l^{-1} (USEPA, 2000). According to the US Geological Survey (2000) 10% of the groundwaters tested have arsenic concentrations >0.01 mg l^{-1}. Arsenic is now almost exclusively used in the manufacture of the wood preservative CCA. This pesticide is used to pressure-treat timber but can leach from the timber to contaminate soils and, being persistent, finds its way into surface and ground waters. The explosion in the use of such timber for garden decking and similar constructions may be putting household wells in particular at risk from arsenic contamination.

The other metals of concern arising from the resource are antimony, cadmium, chromium, mercury, molybdenum, nickel and selenium. Without exception they all have the potential to cause significant health problems when present in groundwaters at levels in excess of the WHO guideline value.

It is extremely difficult to remove arsenic, and other metals, from water without expensive treatment, often involving advanced systems such as activated carbon and membrane processes. This makes small communities that are reliant on a single well, especially in poor countries, and individual households with their own groundwater supply in any country extremely vulnerable to chemical toxicity.

With the exception of iron and manganese all the other metals included in this chapter are listed in the EC Dangerous Substances Directive (76/464/EEC), although only mercury and cadmium are in List I (Appendix 6). This list will be replaced under Article 16 of the EC Water Framework Directive (2000/60/EC) (Appendix 5) with a new list of priority hazardous substances (PHSs) that already includes many of these metals (Gray, 2005). The USEPA have also included a number of these metals on their priority substances list that controls the release of listed substances into the catchment through better regulation and control. The use of many of these priority substances and PHSs will eventually be banned.

References

Afzal, B. M. (2006). Drinking water and women's health. *Journal of Midwifery and Women's Health*, **51**, 12–18.

ATSDR (2000a). *Toxicological Profile for Manganese*. Atlanta, Georgia: Agency for Toxic Substances and Disease Registry (ATSDR), US Department of Health and Human Services.

ATSDR (2000b). *Toxicological Profile for Arsenic (Update)*. Atlanta, Georgia: Agency for Toxic Substances and Disease Registry (ATSDR), US Department of Health and Human Services.

Bagla, P. and Kaiser, J. (1996). India's spreading health crisis draws global arsenic experts. *Science*, **274**, 174–5.

Basu, A., Mathata, J., Gupta, S. and Giri, A. K. (2001). Genetic toxicology of a paradoxical human carcinogen, arsenic: a review. *Mutation Research*, **488**, 171–94.

Chen, C. J., Hsu, L. I., Tseng, C. H., Hsueh, Y. M. and Chiou, H. Y. (1999). Emerging epidemics of arseniasis in Asia. In *Arsenic Exposure and Health Effects: IV*, ed. W. R. Chappell, C. O. Abernathy and R. L. Calderon. Oxford: Elsevier, pp. 113–25.

Chiarle, S., Ratto, M. and Rovatti, M. (2000). Mercury removal from water by ion exchange resins adsorption. *Water Research*, **34**, 2971–8.

Chiou, H. Y., Huang, W. I. and Su, C. L. (1997). Dose-response relationship between prevalence of cerebrovascular disease and ingested inorganic arsenic. *Stroke*, **28**, 1717–23.

Chou, C.-H. S. J. and De Rosa, C. T. (2003). Case studies – arsenic. *International Journal of Hygiene and Environmental Health*, **206**, 381–6.

Clarke, L. (1988). *The Field Guide to Water Wells and Boreholes*. Chichester: John Wiley and Sons.

Criddle, J. (1992). *The Toxicity of Manganese in Drinking Water*. Report FR0313. Marlow: Foundation for Water Research.

Das, D., Chatterjee, A., Mandal, B. K., Samanta, G. and Chakraborty, D. (1995). Arsenic in groundwater in six districts of west Bengal, India: the biggest arsenic calamity in the world. *Analyst*, **120**, 917–24.

EC (1998). Council Directive 98/83/EEC of 3 November 1998 on the quality of water intended for human consumption, *Official Journal of the European Communities*, **L330** (5.12.98), 32–53.

Fonds, A. W., van den Eshof, A. J. and Smit, E. (1987). *Water Quality in the Netherlands*. Report No. 218108004. Bilthoven, the Netherlands: National Institute of Public Health and Environmental Protection.

Gray, N. F. (2005). *Water Technology: An Introduction for Environmental Scientists and Engineers*. Oxford: Elsevier.

Greathouse, D. G. and Osborne, R. H. (1980). Preliminary report on nationwide study of drinking water and cardiovascular diseases. *Journal of Environmental Pathology and Toxicology*, **42**(2–3), 65–76.

Hopenhayn-Rich, C., Browning, S. R., Hertz-Picciotto, I. *et al.* (2000). Chronic arsenic exposure and risk of infant mortality in two areas of Chile. *Environmental Health Perspective*, **108**, 667–73.

IPCS (2003). *Elemental Mercury and Inorganic Mercury Compound: Human Health Aspects*. Concise International Chemical Assessment Document: 50, International Programme on Chemical Safety. Geneva: World Health Organization.

Lewis, D. R., Southwick, J. W., Oullet-Hellstrom, R., Rench, J. and Calderon, R. L. (1999). Drinking water arsenic in Utah. *Environmental Health Perspectives*, **107**, 359–65.

Stetter, D., Dordlemann, O. and Overath, H. (2002). Pilot scale studies on the removal of trace metal contaminations in drinking water treatment using chelating ion-exchange resins. *Water Supply*, **2**, 25–35.

Stilwell, D. E. and Gorny, K. D. (1997). Contamination of soil with copper, chromium and arsenic under decks built from pressure treated wood. *Bulletin of Environmental Contamination and Toxicology*, **58**, 22–9.

Sun, G. F., Pi, J., Li, B. *et al*. (2001). Progresses on researches of endemic arsenism in China: population at risk, intervention actions, and related scientific issues. In *Arsenic Exposure and Health Effects: IV*, ed. W. R. Chappell, C. O. Abernathy and R. L. Calderon. Oxford: Elsevier, pp. 79–85.

Tondel, M., Rahman, M., Magnuson, A. *et al*. (1999). The relationship of arsenic levels in drinking waters and the prevalence rate of skin lesions in Bangladesh. *Environmental Health Perspectives*, **107**, 727–9.

Tseng, C. H., Tai, T. Y., Chong, C. K., Tseng, C. P. and Lai, B. J. (2000). Long term arsenic exposure and incidence of non-insulin dependent diabetes mellitus. A cohort study in arseniasis-hyperendemic villages in Taiwan. *Environmental Health Perspectives*, **108**, 847–51.

Tseng, W. P., Chu, H. M., How, S. W. *et al*. (1968). Prevalence of skin cancer in an endemic area of chronic arsenicism in Taiwan. *Journal of the National Cancer Institute*, **40**, 453–63.

USEPA (2000). *Arsenic Occurrence in Public Drinking Water Supplies*. EPA-815-R-00-023. Washington, DC: Office of Ground Water and Drinking Water, US Environmental Protection Agency.

US Geological Survey (2000). *Arsenic in ground-water resources of the United States*. USGS Fact Sheet: 063–00. Reston, VA: US Geological Survey.

Ware, G. D. (1989). Mercury. *Reviews of Environmental Contamination and Toxicology*, **107**, 93–102.

Welté, B. (2002). Le nickel: 4e Partie. Traitment. *Techniques, Sciences, Methodes*, **97**(5), 61–6.

WHO (2003a). *Iron in Drinking Water*. Report No. WHO/SDE/WSH/03.04/08. Geneva: World Health Organization.

WHO (2003b). *Manganese in Drinking Water*. Report No. WHO/SDE/WSH/03.04/104. Geneva: World Health Organization.

WHO (2003c). *Chromium in Drinking Water*. Report No. WHO/SDE/WSH/03.04/04. Geneva: World Health Organization.

WHO (2004). *Guidelines for Drinking Water Quality*, Vol. 1. *Recommendations*, 3rd edn. Geneva: World Health Organization.

Chapter 10
Hardness and total dissolved solids

10.1 Introduction

The hardness or softness of water varies from place to place and reflects the nature of the geology of the area with which the water has been in contact. In general, surface waters are softer than groundwaters. Hard waters are associated with chalk and limestone catchment areas, whereas soft waters are associated with impermeable rocks such as granite (Section 4.4). Water hardness is a traditional measure of the ability of water to react with soap to produce a lather, and for most consumers the problems associated with washing, and also the scaling of pipes and household appliances that use water, are the two major factors of concern.

An alternative measure of hardness is total dissolved solids (TDS), which is a measure of the total concentration of ions in water. The TDS in groundwater is often an order of magnitude higher than in surface waters. In aquifers the TDS increases with depth due to less fresh recharge water to dilute existing groundwater and the longer period for ions to be dissolved. The older and deeper the groundwater the more mineral rich the water becomes resulting in quite saline water. High concentrations of salts in groundwaters are often due to over-abstraction or to drought conditions when old saline groundwaters may enter boreholes through upward replacement, or due to saline intrusion into the aquifer from the sea. In Europe conductivity is used as a replacement for TDS measurement and is widely used to measure the degree of mineralization of groundwaters (Table 4.5).

10.2 Chemistry of hardness

Hardness is caused by metal cations such as calcium (Ca^{2+}), but in fact all divalent cations cause hardness (Table 10.1). They react with certain anions such as carbonate or sulphate to form a precipitate. Monovalent cations such as sodium (Na^+) do not affect hardness. Strontium, ferrous iron (Fe^{2+}) and manganese are usually such minor components of hardness that they are generally ignored, with the total hardness taken to be the sum of the calcium and

Table 10.1 *Principal metal cations causing hardness and the major anions associated with them*

Cations	Anions
Ca^{2+} (calcium)	HCO_3^- (hydrogencarbonate)
Mg^{2+} (magnesium)	SO_4^{2-} (sulphate)
Sr^{2+} (strontium)	Cl^- (chloride)
Fe^{2+} (iron)	NO_3^- (nitrate)
Mn^{2+} (manganese)	SiO_3^{2-} (silicate)

magnesium concentrations. Aluminium (Al^{2+}) and ferric iron (Fe^{3+}) can affect hardness but their solubility is limited at the pH of natural waters, so that ionic concentrations can be considered negligible. Barium and zinc can also cause hardness, but their concentrations in water are normally extremely small. The majority of divalent cations in water are Ca^{2+} and Mg^{2+}, so hardness is calculated by measuring these cations only.

There are a number of additional terms relating to hardness. *Total hardness* is the direct measurement of hardness ($Ca^{2+} + Mg^{2+}$). *Calcium hardness* is the direct measurement of calcium only. *Carbonate hardness* is the hardness derived from the solubilization of calcium or magnesium carbonate by converting the carbonate to hydrogencarbonate. This hardness can be removed by heating. *Magnesium hardness* is the hardness derived from the presence of magnesium and is calculated by subtracting the calcium hardness from the total hardness. *Non-carbonate hardness* is the hardness attributable to all cations associated with all anions except carbonate, e.g. calcium chloride and magnesium sulphate. *Permanent hardness* is equivalent to non-carbonate hardness, which cannot be removed from water by heating. *Temporary hardness* is equivalent to carbonate hardness and can be removed by heating (Flanagan, 1988). All hardness values are expressed in mg $CaCO_3$ l^{-1}.

The classification of waters into hard and soft is arbitrary, with a number of classifications used (Table 10.2). It is generally accepted that waters with a hardness of <60 mg l^{-1} are soft.

In terms of water quality, hardness can have a profound effect. The hardness of water was originally measured by the ability of the water to destroy the lather of soap, as this is one of the principal problems of very hard water. Although hardness does neutralize the lathering power of soap it does not affect modern detergent formulations. Soft waters are more aggressive than hard waters, enhancing the corrosion of copper and lead pipes. Above a hardness of 150–200 mg l^{-1} scaling becomes a problem, although this is only the case with calcium hardness or temporary hardness. The hydrogencarbonate is removed during heating to form calcium carbonate, which forms a thick scale over the

Table 10.2 *Two examples of classifications used for water hardness*

Classification A		Classification B	
Concentration (mg l^{-1})	Degree of hardness	Concentration (mg l^{-1})	Degree of hardness
0–50	Soft		
50–100	Moderately soft	0–75	Soft
100–150	Slightly hard	75–150	Moderately hard
150–250	Moderately hard	150–300	Hard
250–350	Hard		
350+	Excessively hard	300+	Very hard

surface of pipes, boilers and, of course, kettles and the heating elements in washing-machines and dishwashers. The chemical reaction that occurs is

$$2HCO_3^- \rightarrow H_2O + CO_2 + CO_3^{2-}$$
Hydrogencarbonate

$$CO_3^{2-} + Ca^{2+} \rightarrow CaCO_3$$
Calcium carbonate

The greater the degree of carbonate hardness the more severe the problem becomes, although a moderate level of hardness (150 mg l^{-1}) is useful as it forms a protective film of calcium carbonate over the inside of pipes, preventing the leaching of metals and corrosion.

In the USA TDS is used in place of hardness. It is a broader classification than hardness that measures the total amount of inorganic salts and the small amount of organic matter present. Traditionally TDS is calculated gravimetrically by drying water at $180\,°C$ and measuring the residual in mg l^{-1}. The main constituent cations are calcium, magnesium, sodium and potassium associated with the anions carbonate, hydrogencarbonate, chloride, sulphate and nitrate. However, TDS is closely correlated with hardness. The alternate method of calculating TDS is by using conductivity and multiplying the conductance by a specific factor, usually between 0.55 and 0.75. This conversion factor varies with the type of water source, increasing with anion concentration, although once determined gravimetrically it remains fairly constant for a specific water source. At TDS concentrations >500 mg l^{-1} excessive scaling occurs (WHO, 2003).

10.3 Standards

Hardness is an important factor in the taste of water, although at above 500 mg l^{-1} the water begins to taste unpleasant. The World Health Organization

(WHO, 1984) set a maximum recommended concentration of 500 mg l^{-1} in drinking water on aesthetic, not health, grounds. In the second revision (WHO, 1993) again no health-related standard was felt necessary but a guideline of 200 mg l^{-1} was suggested to avoid scale deposition in distribution systems. In the latest revision the WHO (2004) has made no health-based guideline for hardness but acknowledges that the degree of hardness can affect acceptability to the consumer in terms of taste and scale deposition. They indicate that scale deposition will occur in plumbing, especially when the water is heated, at concentrations >200 mg l^{-1}, and that at <100 mg l^{-1} the water has a low buffering capacity and can increase corrosion in pipework. In the USA, where TDS is used in place of hardness, the National Secondary Drinking Water Regulations set a non-enforceable maximum contaminant level of 500 mg l^{-1} for TDS to prevent taste and scaling problems.

The current EC Drinking Water Directive does not set any specific standards for hardness. In the earlier Directive a standard was set for water that had been softened, where a minimum hardness concentration of 150 mg l^{-1} was required. In the earlier Directive hardness was expressed solely in terms of calcium ions. So 150 mg l^{-1} as $CaCO_3$ is equivalent to 60 mg l^{-1} as Ca^{2+}. To convert ions to $CaCO_3$ equivalent, multiply mg l^{-1} of Ca^{2+} by 2.495, and Mg^{2+} by 4.112 to give mg l^{-1} as $CaCO_3$ equivalent. Alternatively:

$$\text{Hardness} = \text{mg } l^{-1} \text{ of each ion} \times \frac{\text{Equivalent weight of } CaCO_3}{\text{Equivalent weight of each ion}} \quad (\text{mg } CaCO_3 \; l^{-1}).$$

So, for example, for 30 mg l^{-1} of the ion calcium (Ca^{2+}) the hardness is calculated as:

$$30 \text{ mg } l^{-1} \text{ of } Ca^{2+} \times 50/20.04 = 74.9 \text{ mg } CaCO_3 \; l^{-1}$$

where the equivalent weights for the most important ions are Ca^{2+} 20.04, Mg^{2+} 12.16, HCO_3^- 6.1 and SO_4^{2-} 48 and the equivalent weight of $CaCO_3$ is 50. Concentrations of calcium in natural waters of $\leq$ 100 mg $Ca^{2+} l^{-1}$ are common and > 200 mg $Ca^{2+} l^{-1}$ rare. High levels of magnesium in water supplies are far more uncommon with concentrations >100 mg $Mg^{2+} l^{-1}$ unusual. The TDS of natural waters range from 30 to 6000 depending on the solubilities of the minerals within the catchment. Most surface waters have a TDS <500 mg l^{-1} compared to <1000 mg l^{-1} for groundwaters.

10.4 Health aspects

It was observed as early as 1957 that the occurrence of cardiovascular disease in the population was related to the acidity of water (Zoetman and Brinkman, 1976). Since that time a considerable number of studies, mainly in the USA, have indicated a correlation between hardness or TDS and mortality, especially cardiovascular disease. A study carried out in the UK and reported in 1982 confirmed that mortality from cardiovascular disease was closely associated with water hardness, with mortality decreasing as hardness increased. The effect was present for both stroke and ischaemic heart disease, but not for non-cardiovascular disease. The difference in mortality rates between soft water (around 25 mg $CaCO_3$ l^{-1}) areas and hard water areas ranged from 10% to 15%. This is in general agreement with an earlier figure of a 0.8% decrease in mortality with every 10 mg $CaCO_3$ l^{-1} increase in hardness up to 170 mg $CaCO_3$ l^{-1} (Lacey, 1981; Powell *et al.*, 1982).

Hardness is not usually associated with cancer, yet a number of studies have identified just such a relationship. Turner (1962) studied a number of counties in England and Wales and found that gastric cancer was inversely associated with water hardness. Similar results were reported from a study involving 66 cities in Canada (Wigle *et al.*, 1986), while a weak inverse association between total cancer mortality and water hardness in 473 US cities was reported by Morin *et al.* (1985). In contrast elevated stomach and female breast cancer mortalities have been reported in hard water areas in a study of 80 British towns (Stocks, 1973). Most of the research on health effects in relation to the mineral content of water has been done using hardness, although the results of these studies equally apply to TDS.

10.5 Conclusions

Drinking water hardness is a contentious issue with both suppliers and consumers alike. Hardness is very much linked to taste, and many consumers in hard water areas love the unique taste that hardness imparts to their water, likewise those in soft water areas. The taste threshold for calcium hardness is between 100 and 300 mg l^{-1}, depending on the associated anion, with concentrations >500 mg l^{-1} unacceptable for most consumers. In terms of TDS the taste of water can be broadly classified as: excellent >300 mg l^{-1}, good 300–600 mg l^{-1}, fair 600–900 mg l^{-1}, poor 900–1200 mg l^{-1} and unacceptable >1200 mg l^{-1}. If the TDS is extremely low then the water takes on an insipid flat taste that is unpalatable.

It is widely known that very soft waters have an adverse effect on the mineral balance of the body and this has been linked to both stroke and ischaemic heart disease. There is evidence that gastric cancer is also inversely associated with water hardness. However, both calcium and magnesium are common in a wide range of foods with only 5–20% of the daily intake of these ions coming from water. So why these relationships exist remains unclear. It is also a general rule that the toxicity of pollutants and contaminants is significantly less in hard water than soft water.

The general consensus is that moderately hard water is good for you, but causes scaling in distribution pipes and household plumbing, and reduces the efficiency of boilers, resulting in high fuel consumption. So in practice, if water contains more than $300 \, mg \, l^{-1}$ as $CaCO_3$ then the water supply companies will tend to soften the water (Chapter 14), usually by the addition of lime or soda ash. The practice of softening has become more widespread over the past 20 years with point-of-use water softeners increasingly common (Section 29.3).

Climate change, linked with over-abstraction, will increase the incidence of saline intrusion in many groundwaters, while groundwaters in areas of low rainfall but high demand are increasingly becoming more saline as deeper, older groundwaters are utilized. The problem of elevated sodium, sulphate and chloride concentrations is considered in Section 29.2. Many islands are increasingly more dependent on desalination and natural rainfall harvesting, resulting in a greater proportion of their supplies being extremely soft. Bottled waters are usually moderate to very hard as they impart a unique taste to the water (Section 29.2). In the light of reported health effects associated with soft waters, the Drinking Water Inspectorate has commissioned research to examine the need to reintroduce a minimum hardness standard for drinking water that has been either softened or desalinated. It would appear that a minimum hardness of $150 \, mg \, l^{-1}$ would be beneficial to health, while a maximum hardness of $250 \, mg \, l^{-1}$ would prevent excessive scaling.

References

Flanagan, P. J. (1988). *Parameters of Water Quality, Interpretation and Standards.* Dublin: Environmental Research Unit.

Lacey, R. F. (1981). *Changes in Water Hardness and Cardiovascular Death-Rates.* Technical Report 171. Medmenham: Water Research Centre.

Morin, M. M., Sharrett, A. R., Bailey, K. R. and Fabsitz, R. R. (1985). Drinking water source and mortality in US cities. *International Journal of Epidemiology,* **14**, 254–64.

Powell, R., Packham, R. F., Lacey, R. F. and Russell, P. F. (1982). *Water Quality and Cardiovascular Disease in British Towns.* Technical Report 171. Medmenham: Water Research Centre.

Stocks, P. (1973). Mortality from cancer and cardiovascular diseases in the county boroughs of England and Wales classified according to the sources and hardness of their water supplies, 1958–1967. *Journal of Hygiene, Cambridge,* **71**, 237–52.

Turner, R. C. (1962). Radioactivity and hardness of drinking waters in relation to cancer mortality rates. *British Journal of Cancer,* **16**, 27–45.

WHO (1984). *Guidelines for Drinking Water Quality,* Vol. 2. *Health Criteria and Other Supporting Information.* Geneva: World Health Organization.

WHO (1993). *Guidelines for Drinking Water Quality,* Vol. 1. *Recommendation,* 2nd edn. Geneva: World Health Organization.

WHO (2003). *Total Dissolved Solids in Drinking Water.* Report No. WHO/SDE/WSH/03.04/16. Geneva: World Health Organization.

WHO (2004). *Guidelines for Drinking Water Quality,* Vol. 1. *Recommendation,* 3rd edn. Geneva: World Health Organization.

Wigle, D. T., Mao, Y., Semenciw, R., Smith, M. H. and Toft, P. (1986). Contaminants in drinking water and cancer risks in Canadian cities. *Canadian Journal of Public Health*, **77**, 335–42.

Zoetman, B. C. J. and Brinkman, F. J. J. (1976). *Hardness of Drinking Water and Public Health*. Oxford: Pergamon Press.

Chapter 11
Algae and algal toxins

11.1   Introduction

Many impoundment and storage reservoirs are rich in nutrients from farming and the discharge from sewage treatment plants upstream (Chapter 5). Given adequate light, algae can quickly establish themselves and become a problem. Algal development is worse in the summer due to thermal stratification, with the lower zones of the reservoir (i.e. hypolimnion) becoming anaerobic. This causes ammonia, phosphorus and silica, all algal nutrients, to be released from the bottom sediments, which encourages even more algal growth (Sections 4.3 and 5.5).

11.2   Problems associated with algae

There are two specific problem areas for the water supply engineer associated with algal blooms. The first is an operational problem. The physical presence of the algae makes water treatment more difficult because the algae must be removed from the finished water. If small algae do get through the filter, then algae will enter the distribution system and cause several problems. The water will be coloured and turbid, the algae will decay causing taste and odour problems, and finally the algae will either become the source of food for micro-organisms growing on the walls of the supply pipes, or the source of food for larger animals infesting the supply system (Chapter 24). In practice, algae reduce the rate of flow through the treatment plant by blocking microstrainers; the larger algae blocking rapid sand filters and the smaller species blocking the finer slow sand filters. In severe cases the supply may be drastically reduced or stopped altogether.

The second problem area is one of metabolism. The algae do two things. They all produce carbon dioxide during respiration, and this can cause a severe alteration of the pH of the water. This is most often caused by large crops, known as blooms, of blue-green algae. These changes in pH severely disrupt the coagulation process, resulting in the loss of coagulant, often aluminium sulphate, into the finished water and a reduction in quality in terms of colour and turbidity. Certain algae also release extracellular products, some of which are toxic, others that increase the organic matter content of the water or form halogenated by-products that are carcinogenic (Chapter 18). Taste and odour problems can cause treatment plants to close during the summer for up to two months (Section 14.2) (National Rivers Authority, 1990).

11.2.1   Toxins

There has been much interest in toxins that are released by large blooms of algae, not only in reservoirs but also in coastal waters. It is the blue-green algae that are responsible for toxins in drinking water. Despite their name, these algae are actually a group of bacteria capable of photosynthesis. Therefore the blue-green algae are also referred to as cyanobacteria (Hunter, 1991). It is not a new phenomenon for algal blooms to release toxins that can kill livestock, domestic animals and even fish and birds that have drunk contaminated water (Codd *et al.*, 1989), with the earliest recorded report over 110 years ago. In the British Isles three algae are known to produce toxins in freshwater, *Microcystis aeruginos, Anabaena flos-aquae* and *Aphanizomenon flos-aquae* (Falconer, 1989, 1991). Cyanotoxins that affect either the liver or nervous system include microcystins, anatoxin *a*, cylindrospermopsins and saxitoxins. Blue-green algae also produce endotoxins that affect the gastrointestinal tract such as lipopolysaccharides. Two major types of toxins are produced by blue-green algae. Neurotoxins, such as anatoxin A, which cause paralysis of the skeletal and respiratory muscles, can result in death in as little as five minutes. These are mainly alkaloids produced by species of the genera *Anabaena, Aphanizomenon* and *Oscillatoria*. Hepatotoxins, such as microcystins, nodlarines and cylindrospermopsins, cause severe and often fatal liver damage. These are peptides, mainly different types of microcystin, and are produced by many common species of the genera *Microcystis, Oscillatoria, Anabaena* and *Nodularia* (Chorus and Bartram, 1999). The two most important toxin groups that have been associated with human poisoning are the microcystins and cylindrospermopsins (Falconer and Humpage, 2006). Other toxins, associated with different species of blue-green algae, are also produced (Table 11.1). While some species are ubiquitous, others have a more restricted distribution.

When there is sunlight, a nitrogen source (e.g. nitrates from farm runoff or a sewage treatment plant effluent) and a phosphorus source (mainly from sewage treatment plants), then algal growth will develop rapidly to very high densities, i.e. eutrophication. The toxins are thought to be released on decay (lysis), rather than as extracellular products released while active. Decay can occur naturally, but massive releases of toxins only occur when the algae are killed by the addition of chemicals such as copper sulphate.

Cyanobacterial poisoning occurs in humans and animals in three ways: through contact with contaminated water, by the consumption of fish or other species taken from such waters, or by drinking contaminated water (Hunter, 1991). A major occurrence of toxic algae took place at Rutland Water in Leicestershire, which is one of the largest reservoirs in Western Europe. During

Table 11.1 *Details on the four most common toxins produced by blue-green algae in Australia. Reproduced from Department of Natural Resources, Mines and Water (2006) with permission from Department of Natural Resources and Water, Queensland Government*

Toxin type	Potential health effects	Organism producing toxin
Hepatotoxins Microcystins	Damages liver cells, leading to pooling of blood and finally liver failure	*Microcystis aeruginosa* *Microcystis flos-aquae* *Nodularia spumigena*
Neurotoxins Saxitoxins	Interfere with the function of the nervous system. Death by paralysis of the respiratory system as a result of muscle failure	*Anabaena circinalis*
Non-specific toxins Cylindrospermopsin	Relatively slow-acting toxin that damages a number of organs of the body including liver, kidney and thymus	*Cylindrospermopsis raciborskii Aphanizomenon ovalisporum*
Endotoxins Dermatotoxic lipopolysaccharides	Associated with outbreaks of gastroenteritis, skin and eye irritation and hay fever in humans who come into contact with blue-green algae during recreational activities	Potentially produced by all blue-green algae

late August and mid-September 1989, blooms of *Anabaena* and *Aphanizomenon* were followed by *Microcystis aeruginosa*. Twenty sheep and 15 dogs are thought to have been killed by drinking the water, which is part of the supply network for 1.5 million people. Another toxic bloom also occurred in 1989 at Rudyard Lake in Staffordshire. This time army cadets who had been canoeing and swimming through the algal bloom had influenza-type symptoms, two being admitted to hospital (Turner *et al.*, 1990). A survey in Florida found that 75 out of 167 surface waters used for drinking water supply contained toxic cyanobacterial blooms during 1999. This resulted in some finished waters containing toxins with maximum concentrations recorded for microcystins of $12.5\,\mu\mathrm{gl}^{-1}$, anatoxin-A $8.5\,\mu\mathrm{g\,l}^{-1}$ and cylindrospermopsin $97.1\,\mu\mathrm{g\,l}^{-1}$ (Burns, 2003). Cylindrospermopsin had been recorded earlier from Australia where 138 children and 10 adults were poisoned after a *Cylindrospermopsis* bloom in a reservoir (Richardson, 2003). The presence of algae can cause other toxicity problems. For example halogenated acetonitriles are developmental toxicants that are produced during chlorination or chloramination when algae are present.

11.2.2　Taste and odour

The cyanobacterial blooms also cause strong tastes and odours. Pressdee and Hart (1991) report that the Ardleigh Reservoir (Anglian Region) has had to close for supply for two months every year since 1978 because of blooms of *Microcystis aeruginosa*. So at the time when water demand is at its highest and supplies are reaching their lowest, algal blooms can cause havoc with water distribution and supply. Algae produce mainly earthy or musty odours such as geosmin and 2-methylisoborneol (Section 8.4). However, dimethyltrisulphide, dimethyldisulphide and methyl mercaptan are also produced when algae decompose producing a range of odours similar to rotten cabbage to fishy (Table 8.1; Figure 8.2).

11.3   Standards

Microcystin-LR is the most frequently encountered algal toxin at concentrations high enough to be a danger to consumers and water users. As little data exist for the other toxins there is currently insufficient data to derive specific guideline values. While the World Health Organization (WHO, 2004) has set a guideline for microcystin-LR of $0.001\,mg\,l^{-1}$, neither the EC or the USA have set standards for algal toxins; although the US Environmental Protection Agency have included it in their Contaminant Candidate List (Appendix 7). This is in contrast with other countries, for example in Australia a drinking water guideline value for microcystins has been set at $1.3\,\mu g\,l^{-1}$ expressed as microcystin-LR toxicity equivalents.

In the Netherlands, where this is also a serious problem, maximum acceptable algal concentrations have been suggested at $200\,\mu g\,l^{-1}$ of chlorophyll-*a* for blue-green algae and $60\,\mu g\,l^{-1}$ for other algal groups. These levels can be dealt with successfully by conventional treatment, but do not relate to the concentrations of the toxins. Water supply companies monitor reservoirs and other surface water resources daily for blue-green algae. Using an inverted microscope, they scan for the toxin-producing species and when these are identified in significant numbers action is taken. Once the density of blue-green algae exceeds 1000 cells per ml in surface waters then there is a risk of cyanotoxins being present in harmful concentrations.

Immunoassay kits are used to screen suspected water for microcystin-LR, but accurate measurements requires analysis by high-performance liquid chromatography after extraction using 75% aqueous methanol. All toxins are peptide-related, highly polar and have relatively high molecular weights making them difficult to measure in environmental samples. The presence of microcystins appears associated with the presence of algal derived tastes and odours in drinking water caused by the compounds geosmin and 2-methylisoborneol. It may be possible to use these compounds as indicators for the toxin (Section 8.4).

11.4   Treatment

Treatment of algal toxins from water requires filtration to remove whole cells, followed by oxidation of the toxin by ozone or chlorine at sufficient concentrations and contact time. Activated carbon is effective under controlled conditions (Hart *et al.*, 1992). Granular activated carbon and ozone are both particularly effective and the use of activated carbon in conjunction with ozonation is, as expected, very effective (Himberg *et al.*, 1989; AWWA, 2002). This is discussed further in Section 16.2. Cyanotoxins quickly degrade or are oxidized in surface waters once they have been released after cell lysis. The biological component of slow sand filters also removes the toxins, but other processes including coagulation, sedimentation, oxidation and chlorination will also contribute to their removal. In a study of 677 source and finished waters 80% tested positive for microcystin-LR. However, only two of the samples of finished drinking water exceeded the WHO guideline value demonstrating that water treatment was effective in removing algal toxins (Carmichael, 2001).

Control options are currently directed at the blue-green algae rather than at the toxins. None of the current methods are reliable. The obvious method is to reduce the nutrient loading to the reservoir to reduce the crop of algae. Destratification of deeper lakes may eliminate some species but encourage others, whereas biological control methods such as using fish to graze the algae are difficult to operate and unreliable (Parr and Clarke, 1992). Decomposing straw has been shown to inhibit both green and blue-green algae (Gibson *et al.*, 1990), although the feasibility of using barley straw to control algae in small reservoirs is still being examined. Currently preferred operational management techniques include selective abstraction, for example varying the depth of abstraction in reservoirs, and using air flotation to remove cells from raw waters (Section 4.3).

11.5 Conclusions

Algal blooms occur not only in lakes and reservoirs, but also in canals and coastal waters. While their occurrence has increased in recent years the reason for this is unknown, although it is possibly linked to climate change causing rising surface water temperatures and greater activity within the nitrogen cycle (Figure 5.1).

Blue-green algae in water resources is a potentially serious threat to health, although the presence of such species does not necessarily indicate the presence of algal toxins in the water. In fact our understanding of the factors controlling the production of toxins is quite poor, with concentrations of toxins very variable. The risk to health depends on the type of toxin, its concentration in the water and the amount of toxin consumed. A major problem is measuring algal toxins in water samples, although the use of taste and odour compounds associated with odour may provide a simple screening method in the future. The removal of toxins at the treatment plant can be variable so it is current practice not to use water resources, if at all possible, when algae are detected above crital threshold values. However, with increasing pressure on water resources selective abstraction techniques and advanced treatment techiques such as ozone and granular activated carbon are becoming more widely adopted. While camping, water contaminated with blue-green algae should neither be handled or drunk. While boiling contaminated water will kill the algae it also releases the cell-bound toxins into the water making it unsafe to drink. Sterilization tablets also have no effect, so care must be taken when drinking water directly from surface sources.

References

AWWA (2002). *Removal of Algal Toxins From Drinking Water Using Ozone and GAC*. Research Foundation Report. Washington, DC: American Water Works Association.

Burns, J. (2003). *Cyanobacteria and Their Toxins in Florida Surface Waters and Drinking Water Supplies*. Tallahassee, FL: Florida Department of Health.

Carmichael, W. W. (2001). *Assessment of Blue-Green Algal Toxins in Raw and Finished Water*. Denver, CO: American Water Works Association Research Foundation.

Chorus, I. and Bartram, J. (eds.) (1999). *Toxic Cyanobacteria in Water: A Guide to Their Public Health Consequences, Monitoring and Management*. London: E. and F.N. Spon.

Codd, A. G., Brooks, W. P., Lawton, L. A. and Beattie, K. A. (1989). Cyanobacterial toxins in European waters: occurrence, properties, problems and requirements. In *Watershed 89. The Future of Water Quality* in *Europe*, ed. D. Wheeler, M. L. Richardson and J. Bridges. *Proceedings of the International Association of Water Pollution Research and Control*, 17–20 April, 1989, London: Pergamon Press, pp. 211–20.

Department of Natural Resources, Mines and Water (2006). *Harmful Algal Blooms: Effects on Drinking Water*. The State of Queensland, Australia: Department of Natural Resources, Mines and Water.

Falconer, I. R. (1989). Effects on human health of some toxic cyanobacteria (blue-green algae) in reservoirs, lakes and rivers. *Toxicity Assessment*, **4**, 1175–84.

Falconer, I. R. (1991). Tumour promotion and liver damage caused by oral consumption of cyanobacteria. *Environmental Toxicology and Water Quality*, **6**, 177–84.

Falconer, I. R. and Humpage, A. R. (2006). Cyanobacterial (blue-green algal) toxins in water supplies: cylindrospermopsins. *Environmental Toxicology*, **21**(4), 299–304

Gibson, M. T., Welch, I. M., Barrett, P. R. F. and Ridge, I. (1990). Barley straw as an inhibitor of algal growth. II: Laboratory studies. *Journal of Applied Phycology*, **2**, 241–8.

Hart, I., Scott, P. and Carlie, P. R. (1992). *Algal Toxin Removal From Water*. Report FR 0303. Marlow: Foundation for Water Research.

Himberg, K., Keijola, A. M., Hiisvirta, L., Pyyasalo, H. and Sivonen, K. (1989). The effects of water treatment processes on the removal of hepatotoxins from *Microcystis* and *Oscillatoria cyanobacteria*: a laboratory study. *Water Research*, **23**, 979–84.

Hunter, P. R. (1991). An introduction to the biology, ecology and potential public health significance of the blue-green algae. *PHLS Microbiology Digest*, **8**, 13–20.

National Rivers Authority (1990). *Toxic Blue-green Algae*. Water Quality Series 12, Anglian Region. Peterborough: NRA.

Parr, W. and Clarke, S. (1992). *A Review of Potential Methods for Controlling Phytoplankton, with Particular Reference to Cyanobacteria, and Sampling Guidelines for the Water Industry*. Report FR 0248. Marlow: Foundation for Water Research.

Pressdee, J. and Hart, J. (1991). *Algal/Bacterial Toxin Removal From Water*. Report FR 0223. Marlow: Foundation for Water Research.

Richardson, S. D. (2003). Disinfection by-products and other emerging contaminants in drinking water. *Trends in Analytical Chemistry*, **22**(10), 666–84.

Turner, P. C., Gammie, A. J., Hollinrake, K. and Codd, G. A. (1990). Pneumonia associated with contact with cyanobacteria. *British Medical Journal*, **300**, 1440–1.

Chapter 12
Radon and radioactivity

12.1 Introduction

The main source of radiation is from naturally occurring radionuclides. In fact, the combined effect of all these natural sources accounts for, on average, 87% of the total radiation exposure to a person over their lifetime. So man-made sources only account for 13% of radiation exposure, of which 11.5% is from medical sources, 0.5% from fallout, 0.4% due to occupational exposure and 0.1% from nuclear discharges, with various other miscellaneous sources accounting for the remainder. Of the natural sources, radon accounts for 32% of this total exposure. The majority of water supplies have very low levels of radionuclides that represent no health risk. Those that are a cause for concern are due to naturally occurring radioactivity. Many types of rock contain mildly radioactive elements, which are known as parent radionuclides, that decay producing other radioactive contaminants known as daughter radionuclides. Depending on their chemical properties both can accumulate in water sources, often at dangerous concentrations. In the environment the parent radionuclides may behave differently from their daughter radionuclides resulting in very different patterns of occurrence in water resources. For example, groundwater with high radium levels tends to have low uranium levels and vice versa, even though uranium-238 is the parent of radium-226.

The intensity or activity of a radionuclide was originally measured by the curie (Ci), which is still in use in the USA. Elsewhere this has largely been replaced by the becquerel (Bq). In practice the Ci is too large for use with normal environmental monitoring so the picocurie (pCi) is used, so $1\,\text{pCi} = 10^{-12}\,\text{Ci}$. Conversion of the two units is $1\,\text{Bq} = 2.7 \times 10^{-11}\,\text{Ci}$ or $1\,\text{Ci} = 3.7 \times 10^{10}\,\text{Bq}$.

Uranium-238 series radionuclides are the major contributors to natural radiation associated with drinking water, the most important being radium-222 (radon), which is also released from water into the air. Other long-lived radionuclides include uranium-234, radium-226, lead-210 and polonium-210, which are mostly alpha-emitters, as well as their shorter-lived progeny that also emit beta and gamma radiation. Uranium-235 is a minor fraction of natural uranium also found occasionally in water.

12.2   Radon

Radon is a natural radioactive gas that has no taste, smell or colour; in fact, special equipment is required to detect its presence. Analysis of drinking water for radionuclides is normally by an ultra-low liquid scintillation counter equipped with an alpha–beta discrimination device (Forte *et al.*, 2007). Radon is formed in the ground by the decay (breakdown) of uranium, which is found in all soils and rocks to some extent. The highest levels of uranium-238 are found in areas of granite with maximum concentrations in the UK, for example, just below 2 ppm (parts per million) in granite from Devon. The decay of uranium-238 results in the formation of radium-226, which subsequently decays to radon-222(Rn). The radon gas slowly migrates through the soil to the surface and is quickly dispersed into the atmosphere, where it is diluted to safe concentrations. Concern has been expressed, however, over modernized houses that are well draught-proofed. Radon percolates up through the foundations of the building, or occasionally from the granite stone used to build the house, and because there is so little air exchange in the building, the radon can accumulate to dangerous concentrations in the enclosed space of the room. Even where newly constructed buildings have adequate radon-proof membranes laid in the foundations to stop the gas from percolating through, radon gas can still accumulate in the buildings from the use of well water in the bathroom and kitchen. Radon is known to be carcinogenic, and as it is primarily inhaled as a gas it causes lung cancer. The US Public Health Service considers radon to be a major environmental health problem and evidence shows it to be the second major cause of lung cancer after smoking. The US Environmental Protection Agency (USEPA) suggest that when ingested with water radon can also increase the risk of stomach cancer (Cross *et al.*, 1985).

Radon is very soluble in water, so when the gas comes into contact with groundwater it dissolves. It is occasionally seen at elevated levels in bottled mineral waters (Section 29.2). For example, in a study of 28 commercially available mineral waters in Hungary, six had activity levels $>0.1\,\mathrm{Bq\,l^{-1}}$ with one with a mean activity level of $2.9\,\mathrm{Bq\,l^{-1}}$(Somlai *et al.*, 2002). In a national survey of groundwaters in the USA the average radon level was $900\,\mathrm{pCi\,l^{-1}}$ (median $300\,\mathrm{pCi\,l^{-1}}$), although in some states the mean value was considerably higher (e.g. New Hampshire $1716\,\mathrm{pCi\,l^{-1}}$). Some groundwaters have been found to contain in excess of $100\,000\,\mathrm{pCi\,l^{-1}}$, with levels in excess of $10\,000\,\mathrm{pCi\,l^{-1}}$ common in the west and north-east of the USA (Dupuy *et al.*, 1992; Helms and Rydell, 1992). Levels in European groundwaters are largely unknown and it is only relatively recently that radon in drinking water has been taken seriously in Europe.

Radon can be consumed by drinking contaminated water or through the inhalation of radon released from water. Radon is primarily released from water when agitated, with up to 40 and 30 times higher concentrations recorded in the bathroom and kitchen respectively, than the living room, due to the radon released from flushing the toilet and running large volumes of water.

Some groundwaters may contain high levels of radon, but those that are affected will mainly be those on private supplies, especially sealed boreholes. This water will have lost very little of its natural radiation before reaching the house. Having a water storage tank in a vented attic significantly reduces radon levels, and if radon is a problem, then consideration should be given to supplying all the water from the attic storage tank. Installing a large aquarium-type aerator in the tank will help to release the radon, whereas installing a small activated carbon filter will remove the radon from the supply to the kitchen tap, which is used for drinking. Installing even a small aquarium-style aerator in the water storage tank will cause some degree of noise, especially at night, so a time switch is required to turn it off at night. Alternatively the tank and motor can be insulated to reduce the vibration that causes the noise. The implications of doing this should be discussed with, and approved by, the local authority or water inspector. Of course, as the aerator will cause considerable mixing of the water, all sediment and debris must be removed from the tank before installation or two tanks should be used in series, the first for radon removal and the second for settlement. Alternatively, as iron and manganese will be constantly precipitated out in the tank the installation of a point-of-use physical filter will be most effective in most cases.

12.3 Non-radon radionuclides

Apart from radon-222, other radionuclides are occasionally found in water; for example, the alpha-emitting radionuclides radium-226, which is associated with an increased risk of bone and head carcinomas, and uranium-234 and -238 associated with an increased risk of bone cancer (Longtin, 1988). With the exception of radium-228, most beta-emitters are associated with human activity, whereas the alpha-emitters are of natural origin and so are far more likely to be detected in groundwaters (Table 12.1).

Generally the concentration of radionuclides in drinking water is very low compared to the maximum contaminant levels (MCLs) in the USA although some parts of the mid-West have significantly higher average combined radium-226 and -228 levels than the rest of the country. This is also true for some Western states that have elevated uranium levels compared to the national average. In a national survey of groundwater supplies in the USA mean uranium concentrations were $1.86 \mu g \ l^{-1}$ with about 3% of water supplies with concentrations $>10 \mu g \ l^{-1}$ (Longtin, 1988). Elevated concentrations are found in groundwater in the Colorado Plateau, Western Central Plateau, Rocky Mountain System, Basin and Range, and the Pacific Mountain System. Pockets of elevated uranium are found scattered in South Carolina, Connecticut and other eastern states, although in general uranium concentrations are low in groundwaters in eastern USA. For example, in one study of the groundwater used by residents in Greenville County in South Carolina, the mean uranium concentration was found to be $620 \mu g \ l^{-1}$ (Orloff et al., 2004). Other

Table 12.1 *Radionuclides found in drinking water that have a significant health risk*

Alpha-emitting radionuclides					
^{210}Po	^{222}Rn	^{226}Ra	^{232}Th	^{234}U	^{238}U

Beta-emitting radionuclides						
^{60}Co	^{89}Sr	^{90}Sr	^{131}I	^{134}Cs	^{137}Cs	^{210}Pb[a] ^{210}Ra[a]

[a] Also have alpha-emitting daughters.

radionuclides have been reported in a small number of drinking water supplies, although compared to radium-226, radium-228 and uranium their occurrence is thought to be rare.

It is inevitable that some water supplies are located in areas that have potential sources of man-made radioactive contamination from facilities that use, manufacture or dispose of radioactive substances. While water supplies can readily become contaminated through accidental releases of radioactivity or through improper disposal practices, there is also some evidence that water supplies can become contaminated from the normal operation of nuclear power plants and reprocessing facilities. There is understandable concern over the proposal to store radioactive wastes underground at Sellafield in the UK, where there are extensive aquifers in the area important for water supply. Water utilities using supplies vulnerable to radioactive contamination are required to perform extensive monitoring for beta particle and photon radioactivity to ensure the safety of drinking water.

Ionizing radiation is perhaps the most established human carcinogen (IARC, 2001). Alpha and beta radiation have a short range compared to gamma radiation, and so need to enter the body in order to damage internal organs. Radionuclides are largely excreted in the urine and so the kidneys and bladder are particularly at risk (Kurttio *et al.*, 2006). Between 0.8% and 7.8% of uranium ingested via drinking water is absorbed, concentrating in the bones and to a lesser extent in the kidneys, liver and other soft tissue, the rest is rapidly lost from the body via the urine although Orloff *et al.* (2004) found that uranium can go on being excreted for up to ten months after the contaminated water had been replaced. Prolonged exposure to drinking water containing alpha-emitters in excess of the USEPA MCL increases the risk of getting cancer, while uranium, which is also carcinogenic, also has toxic effects on the kidney. Beta- and photon-emitters in excess of the MCL are also carcinogenic.

12.4 Standards and treatment

The EC Drinking Water Directive (98/83/EEC) includes radioactivity under Indicator Parameters in Part C (Appendix 1). This sets maximum values for tritium at $100 \, Bq \, l^{-1}$ and the total indicative dose at $0.1 \, mSv \, yr^{-1}$. Total indicative dose is measured excluding tritium, potassium-40, carbon-14, radon and its decay products, but including all other natural decay series radionuclides. While radon and its decay products in drinking water are not covered by the Directive, an EC Recommendation (2001/928/Euratom) *on the protection of the public against exposure to radon in drinking water supplies* proposes an action level for both public and private water supplies of $1000 \, Bq \, l^{-1}$ (EU, 2001). The action level is considered similar to the risk that would arise from breathing air containing radon at $200 \, Bq \, m^{-3}$. Where radon levels in water are above $100 \, Bq \, l^{-1}$ but below $1000 \, Bq \, l^{-1}$, then the local authority must consider whether this poses a risk to human health. If it is concluded that such a risk exists, then remedial action should be considered. The EC Recommendation makes it clear that the action level does not mark a boundary between safe and unsafe, but rather a level at which action will usually be justified. In Ireland, if the radon level in public water supplies exceeds $500 \, Bq \, l^{-1}$, then remediation is considered justified for the protection of human health.

The World Health Organization (WHO) specifies guideline values for gross alpha and beta activity of 0.5 and $1.0 \, Bq \, l^{-1}$ respectively. If samples exceed these levels of activity then further radiological examination is recommended, although higher values do not necessarily imply that the water is unfit for human consumption. A provision guideline value of $15 \, \mu g \, l^{-1}$ has been set for uranium (WHO, 2004).

The USEPA has proposed a combined MCL for radium-226 and radium-228 of $5 \, pCi \, l^{-1}$ and has set a separate MCL for uranium as well as alpha and beta activity (Table 12.2). Although the National Primary Drinking Water Regulations do not apply to private water supplies, they do apply to small water schemes serving more than 25 people (or with 15 or more service connections) using groundwater.

Individual supplies will require a point-of-entry treatment system where radon is present. In a study carried out by the USEPA, granular activated carbon (GAC) was compared with various bubble aeration systems to treat a groundwater supply containing $35 \, 620 \, pCi \, l^{-1}$ radon. It was found that removal by GAC decreased over time from 99.7% to 79%. This was improved if the activated carbon was preceded by ion-exchange, although the performance still decreased from 99.7% to 85% over time. Ion-exchange was necessary to remove iron, which was impeding radon adsorption by fouling the surface of the GAC. In contrast, the bubble aeration systems were highly efficient (>99%) at removing radon from water without any loss of efficiency over time (Kinner *et al.*, 1990).

There is no risk with mains supply, even if the supply is from an aquifer contaminated by high levels of radon. Water treatment of groundwaters normally includes aeration so that 99% of the radon is removed at this point. The removal efficiency varies with the aeration technique, with packed tower

Table 12.2 *Current maximum contaminant levels (MCLs) for radionuclides in US Drinking Water. Reproduced by permission of the US Environmental Protection Agency*

Contaminant	MCL	Notes
Combined radium-226 and radium-228	5 pCi l^{-1}	Naturally occurs in some drinking water sources
Gross alpha activity (not including radon or uranium)	15 pCi l^{-1}	Naturally occurs in some drinking water sources
Beta particle and photon radioactivity	4 mrem yr^{-1}	May occur due to contamination from facilities using or producing radioactive materials
Uranium	30 µg l^{-1}	Naturally occurs in some drinking water sources

aeration particularly effective (>99% removal), compared with only 60–70% removal when using spray aeration. Water treatment plants also use GAC filters to remove radon and other radioactive isotopes. However, there is a concern that filters could become radioactive and then be a major disposal problem.

12.5 Conclusions

As there is no relationship between parent and daughter radionuclides it is not possible to predict which areas are most at risk from contamination. Therefore all domestic drinking water supplies that originate wholly or partially from groundwater should be tested for radon and other radionuclides.

Radon in drinking water has long been a cause for concern in the USA. However, in Europe it is only since the publication of the *EC Recommendation on the protection of the public against exposure to radon in drinking water supplies* in 2001 that member states were encouraged to determine the scale and nature of exposures due to radon in domestic drinking water supplies. For example, between August 2001 and May 2002 radon activity was measured in tap water from private groundwater supplies collected from 166 houses in Co. Wicklow (Ireland). The underlying geology of the area is predominantly granite which contains high levels of uranium from which radon is derived. Four supplies had activity concentrations in excess of the recommended EC action level of 1000 Bq l^{-1} with the highest concentration recorded 5985 Bq l^{-1}. Fifteen supplies had activity concentrations between 500 and 1000 Bq l^{-1}, 51 were between 100 and 500 Bq l^{-1} and the remainder had activity concentrations <100 Bq l^{-1}. This confirmed that for a small percentage of consumers who depend on private groundwater supplies as their primary source of drinking water, radon in drinking water poses a significant health risk. However, the non-mandatory action level of 1000 Bq l^{-1} proposed by the EC is 5400 times greater than the MCL of 5 pCi l^{-1} in the USA and 2000 times greater than the 0.5 Bq l^{-1} guideline set by the WHO for gross alpha activity. So although drinking water radon is a problem solely of some private and very small groundwater supplies, given the difference in the guidelines and maximum values set it appears to be largely ignored within Europe.

Depleted uranium (DU) is a by-product of the process that enriches natural uranium ore for use in nuclear reactors and nuclear weapons. It is extremely dense (19.05 g cm^{-3}) and hard, making it an ideal material for armour piercing munitions. With over 6×10^5 tons of DU stockpiled in the USA alone, it is also a very cheap material in comparison to other traditional materials used for this purpose. First used by US forces during the Gulf War in 1991, it have been widely used in conflicts since that time (Bleise *et al.*, 2003). There have been some concerns about the effects of used and abandoned DU rounds and shells left on battlefields and that this may lead to contamination of water resources. However, recent studies after the Balkan conflicts of 1995–9, in Serbia and Montenegro, has shown that there has been no significant contamination of water, soil or biological material in these areas (Jiam *et al.*, 2005).

Radionuclides in drinking water are unlikely to cause a problem in the short term unless they arise directly from the nuclear industry. However, prolonged exposure to either radon or uranium above the MCL values set in the USA is a cause for concern. Groundwater supplies should be tested for radionuclides as a matter of routine, and all water resources close to current or former nuclear installations should be checked for beta particle and photon radioactivity. The discrepancy between European standards for radioactivity in drinking water and that in the USA is a serious cause of concern.

References

Bleise, A., Danesi, P. R. and Burkart, W. (2003). Properties, use and health effects of depleted uranium (DU): a general overview. *Journal of Environmental Radioactivity*, **64**, 93–112.

Cross, F. T., Harley, N. H. and Hofmann, W. (1985). Health effects and risks from ^{222}Rn in drinking water. *Health Physics*, **48**, 649–70.

Dupuy, C. J., Healy, D., Thomas, M. *et al.* (1992). A survey of naturally occurring radionuclides in ground water in selected bedrock aquifers in Connecticut and implications for public health policy. In *Regulating Drinking Water Quality*, ed. C. E. Gilbert and E. J. Calabrese. Boca Raton, FL: Lewis, pp. 95–119.

EU (2001). Commission Recommendation of 20 December 2001 on the protection of the public against exposure to radon in drinking water supplies. *Official Journal of the European Commission*, **L344**, (28.12.01), 85–8.

Forte, M., Rusconi, R., Cazzaniga, M. T. and Sgorbati, G. (2007). The measurement of radioactivity in Italian drinking waters. *Microchemical Journal*, **85**(1), 98–102.

Helms, G. and Rydell, S. (1992). Regulation of radon in drinking water. In *Regulating Drinking Water Quality*, ed. C. E. Gilbert and E. I. Calabrese. Boca Raton, FL: Lewis, pp. 77–82.

IARC (2001). *Ionising Radiation,* Part 2: *Some Internally Deposited Radionuclides*. IARC Monograph on the Evaluation of the Carcinogenic Risks to Humans, 78. Lyon, France: International Agency for Research on Cancer, IARC Press.

Jiam, G., Belli, M., Sansone, U., Rosamilia, S. and Gaudino, S. (2005). Concentration and characteristics of depleted uranium in water, air and biological samples collected in Serbia and Montenegro. *Applied Radiation and Isotopes*, **63**, 381–99.

Kinner, N. E., Malley, J. P. and Clement, J. A. (1990). *Radon Removal Using Point of Entry Water Treatment Techniques*. EPA/600/2–90/047. Ohio: USEPA.

Kurttio, P., Salonen, L., Ilus, T. *et al.* (2006). Well water radioactivity and risk of cancers in the urinary organs. *Environmental Research*, **102**(3), 333–8.

Longtin, J. P. (1988). Occurrence of radon, radium, and uranium in groundwater. *Journal of the American Water Works Association*, **80**(7), 84–93.

Orloff, K. G., Mistry, K., Charp, P. *et al.* (2004). Human exposure to uranium in groundwater. *Environmental Research*, **94**, 319–26.

Somlai, J., Horváth, G., Kanyács, B. *et al.* (2002). Concentration of ^{226}Ra in Hungarian bottled mineral water. *Journal of Environmental Radioactivity*, **62**, 235–40.

WHO (2004). *Drinking Water Guidelines*. Geneva: World Health Organization.

Chapter 13
Pathogens

13.1 Introduction

There are three different groups of micro-organisms that can be transmitted via drinking water: protozoa, viruses and bacteria (Section 3.2) (Table 13.1). They are all transmitted by the faecal-oral route and so largely arise either directly or indirectly by contamination of water resources by sewage or, increasingly, animal wastes. It is theoretically possible, but unlikely, that other pathogenic organisms such as nematodes (roundworm and hookworm) and cestodes (tapeworm) may also be transmitted via drinking water (Gray, 2004).

13.2 Protozoa

There are two protozoa frequently found in drinking water that are known to be responsible for outbreaks of disease (Table 13.1). These are *Cryptosporidium* and *Giardia lamblia*.

13.2.1 *Cryptosporidium*

This parasitic protozoan is widely distributed in nature, infecting a wide range of animal hosts including pets and farm animals. However, it was only relatively recently that it was found to be a human pathogen as well. The first recorded case of human infection occurred as recently as 1976. The problem is that the protozoa form protective stages known as oocysts that allow them to survive for long periods in water while waiting to be ingested by a host. They are also able to complete their life cycle in just a single host as well as having an auto-infection capacity (Fayer and Ungar, 1986). Once infected the host is a lifetime carrier and is subject to relapses. In normal patients the protozoa give rise to a self-limiting gastroenteritis that lasts for up to two weeks, with children more at risk than adults. If the patient is immunosuppressed, infection will be life threatening. For example, it is a major cause of death among patients with AIDS (acquired immune deficiency syndrome). Two peaks in the number of infections are seen each year, one in the spring and another in the autumn.

Table 13.1 *Bacterial, viral and protozoan diseases generally transmitted by contaminated drinking water. Reproduced from Singh and McFeters (1992) with permission from John Wiley and Sons Ltd*

Agent	Disease	Incubation time
Bacteria		
Shigella spp.	Shigellosis	1–7 days
Salmonella spp.		
S. typhimurium	Salmonellosis	6–72 h
S. typhi	Typhoid fever	1–3 days
Enterotoxigenic *Escherichia coli*	Diarrhoea	12–72 h
Campylobacter spp.	Gastroenteritis	1–7 days
Vibrio cholerae	Gastroenteritis	1–3 days
Viruses		
Hepatitis A	Hepatitis	15–45 days
Norwalk-like agent	Gastroenteritis	1–7 days
Virus-like particles <27 nm	Gastroenteritis	1–7 days
Rotavirus	Gastroenteritis	1–2 days
Protozoa		
Giardia lamblia	Giardiasis	7–10 days
Entamoeba histolytica	Amoebiasis	2–4 weeks
Cryptosporidium	Cryptosporidiosis	5–10 days

The oocysts of *Cryptosporidium* are only 4–7 μm in diameter and so are difficult to remove from raw waters by conventional treatment. The oocysts appear to be very widespread in drinking water resources, including rivers, streams, lakes and reservoirs, being found in over half the samples tested in the UK and in 97% of surface waters in the USA (USEPA, 1999). Oocysts are widespread in all types of water resources, although worse in those receiving treated or untreated wastewater. Reported ranges are: surface waters 0.001–107 oocysts l^{-1}, groundwater 0.001–0.922 oocysts l^{-1}, sewage 1–120 oocysts l^{-1} and filtered treated wastewater 0.01–0.13 oocysts l^{-1} (Rose, 1999). Studies indicate that a single oocyst may be enough to cause infection (Blewett *et al.*, 1993), although outbreaks of cryptosporidiosis are normally associated with gross contamination (Wilkins, 1993). Low-level exposure to oocysts (1–10) is capable of initiating an infection (Rose, 1990). The main symptoms of cryptosporidiosis are stomach cramps, nausea, dehydration and headaches. There may also be significant weight loss associated with infection.

The first major waterborne outbreak of cryptosporidiosis documented was in 1984 due to sewage contamination of a well in Texas (Table 13.2). An outbreak in Georgia in 1987 affected 13 000 people, even though the drinking water had undergone conventional treatment (i.e. coagulation, sedimentation, filtration and

Table 13.2 *Examples of outbreaks of drinking water cryptosporidiosis*

Year	Location (Country)	No. affected	Suspected cause
1983	Surrey (UK)	16	Source contamination
1984	Texas (USA)	2000	Sewage contamination of well
1985	Surrey (UK)	50	Source contamination
1987	Georgia (USA)	13 000	Operation irregularities
1988	Ayrshire (UK)	27	Post-treatment contamination
1989	Swindon (UK)	500	Source contamination
1989	L. Lomond (UK)	442	Source contamination
1990	Humberside (UK)	140	Source contamination
1991	South London (UK)	44	Source contamination
1993	Milwaukee (USA)	440 000	Post-treatment contamination

disinfection). In 1988, 27 cases were confirmed in Ayrshire (UK), although many more people were thought to have been infected. Of those infected, 63% were less than 8 years of age. The outbreak was most likely due to the finished water being contaminated with runoff from surrounding fields on which slurry had been spread. The oocysts were found in the chlorinated water in the absence of faecal indicator bacteria. A far more serious outbreak occurred early in 1989 affecting over 400 people in the Oxford and Swindon areas in the UK (Richardson *et al.*, 1991), although according to Rose (1990) as many as 5000 people may have been affected. The outbreak was quickly traced to the Farmoor Water Treatment Plant near Oxford, which takes its water from the River Thames. On investigation oocysts were found in the filters, in the filter backwash water and in the treated water, even though it had been chlorinated and the microbiological tests had shown it to be of excellent quality. The oocysts were found in the Farmoor Reservoir and also in a tributary of the River Thames, upstream of the treatment plant. The seasonal presence of the organisms was particularly associated with the grazing of lambs and with the scouring that often arises in their early lives. Although the rapid sand filters at Farmoor were removing 79% of the oocysts after coagulation, the recycling of the backwash water had given rise to exceptionally high concentrations of oocysts (10 000 per litre), with resultant breakthrough. Disinfection with chlorine was not effective, and the first action by Thames Water was to stop recycling the backwash water (up to then a normal practice), which brought the problem under control. There was a decrease in the number of reported cases as the number of oocysts in the water decreased. The outbreak in the Loch Lomond area in 1989 was caused by livestock grazing around the reservoir or within the

Table 13.3 *Key contamination routes identified for waterborne cryptosporidiosis*

Slurry tank leakage
Leaching from solid manure storage
Runoff from farmyards
Road and street runoff
Runoff from agricultural land used for grazing
Direct contamination from animals drinking from streams
Percolation through soil to field drainage system and ultimately to watercourse
Disposal of contaminated sewage sludge or sewage effluents
Disposal of waterworks sludge

catchment. Cattle and infected humans can excrete up to 10^{10} oocysts during the course of infection, so that cattle slurry, wastewater from cattle markets and sewage should all be considered potential sources of the pathogen (Table 13.3). The most important outbreak of cryptosporidiosis occurred in the USA during April 1993. The water distribution system serving 800 000 people on Milwaukee became contaminated by floodwater from a nearby river. Over 440 000 people became infected and of the 4400 that were hospitalized almost 50 died (Jones 1994), making *Cryptosporidium* the most serious waterborne disease in developed countries.

13.2.2   *Giardia*

Like *Cryptosporidium, Giardia lamblia* is found in a wide range of animals, where it lives in the free-living (trophozoite) form in the intestines. In water resources cysts are able to survive for long periods, especially in winter. *Giardia* cysts, unlike those of *Cryptosporidium* which are round, are oval and larger (8–14 μm long and 7–10 μm wide). This intestinal protozoan causes acute diarrhoeal illness and has a worldwide distribution. It is particularly common in the USA where it is now considered to be endemic with a carrier rate of 15–20% of the population, depending on their socio-economic status, age and location. *Giardia* is the most common animal parasite of humans in the developed world, although water is probably not the most common mode of transmission. However, *Giardia* remains one of the most common causes of waterborne diseases. Between 1980 and 1985 there were 502 outbreaks of waterborne disease in the USA; 52% of these were due to *Giardia*. For example, sewage contamination of a groundwater supply at a Colorado ski resort resulted in 123 holidaymakers contracting the illness. A survey of the situation in the USA by LeChevallier (1990) showed that 81% of the raw waters and 17% of the treated waters sampled contained *Giardia*. A similar survey in Scotland showed that

48% of the raw waters and 23% of the treated waters sampled also contained *Giardia* cysts. The number of reported cases in England and Wales has risen from 1000 each year in the late 1960s to over 5000 each year by the late 1980s, although these were largely associated with people travelling overseas. The number of outbreaks associated with drinking water contamination is steadily increasing in the UK. A significant outbreak of giardiasis occurred in south Bristol in the summer of 1985, when 108 cases were reported. It is thought that contamination of the supply occurred in the distribution system and was not due to any failure of the water treatment process (Browning and Ives, 1987).

The symptoms of giardiasis develop between one and four weeks after infection. These include explosive, watery, foul-smelling diarrhoea, gas in the stomach or intestines, nausea and, not surprisingly, a loss of appetite. Unlike *Cryptosporidium, Giardia* can be treated by a number of drugs, but there is no way of preventing infection except by adequate water treatment. Boiling water for 20 minutes will kill the cysts. The New Safe Drinking Water Regulations in the USA require surface waters to be filtered to specifically remove cysts and be sufficiently disinfected to destroy *Giardia* (Chapter 19).

13.2.3 *Other protozoan pathogens*

Protozoan pathogens of humans are almost exclusively confined to tropical and subtropical areas, which is why the increased occurrence of *Cryptosporidium* and *Giardia* cysts in temperate areas is so worrying. However, with the increase in travel, carriers of all diseases are now found worldwide, and cysts of all the major protozoan pathogens occur in European sewage from time to time. All the diseases are transmitted by cysts entering the water supplies. Infection is by direct ingestion, either by drinking the water or by swimming in contaminated water. Two other protozoan parasites, which occur in the UK from time to time, are *Entamoeba histolytica* that causes amoebic dysentery and *Naegleria fowleri*, which causes the fatal disease amoebic meningo-encephalitis. The chance of these organisms breaking through water treatment systems in significant numbers is very remote.

13.3 Viruses

Viruses are made up of a core of nucleic acid (RNA or DNA) surrounded by a protein coat. They cannot reproduce without a host cell, but can survive in the environment for very long periods. Human enteric viruses are produced in very large amounts by infected individuals and are faecally excreted. They generally pass unaffected through the wastewater treatment plant and so will be found in surface waters. Infection takes place when a virus is ingested with contaminated food or by drinking contaminated water. The viruses pass through the stomach and generally infect cells of the lower alimentary canal. Infected subjects shed large numbers of viral particles into the faeces, which are then excreted. Up to 10 000 000 viral particles have been reported per gram of faeces, and can continue to be shed for long periods; for example, the mean excretion period for polio is over 50 days.

Infectious hepatitis, enteroviruses, reovirus and adenovirus are all thought to be transmitted via water (Table 13.1). Of most concern in Britain is viral hepatitis. There are three subgroups, hepatitis A, which is transmitted by water, hepatitis B, which is spread by personal contact or inoculation and which is endemic in certain countries such as Greece, and hepatitis C, which is a non-A or -B type hepatitis virus. Hepatitis A is spread by faecal contamination of food, drinking water and areas that are used for bathing and swimming. The virus cannot be cultivated in vitro so studies are confined to actual outbreaks of the disease. Hepatitis A outbreaks usually occur in a cyclic pattern within the community, as once infected the population is immune to further infection by the virus. Therefore no new cases occur for five to ten years until there is a new generation (mainly of children) that has not previously been exposed. There is no treatment for hepatitis A, with the only effective protection good personal hygiene, and the proper protection and treatment of drinking water. Immunoglobulin is often given to prevent the illness developing in possible contacts, although it is not always successful. Symptoms develop 15–45 days after exposure and include nausea, vomiting, muscle ache and jaundice. Hepatitis A virus accounts for 87% of all viral waterborne disease outbreaks in the USA (Craun, 1986).

Enteroviruses, poliomyelitis virus, coxsackievirus and echovirus, all cause respiratory infections and are present in the faeces of infected people. Poliomyelitis in particular is common in British sewage, but this is due to the vaccination programme within communities and does not indicate actual infection. Reovirus is thought to be associated with gastroenteritis whereas adenovirus 3 is associated with swimming pools and causes pharyngo-conjunctival fever.

Warm-blooded animals appear to be able to carry viruses pathogenic to humans. For example, 10% of beagle dogs have been shown to carry human enteric viruses. Therefore, there appears a danger of infection from waters not contaminated by sewage but by other sources of pollution, especially stormwater runoff from paved areas. Most viruses remain viable for several weeks in water at low temperatures, so long as there is some organic matter present. Viruses are found both in surface and ground water sources, and in the USA as many as 20% of all wells and boreholes have been found to be contaminated with viruses. Two viruses that have caused outbreaks of illness due to drinking water contamination are Norwalk virus and rotavirus (Craun, 1991; Cubitt, 1991).

Norwalk virus results in severe diarrhoea and vomiting. It is of particular worry to the water industry in that it appears not to be affected by normal chlorination levels. Also it seems that infection by the virus only gives rise to short-term immunity, whereas most other enteric viruses confer life-long immunity. In 1986 about 7000 people who stayed at a ski resort in Scotland were infected with a Norwalk-like virus. The private water supply, which was untreated, came from a stream subject to contamination from a septic tank.

Rotavirus is a major contributor to child diarrhoea syndrome. This causes the death of millions of children in developing countries each year. This is not, thankfully, a serious problem in Europe due to better hygiene, nutrition and health care. Outbreaks do occur occasionally in hospitals, and although associated with child diarrhoea, can be much more serious if contracted by an adult. Gerba and Rose (1990) have produced an excellent review on viruses in source and drinking waters.

13.4 Bacteria

Bacteria are the most important group in terms of frequency of isolation in drinking water and reported outbreaks of disease (Table 13.1). The most important bacterial diseases are commonly associated with faecal contamination of water. For example, in temperate regions these include *Salmonella* (typhoid, paratyphoid), *Campylobacter, Shigella* (bacterial dysentery), *Vibrio cholerae* (cholera) and *Escherichia coli*.

13.4.1 *Salmonellosis*

Due to the elimination of other classical bacterial diseases through better sanitation, higher living standards and the widespread availability of antibiotic treatment, the various serotypes that make up the genus *Salmonella* are now the most important group of bacteria affecting the public health of both humans and animals in western Europe. *Salmonella* is commonly present in raw waters but is only occasionally isolated from finished waters, as chlorination is highly effective at controlling the bacteria. The symptoms of salmonellosis are caused by an endotoxin and are typically acute gastroenteritis with diarrhoea, often associated with abdominal cramps, fever, nausea, vomiting, headache and, in severe cases, even collapse and possible death. Compared with farm animals the incidence of salmonellosis in humans is low and shows a distinct seasonal variation. A large number of serotypes are pathogenic to humans and their low frequency of occurrence varies annually from country to country. Low-level contamination of food or water rarely results in the disease developing because between 10^5 and 10^7 organisms have to be ingested before development. Once infection has taken place then large numbers of the organisms are excreted in the faeces ($>10^8 \, g^{-1}$). Infection can also result in a symptomless carrier state in which the organism rapidly develops at localized sites of chronic infection, such as the gall bladder or uterus, and is excreted in the faeces or other secretions. Water becomes contaminated by raw or treated wastewater.

The most serious diseases associated with specific serotypes are typhoid fever *(Salmonella typhi)* and paratyphoid *(Salmonella paratyphi* and *Salmonella schottmuelleri)*. The last major outbreak of typhoid in Britain occurred in Croydon, Surrey during the autumn of 1937 when 341 cases were reported, resulting in over 40 deaths. There have been five minor outbreaks of typhoid and three of paratyphoid since then. The number of reported cases of typhoid fever in the UK has fallen to less than 200 each year, 85% of these being contracted abroad. Of the remainder, few are the result of drinking contaminated water. Although *Salmonella paratyphi* is recorded in surface waters all over the British Isles, there have been less than 100 cases of paratyphoid fever reported annually. Typhoid has also been largely eliminated from the USA, although in 1973 there was an outbreak in Dade County in which 225 people contracted the disease from contaminated well water (Craun, 1986).

13.4.2 *Campylobacter*

Campylobacter are spiral curved bacteria 2000–5000 nm in length consisting of two to six coils. Although discovered in the late nineteenth century they were not isolated from diarrhoetic stool specimens until 1972. It is only since the development in 1977 of a highly selective solid growth medium allowing culture of the bacterium that its nature has been revealed. *Campylobacter* species have been isolated from both fresh and estuarine waters with counts ranging from 10 to 230 campylobacters $100\,ml^{-1}$ in rivers in north-west England. There were 27 000 reported outbreaks of campylobacter enteritis in the UK during 1987, rising to over 30 000 in 1990, causing severe acute diarrhoea. It is now thought that *Campylobacter* is the major cause of gastroenteritis in Europe, being more common than *Salmonella*. The most important reservoirs of the bacterium are meat, in particular poultry, and unpasteurized milk. Household pets, farm animals and birds are also known to be carriers of the disease. Unchlorinated or poorly chlorinated water supplies have been identified as a major source of infection. For example, 3000 of a total population of 10 000 developed campylobacter enteritis from an inadequately chlorinated mains supply in Bennington, Vermont, while 2000 people who drank unchlorinated mains water contaminated with faecally polluted river water in Sweden also contracted the disease. Water is either contaminated directly by sewage, which is rich in *Campylobacter*, or indirectly from animal faeces. There is a definite seasonal variation in the number of campylobacters in river water, with the greatest numbers occurring in the autumn and winter. This is opposite to the seasonal variation of infection in the community, with the number of infections increasing dramatically during May and June (Jones and Telford, 1991). Serotyping of isolates has established that *C. jejuni* serotypes, common in human infections, are found downstream of sewage effluent sites, confirming that sewage effluents are important sources of *C. jejuni* in the aquatic environment. Gulls are known carriers and can contaminate water supply reservoirs while they roost. Dog faeces, in particular, are rich in the bacterium. In a UK study *C. jejuni* was isolated from 4.6% of 260 specimens of dog faeces sampled, whereas *Salmonella* spp. were isolated from only 1.2%. The incidence of *C. jejuni* is low compared with other studies on dog faeces, with infection rates ranging from 7% to 49%. Dog faeces can cause contamination of surface waters during storms as surface runoff removes contaminated material from paved areas and roads. An outbreak affecting 50% of a rural community in northern Norway was traced to contaminated faecal deposits from sheep grazing the banks of a small lake that were washed into the water during a heavy storm that melted the snow on the banks. The water supply for the village came directly from the lake without chlorination. Natural aquatic systems in temperate areas are generally cool and research has shown that *Campylobacter* can remain viable for extended periods in streams and groundwaters. Survival of the bacterium decreases with increasing temperature, but at 4 °C survival in excess of 12 months is possible. The incidence of *Campylobacter* in water can be estimated by an MPN (most probable number) technique. *Campylobacter* infections have been reviewed by Skirrow (1982).

13.4.3 *Shigella*

Shigella causes bacterial dysentery and is the most frequently diagnosed cause of diarrhoea in the USA, where it accounted for 19% of all the cases of waterborne diseases reported in 1973. The bacterial genus is rather similar in epidemiology to *Salmonella*, except that *Shigella* rarely infects animals and does not survive as well in the environment. When the disease is present as an epidemic it appears to be spread mainly by person-to-person contact, especially between children, shigellosis being a typical institutional disease occurring in overcrowded conditions. However, there has been a significant increase in the number of outbreaks arising from poor quality drinking water contaminated by sewage. Of the large number of serotypes (>40) *S. sonnei* and *S. flexneri* account for more than 90% of isolates. The number of people excreting *Shigella* is estimated to be 0.46% of the population in the USA, 0.33% in Britain, but 2.4% in Sri Lanka. In England and Wales notifications of the disease have fallen from between 30 000 and 50 000 each year to <7000 annually in the 1980s (Galbraith *et al.*, 1987). It is a problem in both developed and developing countries but is endemic in Eastern Mediterranean countries.

13.4.4 *Enteropathogenic Escherichia coli*

There are 14 distinct serotypes of *Escherichia coli* which cause gastroenteritis in humans and animals, being especially serious in newborn infants and children under five years of age. It is common throughout Europe and is also thought to be the cause of 'traveller's tummy', the bout of diarrhoea that affects so many tourists who visit the warmer areas of Europe. The symptoms are profuse watery diarrhoea with little mucous, nausea and dehydration. The disease does not cause any fever and is rarely serious in adults. Up to 2.4% of children in England and Wales are thought to be carriers, although much higher percentages are found in people engaged in high-risk occupations such as food handling. Enteropathogenic *E. coli* is commonly isolated from sewage but probably represents less than 1% of the total coliforms present in polluted waters. However, only 100 bacteria are required to cause illness. Survival of the organism is the same as for other serotypes of *E. coli* and under warm, nutrient-rich conditions they are able to multiply in water. An outbreak of gastroenteritis in Worcester in the winter of 1965–6 affected 30 000 people and was thought to be due to contamination of the water supply as a result of flooding. In 1986 the water tanks on a cruise-liner became contaminated by sewage, resulting in an outbreak of the disease that affected 251 passengers and 51 crew members (O'Mahony *et al.*, 1986).

Escherichia coli 0157:H7 causes haemorrhagic colitis, haemolytic uraemic syndrome and is a major cause of kidney disease in children. Like *Campylobacter jejuni*, this organism is generally associated with food, in particular beef and milk, but in recent years has been implicated in a number of waterborne outbreaks. The number of organisms required to initiate infection is thought to be <100, with outbreaks the result of faecal contamination of surface waters and poor disinfection. Contaminated water entered the distribution system in Cabool in Missouri (USA) during the winter of 1990 after the pipework had been disturbed resulting in 240 confirmed cases of 0157:H7 gastroenteritis and four deaths (Geldreich *et al.*, 1992).

13.4.5 *Cholera*

Cholera is thought to have originated in the Far East, where it has been endemic in India for many centuries. In the nineteenth century the disease spread throughout Europe where it was eventually eliminated by the development of uncontaminated water supplies, water treatment and better sanitation. It is still endemic in many areas of the world, especially those that do not have adequate sanitation and, in particular, in situations where the water supplies are continuously contaminated by sewage. Over the past 20 years the incidence and spread of the disease has been causing concern; this has been linked to the increasing mobility of travellers and the speed of travel. Healthy, symptomless carriers of *Vibrio cholerae* are estimated to range from 1.9% to 9.0% of the population. This estimate is now thought to be rather low, with a haemolytic strain of the disease reported as being present in up to 25% of the population. The holiday exodus of Europeans to the Far East, which has been steadily increasing since the mid 1960s, will have led to an increase in the number of carriers in their home countries and an increased risk of contamination and spread of the disease. While there are regularly reported cases of cholera, none are thought to have been waterborne (Galbraith *et al.*, 1987).

Up to 10^8-10^9 organisms are required to cause the illness, so cholera is not normally spread by person-to-person contact. It is readily transmitted by drinking contaminated water or by eating food handled by a carrier, or that has been washed with contaminated water, and is regularly isolated from surface waters in the UK. It is an intestinal disease with characteristic symptoms, i.e. sudden diarrhoea with copious watery faeces, vomiting, suppression of urine, rapid dehydration, lowered temperature and blood pressure, and complete collapse. Without treatment the disease has a 60% death-rate, the patient dying within a few hours of first showing the symptoms, although with suitable treatment the death-rate can be reduced to less than 1%.

The bacteria are rapidly inactivated under unfavourable conditions, such as high acidity or a high organic matter content in the water. In cool, unpolluted waters *V. cholerae* will survive for up to two weeks. Two different somatic serotypes, O1 and O139, are currently responsible for epidemic cholera, with the latter first reported in Madras in 1992 (Mukhopadhyay *et al.*, 1995). Other types of *Vibrio* cause a milder form of the disease or gastroenteritis.

13.4.6 *Opportunistic bacterial pathogens*

Opportunistic bacteria are usually found as part of the normal heterotrophic bacterial flora of aquatic systems and may also exist as part of the normal body microflora (Table 13.4) (Section 19.4) (Reasoner, 1992). These organisms are normally not a threat to healthy individuals but under certain circumstances they can lead to infection in certain segments of the community, in particular new-born babies, the elderly and the immunocompromised. It is thought that numerous hospital-acquired infections are attributable to such organisms (DeZuane, 1990). Some of the organisms listed in Table 13.4 are also considered as primary pathogens, meaning that they are also capable of being primary disease-causing agents, such as *Salmonella, Campylobacter, Shigella, V. cholerae* and *E. coli*, rather than secondary invaders.

13.5 Unusual sources of contamination

There was some concern in the early 1980s over the open policy adopted by the then Regional Water Authorities in allowing more extensive use of their potable supply reservoirs for recreation. One of the potential hazards identified has been the use of bait by fishermen. It has been suggested that such bait could be a potential source of problems due to: (1) the introduction of pathogenic organisms affecting either humans or fish; (2) increasing the nutrient concentration of the water and thus causing eutrophication; and (3) that certain

Table 13.4 *Opportunistic bacterial pathogens normally isolated from drinking waters. Reproduced from Reasoner (1992) with permission from US Environmental Protection Agency*

Acinetobacter spp.
Achromobacter xylosoxidans
Aeromonas hydrophila
Bacillus spp.
Campylobacter spp.[a]
Citrobacter spp.
Enterobacter aerogenes
Enterobacter agglomerans
Enterobacter cloacae
Flavobacterium meningusepticum
Hafnia alvei
Klebsiella pneumoniae[a]
Legionella pneumopila[a]
Moraxella spp.
Mycobacterium spp.
Pseudomonas spp. (non-aeruginosa)[a]
Serratia fonticola
S. liquefaciens
S. marcescens
Staphylococcus spp.[a]
Vibrio fluvialis[a]

[a] Indicates that the organism may be a primary pathogen.

chrysoidine dyes which are used to colour maggots are possibly carcinogenic. In a study carried out by the Water Research Centre, there appeared to be no risk at all, although the report noted that most of the Regional Water Authorities were already adopting restrictions (Solbe and De, 1983). It had been shown that a botulinus toxin, which originated from maggots used as bait and contaminated with *Clostridium botulinum*, had caused the deaths of waterfowl on the River Thames in the summer of 1982. However, it was felt that chlorination was effective in destroying or inactivating the toxin, so that any risk to drinking water was minimal. Restricting baiting to moderate levels, especially on small reservoirs, appears to be sensible, but the risk is negligible.

Although migratory waterfowl that roost at reservoirs have no serious deleterious effect on water quality, gulls that feed on contaminated and faecal material at refuse tips and sewage works have been shown to excrete pathogens. All five British species of *Larus* gulls have been rapidly increasing in numbers in recent years, with the herring gull *(Larus argentatus)* doubling its population size every five to six years. This has resulted in a large increase in the inland population of *Larus* gulls throughout the British Isles, both permanent and those overwintering. These birds are opportunist feeders and have taken advantage of the increase in the human population and its standard of living, feeding on contaminated waste during the day and then roosting on inland water bodies, including reservoirs, at night. Faecal bacteria, especially *Salmonella* spp., have been traced from feeding sites such as domestic waste tips to the reservoir, showing that gulls are directly responsible for the dissemination of bacteria and other human pathogens. Many reservoirs have shown a serious deterioration in bacterial quality due to contamination by roosting gulls, and those situated in upland areas where the water is of a high quality, so that treatment is minimal before being supplied to the consumer, are particularly at risk from contamination. Research has shown that numerous *Salmonella* serotypes as well as faecal coliforms, faecal streptococci and spores of *Clostridium welchii* can be isolated from gull droppings (Gould, 1977). Ova of parasitic worms have also been isolated from gull droppings and birds are thought to be a major cause of the contamination of agricultural land with the eggs of the human beef tapeworm *(Taenia saginata)*. There is a clear relationship between the number of roosting gulls on reservoirs and the concentration of all the indicator bacteria. Gull droppings can also contribute significant amounts of nitrogen and phosphorus to reservoirs, which can lead to eutrophication.

Although contamination of small service reservoirs has been eliminated by covering them, this is impracticable for large upland reservoirs serving major cities and towns (HMSO, 1989). A considerable degree of success has been achieved by using bird-scarers and other techniques developed originally for the control of birds at airports. Roosting can be discouraged by broadcasting species-specific distress calls of *Larus* gulls, which has led to a dramatic reduction in bacterial contamination. Such a control option is far more cost effective than installing and operating more powerful disinfection treatment systems.

Drinking water standards for microbial pathogens and the efficiency of water treatment to remove them from raw waters are considered in Chapter 19. Recent developments in drinking water microbiology have been reviewed by McFeters (1990a, b), Morris *et al.* (1993), Hunter *et al.* (2002) and Bitton (2005).

13.6 Conclusions

The extent of waterborne diseases is measured though a voluntary reporting scheme in most countries, relying on accurate identification of the cause of illness by the medical practitioner. Therefore, estimates of the extent of specific disease-causing pathogens are notoriously inaccurate. It is clear, however, that drinking water is assumed by most consumers in the developed world as being microbially and chemically safe to drink. Yet waterborne pathogens are the major cause of gastroenteritis-type diseases within the population. While bacterial contaminants are effectively controlled by conventional water treatment and disinfection, protozoa and viruses can break through. Private supplies, especially shallow wells in agricultural areas, are particularly at risk from agricultural runoff. Particularly at risk are the young, elderly and immuno-compromised. Both primary and opportunistic bacterial infections that are not life-threatening to healthy individuals, can result in significant mortalities amongst those with AIDS (e.g. cryptosporidiosis and salmonellosis) requiring new management strategies for water utilities (Wong *et al.*, 1994). Most primary bacterial pathogens, although commonly isolated from water resources, rarely cause disease in the developed world. However, major outbreaks do occur in developing countries, or in war zones and disaster areas where normal sanitation and water supplies are damaged. For example, over 500 000 cases of cholera were reported in Peru during 1991–4. New emerging pathogens are a particular concern. For example, there were major drinking water-related outbreaks of *E. coli* 0157:H7 in 1998, 1999 and 2000 in the USA affecting thousands of individuals and resulting in several deaths. Haemolytic uraemic syndrome is a serious complication following infection by this strain of *E. coli* with 2% and 7% of the elderly and under-5-year olds developing it respectively.

Microbial contamination of drinking water resources requires a direct route from carrier to the supply. In the USA, livestock generates 130 times more waste than humans resulting in 70% of all surface waters affected by agricultural runoff (US Senate Committee of Agriculture, Nutrition and Forestry, 1997). Crypto-sporidiosis was originally thought to be a zoonosis contracted by direct contact with faeces from livestock. The early reported outbreaks confirmed this being caused by surface runoff from agricultural land. However, a number of factors made this protozoan a particularly successful waterborne pathogen (Table 13.3) including its lack of host specificity, now being recorded in over 150 different species. It can also be readily spread from person to person and is auto-infective. *Cryptosporidium parvum* is now known to exist as two distinct genotypes or species. Genotype 1, or H for human, is almost exclusively a parasite of humans and has been renamed *C. hominis*. This genotype is most aggressive in humans and was associated with the cryptosporidiosis outbreak in Milwaukee in 1993. Genotype 2 (or C for calf) is now identified as the traditional *C. parvum* that infects a wide range of animals including humans. Both species are now ubiquitous in water resources and the oocysts can survive for up to 18 months depending on the temperature. Its small size and resistance to both chlorination and ultraviolet (UV) radiation has resulted in cryptosporidiosis now being the most frequent and serious cause of drinking water-associated gastroenteritis both in the USA and Europe. Protozoa seem particularly problematic pathogens in drinking water with *Giardia* causing over one million cases of severe and ten million cases of mild giardiasis annually (Smith, 1996).

Due to water scarcity, treated sewage effluents are increasingly being reused for supply purposes (Section 4.5) (Figure 4.9). While such effluents are disinfected in the USA, this is not the general practice elsewhere. Also some viruses and the cysts or oocysts of pathogenic protozoa are resistant to such disinfection. Therefore the water being abstracted for supply may contain elevated numbers of faecal pathogens such as *Campylobacter* spp. and *Salmonella* spp. There is a serious risk to all water resources from both animal and human pathogens and all water resources should therefore be considered as being potentially microbially contaminated and unsafe for human consumption without appropriate treatment and disinfection.

Drinking water is protected from microbial contamination by a mixture of resource protection and treatment strategies. Used water is treated at the wastewater treatment plant to remove pathogens, with the final effluent often disinfected, using UV radiation in Europe or chlorination in the USA, to ensure that pathogens are eliminated if discharged into surface waters that are to be reused for supply. This is particularly important with the rise of antibiotic strains of bacteria associated with effluents and surface waters. Better farm waste management and a range of structural and management practices to control surface runoff minimizes microbial and chemical contamination of water resources used for supply purposes (Gray, 2005). This water then undergoes appropriate treatment before being supplied to consumers as drinking water depending on the degree of contamination. This may include pre-chlorination as well as post-chlorination or other form of disinfection. Continuous microbial monitoring ensures a high microbial quality of water supplied to the consumer (Chapter 19). Water supplies are also vulnerable to terrorism and this is examined in Chapter 30.

References

Bitton, G. A. (2005). *Wastewater Microbiology*, 3rd edn. New York: John Wiley and Sons.

Blewett, D. A., Wright, S. E., Casemore, D. P., Booth, N. E. and Jones, C. E. (1993). Infective dose size studies on *Cryptosporidium parvum* using genotobiotic lambs. *Water Science and Technology*, **27**(3–4), 61–4.

Browning, J. R. and Ives, D. G. (1987). Environmental health and the water distribution system: a case history of an outbreak of giardiasis. *Journal of the Chartered Institution of Water and Environmental Management*, **1**, 55–60.

Craun, G. F. (1986). *Waterborne Diseases in the United States*. Boca Raton, FL: CRC Press.

Craun, G. F. (1991). Cause of waterborne outbreaks in the United States. *Water Science and Technology*, **24**(2), 17–20.

Cubitt, D. W. (1991). A review of the epidemiology and diagnosis of waterborne viral infections. *Water Science and Technology*, **24**(2), 197–203.

DeZuane, J. (1990). *Handbook of Drinking Water Quality: Standards and Control*. New York: Van Nostrand Reinhold.

Fayer, R. and Ungar, B. L. P. (1986). *Cryptosporidium* species and cryptosporidiosis. *Microbiological Reviews*, **50**, 458–83.

Galbraith, N. S., Barrett, N. and Stanwell-Smith, R. (1987). Water disease after Croydon: a review of waterborne and water associated disease in the UK 1937–86. *Water and Environmental Management*, **1**, 7–21.

Geldreich, E. E., Fox, K. R., Goodrich, J. A. *et al.* (1992). Searching for a water supply connection in the Cabool, Missouri disease outbreak of *Escherichia coli* O157:H7. *Water Research*, **26**, 1127–37.

Gerba, C. P. and Rose, J. B. (1990). Viruses in source and drinking water. In *Drinking Water Microbiology*, ed. G. A. McFeters. New York: Springer-Verlag, pp. 380–96.

Gould, D. J. (1977). *Gull Droppings and Their Effects on Water Quality*. Technical Report 37. Stevenage: Water Research Centre.

Gray, N. F. (2004). *Biology of Wastewater Treatment*, 2nd edn. London: Imperial College Press.

Gray, N. F. (2005). *Water Technology: An Introduction for Environmental Scientists and Engineers*, 2nd edn. Oxford: Elsevier.

HMSO (1989). *Guidance on Safeguarding the Quality of Public Water Supplies*. London: Department of the Environment, HMSO.

Hunter, P. R., Waite, M. and Ronchi, E. (eds). (2002). *Drinking Water and Infectious Disease*. Boca Raton, FL: CRC Press.

Jones, K. (1994). Inside science: 73. Waterborne diseases. *New Scientist*, **143**(1993), 1–4.

Jones, K. and Telford, D. (1991). On the trail of the seasonal microbe. *New Scientist,* **130** (1763), 36–39.

LeChevallier, M. W. (1990). Coliform regrowth in drinking water: a review. *Journal of the American Water Works Association,* **82**(11), 74–86.

McFeters, G. A. (ed.) (1990a). *Drinking Water Microbiology*. New York: Springer-Verlag.

McFeters, G. A. (1990b). Enumeration, occurrence, and significance of injured indicator bacteria in drinking water. In *Drinking Water Microbiology*, ed. G. A. McFeters. New York: Springer-Verlag, pp. 478–92.

Morris, R. W., Grabow, W. O. K. and Dufour, A. P. (eds.) (1993). *Health-Related Water Microbiology 1992*. Oxford: Pergamon Press.

Mukhopadhyay, A. K., Saha, P. K., Garg, S. *et al.* (1995). Distribution and virulence of *Vibrio cholerae* belonging to serotypes other than O1 and O139: a nationwide survey. *Epidemiology and Infection,* **114**, 65–70.

O'Mahony, M. C., Noah, N. D., Evans, B., Harper, D. and Rowe, B. *et al.* (1986). An outbreak of gastroenteritis on a passenger cruise ship. *Journal of Hygiene, Cambridge*, **97**, 229–35.

Reasoner, D. (1992). *Pathogens in drinking water – are there any new ones?* Washington, DC: US Environmental Protection Agency.

Richardson, A. J., Frankenberg, R. A. and Buck, A. C. (1991) An outbreak of waterborne cryptosporidiosis in Swindon and Oxfordshire. *Epidemiology and Infection,* **107**, 485–95.

Rose, J. B. (1990). Emerging issues from the microbiology of drinking water. *Water Engineering and Management,* July, 23–8.

Rose, J. B. (1999). Environmental ecology of *Cryptosporidium* and public health implications. *Annual Review of Public Health*, **18**, 135–61.

Singh, A. and McFeters, G. A. (1992). Detection methods for waterborne pathogens. In *Environmental Microbiology*, ed. R. Mitchell. New York: John Wiley and Sons.

Skirrow, M. B. (1982). *Campylobacter enteritus*: the first five years. *Journal of Hygiene, Cambridge*, **89**, 175–84.

Smith, H. V. (1996). Detection of *Giardia* and *Cryptosporidium* in water: current status and future prospects. In *Molecular Approaches to Environmental Microbiology*, ed. R. W. Pickup and J. R. Saunders. Chichester, UK: Ellis Horwood, pp. 195–225.

Solbe, L. F. and De, L. G. (1983). *Water Authority Regulations Concerning the Use of Angler's Bait on Potable Supply Reservoirs*. Technical Report 188. Medmenham: Water Research Centre.

USEPA (1999). *25 years of the Safe Drinking Water Act: History and Trends*. EPA-816-R-99-007. Washington, DC: US Environmental Protection Agency.

US Senate Committee of Agriculture, Nutrition and Forestry (1997). *Animal Waste Pollution in America: An Emerging National Problem*. Washington, DC: Minority Staff of the US Senate Committee of Agriculture, Nutrition and Forestry, US Government Printing Office.

Wilkins, M. (1993). The fight against *Cryptosporidium. Water Bulletin,* **544** (12 February), 12.

Wong, S. S. Y., Yuen, K. Y., Yam, W. C., Lee, T. Y. and Chau, P. Y. (1994). Changing epidemiology of human salmonellosis in Hong Kong, 1982–1993. *Epidemiology and Infection*, **113**, 425–34.

PART III
PROBLEMS ARISING FROM WATER TREATMENT

Chapter 14
Water treatment

14.1 Introduction

Water treatment plants must be able to produce a finished product of consistently high quality regardless of the demand. With the exception of particularly pure groundwaters, all water supplied for drinking requires purification. Although, in theory, the dirtiest water can be purified to drinking water quality, in practice treating even relatively pure water to produce finished water of consistent quality, and of sufficient volume, is technically very difficult. Water treatment consists of a range of unit processes that are usually operated in series (Stevenson, 1998). These are listed in Table 14.1, although it is unusual for all of them to be used at any one particular plant. The cleaner the raw water, the smaller the number of steps or unit processes that are required, and hence the overall cost of the water is less. The most expensive operations are sedimentation and filtration in conventional treatment, whereas specialized processes for softening water or removing specific contaminants such as nitrates or pesticides using membrane processes or activated carbon can be very expensive (Parsons and Jefferson, 2006).

Groundwater is generally much cleaner than surface waters and so does not require the same degree of treatment, apart from aeration and disinfection, before supply. Naturally occurring substances, which may need to be reduced or eliminated in groundwaters, include iron (Section 9.1), hardness (if in excess of $300\,mg\,l^{-1}$) (Chapter 10) and carbon dioxide. Substances originating from humans that are becoming increasingly common in groundwaters and that need to be removed include nitrates (Chapter 5), pathogens (especially bacteria and viruses) (Chapter 13), trace amounts of organic compounds such as pesticides (Chapter 6) and endocrine-disrupting compounds and PPCPs (Chapter 7). The water industry tries to obtain the cleanest water possible for supply, although the volume and consistency of supply are the major factors in the selection of a resource. The cleanest of the suitable resources available is usually selected; however, it may be necessary to blend several resources to dilute unwanted contaminants to below harmful concentrations (Section 5.6).

Table 14.1 *Main unit processes in water treatment in general order of use*

Treatment category	Unit process
Intake	
Pretreatment	Coarse screening
	Pumping
	Storage
	Fine screening
	Equalization
	Neutralization
	Aeration
	Chemical pretreatment
Primary treatment	Coagulation
	Flocculation
	Sedimentation
Secondary treatment	Rapid sand filtration
	Slow sand filtration
Disinfection	
Advanced treatment	Adsorption
	Activated carbon
	Fe and Mn removal
	Membrane processes
Fluoridation	
Distribution	

14.2   Unit processes

The selection of unit processes depends on the quality of the raw water entering the treatment plant and the quality of the finished water required. A general layout of a large urban water treatment plant is shown in Figure 14.1. Only a brief description of the major unit processes has been given below. Detailed reviews of water treatment are published elsewhere by Twort *et al.* (1994) and Vigneswaran and Visvanathan (1995).

14.2.1   Preliminary screening

Large-scale treatment plants, which serve large conurbations, are rarely close to the sources of water except where direct abstraction from rivers is practised. Most upland reservoirs are many miles from the point of consumption, so the raw water must be conveyed to the treatment plant either by pipe or open channel. The raw water is passed through a set of coarse screens to remove gross solids such as weeds, sticks and other large material before starting its journey to the plant. This is mainly carried out to protect pipes from becoming blocked or pumps being damaged.

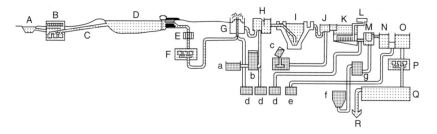

Figure 14.1 Schematic diagram of a large water treatment plant showing all the unit processes in series. A = river, B = intake and pumping station, C = rising main, D = reservoir, E = mechanical strainers, F = raw water pumping station, G = aeration tower, H = coagulant flash mixers, I = sedimentation tanks, J = flash mixer, K = rapid gravity filter, L = air blowers, M = flash mixer, N = chlorine retention chamber, O = clean water tank, P = outgoing pump house, Q = storage reservoir, R = to supply distribution system, a = storage, b = coagulant solution, c = activated carbon slurry, d = chlorine supply, e = sulphonation, f = lime, g = lime slurry.

14.2.2   Storage

Raw water is pumped from the intake to the storage reservoir where it is often stored to improve quality before treatment, as well as ensuring adequate supplies at periods of peak demand. There are a number of natural processes at work during storage that all significantly improve water quality. The filtration process can only deal effectively with suspended solids concentrations of less than 10 mg l^{-1}. For waters with suspended solids concentrations in excess of 50 mg l^{-1}, storage is necessary to allow this particulate matter to settle out of suspension. Ultraviolet radiation is another important natural treatment process. In the upper zones of the reservoir it destroys harmful bacteria and some other pathogenic organisms, bleaches colour and oxidizes some organic impurities that are responsible for taste and odour problems (Gray, 2004). Excessive hardness can also be reduced by the liberation of carbon dioxide by algae present in the reservoir during summer. This converts the hydrogencarbonates into carbonates, which are then precipitated out of solution (Table 14.2).

Storage can largely eliminate variations in water quality that can occur in surface waters, especially due to floods, or variations in the dilution of any pollutants present. Although the advantages largely outweigh the disadvantages, there are a number of important problems in the operation of storage reservoirs; for example, atmospheric pollution and fallout, pollution from birds (especially roosting gulls), algal development (if stored for longer than ten days) and, of course, like all reservoirs they take up considerable areas of land so that their construction can be contentious. In deeper reservoirs the major problem is thermal stratification with the deeper water becoming anaerobic (Section 4.3).

14.2.3   Screening and microstraining

Before treatment the raw water is screened again, this time through fine screens. If considerable amounts of fine solids or algae are present, then microstraining may be used before the next stage. Microstrainers consist of fine stainless-steel mesh drums, with up to 25 000 apertures cm^{-2}. The microstrainer is a rotary drum that is partially immersed in the water. As it rotates the head difference drives the water through the micromesh, which strains out the particles, especially the algae. The apertures normally used in water treatment are either

Table 14.2 *The improvement brought about by storage in the quality of water abstracted from the River Great Ouse at Grafham prior to full treatment. Adapted with permission from Anglian Water*

	Before storage	After storage
Colour (Hazen)	30	5
Turbidity (NTU)	10	1.5
Ammoniacal N (mg l^{-1})	0.3	0.06
Biochemical oxygen demand (mg l^{-1})	4.5	2.5
Total hardness (mg l^{-1})	430	280
Presumptive coliforms (100 ml l^{-1})	6500	20
Escherichia coli most probable number (100 ml l^{-1})	1700	10
Colony counts (ml l^{-1})		
3 days at 20 °C	50 000	580
2 days at 37 °C	5000	140

25 or 35 μm in diameter. When clean, no micro-organisms are retained but the particles and algae retained on the micromesh form a straining layer, allowing some retention of micro-organisms to occur, although this is incidental and cannot be either predicted or relied upon. Microstraining produces a wash-water in which all the strained particles, including the algae, are concentrated. It can represent as much as 3% of the total volume of water passing through the strainer and needs to be treated separately. There is evidence to show that such waters can be potentially rich in pathogens, especially cysts and oocysts of protozoans, and so great care must be taken not to allow this wash-water to contaminate the finished water.

14.2.4   Aeration

Water from groundwater resources, from the bottom of a stratified lake or reservoir, or from a polluted river, will contain very little or no dissolved oxygen. If anaerobic water is allowed to pass through the treatment plant it will damage or affect other unit processes, in particular filtration and coagulation. Therefore the raw water needs to be aerated before it is treated further. This is achieved by bringing the water into contact with air. Although bubbling air through water is the way this is often done on a small scale, this is prohibitively expensive at larger plants because of the vast volumes of water that are treated daily. The simplest method is a cascade or fountain system. In a cascade, the water pours down a tower structure, which ensures reaeration through excessive turbulence. Alternatively, the pressure of the water as it is forced out of the reservoir by the weight of the water above it allows natural jets or fountains of water to spray up into the air. The jets do not rise high into the air as they are normally wide to cope with the large volumes of water being processed. Many such jets are partially enclosed to reduce evaporation losses. These two methods are visually spectacular and are often used to great effect in the overall design of treatment plants. There are many other types of aeration systems used, including packed towers and diffusers (Twort *et al.*, 1994).

Apart from ensuring optimum treatment within the treatment plant, aeration also provides oxygen for purification and significantly improves the quality, especially the taste, of water. Aeration also reduces certain objectionable odours, and reduces the corrosiveness of water by driving off any excess carbon dioxide gas present, thus raising the pH. Aeration cannot, however, reduce the corrosive properties of acid waters alone and neutralization with lime may be needed. Iron and manganese can also be removed from solution by aeration. These metals are only soluble in water with a pH of less than 6.5 and in the absence of dissolved oxygen, and so are common in certain groundwaters. Aeration oxidizes the soluble metal salts into insoluble metal hydroxides, which can then be removed by flocculation or filtration.

14.2.5 Coagulation

After fine screening most of the remaining suspended solids will be very small, usually $<10\,\mu m$. These colloidal solids are so small that they may never settle out of suspension naturally. Colloidal solids are particles of clay, metal oxides, large protein molecules and micro-organisms. All small particles tend to be negatively charged, and, as like charges repel, all the negatively charged colloidal particles in the water tend to repel one another, preventing aggregation into larger particles that could then settle out of suspension. A particle $100\,\mu m$ in diameter will settle 200 000 times faster than a particle $0.1\,\mu m$ in diameter, although the exact settling velocity depends on the specific gravity of the particle as well as its size (Table 14.3). The removal of colloidal matter is a two-step process, coagulation followed by flocculation.

A coagulant is added to the water to destabilize the particles and to induce them to aggregate into larger particles known as flocs. A variety of coagulants are used. The most common salts are aluminium sulphate (alum), aluminium hydroxide, polyaluminium chloride, iron (III) chloride, iron (III) sulphate and lime. Iron (II) sulphate ($FeSO_4.7H_2O$), known as copperas, is also used although it is generally mixed with chlorine to give a mixture of iron (III) chloride ($FeCl$) and iron (III) sulphate [$Fe(SO_4)_3$], known as chlorinated copperas. There is a rapidly increasing range of synthetic organic polymers also available. These include polyacrylamides, polyethylene oxide and polyacrylic acid. The use of coagulants and some of the problems arising from their use are considered in Chapter 15. The actual mechanisms of coagulation are complex and include adsorption, neutralization of charges and entrainment within the physical–chemical matrix formed (Twort *et al.*, 1994).

Table 14.3 *Settling velocity of particles as a function of size only*

Particle size (μm)	Settling velocity (m h^{-1})
1000	600
100	2
10	0.3
1	0.003
0.1	0.00001
0.01	0.0000002

The amount of coagulant added to the water is critical. Too little results in ineffective coagulation so that the filtration apparatus may become blocked too rapidly; while too much coagulant can lead to the excess chemical being discharged with the finished water. The coagulant is added to the process stream at a specific concentration (30–100 mg l^{-1}) using a mixing device. Either a mixing chamber (flash-mixer) is used with a high-speed mixer, or the coagulant is added to the water in a mixing channel (using a hydraulic jump within a measuring flume) to induce mixing. Coagulation is complete within one minute of addition. Metal salts react with the alkalinity in the water to produce an insoluble metal hydroxide precipitate, which enmeshes the colloidal particles. The precipitate formed appears fluffy when seen under a low-power microscope, and apart from enmeshing any particles in the water it also adsorbs some of the dissolved organic matter present. As already stated, all natural particles in water have a very small negative electrical potential on their surfaces, usually less than -20 mV. As the hydroxide flocs carry a small positive electrical potential, there is a mutual attraction to the particles in the water. Coagulant aids, which are organic polymers with either positive or negative charges, are sometimes used to improve coagulation. They are a supplement to the normal coagulant and can cause problems in slow sand filtration.

14.2.6   Flocculation

When small particles collide in a liquid some naturally aggregate to form larger particles. As the larger particles settle they overtake smaller particles that are settling at much slower rates. If they collide, then the smaller particle will aggregate onto the larger particle.

The chance of particles colliding can be significantly increased by gently mixing the water, a process known as flocculation. When there is a high concentration of colloidal particles, then flocculation can be effective on its own. However, at the lower concentrations usually found in water resources a coagulant must be used. In the water treatment process, flocculation therefore follows chemical addition (coagulation). During this mixing larger flocs are produced that are easily removed during clarification. Flocculation occurs naturally by Brownian motion (perikinetic flocculation); however, for particles larger than 1 μm this is very slow and mechanical mixing devices are required (paddle or turbine mixers) to increase the rate of collisions (orthokinetic flocculation).

14.2.7 Clarification

The flocs formed by the addition of a coagulant or by flocculation are removed by settlement. The process is different from normal sedimentation processes found in wastewater treatment and in industry where the water flows slowly along or across a tank allowing the particles to settle. In water treatment the water flows in an upward direction from the base of the tank. The flocs, which are heavier than water, settle towards the bottom, so the operator must balance the rate of settling against the upward flow of water to ensure that all the particles are held within the tank as a thick sludge blanket. There is a layer of clear, clarified water at the surface that overflows a simple weir to the next step of the treatment process. These tanks are referred to as floc blanket clarifiers and are extremely effective. As the flocs rise through the blanket further flocculation occurs, which increases the floc density. To maintain the sludge blanket at the required height, and the more sludge maintained within the tank the more efficient the separation will be, the excess sludge must be discharged from the tank. Sludge removal, or bleeding as it is commonly referred to, may be carried out continuously or intermittently. The sludge is a concentrated mixture of all the impurities found in the water, especially bacteria, viruses and protozoan cysts. It must therefore be handled carefully and disposed of safely. The volume of sludge is fairly large and is equivalent to between 1.5% and 3.0% of the flow through the clarifier. There are many different designs of sedimentation tanks, including the use of inclined plates to encourage settlement and the development of parallel plate and tube settlers (Twort *et al.*, 1994).

14.2.8 Filtration

After clarification, the water contains a small amount of fine solids ($<10\,\mathrm{mg\,l^{-1}}$) and soluble material. Although some of these particles may have been in the natural raw water, many will have been formed during the coagulation process. Another process, filtration, is required to remove this residual material. The filters contain layers of sand (or anthracite) and gravel graded to ensure effective removal. In their simplest form, filters allow the downward passage of water through layers of fine sand, which are supported on layers of coarser gravels. Pipes at the base of the filter, underdrains, collect the filtered water. Particles that are removed by the sand clog the surface and reduce the rate of flow of water through the filter. Therefore the filter must be cleaned intermittently. This is done by either scraping the surface layer of sand containing the retained particles from the surface, or where possible by pumping water through the filter in the reverse direction under pressure. This washes all the small retained particles from the sand, a process known as backwashing. The efficiency of the filters depends on a number of factors such as the nature and quality of the material to be removed from the water, the size and shape of the filter media, and the flow rate of water through the filter. There are two types of filter used in water treatment: rapid and slow sand filters.

Rapid sand filters contain coarse grades of quartz sand (1 mm diameter) so that the gaps between the grains are comparatively large. This ensures that the water passes rapidly through at rates of between 5 and $10 \, m^3 \, m^{-2} \, h^{-1}$, operating at about 50 times the rate of a slow sand filter. Rapid sand filters are usually deeper, between 3.0 and 3.5 m in depth, and consist of sand, anthracite plus sand, or a similar material such as activated carbon and sand. These filters are used for water that has previously been treated by coagulation and sedimentation, and are less effective than slow sand filters in retaining very small solids. Therefore bacteria, taste and odours are less effectively removed than by slow sand filtration. Because of the high loading rates, rapid sand filters are much smaller and more compact than slow sand filters. In use, the amount of water passing through the filter each hour gradually decreases due to the gaps in the sand becoming blocked by the retained solids. When the filtration rate becomes too low the filter must be cleaned. Depending on the design and loading rates this is carried out several times a day or every few days by blowing air up through the layer of sand to scour material free from the grains of sand, and then washing away the solids by backwashing with clean water. Immediately after backwashing there is an increased risk of pathogenic micro-organisms penetrating the filter. It may take up to 20 minutes for rapid sand filters to reach their optimum performance in terms of water quality. The water used for backwashing should not be recycled through the plant. Rapid sand filters are of two main designs, open-topped gravity-fed systems or more efficient closed systems where the filter is enclosed within a metal shell and the water is forced through the sand by pressure.

In contrast, slow sand filters have a layer of much finer quartz sand (0.5–2.0 m depth) overlying coarse sand or gravel (1.0–2.0 m depth) that physically removes fine solids (Visscher, 1988). However, these filters, apart from physical removal, also provide a degree of biological treatment. The top 2 mm of the sand is host to a mixture of algae and nitrifying bacteria (autotrophic layer). Here nitrogen and phosphorus are removed and oxygen is released. Below this autotrophic layer is a thicker layer of sand, up to 300 mm, which is colonized by bacteria (heterotrophic layer) and other micro-organisms that remove the residual colloidal and soluble organic material from the water (Duncan, 1988). Water treatment in sand filters is therefore a combination of physical and biological activity, with pathogenic bacteria, taste and odour (due to algae and organic compounds) largely removed. The quality of the water is excellent, unlike that from a rapid sand filter where further treatment may be required as there is little biological activity and only the larger solids are retained. The rate of filtration in slow sand filters is controlled by gravity alone and so combined with the small gaps between the particles of sand, water only passes through such filters slowly. On average the rate of filtration is between 0.1 and $0.3\,\mathrm{m^3\,m^{-2}\,h^{-1}}$, so a very large area of filters is required. They are expensive to operate because the dirt layer that collects on the surface of the sand and impedes drainage must be mechanically skimmed off after the filter has been drained. Unlike rapid sand filters, slow sand filters cannot be backwashed. After 2–3 months the sand that has been removed during the skimming operations must be replaced to maintain the required depth of fine sand. This makes slow sand filtration labour intensive and operationally expensive.

14.2.9 pH adjustment

The pH of the finished water may require adjusting so that it is neither too acidic, which may corrode metal distribution pipes and household plumbing, or too alkaline, which will result in the deposition of salts within the distribution system causing a reduction in flow. The pH may be adjusted at a number of unit processes, such as coagulation, to ensure maximum efficiency. Alkalis such as lime, sodium carbonate or caustic soda are used to increase the pH, whereas acids are used to decrease it.

14.2.10 Fluoridation

Fluoridation is very controversial and is where trace quantities of fluoride are continuously added to drinking water to improve the community resistance to dental caries. This is fully examined in Chapter 17.

14.2.11 Disinfection

Although slow sand filters are extremely efficient at removing bacteria, and the coagulation process is good at removing viruses, the finished water still contains pathogenic viruses and bacteria that need to be removed or destroyed. In practice it is impossible to sterilize water, to kill off all the micro-organisms present, due to the very high concentration of chemicals required, which would make the water very unpleasant and possibly dangerous to drink. Therefore the water is disinfected, rather than sterilized, by using one of the disinfection methods such as chlorination, ozone or ultraviolet (UV) radiation to ensure that pathogens are kept at safe levels. Of the three methods of disinfection chlorination is by far the most widely used.

Ozone has powerful oxidation properties and tends to be used where the natural water contains materials that would combine with chlorine to form unacceptable odours or tastes. Ozone, which is often used in combination with activated carbon, can eliminate all bacteria at a dose rate of 1 ppm within 10 minutes, and can also reduce colour, taste and odour. Apart from being more expensive than chlorination and that it has to be manufactured on site, the lack of residual disinfection action within the distribution mains is the major drawback (Masschelein, 1982). This allows biological growth to develop which causes taste and odour problems. Therefore low-level chlorination is often used after ozonation to prevent such growth. When waters containing bromide are oxidized, using either ozone or hydrogen peroxide, bromate is formed, which is widely considered to be a genotoxic carcinogen (Bull and Kopfler, 1991) (Chapter 18).

Ultraviolet radiation is emitted from special lamps and is effective in killing all micro-organisms as long as the exposure time is adequate. Ultraviolet radiation is electromagnetic energy in the range 250–265 nm. To be effective this energy must reach the nucleic acid within the largest organism to induce structural changes that will prevent replication of the pathogen. The lamps are enclosed in stainless-steel reaction chambers (Wolfe, 1990, Kruithof *et al.*, 1991). They are used at small plants or for institutions where the chance of contamination after treatment is unlikely. Very effective household UV sterilization units are also available for single dwellings (Chapter 29)

Chlorine and its compounds are readily available in gas, liquid or solid forms. It is easy to add to water, has a high solubility (7000 mg l^{-1}) and is cheap. The residues it leaves in solution continue to destroy pathogens after the water has left the treatment plant and as it travels through the distribution network. While toxic to micro-organisms, chlorine is not thought to be harmful to humans at the concentrations used, although a possible link with bladder cancer has been suggested. In the USA the concentration of chlorine used in drinking waters is significantly higher than in Europe, so at the concentrations used in the UK chlorine should be considered totally safe. Chlorine is, however, a dangerous chemical to handle in its concentrated form and produces a poisonous gas.

The chemistry of chlorine is complex. Essentially chlorine (Cl_2) reacts with water to form hypochlorous acid (HOCl) and hydrochloric acid (HCl):

$$Cl_2 + H_2O \rightarrow HCl \qquad + HOCl$$
$$\updownarrow \qquad\qquad \updownarrow$$
$$H^+ + Cl^- \qquad H^+ + OCl^-.$$

In dilute solution this reaction is very rapid and is normally complete within one second. Hypochlorous acid is a weak acid that readily breaks down (dissociates) into the hypochlorite ion (OCl^-). This occurs almost instantaneously. Both hypochlorous acid and the hypochlorite ion act as disinfectants, although the former is about 80 times more effective than the latter. A chemical equilibrium develops between the two forms, although dissociation is suppressed as the pH decreases. In practice, at about pH 9, 100% of the chlorine is in the hypochlorite form, about 50% at pH 7.5 and at pH 5 or less it is all present as hypochlorous acid. Disinfection is therefore much more effective at acidic pH.

Chlorine is not as aggressive a disinfectant as ozone and there are a number of pathogenic micro-organisms that are resistant to chlorination. Effectively eliminating all the coliforms present does not necessarily indicate that all other pathogenic micro-organisms have also been destroyed (LeChevallier, 1990). Temperature also affects chlorination, its efficiency decreasing at lower temperatures. Many substances will readily combine with chlorine, especially reducing agents and unsaturated organic compounds. These compounds exert an immediate chlorine demand that must be satisfied before chlorine becomes available for disinfection. Excess chlorine must therefore be added to satisfy this demand as well as to leave a residual amount in the water long enough to penetrate and destroy all the micro-organisms present. Suspended organic and inorganic matter absorbs chlorine, whereas iron and manganese neutralize chlorine by forming insoluble chlorides. Thus it is better to remove these problematic substances by appropriate treatment before disinfection rather than increasing the dose of chlorine. Chlorine is very reactive and combines with almost everything in the water. Research has shown that chlorine reacts with organic compounds that occur naturally in water to form disinfection by-products, many of which are toxic and carcinogenic (Chapter 18). Chlorination is thus unsatisfactory when waters are rich in organic acids, such as those from peaty upland areas.

A major problem is the presence of ammonia. This reacts readily with chlorine to form a range of compounds known as chloramines, the exact nature of which depends on the relative concentrations of the two chemicals and the pH (White, 1972). Three chloramines are formed: these are monochloramine (NH_2Cl), dichloramine ($NHCl_2$) and trichloramine or nitrogen trichloride (NCl_3).

$$NH_4^+ + HOCl \rightarrow NH_2Cl + H_2O + H^+$$
Monochloramine

$$NH_2Cl + HOCl \rightarrow NHCl_2 + H_2O$$
Dichloramine

$$NHCl_2 + HOCl \rightarrow NCl_3 + H_2O$$
Trichloramine

When ammonia is present the dose of chlorine must be increased to ensure that sufficient excess chlorine (residual) is left in the water to destroy the pathogens. However, combined chlorine (combined residuals) such as mono- and dichloramines retain some of their disinfection potential and, although less effective than free chlorine present as hypochlorous acid and hypochlorite ions

Figure 14.2 Breakpoint chlorination curve.

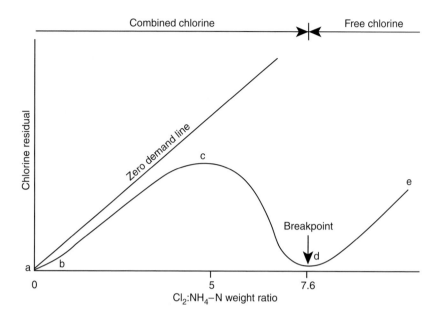

(free residuals), they have long-lasting disinfection properties. Therefore it has become the practice at some treatment plants to add ammonia at the chlorination stage to give a combined rather than a free residual effect (i.e. chloramination). Combined residual chlorine requires a contact time of a hundred times longer than free residual chlorine to achieve the same degree of elimination of pathogens. When chlorine is added to water containing ammonia, which is either present naturally or added deliberately to produce combined chlorine rather than a free chlorine residual, then a breakpoint curve is produced (Figure 14.2). If inorganic reducing agents such as iron and manganese are present then these will exert a chlorine demand that does not produce a residual and so is seen as a flat line at the start of the curve (a-b). Once satisfied the chlorine residual in the water is a function of the chlorine:ammonia weight ratio. When the ratio of chlorine: ammonia is $< 5:1$ at pH 7–8, then the chlorine residual is all monochloramine (b-c). Increasing the chlorine:ammonia ratio results in some of the monochloramine reacting to form small amounts of dichloramine. As the ratio approaches 7.6:1, the chloramines are oxidized by the excess chlorine to nitrogen gas, resulting in a rapid loss of residual chlorine in the water (c-d)

$$NH_2Cl + NHCl_2 \leftrightarrow N_2 + 3H^+ + 3Cl^-$$

After the breakpoint there is no ammonia left to react with the chlorine so the residual concentration increases in proportion to the amount of chlorine applied (d-e). The disinfection potential, before the breakpoint, is therefore reached due to combined residuals (mono- and dichloramines) and after the breakpoint due to free chlorine, although trace amounts of dichloramine and trichloramine may remain at lower pH values (Figure 14.2). This is excellently reviewed by Bryant *et al.* (1992). Excess ammonia will find its way into the distribution main if the chlorine:ammonia ratio is too low, resulting in nitrification and the formation of nitrate, or more importantly, nitrite (Chapter 5).

Another reason for adding ammonia is to prevent the chlorine from reacting with trace amounts of organic compounds in the water, such as phenol, and forming unpleasant tastes (Section 16.1). When ammonia is present in the raw water at concentrations in excess of 0.3 mg l^{-1} it can cause an unpleasant taste and odour if it is allowed to degrade anaerobically during treatment. Therefore it is not unusual for the raw water to be chlorinated as it enters the treatment plant to control the bacteria that cause the problem, although pre-chlorination will upset the biological activity in slow sand filters.

Chlorine is relatively easy to handle and cost effective, although chlorine dioxide is used under alkaline pH conditions, since under these conditions it will not form combined residuals with ammonia. As a general guide, the cleaner the water the larger the residual effect. The amount of chlorine used depends on the rate of flow and the residual chlorine concentration required, which is usually 0.2–0.5 mg l^{-1} after 30 minutes. The problem of inadequate residual chlorine allowing pathogens to survive in distribution systems, and the problems of the contamination of supplies after treatment is dealt with in Chapter 25.

Super-chlorination is used to destroy problematic odours and tastes. Excess chlorine is added to the water to oxidize any remaining organic compounds. The excess chlorine is then removed after the required contact time by the addition of sulphur dioxide, a process known as sulphonation.

14.2.12   Softening and other tertiary (advanced) processes

Conventional water treatment is unable to remove a number of soluble inorganic and non-biodegradable organic substances from water. Soluble inorganic material is removed by precipitation or ion-exchange, whereas organic substances that are not biologically degraded can be removed by adsorption using activated carbon. Membrane filtration, including reverse osmosis is also widely used for the removal of both inorganic and organic contaminants.

Chemical precipitation is more widely known as precipitation softening. It is used primarily to remove or reduce the hardness in water that is caused by excessive salts of calcium and magnesium. The cause of hardness and the problems caused by excessively hard waters are examined in detail in Chapter 10. Precipitation softening converts the soluble salts into insoluble ones, so that they can be removed by subsequent sedimentation. Lime or soda ash is normally used to remove the hardness, although the exact method of addition depends on the type of hardness present. Lime is most widely used but, like soda ash, produces a large volume of sludge that has to be disposed of (Pontius, 1990). Softening using ion-exchange reactions is becoming increasingly common.

Ion-exchange separation uses a resin, usually natural zeolites which are sodium aluminosilicates (Hill and Lorch, 1987). The zeolites exchange sodium ions for calcium and magnesium ions. The hardness is therefore removed and bound to the resin while sodium, which does not cause hardness, takes the place of calcium and magnesium in the water making it softer. Resins are housed in an enclosed metal tank similar to a pressurized rapid sand filter in design. Once the resin is exhausted and no more sodium ions are available for exchange, the exchange column is taken out of service and the resin regenerated. This is done by pumping concentrated salt solution through the resin, which removes all the calcium and magnesium held in the column, replacing them with sodium again ready for further softening. The regeneration process produces a concentrated solution of unwanted calcium and magnesium chlorides that must be disposed of. The reactions are as follows:

Softening

$$Ca^{2+}/Mg^{2+} + Na_2 \text{ (zeolite)} \rightarrow Ca/Mg \text{ (zeolite)} + 2Na^+$$

| Hard water | Resin | Resin | Soft water |

Regeneration

$$Ca/Mg \text{ (zeolite)} + 2NaCl \rightarrow Na_2 \text{ (zeolite)} + CaCl_2/MgCl_2$$

| Exhausted resin | Salt solution | Regenerated resin | Waste solution |

Synthetic ion-exchange materials have been developed to give much higher exchange capacities than natural compounds such as zeolites. Some organic ion-exchange resins can be used to remove anions such as nitrates, sulphates, chlorides, silicates and carbonates. Ion-exchange is increasingly used to remove nitrates from drinking water, although biological denitrification is also widely practised.

Trace concentrations of synthetic organic compounds, especially pesticides and industrial solvents, are found in surface and ground water resources. Activated carbon is used to remove these materials and other complex organic compounds responsible for taste and odour problems. Activated carbon works by adsorption of the organic molecule onto its porous structure. The material is continuously adsorbed until the activated carbon is saturated and its capability to adsorb more is exhausted. The carbon, which is wood ash or another lignin-based material that has been specially conditioned, comes in two forms powdered activated carbon (PAC) or granular (GAC). If powder is used then it must be discarded when it is exhausted; granules, on the other hand, can be regenerated for reuse by heat treatment. Although more expensive, granules are used when a permanent activated carbon capacity is required due to significant concentrations of unwanted organic compounds in the water. The granules are used in a permanent bed similar to a rapid sand filter that incorporates a backwash facility. The powdered form is only used for occasional or intermittent use. There are significant concerns about the formation of dioxins during regeneration, and the need to treat spent activated carbon as a hazardous waste. When plants do not have on-site regeneration facilities, and only the largest water treatment plants do, then the transportation of the spent activated carbon to another site for regeneration may also raise concerns (Mallevialle and Suffet, 1992). Other advanced processes have been reviewed by Lorch (1987) and Pontius (1990).

One of the most important advances in water treatment in the past decades has been the development of membrane filtration. This is a highly sophisticated process that employs synthetic polymeric membranes to physically filter out of solution under pressure minute particles including viruses and some ions. Conventional filtration such as microscreening can only deal effectively with particles larger than 10^{-2} mm; however, by employing a range of different synthetic membranes with very small pores particles of any size down to 10^{-7} mm can be removed from water (Figure 14.3). As with all filtration processes the size of the particles retained is approximately an order of magnitude smaller than the pore size of the filter. Membrane filtration is widely used for the advanced and tertiary treatment of potable waters including desalination and removal of organics (reverse osmosis), softening (nanofiltration), disinfection, removal of colour and humic substances (ultrafiltration) and removal of protozoan cysts (microfiltration) (Madaeni, 1999; Parsons and Jefferson, 2006).

Apart from the routine monitoring of finished water quality, continuous monitoring equipment is often used at water treatment plants and fitted with alarm systems to warn operators of problems. Those most widely used are for

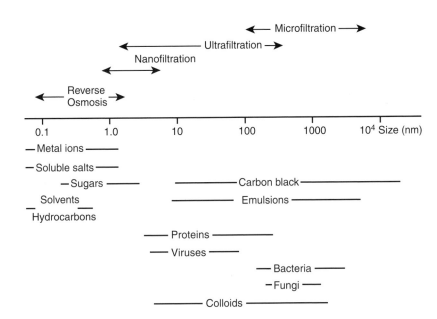

Figure 14.3 Application size range of membrane filtration processes. Reproduced from Scott and Hughes (1996) with permission from Springer Publishing Company.

pH, residual chlorine, fluoride, aluminium, iron, dissolved oxygen, colour, ammonia, turbidity, total organic carbon, nitrate and, of course, flow. Although these are excellent for specific parameters they do not measure the overall quality of the water, or warn of contamination of the raw water by trace contaminants such as pesticides. Fish monitors are now widely used to test both the raw water entering the plant and the finished water. They are generally based on the avoidance responses of fish (DeGraeve, 1982) or their ventilation frequency (Cairns and Garton, 1982) in relation to overall water quality. These responses can be monitored remotely and used to activate alarms to warn of a possible water quality problem.

14.2.13   Sludge

Water treatment produces considerable amounts of waste sludge in the form of a thin slurry. This is mainly gelatinous hydroxide sludge from coagulation and clarification, and the precipitation sludge from water softening. The water used to backwash sand filters is rich in solids, up to $100\,mg\,l^{-1}$, whereas the wash-water from the microstrainers is rich in organic matter, especially algae. These solids are allowed to settle before the sludge is pumped to shallow lagoons to slowly solidify, or is dewatered using a filter press. The solid cake from the lagoons or filter press is then safely disposed of to a landfill site, spread on land, or incinerated. Care must be taken with the disposal of water treatment sludges and wash-waters due to the potential for the transfer of pathogens.

14.3   Process selection

The cost of water treatment is dependent on three factors: (1) the quality of the raw water, with costs increasing as raw water quality deteriorates; (2) the degree of treatment required, so that the purer the finished water required, the more it will cost to produce it; and finally (3) the volume of water required and hence the size of the treatment plant, with the cost of water per unit volume decreasing as the capacity of the treatment plant increases.

There are three broad categories of resources. These are, in decreasing order of raw water quality: groundwaters, impounded surface waters such as reservoirs and lakes, and finally upland and lowland rivers. As demand has increased, surface waters have been recycled, especially in the south of England where resources are short (Section 4.5).

Groundwaters are chemically, and bacteriologically, of good quality and so only minimum treatment is required. This makes groundwater supplies cheap compared with water from other resources (Figure 14.4). The water contains carbon dioxide and iron, which must be removed by aeration, followed by rapid sand filtration. If the water is excessively hard it may be necessary to soften it using either precipitation or ion-exchange. All that is then required is to

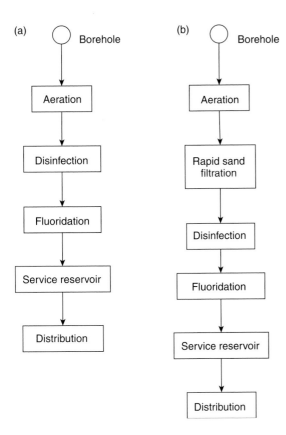

(a)

Borehole

Aeration

Disinfection

Fluoridation

Service reservoir

Distribution

(b)

Borehole

Aeration

Rapid sand filtration

Disinfection

Fluoridation

Service reservoir

Distribution

Figure 14.4 Normal sequence of treatment for a groundwater source to be used for supply where (a) groundwater is of excellent quality and (b) where groundwater has a moderate concentration of iron.

disinfect the water before distribution. Water from reservoirs may be exceptionally clean and as the raw water is aerobic, there is no carbon dioxide or iron to be removed. Such waters may only require microstraining and then disinfection (Figure 14.5a). In contrast, river water supplies will require either storage followed by microstraining, then filtration, flocculation and clarification followed by rapid sand filtration before disinfection and subsequent supply (Figure 14.5b and Figure 14.6). Table 14.4 summarizes the unit processes normally selected for the removal of specific compounds and materials.

14.4   How problems arise

It is during coagulation and flocculation that most problems of contamination of drinking water during treatment arise. The rate of addition of coagulant is governed by many factors that can alter very quickly so that constant control over coagulant addition is required. The optimum conditions for coagulation are determined as often as possible using a simple procedure known as the jar test (Solt and Shirley, 1991). This measures the effect of different combinations of the coagulant dose and pH, which are the two most important factors in the process. The jar test allows a comparison of these different combinations under

Figure 14.5 Typical sequence of treatment for water taken from (a) an upland storage reservoir and (b) an upland river, both of good quality.

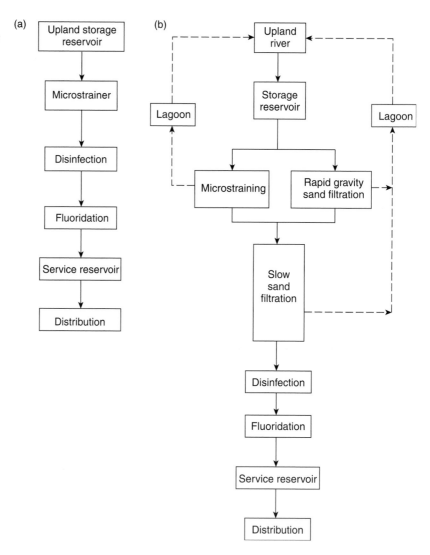

standardized conditions, after which the colour, turbidity and pH of the supernatant (clarified water) are measured. This involves three separate sets of jar tests, using the following approach:

1. The test is first carried out on the raw water without altering the pH. The coagulant dose is increased over a suitable range, so if five jars are used then five different doses can be tested. From this a simple curve can be drawn which shows the turbidity against the coagulant dose (Figure 14.7). This permits the best coagulant dose to be calculated.

2. The next stage is to alter the pH of the raw water by adding either an alkali or acid, and then to repeat the test using the best coagulant dose determined in the first set of tests. The range selected is normally between 5.5 and 8.5, and if possible 0.5 pH

increments are used. From this a plot of final colour and turbidity is plotted against pH, allowing the optimum pH for coagulants to be selected (Figure 14.8).

3. Finally the test is repeated a third time using fresh raw water, but this time corrected to the optimum pH and tested at various coagulant doses again. This determines the exact coagulant dosage at the optimum pH. These values are then used to operate the coagulant process.

On-line, automated systems are now available, although the manual test is still widely used. The curve produced from a lowland river in Figure 14.8 is typical for hard waters of its type, and from this it is clear that few problems should occur in operational management. However, soft upland waters, which are usually highly coloured as a result of humic material, are far more difficult to treat. Figure 14.9 shows that there is only a very narrow band of pH at which to achieve the optimum removal of colour and turbidity. Residual colour and turbidity will always be much higher in soft upland waters than in lowland water or groundwater supplies. The hydroxide flocs formed are small and weak, causing problems at later treatment stages. Another problem is that as conditions change fairly rapidly, these tests need to be repeated, certainly daily and sometimes more often. At small plants where such waters are often treated this is just not possible, so quality in terms of turbidity and colour is more likely to fluctuate.

Changes in quality can occur rapidly, so if the operator sees a change in the colour or turbidity he or she may increase the dosage rate of coagulant, but optimum conditions may not be maintained so that excess coagulant will enter the distribution system. One of the major causes of seasonal and often daily changes is the presence of algae in reservoirs, which can significantly alter the pH of the water.

The jar test does not indicate how much insoluble coagulant will pass through the clarification and filtration stages, as we shall see later. It does, however, give an idea of how much soluble coagulant will be in the final water as this depends solely on pH and the dosage rate. If the curves obtained from the jar test are overlaid by the amount of aluminium that has been taken into solution, a new and more serious problem arises (Figures 14.8 and 14.9). The curve of soluble aluminium in water that is abstracted from the hard lowland river follows the same curve as both turbidity and colour, so that the amount of soluble aluminium coagulant in the treated water will be acceptably low at the optimum conditions for the removal of turbidity and colour (Figure 14.8). With soft coloured waters there is the dilemma that the best conditions for the removal of turbidity and colour do not coincide with those for minimum coagulant solubility. In the example shown (Figure 14.9) the optimum removal of turbidity and colour occurred at a coagulant dose of 3 mg Al l^{-1} at pH 5.3, which results in 0.7 mg l^{-1} of soluble aluminium in the water compared with the EU limit of 0.2 mg l^{-1}. The water is therefore clear and colourless, but contains three and a half times more than the legal limit of aluminium. Where this occurs

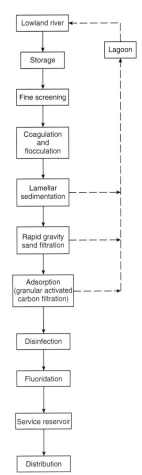

Figure 14.6 Typical sequence of unit processes to treat water from a lowland river supply of moderate to poor quality.

Table 14.4 *Selection of unit processes for the removal of specific parameters*

Parameter	Water treatment process options
Algae	Powdered activated carbon adsorption, microscreens, rapid filtration
Colour	Activated carbon adsorption, coagulation, flocculation, filtration
Floating matter	Coarse screens
Hardness	Coagulation, filtration, lime softening
Coliforms	
>100 per $100\,ml^{-1}$	Pre-chlorination, coagulation, filtration post-chlorination
>20 per $100\,ml^{-1}$	Coagulation, filtration post-chlorination
<20 per $100\,ml^{-1}$	Post-chlorination
Hydrogen sulphide	Aeration
Fe and Mn	Pre-chlorination, aeration, coagulation, filtration
Odour and taste	Aeration, activated carbon adsorption
Suspended solids	Fine screens, microscreens
Trace organics	Activated carbon adsorption
Turbidity	Coagulation, sedimentation, post-chlorination

Figure 14.7 Typical coagulation curve showing the effect of the concentration of coagulant on turbidity (T). Reproduced from Ainsworth *et al.* (1981) with permission from WRc Plc.

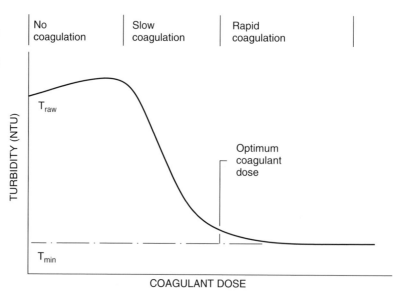

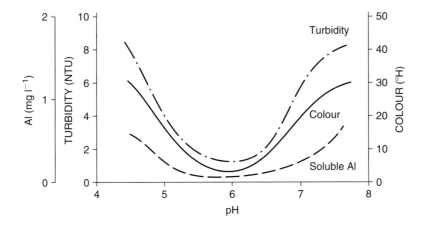

Figure 14.8 Example of a pH optimization curve for a hard lowland water showing the effect on turbidity, colour and the soluble aluminium concentration. Reproduced from Ainsworth *et al.* (1981) with permission from WRc Plc.

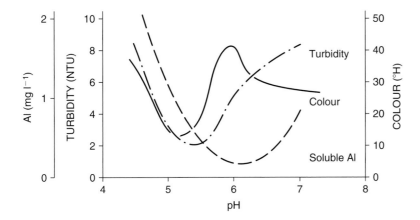

Figure 14.9 Example of a pH optimization curve for a soft, coloured water showing the effect on turbidity, colour and the soluble aluminium concentration. Reproduced from Ainsworth *et al.* (1981) with permission from WRc Plc.

more careful analysis of the dosage rates must be carried out to determine the optimum conditions to achieve the minimum amount of aluminium in the water. In this instance it was a coagulant dose of 5 mg Al l^{-1}, nearly twice as much as before at pH 6.6. This resulted in satisfactory aluminium (0.06 mg l^{-1}), colour (5.2 mg l^{-1}) and turbidity (1.4 FTU), although it was far more expensive and resulted in larger amounts of aluminium-rich sludge being produced. Because the operational range is so narrow, any minor change in water quality will result in a rapid deterioration in colour, turbidity and, of course, soluble aluminium. With higher residual levels of aluminium in soft waters anyway, any random adjustment by the operator, no matter how minor, could cause a significant increase in the amount of aluminium in the water (Ainsworth *et al.*, 1981). Synthetic organic polymers, in particular polyacrylamides, are used extensively to produce stronger flocs and more stable floc blankets at water treatment plants. In the UK the use of polyelectrolytes is widespread, with up to 70% of all plants using such chemicals. Polyacrylamide is primarily used in the treatment of sludge, but as the polymer-rich supernatant is normally returned to the inlet of the plant it is inevitable that polyelectrolytes will find their way into drinking water (Chapter 15).

Sedimentation and filtration are the stages that particulate matter is removed, including the hydroxide flocs formed during coagulation. The efficiency of the sedimentation tank can be monitored continuously using turbidity meters that activate an alarm system if the turbidity exceeds the set maximum. Correct operation of sedimentation tanks is vital to minimize particulate matter passing through the plant. The most serious problem is fluctuating flow rates, which cause the sludge blanket, through which the water flows, to expand too much. Remember that most sedimentation tanks used for water treatment operate in an upward flow direction (Section 14.2.7), so if the sludge blanket expands too much then particles are lost from the tank with the treated water. Infrequent desludging can also have a similar effect. The operation of filters, both rapid and slow sand filters, is complex and poor operation can lead to problems, perhaps the most serious of which is when the sand bed develops cracks, allowing the passage of unfiltered water. The polyelectrolytes used as a coagulant aid can also cause the sand in the filter to crack. The layer of micro-organisms at the surface of slow sand filters reduces the amount of organic matter in the finished water, although taste and odour problems can be caused by their activity. A similar situation can occur with micro-organisms colonizing GAC filters (Camper et al., 1985). Recycling of process water is also a major potential source of pathogen breakthrough.

Nitrogen is usually found in its fully oxidized form as nitrate or in its reduced form as ammonia. Nitrite is an intermediate stage in the oxidation of ammonia to nitrate, which occurs naturally in soil and water. In general, the concentration of nitrite in water is very low, but it can occasionally occur in unexpectedly high concentrations in surface waters due to industrial and sewage pollution.

It is nitrite, not nitrate, which is the actual toxicant associated with methaemoglobinaemia, the formation of N-nitroso compounds and carcinogenic effects, all of which have been discussed fully in Section 5.3. In their detailed report of water quality, the Drinking Water Inspectorate (DWI) (HMSO, 1991) reported increased nitrite levels in many finished waters arising from chloramination disinfection processes at water treatment plants. Ammonia is not completely reacted at chlorine to ammonia ratios of 3–4:1 by weight (Figure 14.2), resulting in excess ammonia reaching the distribution system. Here it acts as a source nutrient for nitrification producing nitrate or nitrite. For this reason a number of treatment works have had to replace their chloramination process due to partial nitrification leading to increased nitrite in the drinking water. For full conversion of ammonia to monochloramine a ratio of >5:1 is necessary (Bryant et al., 1992).

One of the most difficult problems arising from water treatment is the production of excessive colour and turbidity. Discolouration of water is a major cause of complaints, with the difficulty in maintaining optimum coagulation the prime cause. This is particularly difficult where water is abstracted from upland reservoirs and rivers where the water is usually acidic and highly coloured. With

Table 14.5 *Typical values used to avoid discolouration in water leaving the treatment plant*

Parameter	Value to avoid discolouration
Turbidity (NTU)	<0.5
Colour (mg l^{-1} Pt-Co)	<10
Iron (μg Fe l^{-1})	<50
Aluminium (μg Al l^{-1})	<100
Manganese (μg Mn l^{-1})	<50
Dissolved oxygen (%)	>85
pH (at 25 °C)	6.5–9.0

Pt-Co = elements electrode is comprised.

such soft water there is only a very narrow operational band in terms of pH to obtain good colour removal, so that any changes in water quality, for example, due to algal growth, which affects the pH of the water quite significantly during a 24-hour period, will result in discolouration (Chapter 11).

Particulate matter also arises from water treatment and can cause a serious deterioration in water quality, including discolouration, as can dissolved organic matter. Excessive amounts of suspended particulate matter and/or coagulant residual, which may be in an insoluble or soluble form, can be released from the treatment plant and enter the distribution system. These materials quickly settle out of solution by various means and accumulate in areas of low flow. If resuspended they will affect the quality of the water supplied to the consumer. These deposits are composed of aluminium coagulant, iron coagulant, manganese and iron oxides, organic matter and algae. They also harbour a variety of micro-organisms and larger animals which use them as a food source. The Department of the Environment issued guidelines to the water supply companies before privatization giving suggested values in finished water as it leaves the treatment plant to avoid discolouration problems (Table 14.5). These guidelines were in addition to the regulations arising from the EC Drinking Water Directive (98/83/EEC) (Appendix 1). All are related to the precipitation of solids, except for dissolved oxygen and pH, which are related to corrosion (Chapter 27).

Problems associated with the addition of fluoride and the formation of disinfection by-products are dealt with in detail in Chapter 17 and 18 respectively.

14.5 Conclusions

The quality of surface waters intended for abstraction for public supply has to be classified under the EC Surface Water Directive (75/440/EEC) into three categories (A1, A2 and A3) for which mandatory minimum treatment is specified. For example, A1 resources are the cleanest (e.g. groundwaters) and only require minimum treatment whereas A3 waters (e.g. lowland rivers) are the poorest requiring advanced treatment (Table 2.9) (Section 2.5). This Directive will be replaced in 2012 when River Basin Management Plans come into force under the new Water Framework Directive (00/60/EC) that ensures minimum quality standards for all water resources. The introduction of the new EC Drinking Water Directive (Appendix 1) and the revision of the US National Primary Drinking Water Standards (Appendix 2), following the revision of the World Health Organization drinking water guidelines (Appendix 3), have required more advanced treatment processes to achieve even lower permissible concentrations of a wide variety of inorganic and organic compounds. This has led to a significant investment in membrane filtration technology, especially microfiltration to replace coagulation, flocculation, clarification and filtration processes, and reverse osmosis to replace activated carbon adsorption (Scott and Hughes, 1996; Parsons and Jefferson, 2006). Membrane technology can also be used for water disinfection, although like UV radiation and ozonization there is no residual disinfection effect (Madaeni, 1999).

References

Ainsworth, R. G., Calcutt, T., Elvidge, A. F. *et al.* (1981). *A Guide to Solving Water Quality Problems in Distribution Systems*. Technical Report 167. Medmenham: Water Research Centre.

Bryant, E. A., Fulton, G. P. and Budd, G. C. (1992). *Disinfection Alternatives for Safe Drinking Water*. New York: Van Nostrand Reinhold.

Bull, R. J. and Kopfler, F. C. (1991). *Health Effects of Disinfectants and Disinfection By-Products*. Denver: AWWA Research Foundation and the American Water Works Association.

Cairns, M. A. and Garton, R. R. (1982). Use of fish ventilation frequency to estimate chronically safe toxicant concentration. *Transactions of the American Fisheries Society*, **111**, 70–7.

Camper, A. K., LeChevallier, M. W., Broadway, S. C. and McFeters, G. A. (1985). Growth and persistence of pathogens on granular activated carbon filters. *Applied Environmental Microbiology*, **50**, 1178–82.

DeGraeve, G. M. (1982). Avoidance response of rainbow trout to phenol. *Progress in Fish Culture*, **44**, 82–7.

Duncan, A. (1988). The ecology of slow sand filters. In *Slow Sand Filtration: Recent Developments in Water Treatment Technology*, ed. N. J. D. Graham. Chichester: Ellis Horwood, pp. 163–80.

Gray, N. F. (2004). *Biology of Wastewater Treatment*. London: Imperial College Press.

Hill, R. and Lorch, W. (1987). Water purification. In *Handbook of Water Purification*, ed. W. Lorch. Chichester: Ellis Horwood, pp. 226–302.

HMSO (1991). *Private Water Supplies Regulations* 1991. Statutory Instrument 1991/2790. London: HMSO.

Kruithof, J. C., Van Eekeren, M. W. M. and Schippers, J. C. (1991). Membraanfiltratie, geavanceerde oxydatie en UV-desinfectie in de processes, advanced oxidation and UV-USA en Canada. *H₂O*, **4** (19), 537–42.

LeChevallier, M. W. (1990). Coliform regrowth in drinking water: a review. *Journal of the American Water Works Association*, **82**(11), 74–86.

Lorch, W. (ed.) (1987). *Handbook of Water Purification*. Chichester: Ellis Horwood.

Madaeni, S. S. (1999). The application of membrane technology for water disinfection. *Water Research*, **33**, 301–8.

Mallevialle, J. and Suffet, I. H. (1992). *Influence and Removal of Organics in Drinking Water*. Boca Raton, FL: Lewis.

Masschelein, W. J. (1982). *Ozonization Manual for Water and Wastewater Treatment*. Chichester: Wiley.

Parsons, S. and Jefferson, B. (2006). *Introduction to Potable Water Treatment Processes*. London: Blackwell.

Pontius, F. W. (ed.) (1990). *Water Quality and Treatment: a Handbook of Community Water Supplies*. New York: McGraw-Hill.

Scott, K. and Hughes, R. (eds.) (1996). *Industrial Membrane Separation Technology*. London: Blackie Academic and Professional.

Solt, G. S. and Shirley, C. B. (1991). *An Engineer's Guide to Water Treatment*. Aldershot: Avebury Technical Press.

Stevenson, D. G. (1998). *Water Treatment Unit Processes*. London: Imperial College Press.

Twort, A. C., Crowley, F. C., Lawd, F. M. and Ratnayaka, P. P. (1994). *Water Supply*, 4th edn. London: Arnold.

Vigneswaran, S. and Visvanathan, C. (1995). *Water Treatment Processes: Simple Options*. Boca Raton, FL: CRC Press.

Visscher, J. T. (1988). Water treatment by slow sand filtration: considerations for design, operation and maintenance. In *Slow Sand Filtration: Recent Developments in Water Treatment Technology*, ed. N. J. D. Graham. Chichester: Ellis Horwood, pp. 1–10.

White, G. C. (1972). *Handbook of Chlorination*. New York: Van Nostrand Reinhold.

Wolfe, R. L. (1990). Ultraviolet disinfection of potable water. *Environmental Science and Technology*, **24**, 768–73.

Chapter 15
Aluminium and acrylamide

15.1 Aluminium

Aluminium is a widespread and abundant element that is found as a normal
constituent of all soils, plant and animal tissue. It is especially common in food,
resulting in a typical daily intake of between 5 and 20 mg depending on individual
variations in eating and drinking habits. In the UK the mean dietary aluminium
intake is approximated at 3.9 mg d^{-1} (Ysart *et al.*, 2000) (Table 15.1). It is
recommended that the dietary intake should not exceed 6 mg d^{-1} if potential
toxicity problems are to be avoided (Soliman and Zikovsky, 1999). Diet is a
major factor in aluminium uptake. For example, aluminium is taken up in large
amounts by tea plants, so that drinking tea significantly enhances aluminium
uptake (Flaten and Ødegård, 1988). In fact, tea may contain anything from 20 to
200 times more aluminium than the water it is made with. Aluminium can also
be leached from cooking utensils (Jagannatha and Valeswara, 1995), and
cooking acidic foods such as citric fruits, rhubarb or tomatoes can lead to
enhanced leaching from aluminium pots and pans. Enhanced leaching has also
been reported for spinach and other green vegetables. Aluminium leaching also
occurs from coffee percolators made from aluminium. In new percolators coffee
contains on average 4.1 mg Al l^{-1}, of which 85% comes from aluminium
leached from the metal pot. This reduces as the percolator ages, although more
than 70% of the aluminium in the coffee will still be leached from the
pot. Elevated aluminium concentrations have also been noted in soft drinks
(pH <3.0) when supplied in aluminium cans (Table 15.2) (López *et al.*, 2002).
While aluminium packaging materials are normally coated with lacquers to
reduce leaching, aluminium cartons and packaging can contribute to the amount
of aluminium in the diet. It is difficult to estimate the exposure to aluminium
from containers and cooking utensils, and in most cases they are likely to be
small compared with the total dietary intake. However, aluminium pots and pans
should be replaced with steel as the old ones wear out. It has been suggested that
leaching of aluminium from pots and pans is enhanced if the water has been
fluoridated, or if the food contains fluoride. However, recent work has shown
that this does not appear to be so, and those in fluoridated areas are no more
at risk from aluminium leaching than those receiving low-fluoride water,

Table 15.1 *Reported mean or range of dietary intake of aluminium by adults* (mg d^{-1}). *Infant dietary intake of aluminium is much lower at between 0.03 and 0.7 mg d^{-1}*

Australia	1.9–2.4	Finland	6.7	Germany	8–11
Japan	4.5	Netherlands	3.1	Sweden	13.0
Switzerland	4.4	UK	3.9	USA	7.1–8.2

Table 15.2 *Aluminium concentrations in soft drinks. Adapted from López et al.* (2002) *with permission from Elsevier Ltd*

Sample	Type of container	n	Aluminium (μg l^{-1})	
			Mean	Range
Orange	Plastic	6	250.6	205.0–335.9
	Can	5	686.7	320.0–1053.3
	Glass	6	310.0	220.2–470.4
Lemon	Plastic	6	260.7	121.1–273.5
	Can	4	660.0	305.0–1015.1
	Glass	5	325.0	315.2–420.3
Tonic water	Plastic	5	256.0	250.2–310.7
	Can	4	410.3	320.7–510.3
	Glass	5	310.0	300.7–420.8
Cola	Plastic	7	163.8	60.7–301.7
	Can	5	580.7	279.5–985.6
	Glass	5	200.2	190.7–310.0
Soda water	Plastic	6	122.8	44.6–264.8
	Can	6	341.7	47.9–936.6
Several fruit[a]	Glass	4	890.1	869.5–970.1
	Can	9	748.2	290.3–1019.8

[a] Mix of tropical fruit.

especially where the water is soft and slightly acidic. Aluminium is a common food additive and is used in colourings, emulsifiers, stabilizers and anti-caking agents. Pharmaceuticals, in particular antacids and buffered analgesics, can contribute substantially to the total body burden of aluminium; potentially up to 5000 mg d^{-1} in users of antacids. Absorption of the metal through the skin from deodorants containing aluminium has also been reported. So the exposure to, and absorption of, aluminium is hard to control.

Aluminium in water may be present due to leaching from soil and rock. Surveys in the USA and UK have found aluminium levels in natural water sources ranging from $14 \mu g \, l^{-1}$ to $1200 \mu g \, l^{-1}$. For example, Northern Ireland have derogated levels of $500 \mu g \, l^{-1}$ for some supplies derived from catchments where aluminium is naturally elevated. In other parts of the world, such as Australia, aluminium levels have been found in water sources as high as $18000 \mu g \, l^{-1}$ due to clay minerals containing alumino-silicates. Acid rain in poorly buffered regions has led to increased leaching from soils so that the acidified surface waters can have much higher aluminium concentrations than normal (Garhardsson *et al.*, 1994). In acidic soft water areas, surface waters may contain as much as 200–$300 \mu g \, l^{-1}$ of aluminium, increasing to $600 \mu g \, l^{-1}$ where these areas have been afforested. Such catchments are especially at risk if they are near the coast and subject to prevailing winds as they receive precipitation carrying higher salt burdens, which also increases acidity in the soil and leaching of aluminium.

The objective of water treatment is to provide a supply of water that is chemically and microbiologically safe for human consumption. To achieve this chemicals such as aluminium sulphate or iron (III) sulphate are added to the water during treatment to help remove fine particulate matter, including bacteria, a process known as coagulation (Chapter 14). There has been increasing concern about the amount of residual chemicals left in the water after treatment, both in a soluble and insoluble form, and the possible effects on the health of consumers. The chemical causing most concern is aluminium sulphate.

The dose of aluminium sulphate (alum) added to raw water depends on a number of factors, the most critical being the pH, although temperature is also important. The solubility range of aluminium is shown in Figure 15.1. To achieve the EC limit value of $200 \mu g \, Al \, l^{-1}$ a coagulant pH must be between 5.2 and 7.6. If the pH goes outside this range the solubility increases rapidly.

The addition of alum to drinking water as a coagulant will therefore leave some soluble or insoluble residual aluminium (Section 14.4). Subsequent filtration processes should remove the insoluble particulate aluminium, although the effectiveness of removal depends on filtration efficiency. Where residual concentrations are high, deposition of aluminium may occur in the distribution system. Any disturbance of these sediments will increase the concentration of aluminium in drinking water.

Reasons for increased levels of aluminium in drinking water are normally associated with operational problems at the treatment plant. Aluminium residuals can be reduced during the coagulation process by (1) optimization of the pH during coagulation, (2) avoiding excessive aluminium dosage, (3) good mixing at the point of application of the coagulant, (4) optimization of paddle speeds for flocculation, and (5) efficient filtration of the aluminium flocs (Section 14.4). However, accidents have occurred with aluminium sulphate spilt or accidentally discharged into the finished water and discharged into the distribution network. The worst such incident happened at the Lowermour Water Treatment Plant in Cornwall on 6 July 1988. This plant serves over

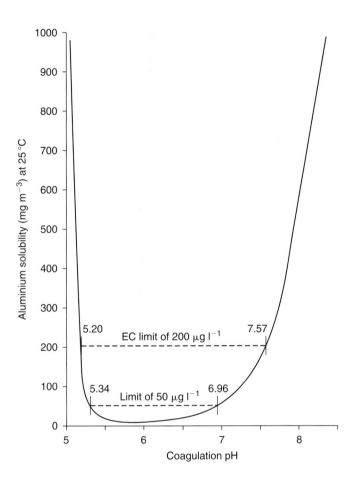

Figure 15.1 Solubility of aluminium with pH. The pH causes severe problems during coagulation. The effective pH range to ensure the EC Drinking Water Quality Directive standard for aluminium, and the smaller effective range of 5.3–7.0 for a 50 µgl^{-1} limit are shown.

20 000 people around Camelford, but on that day a total delivery of alum in solution was put into the contact tank rather than a storage tank. In consequence, concentrations of aluminium several thousand times greater than the limit set by the EC flowed out of the local taps, as well as killing over 30 000 fish in the Rivers Camel and Allen (Clayton, 1991). Consumers who used the water before realizing that it was contaminated suffered a variety of effects such as blisters, mouth ulcers, skin irritations, sore throats, lassitude, diarrhoea and a host of other symptoms. Also, people who washed their hair in the contaminated water developed scalp rashes, whereas those with bleached hair suffered severe discolouration.

In general, absorption of aluminium by the human body is poor with uptake mainly from dietary sources with <5% uptake due to drinking water (Gourier-Fréry and Fréry, 2004). However, in practice it is difficult to estimate the contribution aluminium in drinking water makes to the daily intake of the metal. If we take, for example, a daily intake of 20 mg for an adult who drinks about 2 litres of water each day at 200 µg Al l^{-1}, then the drinking water is only contributing 2% of the total aluminium intake for that person. Conversely, if we assume a lower daily intake of just 5 mg, then 2 litres of water at $200\,\mu g$ Al l^{-1} would represent 8%. It is easy to see that given certain conditions, increased concentrations of aluminium in drinking water could contribute a significant amount of aluminium to the daily intake of the metal. Those who are very young or very old have a far higher intake of water. The elderly are clearly more at risk, and tend to use aluminium for packaging food and cooking utensils, because of its lightness. Any contribution from drinking water, no matter how low, is therefore significant, and can add to the elevated levels found in tea, soft drinks and fruit juices. The forms of aluminium found in drinking water are particularly bioavailable although this is not so with food. In practice a very small proportion of aluminium in water, possibly less than 1%, appears to be absorbed; the rest is excreted with the faeces. Of the absorbed aluminium, most will be excreted via the kidneys with a very small amount accumulating in bone, liver and brain tissues. However, even at these small absorption rates, the aluminium in drinking water may well contribute a much greater proportion of the total aluminium absorbed by the body than currently thought.

15.1.1   Levels in water

The World Health Organization (WHO) no longer classifies aluminium as a chemical of significance to health in drinking water. It is now classed as a substance that may give rise to consumer complaints, and has a revised standard of $200\,\mu g\,l^{-1}$. The US Environmental Protection Agency (USEPA) has not included aluminium in its National Primary Drinking Water Regulations, although it has been included in the list of 51 contaminants on its revised Drinking Water Contaminant Candidate List for further evaluation (Appendix 7) (USEPA, 2005). It is included in the National Secondary Drinking Water Regulations, which deals with aesthetic quality of water, where a non-enforceable guideline of $50-200\,\mu g\,l^{-1}$ has been set. The EC Drinking Water Directive sets a limit value of $200\,\mu g\,l^{-1}$, which has been adopted by both the UK and Ireland. These levels have been repeatedly exceeded in the past on a day-to-day basis in both countries. However, better operational management has led to a high level of compliance, although smaller supplies remain at risk. In their most recent guidelines, the WHO (2004) have stated that 'owing to the limitations in the animal data as a model for humans and the uncertainty surrounding the human data, a health-based guideline value for aluminium cannot be derived'. However, practicable levels based on the optimization of the coagulation process during treatment where aluminium-based coagulants are used have been recommended at $\leq 100\,\mu g\,l^{-1}$ for large treatment plants and $\leq 200\,\mu g\,l^{-1}$ for smaller treatment plants (i.e those serving $< 10\,000$ people). Smaller plants are less able to achieve lower residual values as their size provides little buffering for fluctuations in operation. Also smaller plants have limited resources for process optimization and restricted access to expertise to solve specific operational problems.

This is a worldwide problem, with increased levels of aluminium found in drinking water wherever alum is used as a coagulant (Sollars *et al.*, 1989). Some water companies have indicated a willingness to phase out the use of alum as a coagulant, although due to technical reasons and cost this is less likely to occur in areas where raw waters are soft. It has been suggested that iron (III) sulphate could be used instead of alum; however, it is more expensive, more difficult to handle and causes corrosion problems. Residual aluminium is tasteless, whereas residual iron is not and causes other consumer problems as well (Section 9.1). Iron salts used for coagulation are often contaminated with other metals, in particular chromium, nickel, manganese and lead, which find their way into the finished water (Section 9.4).

15.1.2 Health effects

Aluminium was previously regarded as non-toxic, only causing aesthetic problems if occurring in high concentrations in drinking water. However, over the past 20 years more and more evidence has come to light to link aluminium with certain neurodegenerative diseases (Flaten, 2001).

Most of the information about the effects of aluminium in drinking water comes from patients undergoing kidney dialysis (haemodialysis). It was first described in the early 1970s: a syndrome that was characterized by the onset of altered behaviour, dementia, speech disturbance, muscular twitching and convulsions. The outcome for patients with dialysis dementia, as the syndrome became known, was usually fatal. Aluminium concentrations in the water used to prepare dialysate fluid and the incidence of dialysis dementia were proved, although aluminium from other sources also contributed. Aluminium was added to the dialysis fluid to remove phosphorus from the blood, while antacids containing aluminium were also given orally to patients to bind the phosphorous in the intestines to prevent its absorption into the bloodstream. People receiving long-term dialysis treatment through artificial kidneys are normally exposed to 150–200 litres of dialysis fluid two or three times each week. The water used to make up the fluid must be of extremely high quality as aluminium is readily transferred from the dialysate to the patient's bloodstream during haemodialysis. Waterborne aluminium is also thought to be an important factor in the development of osteomalacia in dialysis patients.

Aluminium also appears to be an important factor in two other severe neurodegenerative diseases, amyotrophic lateral sclerosis and Parkinsonian dementia. These diseases are extremely common in the Western Pacific, especially on the island of Guam, where the drinking water and soil is low in calcium and magnesium, but high in aluminium, iron and silicon. There is also considerable evidence that aluminium plays a role in the development of Alzheimer's disease.

Alzheimer's disease, which is dementia in younger people (less than 45–50 years of age), is a slowly developing disease beginning with learning and

Table 15.3 *Key epidemiological studies of aluminium in drinking water and Alzheimer's disease or dementia. Adapted from Flaten (2001) with permission from Elsevier Ltd*

Significant association reported	No significant association reported
Martyn *et al.*, 1989	Wood *et al.*, 1988
Flaten, 1990	Wettstein *et al.*, 1991
Frecker, 1991	Taylor *et al.*, 1995
Neri and Hewitt, 1991	Martyn *et al.*, 1997
Forbes and McLachlan, 1996	
Jacqmin-Gadda *et al.*, 1996	
McLachlan *et al.*,1996	
Gauthier *et al.*, 2000	
Rondeau *et al.*, 2000	

memory defects and slowly progressing to affect all aspects of intellectual activity including judgement, calculation and language. As the terminal stages approach, urinary incontinence and the loss of useful motor activity results in the need for total nursing care. The disease may take from 18 months to 19 years to take its course, although the average is around 8 years. It is a savage illness, resulting in an untreatable encephalopathy destroying the mind and body. In the USA alone, 1.2 million people have severe dementia, with another 2.5 million having mild to moderate dementia, with the cumulative lifetime risk to each individual of becoming severely demented as high as 20%. An estimated 15% of all elderly people in the British Isles are suffering from dementia. Aluminium has been implicated in Alzheimer's disease because of its assumed pathogenic role in the formation of senile plaques in the brain (Forbes and MacAiney, 1992). Certainly there are significantly increased levels of the metal in the brains of patients with Alzheimer's disease, with 1.5–2.0 times more aluminium present in the grey matter than in non-demented subjects. However, although it is clear that aluminium is associated with pathological changes in Alzheimer's disease, it is not yet clear whether its role is causal or incidental. The effects on the brain of dialysis dementia and Alzheimer's patients are different, although aluminium is associated with both.

Nine of the 13 published epidemiological studies that have examined the relationship between the concentration of aluminium in drinking water and the incidence of Alzheimer's disease have shown significant positive relationships (Table 15.3) (Flaten, 2001). The major problem with such studies is that drinking water only contributes a fraction of the total dietary intake of the metal. Less than 1% of aluminium in water is absorbed with the remainder excreted with the faeces. Of the aluminium absorbed, most is excreted via the kidneys and only a very small amount accumulates in bone, liver and brain tissues. So the general consensus amongst scientists, including the WHO, is that the link between drinking water and the disease is tenuous (Allen and Cumming, 1998; Stauber *et al.*, 1999). Interest is now focussing on the relationship between the disease and the total dietary intake of aluminium, especially those foods that contain large amounts of aluminium additives, which include soft drinks and certain processed foods (Rogers and Simon, 1999). The form of aluminium in food and drinking water (i.e. speciation) may also be critical in controlling its uptake by the body (Gauthier *et al.*, 2000).

15.2 Acrylamide

Anionic, cationic and non-ionic polyacrylamide flocculants are widely used in both water and wastewater treatment with acrylamide monomer contents of around 0.05%. Polyacrylamides are also used as grouting agents in the construction of reservoirs, tunnels and wells. So acrylamide can find its way into water supplies from reuse of sewage effluents, during water treatment or during storage in service reservoirs. It is also used in the sizing of paper and textiles, the manufacture of organic chemicals, plastics, adhesives and dyes, and in oil extraction and ore processing. The main contamination route arises where dosing conditions with the flocculant are not ideal due to changes in either water chemistry (Section 14.4) or due to poor operational practice.

Acrylamide is both a neurotoxin and impairs reproductive function. The body readily absorbs it during digestion and long-term exposure has been associated with scrotal, thyroid and adrenal tumours in male rats, and mammary, thyroid and uterine tumours in female rats (Johnston *et al.*, 1986; Rice, 2005). The International Agency for Research on Cancer have classed it as a Group 2A carcinogen in relation to its presence in drinking water (Section 3.2). When exposed to acrylamide for short periods at concentrations exceeding the maximum contaminant level (MCL) then damage to the nervous system, weakness and coordination problems in the legs have been reported. The USEPA list long-term exposure problems, at concentrations exceeding the MCL, as damage to the nervous system, leading to paralysis, and cancer.

There have been problems of accurately measuring acrylamide at health-related concentrations making routine monitoring difficult and expensive, although new developments using gas chromatography-mass spectrometry are more reliable (Cavalli *et al.*, 2004). For this reason control of the compound has been through product control rather than normal water quality monitoring.

Limiting either the acrylamide content of the polyacrylamide flocculant, reducing the dose, or both, are the only methods to control acrylamide in drinking water. Acrylamide was first introduced into the 1993 WHO drinking water guidelines at $0.5\,\mu g\,l^{-1}$, the revised EC Drinking Water Directive followed suit and set a standard of $0.1\,\mu g\,l^{-1}$ based on product control (Section 2.5). Under the National Primary Drinking Water Standards in the USA, the maximum permissible level of acrylamide monomer permitted in flocculants must not exceed 0.05% dosed at $1 \, mg \, l^{-1}$, with a maximum contaminant level goal set at zero due to its potential carcinogenicity. When the new EC acrylamide standard came into force the Drinking Water Inspectorate for England and Wales originally specified a national limit of free monomer in flocculants $\leq 0.025\%$. Used at the maximum permitted dosing concentration of $0.5 \, mg \, l^{-1}$, this results in a maximum concentration of $0.125 \, \mu g \, l^{-1}$ in drinking water. A stricter purity limit of 0.02% free acrylamide was introduced on 25 December 2003 in order to reduce the concentration in drinking water to below the EC standard.

15.3   Conclusions

Aluminium is not an acute toxicant, as was confirmed at Camelford when 20 000 people were exposed to elevated levels for five days with the resulting symptoms both mild and short-lived. However, the intake of large amounts of aluminium over longer periods is known to cause anaemia, osteomalacia (brittle or soft bones), glucose intolerance and cardiac arrest in humans. There are many questions still to be answered with regard to aluminium toxicity especially the effects in humans exposed to low levels of aluminium over a long period; the earlier onset or progression of a wide range of diseases of the nervous system is one distinct possibility. So it is not possible at present to dismiss aluminium as having some kind of causal role in Alzheimer's disease.

Therefore, using the precautionary approach, aluminium should not be treated as a safe chemical, with stricter operational practice required to control dosage rates and handling of the chemical during water treatment to prevent accidents. The prescribed limit values are really a compromise because residual levels in excess of $100 \, \mu g \, l^{-1}$ may cause discolouration problems anyway. With correct operational practice aluminium concentrations in finished water can be kept very low indeed, so more investment is needed in water treatment in areas where the use of aluminium sulphate is made difficult by the nature of the raw water. Given the other sources of aluminium in food and drinks, such as tea and canned soft drinks, it is unlikely that aluminium in drinking water at the EC limit of $200 \, \mu g \, l^{-1}$ will be problematic. The wider use of aluminium as a food additive, its use in pharmaceuticals and cosmetics, the increasing use of aluminium packaging especially for acidic soft drinks, is resulting in greater exposure in forms that may be more bioavailable. For this reason, younger generations may be at greater risk from long-term exposure to aluminium than either their parents or grandparents. Therefore suppliers should not be complacent about levels of aluminium in drinking water.

Acrylamide has proved to be a difficult contaminant both to monitor and control in drinking water, with water utilities relying on controlling the level of active acrylamide monomer in the flocculants that they use. Safe exposure levels for a 10 kg child drinking one litre of water daily is estimated to be $1.5 \, mg \, l^{-1}$

for a single day exposure, 0.3 mg l^{-1} for 10 days exposure, or 0.002 mg l^{-1} over 7 years exposure. This indicates that there is a potential danger that serious health effects could occur if an industrial discharge of acrylamide entered into a supply resource as it is not removed by conventional treatment and not routinely monitored. Acrylamide is biodegradable and so can be broken down in surface waters (4–12 days) and soil (14 days). Recent work in Sweden has shown that acrylamide is also found in many foods and drinks (Coghlan, 2006). It is formed when food is heated above 120 °C enabling the amino acid asparagine to react with sugars such as glucose and fructose. Acrylamide is especially high in potato products that have been fried although up to 39% of total exposure has been associated with coffee. However, no relationship between the consumption of these products and the known clinical effects associated with the chemical has been established, which is reassuring as it appears that acrylamide in food is almost impossible to avoid. It is estimated that the average daily intake of acrylamide from food is one microgram per kilogram of body weight, so any contribution from drinking water may push the recommended daily intake of the chemical over the safe limit.

References

Allen J. L. and Cumming F. J. (1988). *Aluminium in the Food and Water Supply: an Australian Perspective*. Water Services Association, Urban Water Research Association of Australia Research Report: No. 202.

Cavalli, S., Polesello, S. and Saccani, G. (2004). Determination of acrylamide in drinking water by large volume direct injection and ion-exclusion chromatography-mass spectrometry. *Journal of Chromatography: A*, **1039**(1–2), 155–9.

Clayton, B. (1991). *Water Pollution at Lowermoor, North Cornwall*. Second Report of the Lowermoor Incident Health Advisory Group. London: Department of Health, HMSO.

Coghlan, A. (2006). Acrylamide: the food scare the world forgot. *New Scientist*, **189**(2548), 8–10.

Flaten, T. P. (1990). Geographical associations between aluminium in drinking water and death rates with dementia (including Alzheimer's disease), Parkinson's disease and amyotrophic lateral sclerosis in Norway. *Environmental Geochemistry and Health*, **12**, 152–67.

Flaten, T. P. (2001). Aluminium as a risk factor in Alzheimer's disease, with emphasis on drinking water. *Brain Research Bulletin*, **55**(2), 187–96.

Flaten, T. P. and Ødegård, M. (1988). Tea, aluminium and Alzheimer's disease. *Food Chemistry and Toxicology*, **26**, 959–60.

Forbes, W. F. and MacAiney, C. A. (1992). Aluminium and dementia. *The Lancet*, **340**, 668–9.

Forbes, W. F. and McLachlan, D. R. C. (1996). Further thoughts on the aluminium-Alzheimer's disease link. *Journal of Epidemiology and Community Health*, **50**, 401–3.

Forbes, W. F., Gentleman, J. F., Agwani, N., Lessard, S. and McAiney, C. A. (1997). Geochemical risk factors for mental functioning, based on the Ontario Longitudinal Study of Aging (LSA). VI: The effects of iron on the associations of aluminium and fluoride water concentrations and of pH with mental functioning, based on results obtained from the LSA and from death certificates mentioning dementia. *Canadian Journal of Aging*, **16**, 142–59.

Frecker, M. F. (1991). Dementia in Newfoundland: identification of a geographical isolate? *Journal of Epidemiology and Community Health*, **45**, 307–11.

Garhardsson, L., Oskarsson, A. and Skerfring, S. (1994). Acid precipitation-effects on trace elements and human health. *Science of the Total Environment*, **153**, 237–45.

Gauthier, E., Fortier, I., Courchesne, F. *et al.* (2000). Aluminium forms in drinking water and risk of Alzheimer's disease. *Environmental Research*, **84**, 234–46.

Gourier-Fréry, C. and Fréry, N. (2004). Aluminium. *EMC – Toxicologie Pathologie*, **1**, 79–95.

Jacqmin-Gadda, H., Commenges, D., Letenneur, L. and Dartigues, J.-F. (1996). Silica and aluminium in drinking water and cognitive impairment in the elderly. *Epidemiology*, **7**, 281–5.

Jagannatha, K. S. and Valeswara, G. (1995). Aluminium leaching from utensils – a kinetic study. *International Journal of Food Science and Nutrition*, **46**, 31–8.

Johnston, K. A., Gorzinski, S. J., Bodner, K. M. *et al.* (1986). Chronic toxicity and oncogenicity study on acrylamide incorporated in the drinking water of Fishner 344 rats. *Toxicology and Applied Pharmacology*, **85**(2), 154–68.

López, F. F., Cabrera, C., Lorenzo, M. L. and López, M. C. (2002). Aluminium content of drinking waters, fruit juices and soft drinks: contribution to dietary intake. *Science of the Total Environment*, **292**, 205–13.

Martyn, C. N., Barker, D. J., Osmond, C. *et al.* (1989). Geographical relation between Alzheimer's disease and aluminium in drinking water. *The Lancet*, (**i**), 59–62.

Martyn, C. N., Coggon, D. N., Inskip, H., Lacey, R. F. and Young, W. F. (1997). Aluminium concentrations in drinking water and risk of Alzheimer's disease. *Epidemiology*, **8**, 281–6.

McLachlan, D. R. C., Bergeron, C., Smith, J. E., Boomer, D. and Rifat, S. L. (1996). Risk for neuropathologically confirmed Alzheimer's disease and residual aluminium in municipal drinking water employing weighted residential histories. *Neurology*, **46**, 401–5.

Neri, L. C. and Hewitt, D. (1991). Aluminium, Alzheimer's disease, and drinking water. *The Lancet*, **338**, 390.

Rice, J. M. (2005). The carcinogenicity of acrylamide. *Mutation Research*, **580**(1–2), 3–20.

Rogers, M. A. and Simon, D. G. (1999). A preliminary study of dietary aluminium intake and risk of Alzheimer's disease. *Age and Ageing*, **28**, 205–9.

Rondeau, V., Commenges, D., Jacqmin-Gadda, H. and Dartigues, J.-F. (2000). Relation between aluminium concentrations in drinking water and Alzheimer's disease: an 8 year follow-up study. *American Journal of Epidemiology*, **152**, 59–66.

Soliman, K. and Zikovsky, L. (1999). Concentrations of Al in food sold in Montreal, Canada, and its daily dietary intake. *Journal of Radioanalytical Nuclear Chemistry*, **242**, 807–9.

Sollars, C. J., Bragg, S., Simpson, A. M. and Perry, R. (1989). Aluminium in European drinking water. *Environmental Technology Letters*, **10**, 131–50.

Stauber, J. L., Florence, T. M., Davies, C. M., Adams, M. S. and Buchanan, S. J. (1999). Bioavailability of Al in alum-treated drinking water. *Journal of the American Water Works Association*, **91**, 84–93.

Taylor, G. A., Newens, A. J., Edwardson, J. A., Kay, D. W. K. and Forster, D. P. (1995). Alzheimer's disease and the relationship between silicon and aluminium in water supplies in northern England. *Journal of Epidemiology and Community Health*, **49**, 323–4.

USEPA (2005). *Drinking Water Contaminant Candidate List (CCL)*. Washington DC: Office of Water, US Environmental Protection Agency.

Wettstein, A., Aeppli, J., Gautschi, K. and Peters, M. (1991). Failure to find a relationship between mnestic skills of octogenarians and aluminium in drinking water. *International Archives of Occupational and Environmental Health*, **63**, 97–103.

WHO (2004). *Guidelines for Drinking Water Quality*, Vol.1. *Recommendations*, 3rd edn. Geneva: World Health Organization.

Wood, D. J., Cooper, C., Stevens, J. and Edwardson, J. A. (1988). Bone mass and dementia in hip fracture patients from areas with different aluminium concentrations in water supplies. *Age and Ageing*, **17**, 415–19.

Ysart, G., Miller, P., Croasdale, M. *et al.* (2000). 1997 UK total diet study – dietary exposures to aluminium, arsenic, cadmium, chromium, copper, lead, mercury, nickel, selenium, tin and zinc. *Food Additives and Contaminant*s, **17**, 775–86.

Chapter 16
Odour and taste

16.1 Source of odour and taste problems

Most odour and taste problems occur at the water treatment stage and are linked to chlorination (Levi and Jestin, 1988). Chlorine itself has a distinctive odour with a reported taste threshold of 0.16 mg l^{-1} at pH 7 and 0.45 mg l^{-1} at pH 9. Although consumers generally accept a slight chlorine odour as a sign that the water is microbially safe, excessive concentrations of chlorine can make the water most objectionable. In recent years a number of outbreaks of diarrhoeal diseases have occurred when water companies have reduced the level of disinfection due to complaints of chlorinous odours. Clearly a balance must be struck between the protection of public health and wholesomeness in terms of taste and odour. This is an operational problem which can be solved in a number of ways, such as using more sensitive disinfection equipment at the treatment plant, disinfecting within the supply zone or distribution network or by using alternative disinfectants.

Ammonia reacts with chlorine to produce three chloramines (monochloramine, dichloramine and trichloramine or nitrogen trichloride). These compounds are more odorous than free chlorine and become progressively more offensive with increasing numbers of chlorine atoms, with trichloramine by far the worse. The proportion of each compound formed depends on the relative proportions of chlorine and ammonia present, the chlorine demand exerted by other substances and the pH (Montgomery, 1975). This is usually avoided by using breakpoint chlorination in which a high chlorine to ammonia ratio is used (i.e. >7.6:1 at pH 7–8) so that the residual chlorine present is mainly in the free form (Section 14.2). It is not only ammonia that reacts in this way to produce odorous compounds, although the reactions are considerably slower and continue within the distribution system (Section 21.1).

Phenolic compounds have already been mentioned in Chapter 8. They produce odorous compounds during chlorination. Phenol itself has very little odour, but its chlorinated forms monochlorophenol and dichlorophenol have an intense odour and are difficult to remove by subsequent treatment. Trichlorophenol is the least objectionable, causing little odour, so by adjusting the chlorine to phenol ratio the formation of the more odorous compounds can be avoided. Alternative disinfectants are usually used if phenol is detected in the raw water until its source can be identified and eliminated. However, the chlorination of phenol can cause odour problems due to the formation of chlorophenols even when the phenol in the raw water is below the detection limit. For example, the phenolic compound *p*-cresol has an odour threshold concentration of $55 \, \mu g \, l^{-1}$ compared with just $2 \, \mu g \, l^{-1}$ for dichlorophenol. Many phenolic compounds occur naturally and are found in lowland rivers, making odour problems in the treated water inevitable at certain times of the year.

Algae such as *Synura* produce extracellular products that are themselves odorous. However, such odours can be enhanced by chlorination, thereby intensifying the original problem. Algae and actinomycetes accumulate and even grow on the filter media of sand filters and can impart associated odours and tastes if not kept under control. They produce a range of odorous compounds, in particular geosmin and 2-methylisoborneol (Table 8.1). Regular back washing controls the development of these organisms, but as some odorous compounds that can be produced, such as dimethyldisulphide, have an odour threshold concentration of just $10 \, \mu g \, l^{-1}$ the water used for backwashing should not be returned to the start of the plant for treatment if such compounds are suspected of being present as this would cause odour problems in the finished water (Sections 8.4 and 11.2). Research in this area is reported by Pearson *et al.* (1992).

16.2   Removing odours and tastes

Conventional water treatment, apart from possibly adding odours to drinking water, is not very effective at removing them, although the biological activity that occurs in slow sand filters may remove (oxidize) some odorous compounds. It may be possible to prevent odour problems arising by carefully managing the raw water source more effectively by ensuring it is kept free from industrial pollution, that phenolic and nitrogenous compounds that could react with chlorine are eliminated, and that the growth of algae and other organisms, which could produce odours, are controlled. Access to an alternative water resource, free of odour problems, is less likely, but would be effective.

Where odours and tastes are a problem, specific treatment methods must be used (Mallevialle and Suffet, 1987). There are three options available: (1) *Adsorption*, either physical adsorption onto activated carbon or biological adsorption using a biofilm that grows on a natural humus material such as peat; (2) *Aeration*, which strips any volatile odorous compounds from the water; (3) *Chemical oxidation* using ozone, potassium permanganate, chlorine or chlorine dioxide to chemically break down odorous compounds into non-odorous forms.

The most effective method is adsorption using activated carbon, as this removes a wide range of odours and is especially effective against the two most problematic odours. These are the musty–earthy odours produced by actinomycetes, the compounds geosmin and 2-methylisoborneol. In most instances odour problems only occur once or twice a year due to seasonal peaks in algal or actinomycete growth. Therefore activated carbon is normally used in a powdered form and added to the water stream as a slurry before sedimentation or rapid sand filtration (Najm *et al.*, 1991). The powdered activated carbon (PAC) is disposed of with the normal waterworks sludge. The use of PAC is more cost effective than using granular activated carbon (GAC) in the short term as it does not require the expensive construction of a special bed or filter to house the GAC nor facilities to regenerate it. However, if the odour problem is permanent then a GAC system will be required as it is more cost effective in the long term due to low operating costs, even though the initial capital costs are high. Therefore PAC systems are only suitable for short-term, intermittent use. The efficiency of activated carbon in removing odorous compounds is severely affected by the presence of other organic compounds such as natural humic substances. This can be a particular problem with GAC as it increases the operational costs significantly as the bed must be taken out of service more often.

There is currently much interest in the use of biofilms grown on natural organic polymers to remove odours. The action is the same as activated carbon, by adsorption of the odour-producing compound, but instead of having to reactivate the carbon by heat, the biofilm, which is composed of bacteria and other micro-organisms, degrades the odorous chemicals so that the system never needs to be shut down. This makes biofilm systems much cheaper to operate than conventional activated carbon units. However, there is the possibility that odour-producing organisms may themselves colonize these biofilms.

Aeration can be successful in controlling odours caused by volatile organic compounds or dissolved gases such as hydrogen sulphide. It is also effective in controlling trichloramine. However, it is unable to remove the main odorous compounds and so is of limited use.

Chemicals can be used to oxidize organic compounds as they disinfect the water. None is effective against all odours, with each particularly effective against a specific range of odorous compounds. For example, chlorine can control sulphides in groundwater, but will generally make the odour problem worse for the reasons already discussed. It is also effective against organic sulphur compounds such as dimethyltrisulphide, which causes a fishy odour. Chlorine dioxide can be used under certain circumstances instead of chlorine to disinfect raw water. If phenolic compounds are suspected then chlorine dioxide can be used without chlorophenolic odours being formed. It is also effective at reducing all other odours except those caused by hydrocarbons in the water. Potassium permanganate is used widely in the USA for odour control, but not in the UK. It is applied before filtration, when it is reduced to manganese oxide, which is insoluble and so forms a precipitate which helps to remove odours.

The precipitate is removed by filtration, but this process is now generally considered to be inefficient and has been largely replaced with ozone, which can reduce or change the nature of odours fairly effectively. The problem, however, is that the ozone literally breaks up the large organic molecules into smaller ones, which are much easier for micro-organisms to use as a food source. This tends to encourage microbial growth in the distribution system. Ozonation has also been reported to produce rather intense fruity odours.

16.3 Conclusions

Taste and odour problems normally originate from the resource with chlorination the major source of odour derived from treatment. However, a balance has to be struck between the critical issue of microbial safety of drinking water and the lesser problem of taste. Careful operational management in relation to chlorination is necessary to overcome taste problems, with ozone and other advanced treatment processes required where problems are persistent or extreme.

References

Levi, Y. and Jestin, J. M. (1988). Offensive tastes and odors occurring after chlorine addition in water treatment processes. *Water Science and Technology*, **20**(8–9), 269–74.

Montgomery, J. M. (1975). *Water Treatment Principles and Design*. New York: Wiley.

Najm, I. N., Snoeyink, V. L., Lykins, B. W. and Adams, I. Q. (1991). Using powered activated carbon: a critical review. *Journal of the American Water Works Association*, **83**(1), 65–76.

Pearson, P. E., Whitefield, F. B. and Krasner, S. W. (1992). *Off Flavours in Drinking Water and Aquatic Organisms*. Oxford: Pergamon Press.

Mallevialle, J. and Suffet, I. H. (eds.) (1987). *Identification and Treatment of Tastes and Odours in Drinking Water*. Denver: American Water Works Association Research Foundation and Lyonnaise des Eaux.

Chapter 17
Fluoridation

17.1 Introduction

The fluoridation of water supplies has been as controversial in the past as trace organics in water is today. It was introduced in the 1940s to reduce the incidence of tooth decay in the population after a number of surveys in the USA had shown that it had a beneficial effect. Three relationships were identified: (1) fluoride levels in excess of $1.5\,\text{mg l}^{-1}$ led to an increase in the occurrence and severity of dental fluorosis (i.e. teeth became mottled and brittle) without decreasing the incidence of decay, missing or filled teeth; (2) at $1.0\,\text{mg l}^{-1}$ there was the maximum reduction of decay with no fluorosis; and (3) at concentrations less than $1.0\,\text{mg l}^{-1}$ some benefit was observed. With decreasing concentrations of fluoride in water there was an increase in the incidence of tooth decay.

These studies therefore showed that the addition of fluoride to water supplies to bring the level above $0.6\,\text{mg l}^{-1}$ led to a reduction in tooth decay in growing children, and that the optimum beneficial effect occurred around $1.0\,\text{mg l}^{-1}$. Other studies have indicated that fluoride is also beneficial to older people in reducing the hardening of the arteries and, as fluoride stimulates bone formation, in the treatment of osteoporosis, although the most recent evidence contradicts this (Cooper *et al.*, 1990). These early studies were so convincing that fluoridation was adopted around the world, so that today over 250 million people drink artificially fluoridated water. In the USA, for example, over half of the water supplies are fluoridated. It is widely used in New Zealand, Canada and Australia; in fact, it is widespread in most English-speaking countries. Brazil and the former Soviet Union are two other countries with a strict fluoridation policy. In the Republic of Ireland all water supplies are fluoridated where the natural levels are less than $1.0\,\text{mg l}^{-1}$, none are fluoridated in Northern Ireland and only about 10% of supplies are fluoridated in the UK. In western continental Europe its use has been terminated or never implemented, primarily over health concerns and less than 1% of Europeans drink artificially fluoridated water (Table 17.1). For example, in the Netherlands fluoridation has been phased out and it is also banned in Sweden. In contrast, Spain has been gradually introducing fluoridation since 1985.

Table 17.1 *The number of decayed, missing and filled teeth (DMFT) in 12 year olds in some European countries and whether fluoridation of drinking water has been adopted*

Country	DMFT at 12 years	Fluoridation
Austria	1.70	No
Belgium	2.70	No
Denmark	1.01	No
Finland	1.10	No
France	1.94	No
Germany	1.70	No
Greece	2.70	Partial
Iceland	2.70	No
Ireland	1.10	Yes
Italy	2.12	No
Luxembourg	2.30	No
Netherlands	0.60	No
Norway	1.50	No
Portugal	3.08	No
Spain	2.30	Yes
Sweden	1.00	No
United Kingdom	1.40	Partial

Many areas have naturally occurring fluoride in water (Figure 17.1). In some areas high natural levels of fluoride have to be reduced either by treatment or mixing with low-fluoride water to bring the concentration to less than $0.9 \, mg \, l^{-1}$. The EC Drinking Water Directive (98/83/EEC) sets a maximum value for fluoride at $1.5 \, mg \, l^{-1}$. This value has been adopted in the UK, although in the Republic of Ireland a national limit of $1.0 \, mg \, l^{-1}$ has been set. The effects of fluoride vary with temperature, so as the temperature increases the concentration permitted in drinking water decreases (i.e. $1.5 \, mg \, l^{-1}$ at 8–$12 \, °C$, or $0.7 \, mg \, l^{-1}$ at 25–$30 \, °C$). Some bottled mineral waters may also contain high levels of fluoride (Section 29.2). The US Environmental Protection Agency (USEPA) has set two standards for fluoride: a Primary Drinking Water Standard of $4.0 \, mg \, l^{-1}$ to protect against bone disease, and a non-mandatory Secondary Drinking Water Standard of $2.0 \, mg \, l^{-1}$ to prevent dental fluorosis (Appendix 2). The World Health Organization (WHO) has set a revised guide value of $1.5 \, mg \, l^{-1}$, although they state that the volume of water consumed and intake from other sources should be considered when setting national standards.

17.2 Fluoride addition

Fluoride must be added after coagulation, lime softening and activated carbon treatment, as it can be lost during any of these unit processes. Filtration will not

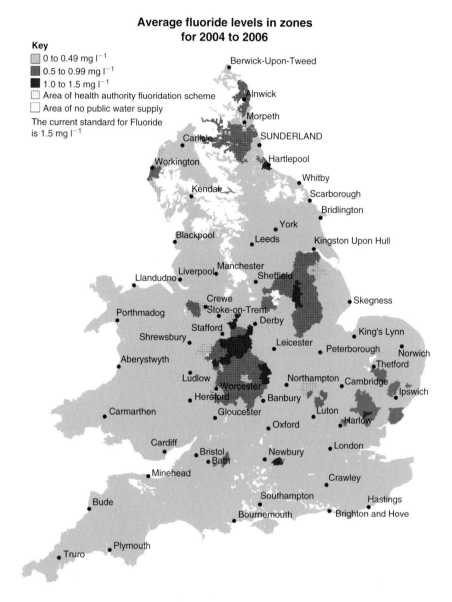

Average fluoride levels in zones for 2004 to 2006

Key
- 0 to 0.49 mg l^{-1}
- 0.5 to 0.99 mg l^{-1}
- 1.0 to 1.5 mg l^{-1}
- Area of health authority fluoridation scheme
- Area of no public water supply

The current standard for Fluoride is 1.5 mg l^{-1}

reduce the fluoride concentration provided it is fully dissolved before reaching the filter. As chlorination has no effect on fluoride, fluoridation is normally carried out on the finished water either before or after chlorination.

Fluorine in its pure elemental state is very reactive and so is always used as a compound. Those most widely used for the fluoridation of supplies are: ammonium fluorosilicate [$(NH_4)_2SiF_6$], which is supplied as white crystals; fluorspar or calcium fluoride (CaF_2) as a powder; fluorosilicic acid (H_2SF_6) as a highly corrosive liquid, which must be kept in rubber-lined drums or tanks; sodium fluoride (NaF), which is either a blue or white powder; and sodium

silicofluoride (Na_2SiF_6), which is also either a white or blue powder. All these compounds dissociate (break up) in water to yield fluoride ions, with the release of the fluoride being almost complete. Silicofluorides are generally used in water treatment in the British Isles, with sodium silicofluoride the cheapest and most convenient to use. The reaction that takes place when it is added to water is:

$$SiF_6 + 3H_2O \leftrightarrow 6F^- + 6H^+ + SiO_3$$

Silicofluoride   Fluoride   Silicate

The compound is emptied into hoppers and then slowly discharged into a tank by a screw feed where it is mixed with water to dissolve it. After correction of the pH the fluoride solution is added to the finished water. Fluoride is removed by precipitation with calcium and excess aluminium coagulant in the finished water, with the precipitate forming in the distribution system. Care is taken to ensure that the dosage is kept at the correct level. This is done by monitoring the weight of chemical used each day and the volume of water supplied. Also fluoride concentration is tested daily by the operator, using a simple test kit, or continuously using an ion-specific electrode and meter to ensure the final concentration is as near to $1.0\,mg\,l^{-1}$ as possible. Fluoride in the home can be removed using a point-of-use treatment system employing either reverse osmosis or activated alumina defluoridation filters (Section 29.3) (Table 3.6).

17.3   Fluoridation and health

Excess fluoride causes teeth to become discoloured (fluorosis) and long-term exposure results in permanent grey to black discolouration of the enamel. Children who drink water containing fluoride in excess of $5\,mg\,l^{-1}$ also develop severe pitting of the enamel. Where excess fluoride is a problem, defluoridation is required. Leaks and spillages of fluoride can be very serious and even fatal. Accidents have occurred resulting in overdosing of fluoride, with reported levels at the consumer's tap of between 30 and $1000\,mg\,l^{-1}$. Excessive fluoride will lower the pH of the water increasing its corrosiveness and thereby increasing the concentration of lead and copper in the water. The general effect of moderately increased fluoride (30–$50\,mg\,l^{-1}$) will be mild gastroenteritis and possible skin irritation (Peterson et al., 1988).

Fluoridation, apart from reducing caries, has been implicated in dental fluorosis, bone health, cancer, immunological effects, reproductive problems, birth defects, renal and gastrointestinal problems, as well as neurological effects. However, it is dental fluorosis and bone health that have received most attention.

Dental fluorosis is a developmental defect of the tooth enamel. It ranges from barely visible white striations on the teeth to severe staining of the enamel and gross defects (Table 17.2). It is widely held in the dental profession that minor forms of dental fluorosis do not represent an aesthetic problem, and may even

Table 17.2 *The standard classification of mottling (i.e. fluorosis) of tooth enamel. Adapted from Dean (1932) with permission.*

Classification	Description of enamel
Normal	Smooth, glossy, pale creamy-white translucent surface
Questionable	A few white flecks or spots
Very mild	Small, opaque, paper-white areas covering <25% of tooth surface
Mild	Opaque white areas covering <50% of tooth surface
Moderate	All tooth surfaces affected; marked wear on biting surfaces; brown stain may be present
Severe	All tooth surfaces affected; discrete or confluent pitting; brown stain present

enhance the appearance of teeth (Hawley *et al.*, 1996). A review of 88 studies showed a total prevalence of fluorosis of 48% in fluoridated areas compared to 15% in non-fluoridated areas; and for what is considered by dentists as aesthetically important fluorosis of 12.5% in fluoridated areas compared to 6.3% in non-fluoridated areas (NHS, 2000). This data also included naturally occurring fluoride, so when the figures are adjusted to remove any fluoride water values $>1.5\,\text{mgl}^{-1}$, then total fluorosis has a prevalence of 46% and 10% in fluoridated and non-fluoridated areas respectively, and 18% and 6% for aesthetically important dental fluorosis (Section 17.5). However, the degree of fluorosis is really a factor of total exposure to fluoride rather than the level in drinking water (WHO, 2004). Although Harrison (2005) has estimated that drinking water in artificially fluoridated areas represents 81% and 66% of the daily intake of fluoride in adults and children respectively, tooth-brushing represents a further 5% and 25% for the two groups respectively. The high levels of fluorosis reported in the USA may be due to the use of discretionary fluoride products. Also low-fluoride toothpaste is not available in the USA (USDHHS, 2001).

Half of the fluoride ingested is taken up by the bone, and water fluoridation can increase the dietary intake of the mineral by 50%. The fluoride ion is known to replace hydroxyl ions in the hydroxyapatite lattice that forms bone. It is suspected that this results in loss of mechanical strength leading to fractures, especially of the hip, and otosclerosis, although the evidence is not strong. The WHO (2004) has indicated that a dietary intake of $6\,\text{mg d}^{-1}$ leads to adverse skeletal effects.

Bone cancer, as well as cancer of the stomach, kidney and thyroid, have all been studied in relation to water fluoridation (MRC, 2002). The results so far suggest that there is no evidence exposure to artificially fluoridated water

increases the risk for any cancers in humans, however, the possibility of some genotoxic effect can not be ruled out completely (NRC, 1993; USDHHS, 2001). There are some reports that fluoride causes an allergic or hypersensitivity reaction in sensitive individuals (Spittle, 1993), although more recent studies have failed to confirm this (Challacombe, 1996). Likewise there is no strong evidence to support other reports that fluoridated water affects the kidneys, thyroid, fertility, pregnancy or causes birth defects (Harrison, 2005). The implication that fluoride affects intelligence levels in children is most likely associated with increased plumbosolvency of fluoridated water releasing lead into solution (Section 27.3).

17.4 Fluoridation and public opinion

Fluoridation has been universally heralded as being a useful and positive process. However, new studies have shown that when similar communities in developed countries are compared there is little difference in the levels of tooth decay between communities using fluoridated and unfluoridated water. The primary explanation given is the now widespread use of fluoridated toothpaste and better dental hygiene. However, where significant differences are seen is when more deprived social groups are compared to their more affluent counterparts.

The major force against fluoridation in the UK was the National Anti-fluoridation Campaign, which was formed in 1960, although they changed their name in 1990 to the National Pure Water Association and are now concerned with all aspects of water quality, although fluoride is still one of their major issues. Their opposition to fluoridation focusses on the following points: (1) fluoride is one of the most toxic inorganic poisons known; (2) it is a precipitative and cumulative drug as well as being an industrial waste product; (3) as fluoride is in the drinking water you have no control over your daily intake; (4) it does not boil off and so it is both difficult and expensive to avoid; (5) fluoride has been rejected because of its damage to health by 13 western European countries; (6) fluoride does not reduce the rate of tooth decay, which can be attributed to other factors including a better diet; (7) the function of public water is to provide safe drinking water, not to serve as a vehicle for dispensing drugs to treat people.

In Ireland, those campaigning against fluoridation have raised concerns that many of the 460 fluoridating schemes in the country fail to monitor natural background concentrations sufficiently leading to concerns that fluoride levels may be regularly breeched. They are also concerned about potentially toxic compounds formed with metals by fluoride in water.

You must make up your own mind about these claims. What is certain is that: there is evidence to show that water fluoridation no longer makes a significant difference to the incidence of tooth decay due to the widespread availability of other sources of fluoride (USPHS, 1991); the incidence of dental fluorosis has

increased since fluoridation in some areas; sensitivity to fluoride, which is well known from patients receiving fluoride treatment for osteoporosis, is also seen occasionally in people who drink large amounts of fluoridated water (e.g. avid tea and coffee drinkers); and preliminary studies carried out by the USEPA have indicated that fluoride may be a carcinogen, although the available epidemiological evidence has not found a causal link between the fluoridation of drinking water and cancer (Chilvers and Conway, 1985). Fluoride is cumulative and so any long-term effects of low-dose exposure may not yet be realized. All these findings need more research, as do the claims of the anti-fluoridation campaigners who are found all around the world.

17.5   Who is at risk

Worldwide many millions of people are exposed to elevated fluoride through their drinking water, which is found naturally elevated in their groundwater. This leads to a wide range of effects from mild dental fluorosis to skeletal fluorosis, which is a crippling condition. Affected areas include Africa, Eastern Mediterranean and South-east Asia. In many arid areas, such as the Rift Valley in Africa, there are no alternative uncontaminated water supplies so that the fluoride cannot be avoided or diluted. The most seriously affected area extends from Turkey through Iraq, Iran, Afghanistan, India, Northern Thailand and China. Removal of fluoride is expensive and technically difficult, although at the local level low-cost solutions are being developed including adsorption onto crushed clay pots and bone charcoal, contact precipitation or the use of activated alumina filters (Fawell *et al.*, 2006).

Skeletal fluorosis, a condition where excess fluoride is stored in bones, has been associated with avid tea drinkers (Whyte *et al.*, 2005) as has dental fluorosis in children in Tibet (Cao *et al.*, 1996). In a detailed study of 37 commercially available black teas, the fluoride content was found to range between 0.7 and 6.0 mg l^{-1} when the tea was made using fluoride-free water (Cao *et al.*, 2006). Tea bags had the highest values being made of older tealeaves. The WHO (1984) has recommended a maximum daily fluoride intake of 2 mg for children and 4 mg for adults. Based on a daily consumption of tea of 800 ml for children and 1500 ml for adults 56% of the brands were considered unsafe for consumption by children and 44% for adults. When fluoridated water is used to make tea then the situation is made significantly worse.

Those most at risk are formula-fed infants and very young children who drink mainly tap water. Other people such as outdoor workers, long-distance runners, people with diabetes insipidus and in fact anyone who drinks large amounts of tap water are at risk. Other groups who are at even greater risk are those with malfunctioning kidneys who store excess fluoride in their bones. The reason for the fluoridation of drinking water was to reduce the level of dental decay in children. However, after the age of 12 years calcification of teeth

stops so that fluoride is of much less benefit. Therefore it does seem more appropriate to prescribe fluoride drugs solely for children rather than this mass-medication approach. Fluoride is widely available in toothpaste, as well as in other forms, so there is now a ready source of fluoride for children.

Giving children fluoride supplements and then allowing them to use fluoride toothpaste can result in serious dental fluorosis. Some medical officers have indicated that fluoride toothpaste is not necessary when water has been fluoridated, or if it is used then only small amounts of toothpaste should be placed on brushes to avoid the possibility of dental fluorosis, albeit unlikely. However, manufacturers of toothpaste are advising parents to use only a pea-sized amount of fluoride toothpaste when other fluoride supplements are taken, which includes access to fluoridated water. Better control of such supplements has resulted in a decline in dental fluorosis within the community (Riordan, 2002).

17.6 Conclusions

There is clear evidence that fluoride benefits children during the development of their teeth by reducing the incidence of dental decay. However, dental fluorosis is an inevitable consequence of water fluoridation. The prescribing of fluoride supplements by doctors as well as dentists increases the risk of fluorosis. Also many countries have school-based systems for fluoride medication, where parents are required to opt out rather than opt into such schemes. While the general consensus of dentists and health practitioners is that fluoride is a good thing, the problem is that there are now so many discretionary sources of fluoride available, overexposure to the mineral is inevitable, especially where the water is fluoridated. The advice of the dental profession to avoid fluoride toothpastes when supplements are given is unhelpful when in many countries, including the USA, low-fluoride toothpaste is not available. Also the advice given by expert panels on the use of fluoridated toothpaste, such as the Irish Forum on Fluoridation (Department of Health and Children, 2002), requires: (1) that toothpaste should not be used when brushing children's teeth up to the age of 2; rather a toothbrush and tap water only should be used and (2) parents should supervise children aged 2 to 7 years when brushing to ensure that only a small pea-size amount of fluoride toothpaste is used and that swallowing of toothpaste is avoided. The Forum concludes this advice by saying that the use of low-fluoride toothpaste cannot be recommended without further research. One of the critical issues here is the level of fluorosis that is acceptable to young people, especially given the huge increase in expenditure on cosmetic dentistry. The system of classification of fluorosis is based on a visual index developed over 70 years ago (Dean, 1932) (Table 17.2). Visual examples of dental fluorosis corresponding to the index devised by Dean can be accessed at www.fluoridealert.org/health/teeth/fluorosis/criteria.html#tf. The concept that even the lowest categories of fluorosis are acceptable to a modern generation, and indeed felt by some dentists as adding to the aesthetic quality of teeth (Department of Health and Children, 2002), is quite surprising. A major review on fluoridation in Ireland concluded that the prevalence and severity of dental fluorosis in the country was increasing with 24% of all children tested in 1997 showing some degree of fluorosis (questionable 14%, very mild 6%, mild 3%, moderate 1% and severe 0%) (Department of Health and Children, 2002). This level of fluorosis is lower than that shown in earlier studies, which predicts that at a fluoride concentration of 1 mg l^{-1} approximately 47% of the population would show some signs of fluorosis, of which approximately 18% would be classed as very mild or worse (Figure 17.2) (NHS, 2000). This increasing trend in dental fluorosis has been reported in many other countries including the USA (Kumar *et al.*, 1989) and Canada (Ismail *et al.*, 1990). The situation is so serious in the USA that new strategies are being developed to reduce the intake of fluoride in infants and young children (USDHHS, 2001). So fluorosis appears to be the price we pay for fluoridated water, although other reported risks appear low.

The mass-medication approach of water fluoridation to benefit a very small minority of children while at the same time unnecessarily exposing the whole community to elevated fluoride levels that causes fluorosis, as well as may being implicated in more serious direct and indirect health-related problems, appears perverse. Either discretionary sources of fluoride must be banned, including fluoride toothpaste, or water fluoridation curtailed and more effort made to offer the dental support to those at most risk. What are children drinking anyway? Consumer studies have shown that the average intake of water by children has steadily declined over the past 20 years from the often-quoted 1.4 litres per day,

Figure 17.2 Relationship between fluoride levels in water and dental fluorosis. The explanation of the degree of fluorosis is given in Table 17.2. Reproduced from Dunning (1986) with permission from the President and Fellows of Harvard College and Harvard University Press, Cambridge, Massachusetts.

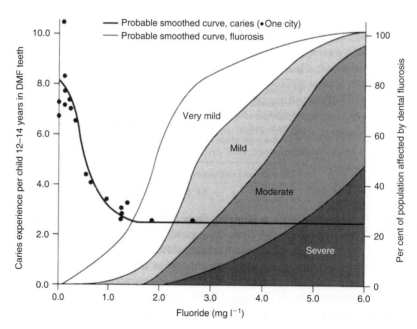

replaced by acidic, sugar-based drinks. Tea can add another $6 \, \text{mg} \, \text{l}^{-1}$ of fluoride to the water it is made with, so a single 250 ml mug of tea could be equivalent of drinking a 1.25 litres of water if fluoridated to $1.5 \, \text{mg} \, \text{l}^{-1}$. The current assumed consumption of water per child is 1.4 litres and 2.0 litres for an adult. So controlling fluoride intake is very difficult if not impossible, especially where water consumption is concerned.

Countries such as Ireland have an evangelical desire to fluoridate their water and have taken extraordinary measures to reassure the public including on-line analysis of fluoride concentrations in finished water and fluoride purity of the hydrofluorosilic acid used at treatment plants (www.fluoride.ie). In contrast the UK water companies have quietly been phasing it out. However, the Water Act 2003 will reverse this trend by placing water companies in England and Wales under a duty to fluoridate water supplies if requested to do so by a Strategic Health Authority. Water companies that fluoridate are required to ensure that the concentration of fluoride in the water supplied is, so far as is reasonably practicable, maintained at $1.0 \, \text{mg} \, \text{l}^{-1}$. However, as we have seen this leads to fluorosis. But in the short term, at least, it appears that fluoridation will continue in many countries.

Fluoride is one of the rare chemicals found in drinking water that can have significant but opposite effects at both low and high concentrations. At low concentrations it has beneficial effects on teeth but excessive exposure in drinking water can lead to a number of adverse effects including skeletal fluorosis. Naturally elevated fluoride concentrations in groundwater supplies have left tens of millions of people with this crippling condition, which is now acknowledged as a major drinking water emergency of global proportions.

References

Cao, J., Bai, X. X., Zhao, Y. *et al.* (1996). The relationship of fluorosis and brick tea drinking in Chinese Tibetans. *Environmental Health Perspectives*, **104**, 1340–3.

Cao, J., Zhao, Y., Li, Y. *et al.* (2006). Fluoride levels in various black tea commodities: measurement and safety evaluation. *Food and Chemical Toxicology*, **44**(7), 1131–7.

Challacombe, S. J. (1996). Does fluoridation harm immune function? *Communications in Dental Health*, **13**(Suppl. 2), 69–71.

Chilvers, C. and Conway, D. (1985). Cancer mortality in England in relation to levels of naturally occurring fluoride in water supplies. *Journal of Epidemiology and Community Health*, **39**, 44–7.

Cooper, C., Wickham, C., Lacey, R. F. and Barker, D. J. P. (1990). Water fluoride concentration and fracture of the proximal femur. *Journal of Epidemiology and Community Health*, **44**, 17–19.

Dean, H. T. (1932). Classification of mottled enamel diagnosis. *Journal of the American Dental Association*, **21**, 1421–6.

Department of Health and Children (2002). *Forum on Fluoridation 2002*. Dublin: Department of Health and Children, Stationery Office.

Dunning, J. M. (1986). *Principles of Dental Public Health*, 4th edn. Massachusetts: Harvard University Press.

Fawell, J., Bailey, K., Chilton, J. *et al.* (2006). *Fluoride in Drinking Water*. Geneva: WHO.

Harrison, P. T. C. (2005). Fluoride in water: a UK perspective. *Journal of Fluorine Chemistry*, **126**, 1448–56.

Hawley, G. M., Ellwood, R. P. and Davies, R. M. (1996). Dental caries, fluorosis and the cosmetic implications of different TF scores in 14-year-old adolescents. *Communications in Dental Health*, **13**, 189–92.

Ismail, A. L., Brodeur, J. M., Kavanagh, M. *et al.* (1990). Prevalence of dental caries and dental fluorosis in students 11–17 years of age in fluoridated and non-fluoridated cities in Quebec. *Caries Research*, **24**, 290–7.

Kumar, J. V., Green, E. L. and Wallace, W. (1989). Trends in dental fluorosis and dental caries prevalences in Newburgh and Kingston, N.Y. *American Journal of Public Health*, **79**, 565–9.

MRC (2002). *Water Fluoridation and Health*. Working Group Report. London: Medical Research Council.

NHS (2000). *A Systematic Review of Public Water Fluoridation*. CRD Report No. 18. University of York: NHS Centre for Review and Dissemination.

NRC (1993). *Health Effects of Ingested Fluoride*. Washington, DC: National Academy Press.

Peterson, L. R., Denis, D., Brown, D., Hadler, J. L. and Helgerson, S. D. (1988). Community health effects of a municipal water supply hyper-fluoridation accident. *American Journal of Public Health*, **78**(6), 711–13.

Riordan, P. J. (2002). Dental fluorosis decline after changes to supplement and toothpaste regimens. *Community Dentistry and Oral Epidemiology*, **30**(3), 233–40.

Spittle, B. (1993). Allergy and hypersensitivity to fluoride. *Fluoride*, **26**(4), 267–73.

USDHHS (2001). *Recommendations for Using Fluoride to Prevent and Control Dental Caries in the United States*. US Department of Health and Human Services Centres for Disease Control and Prevention, Morbidity and Mortality Weekly Report 50 (RR-14). Atlanta, GA: US Department of Health and Human Services.

USPHS (1991). *Review of Fluoride Benefits and Risks*. Report of the Ad-Hoc Sub-Committee on Fluoride of the Committee to Co-ordinate Environmental Health and Related Programs. Washington, DC: US Public Health Service.

WHO (1984). *Fluorine and Fluoride*. Environmental Health Criteria 26. Geneva: World Health Organization.

WHO (2004). *Guidelines for Drinking Water Quality*, Vol. 1. *Recommendations*, 3rd edn. Geneva: World Health Organization.

Whyte, M. P., Essmyer, K., Francis, H., Gamon, F. H. and Reimas, W. R. (2005). Skeletal fluorosis and instant tea. *American Journal of Medicine*, **118**, 78–82.

Chapter 18
Disinfection by-products

18.1 Formation

Since the beginning of the twentieth century chlorine has been used to disinfect drinking water. Although it has provided an effective barrier to the spread of waterborne diseases, chlorine is also very reactive towards the natural compounds present in water. Humic acids, which give peaty water its clear brown colour, and ammonia interfere with the disinfection process, whereas other compounds such as phenol react with chlorine to affect both taste and odour (Section 14.2).

It was not until the development of new analytical methods such as gas chromatography coupled with mass spectrometry, which can identify the organic compounds present in water, in the early 1970s that another problem related to chlorination was identified. This new technology revealed that water could contain hundreds of natural and man-made organic compounds, most of which were present only at very low concentrations of less than $1 \mu g \, l^{-1}$. It was discovered that some of these organic compounds could react with the chlorine during the disinfection process at the treatment plant to form new, complex and often dangerous chemicals known generically as disinfection by-products (DBPs). Chloroform and other trihalomethanes (THMs) were first identified in chlorinated drinking water taken from the River Rhine (Rook, 1974). Trihalomethanes (THMs) and haloacetic acids (HAAs) are the commonest DBPs in drinking water although other groups and compounds are also formed including haloacetonitriles (HANs), haloketones (HAKs) chloral hydrate (CH) and chloropicrin (CP) (Nikolaou $et \, al.$, 1999).

Trihalomethanes are simple, single-carbon compounds that have the general formulae CHX_3, where X may be any halogen atom (i.e. chlorine, bromine, fluorine, iodine, or a combination of several of these). They are all considered to be possible carcinogens and are therefore undesirable in drinking water. There are four common THMs found in drinking water: chloroform ($CHCl_3$), bromodichloromethane ($CHBrCl_2$), dibromochloromethane ($CHBr_2Cl$) and bromoform ($CHBr_3$). The THMs and other chlorinated by-products are only found in raw or treated waters that are disinfected using chlorine. For a specific chlorine dose, the rate and degree of THM formation increases at higher concentrations of organic substrate, higher temperatures and high pH values. Where free chlorine residuals exist in the water THM formation will continue until either the chlorine or organic substrate is exhausted. The chlorine to substrate concentration ratio (i.e. the amount of chlorine added in relation to the amount of organic matter present) is also an important factor in determining which by-products are formed. For example, at low chlorine dose rates phenol is converted to taste-producing chlorophenols, whereas at higher dosages these are converted to tasteless chlorinated quinones.

The presence of other halogens, especially bromine, in the water is also important. Bromide is oxidized by chlorine to hypobromous acid, which then results in brominated analogues of the chlorinated by-products. For example, bromoform is the analogue of chloroform (Fawell *et al.*, 1987). The reaction pathways are extremely complex (Peters *et al.*, 1978). Chloroform is the major THM found in chlorinated drinking waters and concentrations of 30–60 μg l^{-1} are common, with maximum levels of up to 500 μg l^{-1} recorded. In contrast, bromoform is typically found at much lower concentrations of <10 μg l^{-1}. The occurrence and concentrations of by-products are reviewed by Cable and Fielding (1992).

Since the early 1970s a huge range of compounds has been identified which are formed by chlorine reacting with natural humic material and other organic compounds such as amino acids that are found in water (Table 18.1). Other commonly occurring chlorinated by-products include di- and trichloroacetic acids, chloral hydrate, dichloroacetonitrile, chloropicrin, chlorinated acetones, 2-chloropropenal and in most instances their brominated analogues. Although many of these by-products are mutagenic they are generally found in very low concentrations of <1 μg l^{-1}, with the major compounds such as the haloforms (chloroform and bromoform) and di- and trichloroacetic acids normally present in water at 1–50 μg l^{-1}. However, some by-products are very potent mutagens; for example 3-chloro-4-(dichloromethyl)-5-hydroxy-2(*5H*)-furanone (or MX for short). The presence of MX in drinking water at any concentration, no matter how low, is therefore undesirable. It is currently found in UK drinking waters at concentrations ranging from 1–60 ng l^{-1}, depending on the humic material present. It is responsible for between 30% and 60% of the total mutagenic potential of water (Horth *et al.*, 1991).

Unlike THMs, dihaloacetonitriles (DHANs) such as dichloroacetonitrile (DCAN) are unstable and readily broken down (hydrolysed) in aqueous solutions at increased temperatures and pH values, which can occur during storage or analysis. Details of concentrations in drinking water are therefore rare. The most common DHAN is DCAN, which was first reported in drinking water in 1976 (McKinney *et al.*, 1976), with its brominated analogues bromochloroacetonitrile and dibromoacetonitrile discovered several years later. Studies in the Netherlands have shown that all chlorinated drinking waters contain DHANs as well as THMs, but in much lower concentrations within the range 0.04–1.05 μg l^{-1}.

Table 18.1 *Common disinfection by-products found in drinking water*

Chlorine
  Trihalomethanes
    Chloroform (trichloromethane)[a]
    Bromoform (tribromomethane)[a]
    Bromodichloromethane (BDCM)[a]
    Dibromochloromethane (DBCM)[a]
  Other trihalomethanes: chlorinated acetic acids
    Monochloroacetic acid (MCA)[a]
    Dichloroacetic acid (DCA)[a]
    Trichloroacetic acid (TCA)[a]
  Other trihalomethanes: haloacetonitriles
    Dichloroacetonitrile (DCAN)[a]
    Dibromoacetonitrile (DBAN)[a]
    Bromochloroacetonitrile (BCAN)[a]
    Trichloroacetonitrile (TCAN)[a]
  Other
    Chloral hydrate[a]
    3-Chloro-4-(dichloromethyl)-5-hydoxy-2(*5H*)-furanone (MX)[a]
    Chloropicrin[a]
    Chlorophenols[a]
    Chloropropanones[a]
Ozone
  Bromate[a]
  Formaldehyde[a]
  Acetaldehyde[a]
  Non-chlorinated aldehydes
  Carboxylic acids
  Hydrogen peroxide
Chlorine dioxide
  Chlorite[a]
  Chlorate[a]
Chloramination
  Cyanogen chloride[a]

[a] Included in the 1993 World Health Organization drinking water guidelines.

There appears to be a direct correlation between THMs and DHANs, with the average DHAN concentration about 5% of the average THM concentration (Peters *et al.*, 1990). However, in the Netherlands chlorine dosage rates are five to ten times lower than in the USA or Canada, typically about 0.2–1.0 mg l^{-1}. This is why both THM and DHAN concentrations are so much higher in North America, with DHANs found at concentrations of 1–10 μg l^{-1}.

Two HAAs are commonly recorded in drinking waters, dichloroacetic and trichloroacetic acids. In a study of DBPs in drinking water in Athens (Greece) supplied by four water treatment plants all samples contained DBPs, the most commonest being chloroform (annual means for the four plants ranging from 8.0 to 42.5 µg l^{-1}), trichloroacetic acid (3.5 to 18.1 µg l^{-1}), dichloroacetic acid (2.3 to 24.5 µg l^{-1}) and monochloroacetic acid (1.1 to 61.8 µg l^{-1}). Total THMs never exceeded the EC maximum, with higher concentrations of both THMs and HAAs recorded during the summer and autumn (Golfinopoulos and Nikolaou, 2005). Rodriguez *et al.* (2004) also recorded seasonal variations in DBPs. Working in Canada, where both surface water quality and temperature varies significantly over the year, THM levels were on average five times higher in the summer and autumn than the winter. However, average HAAs in spring were four times higher than winter. Disinfection by-products continue to develop as the finished water flows through the distribution network, and Rodriquez and his co-workers showed that as the water travelled through the system the concentration of THMs increased with time and finally stabilized at the farthest extremities of the system. Haloacetic acids also increased with time, but in the warmer months began to fall again due to microbial degradation of dichloroacetic acid in particular. The study showed that contact time and temperature were important factors in the development of both THMs and haloacetic acids within the distribution network. The importance of contact time is also supported by Kim *et al.* (2002) who also found maximum DBP formation occurred at pH 7.0, although the formation potential of THMs and HAAs increased and decreased with pH respectively.

When waters containing bromide are oxidized, especially using ozone or hydrogen peroxide, bromate is formed. The inclusion of bromate in the World Health Organization (WHO) drinking water guidelines poses a number of problems to the water industry. Bromide is common in surface and ground waters, and although it can arise from industrial discharges, bromide is particularly common in rainwater and certain groundwaters as a residual from seawater. It is particularly common in Dutch surface waters. Bromate formation will therefore occur whenever ozone is used for disinfection, or during removal of pesticides with activated carbon.

18.2 Standards

The EC Drinking Water Directive (98/83/EEC) sets a limit for total THMs of 100 µg l^{-1}, which is calculated as the sum of the four commonest THMs, chloroform, bromoform, dibromochloromethane and bromodichloromethane. Some European countries have set more stringent maximum levels for total THMs, for example 30 µg l^{-1} in Italy . The WHO (2004) has set individual guideline values for each these four compounds (Table 18.2). A guideline value for THMs has been set as the sum of the ratio of the concentration of each to its

Table 18.2 *The WHO (2004) standards for the common THMs*

Chloroform	$CHCl_3$	$200\,\mu g\,l^{-1}$
Bromoform	$CHBr_3$	$100\,\mu g\,l^{-1}$
Dibromochloromethane	$CHBrCl$	$100\,\mu g\,l^{-1}$
Bromodichloromethane	$CHBrCl_2$	$60\,\mu g\,l^{-1}$

Table 18.3 *Typical concentrations of the common DBPs found in treated drinking water*

Trihalomethanes	
Chloroform	$10–100\,\mu g\,l^{-1}$
Chlorodibromomethane	$0–5.0\,\mu g\,l^{-1}$
Bromodichloromethane	$1–50\,\mu g\,l^{-1}$
Bromoform	$1–10\,\mu g\,l^{-1}$
Halogenated acetic acids	
Trichloroacetic acid	$1–50\,\mu g\,l^{-1}$
Dichloroacetic acid	$1–50\,\mu g\,l^{-1}$
Chlorinated furanones	
3-Chloro-4-(dichloromethyl)-5-hydroxy-2(*5H*)-furanone (MX)	$0–80\,ng\,l^{-1}$
3-Chloro-4-(chloromethyl)-5-hydroxy-2(*5H*)-furanone (CMCF)	$0–17\,ng\,l^{-1}$
3-Chloro-4-methyl-5-hydroxy-2(*5H*)-furanone (MCF)	$0–78\,ng\,l^{-1}$
3,4-Dichloro-5-hydroxy-2(*5H*)-furanone (MCA)	$0–12\,ng\,l^{-1}$

representative guidelines value should not exceed 1. The DBPs for which the WHO has set a guideline are listed in Table 18.1 and the actual guideline values can be found in Appendix 3. On 31 December 2003 the US Environmental Protection Agency set a new maximum contaminant level (MCL) of $80\,\mu g\,l^{-1}$ for total THMs. The existing MCLs of $60\,\mu g\,l^{-1}$ for the sum of five HAAs, $100\,\mu g\,l^{-1}$ for chlorite and $10\,\mu g\,l^{-1}$ for bromate all remain unchanged (Appendix 2). Typical concentrations of DBPs in drinking water are given in Table 18.3. The EC limit for bromate is $10\,\mu g\,l^{-1}$, equivalent to the WHO guideline.

18.3 Health risks

Most of the information relating to the effects of DBPs, and in particular THMs, on human health is based largely on chloroform. The lethal dose of chloroform is about $630\,mg\,kg^{-1}$ body weight, so about 44 g is required to kill an adult man weighing approximately 70 kg. Laboratory studies on animals have shown that smaller doses are mutagenic and can induce cancer, although epidemiological studies have not positively identified a strong causal relationship between THMs and cancer. There is some evidence to suggest an association between long-term, low-level exposure to THMs in drinking water with rectal, intestinal and bladder cancer (WHO, 1984). The concentration of the chlorinated furanone MX in drinking water is only 1/100 to 1/1000 of that of THMs or HAAs, yet MX represents a significant proportion (60%) of the total mutagenicity of chlorinated drinking water (Kronberg and Vartiainen, 1988; Smeds *et al.*, 1997). Little is known of the effect of mixtures of DBPs on the carcinogenicity of drinking water. An additive effect has generally been assumed, although laboratory studies have been conflicting with inhibitive and synergistic effects also reported for different combinations of DBPs (Komulainen, 2004). The health effects, and in particular the carcinogenicity, of DBPs are reviewed by Bull and Kopfler (1991) and Komulainen (2004).

18.4   Prevention of by-product formation

A high organic matter content in water supplies is not only due to waters rich in humic acids which drain from peaty moorlands and feed upland reservoirs and supply streams. It is also a problem in nutrient-rich lowland rivers. These supplies are most at risk from THM formation due to chlorination and the greater the chlorine dose, the greater the risk of producing THMs. Although some advances have been made in removing chlorination by-products using granular activated carbon filters (Lykins *et al.*, 1988) and other advanced techniques such as membrane filtration, it is expensive. Also, like all other water treatment processes it can result in undesirable changes in water quality.

The risk from THMs appears to be very low. The WHO (1984) estimates that the risk of an additional cancer due to a lifetime exposure to a carcinogen such as chloroform in drinking water, assuming a concentration of $30\,\mu g\,l^{-1}$ and a daily consumption rate of 2 litres, is less than 1 in 100 000. Clearly the best option is to control chlorination rates as carefully as possible. In those areas where there is a high concentration of organic matter which can react with chlorine to form THMs, then the use of chlorine dioxide or chloramination should be considered, or alternative methods of disinfection such as ozonation (Jacangelo *et al.*, 1989) or ultraviolet radiation examined. The risk associated with inadequate disinfection would be much greater than that resulting from concentrations of chloroform and other THMs well in excess of the WHO's guideline value.

Alternative disinfectants to chlorine include ozone, chlorine dioxide and chloramine; however, they also produce various by-products (Table 18.1). While ozone does not form THMs, it does not have a residual disinfection potential. It is so reactive that oxidation products are also formed. Those causing most concern are formaldehyde, acetaldehyde, other non-chlorinated aldehydes and carboxylic acids. In water with $<10\,\mu g\,l^{-1}$ of bromide present chloroform will be the main THM produced; when present in greater concentrations then bromide will be oxidized during ozonation to bromate, which is thought to be carcinogenic, and other brominated compounds such as bromoform (Jancangelo *et al.*, 1989). The reaction is shown below where hypobromous acid (HBrO), which is in equilibrium with hypobromite ions (BrO$^-$) during bromide oxidation by ozone, reacts with the natural organic matter present to produce brominated organic by-products (THM-Br) such as bromoform (Sorlini and Collivignarelli, 2005).

$$Br^- \overset{O_3}{\longrightarrow} BrO^- \longrightarrow HBrO \overset{\text{organic matter}}{\longrightarrow} THM\text{-}Br.$$

If chlorination is used after ozonation then other THMs will be formed such as bromodichloromethane and dibromochloromethane. Chlorine dioxide is an excellent disinfectant and does not generally react with organic matter to form chlorinated by-products (Lykins and Griese, 1986). However, it is often contaminated with chlorine, which forms THMs. In practice, as the chlorine dioxide oxidizes any organic matter in the water it is reduced to chlorite and chlorate, both of which are toxic. The same by-products are formed as with chlorine, except as chlorine dioxide is far less reactive significantly lower concentrations of by-product are formed. Although less efficient than free chlorine, chloramination is becoming more widely adopted as a disinfection method. It uses chlorine and ammonia to produce primarily monochloramine. A significant problem is the excessive use of ammonia to form monochloramine, resulting in the excess ammonia being converted to nitrite in the distribution system. Cyanogen chlorine is possibly the major by-product of chloramination, which is metabolized rapidly in the body to cyanide and thiocyanate (Krasner *et al.*, 1989).

As by-products are not readily removed by existing treatment processes, attention is being paid to their formation by the more effective removal of dissolved organic matter before disinfection. This is primarily achieved by improved coagulation, which can remove substantial amounts of organic matter. Treatment options include air stripping, although this can lead to air contamination and loss of residual disinfection, and reverse osmosis, which is currently the favoured option along with granular activated carbon both of which require high levels of maintenance. The use of models and risk assessment is also being increasingly employed to minimize DBP formation (Montgomery Watson Inc., 1993; Milot *et al.*, 2000; Von Gunten *et al.*, 2001).

18.5 Conclusions

In a survey carried out by Friends of the Earth and published in the *Observer Magazine* in August 1989, data on THM concentrations from water supplies throughout England and Wales showed that the limit for THMs in place at that time of $100 \mu g\ l^{-1}$ was exceeded at some time during the previous 2 years in no less than 82 local supply areas. Where drinking water is predominately obtained from surface water, which is the case in the UK, then the presence of natural organic matter is inevitable. While water treatment is designed to remove this organic matter, in particular humic material, prior to disinfection, it is extremely difficult to achieve consistently high removal. Therefore, while compliance for THMs in the UK has improved enormously since 1986, due to the unpredictable nature of organic material in drinking water resources and the need to chlorinate supplies, it remains a constant challenge to suppliers. Thus of all the listed parameters within the EC Drinking Water Directive (EC, 1998), compliance with THMs continues to be the most difficult, with the lowest levels of compliance of any parameter. For example in Northern Ireland 358 of the 1057 tests on drinking water carried out in 2004 exceeded the $100 \mu g\ l^{-1}$ limit. The Northern Ireland Drinking Water Inspectorate (www.ehsni.gov.uk/water/drinkwater.htm) noted that while the level of compliance remained more or less the same from previous years, the concentration of THMs in drinking water was continuing to decline overall (Northern Ireland Drinking Water Inspectorate, 2005). This pattern is reflected throughout the UK, wherever drinking water is disinfected, although the control of DBPs remains one of the biggest challenges in water treatment.

It is clear that many THMs are carcinogenic at high doses in laboratory animals and also that certain chlorinated by-products, such as MX, are more dangerous than others. It therefore appears prudent to try to reduce the levels of these compounds in our drinking water to as low a concentration as possible.

References

Bull, R. J. and Kopfler, F. C. (1991). *Health Effects of Disinfectants and Disinfection By-Products*. Denver, CO: AWWA Research Foundation and the American Water Works Association.

Cable, C. J. and Fielding, M. (1992). *Review of Transformation Products in Water Sources/Supplies*. Report FR 0286. Marlow: Foundation of Water Research.

EC (1998). Council Directive 98/83/EC of 3 November 1998 on the quality of water intended for human consumption, *Official Journal of the European Communities*, **L330** (5.12.98) 32–53.

Fawell, J. K., Fielding, M. and Ridgeway, J. W. (1987). Health risks of chlorination: is there a problem? *Water and Environmental Management*, **1**, 61–6.

Friends of the Earth (1989). Poison on tap. *Observer Magazine*, 6 August, 15–24.

Golfinopoulos, S. K. and Nikolaou, A. D. (2005). Survey of disinfection by-products in drinking water in Athens, Greece. *Desalination*, **176**, 13–24.

Horth, H., Fawell, I. K., James, C. P. and Young, W. (1991). *The Fate of the Chlorinated-Derived Mutagen MX in vivo*. Report FR 0068. Marlow: Foundation for Water Research.

Jacangelo, J. G., Patania, N. L., Reagan, K. M. *et al.* (1989). Ozonation: assessing its role in the formation and control of disinfection by-products. *Journal of the American Water Works Association*, **81**(8), 74–84.

Kim, J., Chung, Y., Shin, D. *et al.* (2002). Chlorinated by-products in surface water treatment process. *Desalination*, **151**, 1–9.

Komulainen, H. (2004). Experimental cancer studies of chlorinated by-products. *Toxicology*, **198**, 239–48.

Krasner, S. W., McGuire, M. J., Jacangelo, J. G. *et al.* (1989). The occurrence of disinfection by-products in US drinking water. *Journal of the American Water Works Association*, **81**(8), 41–53.

Kronberg, L. and Vartiainen, T. (1988). Ames mutagenicity and concentration of the strong mutagen 3-chloro-4-(dichloromethyl)-5-hydroxy-2(*5H*)-furanone and of its geometric isomer E-2-chloro-3-(dichloromethyl)-4-oxo-butenoic acid in chlorine-treated tap waters. *Mutation Research*, **206**, 177–82.

Lykins, B. W. and Griese, M. H. (1986). Using chloride dioxide for trihalomethane control. *Journal of the American Water Works Association*, **78**(6), 88–93.

Lykins, B. W., Clark, R. M. and Adams, J. A. (1988). Granular activated carbon for controlling THMs. *Journal of the American Water Works Association*, **80**(5), 85–92.

Milot, J., Rodriguez, M. J. and Sérodes, J. B. (2000). Modelling the susceptibility of drinking water utilities to form high concentrations of trihalomethanes. *Journal of Environmental Management*, **60**, 155–71.

Montgomery Watson Inc. (1993). *Mathematical Modelling of the Formation of THMs and HAAs in Chlorinated Natural Waters*. Denver, CO: American Water Works Association.

Nikolaou, A. D., Kostopoulou, M. N. and Lekkas, T. D. (1999). Organic by-products of drinking water chlorination. *Global Nest*, **1**(3), 143–56.

Northern Ireland Drinking Water Inspectorate (2005) *Drinking Water Quality in Northern Ireland 2004*. Belfast: Northern Ireland Drinking Water Inspectorate, Environment and Heritage Service.

Peters, C. I., Young, R. I. and Perry, R. (1978). *Chemical Aspects of Aqueous Chlorination Reactions: a Literature Review*. Medmenham: Water Research Centre.

Peters, R. I. B., De Leer, E. W. B. and De Galan, L. (1990). Dihaloacetonitriles in Dutch drinking waters. *Water Research*, **24**, 797–800.

Rodriguez, M. J., Sérodes, J. -B. and Levallois, P. (2004). Behaviour of trihalomethanes and haloacetic acids in a drinking water distribution system. *Water Research*, **38**, 4367–82.

Rook, J. J. (1974). Formation of haloforms during chlorination of natural waters. *Water Treatment and Examination*, **23**, 234–243.

Smeds, A., Vartiainen, T., Maki-Pakkanen, J. and Kronberg, L. (1997). Concentrations of Ames mutagenic chlorohydroxyfuranones and related compounds in drinking waters. *Environmental Science and Technology*, **31**, 1033–9.

Sorlini, S. and Collivignarelli, C. (2005). Trihalomethane formation during chemical oxidation with chlorine, chlorine dioxide and ozone of ten Italian natural waters. *Desalination*, **176**, 103–11.

Von Gunten, U., Driedger, A., Gallard, H. and Salhi, E. (2001). By-products formation during drinking water disinfection: a tool to assess disinfection efficiency. *Water Research*, **35**(8), 2095–9.

WHO (1984). *Guidelines for Drinking Water Quality*, Vol. 2. *Health Criteria and Other Supporting Information*. Geneva: World Health Organization.

WHO (1993). *Revision of the WHO Guidelines for Drinking Water Quality*, 2nd edn. Geneva: World Health Organization.

WHO (2004). *Guidelines for Drinking Water Quality,* Vol. 1. *Recommendations*. 3rd edn. Geneva: World Health Organization.

Chapter 19
Monitoring and removal of pathogens

19.1 Introduction

Chemical analysis can only be used for the assessment of water treatment efficiency and to monitor compliance to legal standards, while biological examination of water is used to detect the presence of algae and animals that may affect treatment or water quality, and to identify possible defects in the distribution network. Microbial safety of water is the historical and fundamental reason for water treatment, with pathogens monitored using complex, and often costly, techniques. The objective of water treatment is to remove pathogens rather than to act as a reservoir or source of micro-organisms, although the very act of treatment can inadvertently result in periods of high contamination as seen with *Cryptosporidium* (Section 13.2).

19.2 Monitoring pathogens

Pathogens are outnumbered by the normal commensal bacterial flora in both human and animal intestines, so they can only be recovered by filtering relatively large volumes of water (≤ 2 l). Isolation of pathogens requires specific and complicated tests using special equipment; with positive identification often only possible after further biochemical, serological or other analysis. This makes routine monitoring of individual pathogenic micro-organisms in drinking water both difficult and expensive. It is currently impracticable to examine all water supplies on a routine basis for the presence or absence of all pathogens, although the developments in gene probe technology may well make this possible in the future. To overcome this problem a rapid and preferably a single test is required, the theory being that it is more effective to examine a water supply frequently using a simple general test, as most cases of contamination of water supplies occur infrequently, than only occasionally by a series of more complicated tests. This has led to the development of the use of indicator organisms to determine the likelihood of contamination by faeces.

The main criteria for selection of a suitable indicator organism are that (1) they should be a member of the normal intestinal flora of healthy people;

(2) they should be exclusively intestinal in habit and therefore exclusively faecal in origin if found outside the intestine; (3) ideally they should only be found in humans; (4) they should be present when faecal pathogens are present and only when faecal pathogens are expected to be present; (5) they should be present in greater numbers than the pathogen they are intended to indicate; (6) they should be unable to grow outside the intestine with a die-off rate slightly less than the pathogenic organisms; (7) they should be resistant to natural environmental conditions and to water and wastewater treatment processes in a manner equal to or greater than the pathogens of interest; (8) they should be easy to isolate, identify and enumerate; and finally (9) that they should be non-pathogenic. In temperate regions most of these requirements are fulfilled only by *Escherichia coli* although other coliform organisms, *Enterococci* and *Clostridium perfringens* are also used as indicators of faecal contamination in drinking water.

These three groups are able to survive for different periods of time in the aquatic environment. Faecal streptococci (*Enterococci*) die fairly quickly outside the host and their presence is an indication of recent pollution. *Escherichia coli* (faecal coliforms) can survive for several weeks under ideal conditions and are far more easily detected than the other indicator bacteria. Because of this it is the most widely used test although the others are often used to confirm faecal contamination if *E. coli* is not detected. Sulphate-reducing clostridia (*C. perfringens*) can exist indefinitely in water. When *E. coli* and faecal streptococci are absent, its presence indicates remote or intermittent pollution. It is especially useful for testing lakes and reservoirs, although the spores do eventually settle out of suspension. The spores are more resistant to industrial pollutants than the other indicators and it is especially useful in waters receiving both domestic and industrial wastewaters. It is assumed that these indicator organisms do not grow outside the host and, in general, this is true. However, in tropical regions *E. coli* in particular is known to multiply in warm waters and there is increasing evidence that *E. coli* is able to reproduce in enriched waters generally, thus indicating an elevated health risk. Therefore, great care must be taken in the interpretation of results from tropical areas, so the use of bacteriological standards designed for temperate climates are inappropriate for those areas. Regrowth of coliform in distribution networks is now well documented (Section 25.2).

19.3   Measurement and standards

Two techniques are principally used to enumerate indicator bacteria, the membrane filtration and the multiple tube methods. The EC Drinking Water Directive specifies that *E. coli*, *Enterococci* (faecal streptococci), *Pseudomonas aeruginosa* and *C. perfringens* must all be isolated using the membrane filtration method, although the multiple tube method is still widely used for clostridia because of the need to incubate under anaerobic conditions (Gray, 2004). Details of all microbial methods are given in *Standard Methods* for the

USA (APHA, 1992) and for the UK in *Report on Public Health and Medical Subjects No. 71* (Department of the Environment, 1994), although the latter is now regularly updated on-line by the Environment Agency (Table 19.1).

Coliforms not only occur in faeces, but are also normal inhabitants of water and soil. The presence of coliforms in a water sample does not necessarily indicate faecal contamination, although in practice it must be assumed that they are of faecal origin unless proved otherwise. The total coliform count measures all the coliforms present in the sample. However, only *E. coli* is exclusively faecal in origin with numbers in excess of $10^8 \, g^{-1}$ of fresh faeces. So it is important to confirm *E. coli* is present. Routine coliform testing comprises of two tests giving the total coliform count and faecal coliform (*E. coli*) count.

Water supplied under the EC Drinking Water Directive (98/83/EEC) must be free of pathogenic micro-organisms and parasites in numbers constituting a danger to public health, although no parametric values are given for specific viruses, protozoans or bacterial pathogens. The Directive specifies numerical standards for *E. coli* and *Enterococci* in drinking water; and numerical standards for *E. coli*, *Enterococci*, *P. aeruginosa* and total viable counts of heterotrophic bacteria (heterotrophic plate count) at 22 and 37 °C in bottled waters (Table 19.2). Routine (indicator) monitoring is restricted to coliforms, *C. perfringens* and heterotrophic bacteria at 22 °C only (Table 19.2) (Appendix 1). Specific and detailed notes are given on analysis, including the composition of all recommended media. The recommended analyses in the proposed Directive are given in Table 19.3. Bottled natural mineral waters have their own Directive (80/777/EEC) (Section 29.2).

The US Environmental Protection Agency (USEPA) set a maximum contaminant level (MCL) based on the presence or absence (P–A) concept for coliforms, which is linked to the recommended sampling frequency of water supplies that is dependent on the population served. For systems requiring more than 40 samples per month, <5% of samples must be total coliform positive. For systems where the frequency of analysis is less than 40 samples per month, then no more than one sample per month may be total coliform positive. If a sample is found to be total coliform positive then repeat samples must be taken within 24 hours. Repeat samples must be taken at the same tap and also at adjacent taps within five service connections, both above and below the original sample point. If these repeated samples are also total coliform positive, then the samples must be immediately tested for faecal coliforms or *E. coli*. If these prove positive then the public must be informed. Full details of the revised coliform rule is given by Berger (1992). The World Health Organisation (WHO, 2004) guidelines are based solely on the presence of *E. coli* or thermotolerant coliforms, which must not be detectable in any 100 ml sample of water entering or within the distribution system, or in the drinking water delivered to the consumer (Table 19.4).

There are many emerging new technologies for microbial water testing, primarily to detect coliforms. These are focussed on complex biochemical

Table 19.1 *The standard methods (blue books) associated with the sampling, isolation and enumeration of microbial pathogens in drinking water in the UK. These can be downloaded and updated on-line from (www.environment-agency.gov.uk/commercial/1075004/399393/401849/?version=1&lang=_e)*

The Assessment of Taste, Odour and Related Aesthetic Problems in Drinking Waters (1998). Blue Book No. 171. London: Standing Committee of Analysts, Environmental Agency. (See also Blue Book Nos. 196 and 197)

Isolation and Identification of Cryptosporidium *Oocysts and* Giardia *Cysts in Waters (1999).* Blue Book No. 172. London: Standing Committee of Analysts, Environmental Agency.

The Microbiology of Recreational and Environmental Waters (2000). Blue Book No. 175. London: Standing Committee of Analysts, Environmental Agency.

The Microbiology of Drinking Water (2002) – Part 1 – *Water Quality and Public Health. Blue Book No. 176. London: Standing Committee of Analysts, Environmental Agency.*

The Microbiology of Drinking Water (2002) – Part 2 – *Practices and Procedures for Sampling.* Blue Book No. 177. London: Standing Committee of Analysts, Environmental Agency.

The Microbiology of Drinking Water (2002) – Part 3 – *Practices and Procedures for Laboratories.* Blue Book No. 178. London: Standing Committee of Analysts, Environmental Agency.

The Microbiology of Drinking Water (2002) – Part 4 – *Methods for the Isolation and Enumeration of Coliform Bacteria and* Escherichia coli *(Including* E. coli *O157:H7).* Blue Book No. 179. London: Standing Committee of Analysts, Environmental Agency.

The Microbiology of Drinking Water (2006) – Part 5 – *The Isolation and Enumeration of* Enterococci *by Membrane Filtration.* Blue Book No. 202. London: Standing Committee of Analysts, Environmental Agency.

The Microbiology of Drinking Water (2004) – Part 6 – *Methods for the Isolation and Enumeration of Sulphite-Reducing* Clostridia *and* Clostridium perfringens *by Membrane Filtration.* Blue Book No. 192. London: Standing Committee of Analysts, Environmental Agency.

The Microbiology of Drinking Water (2002) – Part 7 – *The Enumeration of Heterotrophic Bacteria by Pour and Spread Plate Techniques.* Blue Book No. 182. London: Standing Committee of Analysts, Environmental Agency.

The Microbiology of Drinking Water (2002) – Part 8 – *Methods for the Isolation and Enumeration of* Aeromonas *and* Pseudomonas aeruginosa *by Membrane Filtration.* Blue Book No. 183. London: Standing Committee of Analysts, Environmental Agency.

The Microbiology of Drinking Water (2006) – Part 9 – *Methods for the Isolation and Enumeration of* Salmonella *and* Shigella *by Selective Enrichment, Membrane Filtration and Multiple Tube-most Probable Number Techniques.* Blue Book No. 206. London: Standing Committee of Analysts, Environmental Agency.

The Microbiology of Drinking Water (2002) – Part 10 – *Methods for the Isolation of* Yersinia, Vibrio *and* Campylobacter *by Selective Enrichment.* Blue Book No. 185. London: Standing Committee of Analysts, Environmental Agency.

The Microbiology of Drinking Water (2004) – Part 11 – *Taste, Odour and Related Aesthetic Problems.* Blue Book No. 196. London: Standing Committee of Analysts, Environmental Agency.

Table 19.1 (cont.)

The Microbiology of Drinking Water (2004) – Part 12 – *Methods for the Isolation and Enumeration of Micro-organisms Associated with Taste, Odour and Related Aesthetic Problems.* Blue Book No. 197. London: Standing Committee of Analysts, Environmental Agency.

The Determination of Legionella *Bacteria in Water and Other Environmental Samples (2005)* – Part 1 – *Rationale of Surveying and Sampling.* Blue Book No. 200. London: Standing Committee of Analysts, Environmental Agency.

Table 19.2 *Microbiological and microbial indicator values in EC Drinking Water Directive (98/83/EEC)*

Parameter	Parametric value	Notes
Part A Microbiological parameters		
Escherichia coli	0/100 ml	
Enterococci	0/100 ml	
In water offered for sale in bottles or containers		
Escherichia coli	0/250 ml	
Enterococci	0/250 ml	
Pseudomonas aeruginosa	0/250 ml	
Colony counts at 22 °C	100/100 ml	
Colony counts at 37 °C	20/100 ml	
Part C Indicator parameters		
Clostridium perfringens	0/100 ml	From, or affected by, surface water only
Coliform bacteria	0/100 ml	For bottled waters 0/250 ml
Colony counts at 22 °C	0/100 ml	

techniques such as hybridization and polymerase chain reaction (PCR), gene probe technology and monoclonal antibody methods, which allow single bacterial cells to be detected (Gleeson and Gray, 1997). Enzyme detection methods are now widely in use, especially in field situations. Total coliform detection is based on the presence of β-galactosidase, an enzyme that catalyzes the breakdown of lactose into galactose and glucose; while *E. coli* is based on the detection of β-glucuronidase activity. These tests are known as ONPG and MUG methods, respectively, after the substrates used in the tests and are now accepted standard monitoring methods in the USA. These ONPG–MUG tests can be used to give an MPN (most probable number) value or simply indicate the P–A of coliforms or *E. coli.* These tests will become increasingly important as there is a general swing away from standards based on microbial density to those simply based on

Table 19.3 *Microbial methods recommended by the EC Drinking Water Directive (98/83/EEC)*

Total coliforms
Membrane filtration followed by incubation on membrane lauryl broth for 4 h at 30 °C followed by 14 h at 37 °C. All yellow colonies are counted regardless of size.

E. coli
Membrane filtration followed by incubation on membrane lauryl broth for 4 h at 30 °C followed by 14 h at 44 °C. All yellow colonies are counted regardless of size.

Faecal streptococci (Enterococci)
Membrane nitration followed by incubation on membrane *Enterococcus* agar for 48 h at 37 °C. All pink, red or maroon colonies which are smooth and convex are counted.

Sulphite-reducing clostridia
Maintain the sample at 75 °C for l0 min prior to membrane filtration. Incubate on tryptose-sulphite-cycloserine agar at 37 °C under anaerobic conditions. Count all black colonies after 24 and 48 h incubation.

Pseudomonas aeruginosa
Membrane filtration followed by incubation in a closed container at 37 °C on modified Kings A broth for 48 h. Count all colonies which contain green, blue or reddish-brown pigment and those that fluoresce.

Total bacteria counts
Incubation in a YEA for 72 h at 22 °C and for 24 h at 37 °C. All colonies to be counted.

Table 19.4 *The WHO drinking water guide values for microbial quality. Reproduced from WHO (2004) with permission from the World Health Organization, Geneva*

Organisms	Guideline value
All water directly intended for drinking	
E. coli or thermotolerant coliform bacteria[a,b,c]	Must not be detectable in any 100-ml sample
Treated water entering the distribution system	
E. coli or thermotolerant coliform bacteria[a,b]	Must not be detectable in any 100-ml sample
Treated water in the distribution system	
E. coli or thermotolerant coliform bacteria[a,b]	Must not be detectable in any 100-ml sample

[a] Immediate investigative action must be taken if *E. coli* are detected.
[b] Although *E. coli* is the more precise indicator of faecal pollution, the count of thermotolerant coliform bacteria is an acceptable alternative. If necessary, proper confirmatory tests must be carried out. Total coliform bacteria are not acceptable indicators of the sanitary quality of water supplies, particularly in tropical areas, where many bacteria of no sanitary significance occur in almost all untreated supplies.
[c] It is recognized that in the great majority of rural water supplies, especially in developing countries, faecal contamination is widespread. Especially under these conditions, medium-term targets for the progressive improvement of water supplies should be set.

presence or absence of coliforms in a sample. The best known commercial ONPG–MUG preparations are currently Colilert® (Access Analytical, Branford, CT), Coliquick® (Hach Co., Loveland, CO) and Colisure® (Millipore Co., Bedford, MA). The ingredients for these new tests come in powder form (in test tubes for the quantitative MPN method and in containers for P–A analysis). A measured amount of water is added to each tube or container and the powder dissolves into a colourless solution. The tubes are placed in an incubator for 24 hours at 35 °C. The solution in tubes with total coliforms will be yellow, which is then exposed to a hand-held fluorescent light. If the tube contains *E. coli* the solution will fluoresce brightly. The specificity of this method eliminates the need for confirmatory and completed tests.

19.4   Heterotrophic plate counts

Heterotrophic plate counts (HPCs) represent the aerobic and facultatively anaerobic bacteria that derive their carbon and energy from organic compounds (Table 25.1). Certain HPC organisms are considered to be opportunistic pathogens (Table 13.4) and have been implemented in gastrointestinal illness (Section 13.4).

Heterotrophic bacteria are commonly isolated from raw waters and are widespread in soil and vegetation and can survive for long periods in water and rapidly multiply, especially at summer temperatures. There is also concern that these organisms can rapidly multiply in bottled waters, especially if not stored properly once opened. The EC Drinking Water Directive requires that there is no significant increase from background levels of HPC bacteria in either tap or bottled waters. While HPC bacteria are not a direct indicator of faecal contamination, they do indicate variation in water quality and the potential for pathogen survival and regrowth. Heterotrophic plate counts are carried out normally with the spread plate method using yeast extract agar (YEA) and incubated at 22 °C for 72 hours and 37 °C for 24 hours. Results are expressed as colony forming units (cfu) per ml. Counts at 37 °C are especially useful as they can provide rapid information of possible contamination of water supplies (Department of the Environment, 1994).

Heterotrophic plate counts have long been employed to evaluate water quality, although less importance is currently placed on HPCs for assessing the potability of drinking water. It is considered that their value lies mainly in indicating the efficiency of various water treatment processes including disinfection, as well as the cleanliness and integrity of the distribution system. They are more useful in assessing the quality of bottled waters, which may be stored for long periods before being sold for consumption. The US National Primary Drinking Water Regulations now include MCLs of no more than 500 cfu ml^{-1} for HPCs (USEPA, 1990) although this is primarily to reduce possible interference with the detection of coliforms.

19.5 Removal of pathogens

19.5.1 Bacteria

One of the most effective ways of removing bacteria from water is by storage in reservoirs. During the spring and summer, sunlight, increased temperatures and biological factors ensure that between 90% and 99.8% reductions of *E. coli* occur. The percentage reduction is less during the autumn and winter due to the main removal mechanisms being less effective, so expected reductions fall to between 75% and 98%. The lowest reductions occur when reservoirs are mixed to prevent stratification. The greatest decrease in E. *coli* and *Salmonella* bacteria occurs over the first week, although the longer the water is stored the greater the overall reduction (Denny *et al.*, 1990). *Salmonella,* faecal streptococci and *E. coli* are excreted in large numbers by gulls. The presence of gulls, especially in high numbers, on storage reservoirs may pose a serious problem either from direct faecal discharge and/or from rainfall runoff along contaminated banks (Denny, 1991) (Section 13.5).

Bacteria are removed by a number of unit processes in water treatment, especially coagulation, sand filtration and activated carbon filtration. Coagulation using alum removes about 90% of faecal indicator bacteria and about 60–70% of the total plate count bacteria, although these figures vary widely from treatment plant to treatment plant. Rapid sand filtration is not dependable and is erratic unless the water is coagulated first. In contrast, slow sand filtration is able to remove up to 99.5% of coliforms, although its performance is generally worse in the winter. Activated carbon is being widely used to control taste and odour in drinking water, as well as removing a wide range of organic compounds. However, these filters can become heavily colonized by heterotrophic micro-organisms. Minute fragments are constantly breaking off the GAC and each is heavily coated with micro-organisms. These micro-organisms are not affected to any great extent by disinfection, and so can be introduced in large numbers into the distribution system (Section 13.6) (LeChevallier and McFeters, 1990). Microfiltration (MF) can remove particles down to between 0.05 and 5 μm and so are effective in removing bacteria (Figure 14.3) (Gray, 2004).

19.5.2 Viruses

Like bacteria, viruses are also significantly reduced when water is stored. For example, the number of viruses in River Thames water decreases from between 12 and 49 pfu l^{-1} (plaque forming units per litre of water) to 1.9 pfu l^{-1} after storage. Similar results have been reported for water from the River Lee. Temperature is also an important factor. Using water from the River Lee, researchers have found that polio virus was reduced by 99.8% in less than 15 days at 15–16 °C compared with 9 weeks for a similar reduction at 5–6 °C. For the optimum removal of all micro-organisms of faecal origin, 10 days of retention should achieve between 75% and 99% reduction regardless of temperature.

Coagulation using alum achieves 95–>99% removal of all viruses, with other coagulants such as iron (III) chloride and iron (III) sulphate not quite as efficient. The use of polyelectrolytes as coagulant aids does not improve the removal of viruses. As with bacteria, rapid sand filtration is largely ineffective in removing viruses unless the water has been coagulated before filtration. Slow sand filtration is effective in removing the bulk of viral particles in water, with removal rates between 97% and 99.8%. However, there is always some measurable contamination left in the water and so disinfection is required (Logsdon, 1990). Activated carbon can remove viruses; these are adsorbed onto the carbon by electrostatic attraction between positively charged amino groups on the virus and negatively charged carboxyl groups on the surface of the carbon. The removal efficiency is very variable and depends on the pH (maximum removal occurs at pH 4.5), the concentration of organic compounds in the water and the time the filter has been in operation. Removal rates of 70–85% are common. Ultrafiltration is able to remove particles between 0.001 and 0.02 μm and so is widely adopted for the removal of viruses, however, membrane processes are excellent at removing micro-organisms but offer no protection to contamination of water while in the distribution system.

19.5.3   Protozoan cysts

Protozoan cysts are not effectively removed by storing the water in reservoirs due to their small size and density. *Cryptosporidium* oocysts have a settling velocity of 0.5 μm s^{-1}. Therefore, if the reservoir is 20 m deep it will take the oocyst 463 days to settle to the bottom, assuming that there is no water current or water movement to disturb this quiescent settlement. *Giardia* cysts are much larger and have a greater settling velocity of 5.5 μm s^{-1}. So in the same 20 m deep reservoir it will take *Giardia* 38 days to reach the bottom. It would be feasible in a large storage reservoir to remove a percentage of the *Giardia* cysts if the retention time was greater than six weeks and if mixing and currents were minimal (Denny, 1991).

Cysts should be removed effectively by coagulation and the addition of polyelectrolyte coagulant aids should also enhance removal. Using optimum coagulant conditions (as determined by the jar test) 90–95% removal should be possible. Rapid sand filtration is not an effective barrier for cysts unless the water is coagulated before filtration. When used after coagulation effective removal of *Giardia* is achieved (99.0–99.9%). The only proved effective method of removing both *Cryptosporidium* and *Giardia* cysts is by slow sand filtration, with 99.98 and 99.99% of cysts removed, respectively (Hibler and Hancock, 1990) or by membrane filtration. *Cryptosporidium* oocysts can be found in the raw water after slow sand filtration, so it must be assumed that water treatment is not able to remove all the protozoan cysts that may be present in raw water.

19.6   Disinfection

The efficiency of water treatment in removing pathogenic micro-organisms varies from month to month, and even when the treatment plant is achieving 99.9% removal there will always be some pathogens remaining in the water. This means that disinfection is absolutely vital to ensure that any micro-organisms arising from faecal contamination of the raw water are destroyed. Chlorination is by far the most effective disinfectant for bacteria and viruses because of the residual disinfection effect that can last throughout the water's journey through the distribution network to the consumer's tap (Section 14.2). The most effective treatment plant design to remove pathogens is rapid and slow sand filtration, followed by chlorination or treatment by pre-chlorination followed by coagulation, sedimentation, rapid sand filtration and post-chlorination. Both of these systems give >99.99% removal of bacterial pathogens including *C. perfringens*. Chlorine and monochloramine are ineffective against *Cryptosporidium* oocysts, although ozone and chlorine dioxide may be suitable alternative disinfectants (Korich *et al.*, 1990). Ozone at a concentration of about 2 mg l^{-1} is able to achieve a mean reduction in the viability of oocysts of between 95% and 96% over a 10-minute exposure period. Disinfection is considered in more detail in Section 14.2. Chlorine is only effective against *Giardia* under very controlled conditions and in practice is unlikely to be an effective barrier on its own (Hibler and Hancock, 1990). Advanced disinfection technologies for treated wastewaters are compared in Table 19.5 where the effectiveness of MF and ultrafiltration (UF) are evident. This summary can be equally applied to water treatment, although neither membrane processes have a residual disinfection effect.

Wash-water and water treatment sludge contain all the pathogens removed from the water and so must be handled as carefully as any other microbiologically hazardous waste. It is important that the sludge is not disposed of in such a way as to recontaminate the raw water source (Section 14.2).

Conventional water treatment cannot guarantee the safety of drinking water supplies at all times. Outbreaks of waterborne diseases can and do happen, although infrequently. With correct operation the chances of pathogenic micro-organisms causing problems to consumers will be reduced even further. The greatest risks come from private supplies that are not treated. Finding dead sheep decomposing in small streams just above a water intake point for an upland cottage is not an unusual occurrence, and the risk to shallow wells from septic tank drainage or surface contamination by faeces from animals that are grazing nearby is high. It may be advisable for everyone on a private supply to install an ultraviolet sterilizer to disinfect their water as it comes into the house (Section 29.3). However, the only effective way to destroy protozoan cysts is to

Table 19.5 *Comparison of technical-economic characteristics of advanced disinfection technologies. Reproduced from Lazarova et al. (1999) with permission from IWA*

Characteristics/ Criteria	Chlorination/ Dechlorination	UV	Ozone	MF	UF
Safety	+	+++	++	+++	+++
Bacteria removal	++	++	++	+++	+++
Virus removal	+	+	++	+	+++
Protozoa removal[a]	−	−	++	+++	+++
Bacterial regrowth	+	+	+	−	−
Residual toxicity	+++	−	+	−	−
By-products	+++	−	+	−	−
Operating costs	+	+	++	+++	+++
Investment costs	++	++	+++	+++	+++

− none; + low; ++ middle; +++ high.
[a] *in vitro* analysis of *Cryptosporidium*.

physically remove them. This can be done using a 1 µm pore size fibre cartridge filter, which should be placed upstream of the ultraviolet sterilizer. This will also increase the efficiency of the sterilization by removing any particulate matter that may harbour and shield pathogens from the ultraviolet radiation. Those on a mains supply should be reassured that their supplies are safe.

If the sand filters at the water treatment plant are by-passed for any reason, or the disinfection system fails, then the water should not be consumed before boiling. It is the responsibility of the water supply company to ensure that notices informing the public of the need to boil supplies are issued immediately (Water Authorities Association, 1985; HMSO, 1989).

19.7 Conclusions

In an ideal world, water destined for human consumption should be free from all micro-organisms, although in practice this is an unattainable goal. However, much can be done to minimize pathogens in drinking water, by using multiple barriers to the transfer of pathogens and setting strict limits for bacteria in water. Minimizing the health risks from microbial pathogens in drinking water requires a holistic approach to water resource management. It requires the prevention of contamination of water resources from agricultural point and diffuse sources and from wastewater effluents. So the treatment of wastewater to prevent the release of pathogens into the environment is as important as treatment of raw water for supply; forming important barriers preventing pathogen transfer from treatment plants to raw water, and from raw water into the distribution network respectively (Figure 19.1). Unlike the USA, which has a long tradition of

Figure 19.1 The use of barriers is vital in the control of pathogens in water resources. Reproduced from Geldreich (1991) with permission from John Wiley and Sons Ltd.

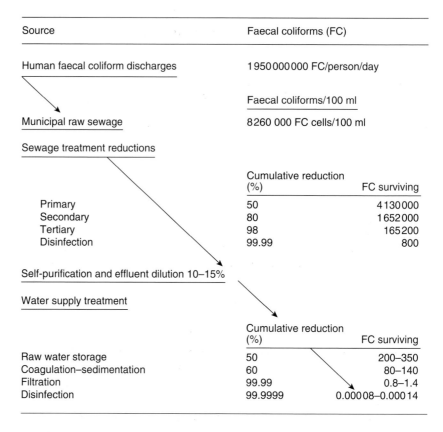

Source	Faecal coliforms (FC)
Human faecal coliform discharges	1 950 000 000 FC/person/day
	Faecal coliforms/100 ml
Municipal raw sewage	8 260 000 FC cells/100 ml

Sewage treatment reductions

	Cumulative reduction (%)	FC surviving
Primary	50	4 130 000
Secondary	80	1 652 000
Tertiary	98	165 200
Disinfection	99.99	800

Self-purification and effluent dilution 10–15%

Water supply treatment

	Cumulative reduction (%)	FC surviving
Raw water storage	50	200–350
Coagulation–sedimentation	60	80–140
Filtration	99.99	0.8–1.4
Disinfection	99.9999	0.000 08–0.000 14

disinfecting treated wastewater effluents, its adoption in Europe has been linked primarily, although not exclusively, with meeting the microbial requirements of the Bathing Water Directive. The importance of isolating sources of microbial contamination in the catchment and preventing their dissemination into, and contamination of, water resources is paramount in protecting the health of the general public.

The universally adopted use of indicator organisms as a means of controlling pathogens is based on the assumption that there is a quantifiable relationship between indicator density and the possible health risks involved. From this relationship a guideline value of indicator organism density can be derived above which there is an unacceptable risk (Figure 19.2). The use of indicator organisms as drinking water standards, although often criticized, is unlikely to be replaced in the foreseeable future; even though gene probes already exist that allow a screening of water for individual pathogen species and genera.

The coliform test is still widely considered the most reliable indicator for potable water, although there is growing dissatisfaction due to waterborne outbreaks, largely as a result of protozoan and viral agents, in drinking water that was tested as fit for consumption under the old coliform standards. The major deficiencies identified with the use of coliforms as indicators for drinking

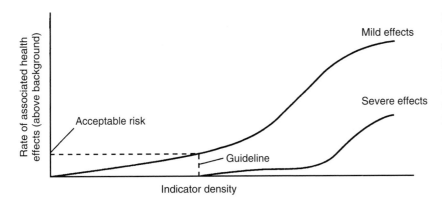

Figure 19.2 The desired water quality criteria and the development of guidelines from them. Reproduced from Cabelli (1978) with permission from John Wiley and Sons Ltd.

water quality assessment are: (1) the regrowth of coliforms in aquatic environments; (2) the regrowth of coliforms in distribution networks; (3) suppression by high background bacterial growth; (4) they are not directly indicative of a health threat; (5) a lack of correlation between coliforms and pathogen numbers; (6) no relationship between either protozoan or viral numbers; and finally (7) the occurrence of false positive and false negative results (Gleeson and Gray, 1997). The adoption of a zero approach to coliforms ensures that water is rejected if a route of faecal contamination is identified, until it can be dealt with. Therefore, the current trend is to use new tests such as the P–A coliform test or the minimal-media ONPG–MUG test to rapidly screen water for the presence of either coliforms or *E. coli*. The P–A concept has a number of potential advantages: (1) sensitivity is improved because it is more accurate to detect coliform presence than to make quantitative determinations; (2) the concept is not affected by changes in coliform density during storage; and (3) data manipulation is much improved.

The WHO endorse this simple P–A approach as it makes water safety easier to manage with the presence of either *E. coli* or thermotolerant coliforms indicative that faecal contamination is able to enter the supply system. The analytical methods for specific bacteria, viruses or parasites are considered too costly, complex and time-consuming for routine laboratory use and hence surveillance. The guideline criteria used are based on the likely viral content of source waters and the degree of treatment necessary to ensure that even large volumes of water have a negligible risk of containing viruses. It is considered that 'the attainment of the bacteriological criteria and the application of treatment for virological reduction should ensure the water presents a negligible health risk' (WHO, 2004).

The increasingly strict microbial water quality standards linked with better controls within catchments to prevent the transfer of pathogens has resulted in a dramatic decline in epidemic and endemic waterborne bacterial diseases in the developed world (Sobsey *et al.*, 1993). Viruses, protozoans and emerging pathogens, including the increasing risk from exotic and antibiotic-resistant strains, constantly renew this challenge. The membrane treatment technologies, such as microfiltration and ultrafiltration that are able to physically remove all micro-organisms including viruses and pyrogens, will play an increasingly important role in the protection of drinking water well into this century (Scott and Hughes, 1996). However, membrane filtration, like ozone and ultraviolet sterilization, offers no protection to treated water from contamination while in the distribution system, so chlorination will still be required (Chapter 25).

References

APHA (1992). *Standard Methods for the Examination of Water and Waste Water*, 18th edn. Washington, DC: American Public Health Association.

Berger, P. S. (1992). Revised coliform rule. In *Regulating Drinking Water Quality*, eds. C. E. Gilbert and E. J. Calabrese. Boca Raton: Lewis, pp. 161–6.

Cabelli, V. (1978). New standards for enteric bacteria. In *Water Pollution Microbiology*, Vol. 2, ed. R. Mitchell. New York: Wiley-Interscience. pp. 233–73.

Denny, S. (1991). *Microbiological Efficiency of Water Treatment*. Report FR 0219. Marlow: Foundation for Water Research.

Denny, S., Broberg, P. and Whitemore, T. (1990). *Microbiological Hazards in Water Supplies*. Report FR 0114. Marlow: Foundation for Water Research.

Department of the Environment (1994). *The Microbiology of Water 1994: Part 1 Drinking Water*. Reports on Public Health and Medical Subjects No. 71. Methods for the Examination of Water and Associated Materials. London: HMSO.

Geldreich, E. E. (1991). Microbial water quality concerns for water supply use. *Environmental Toxicology and Water Quality*, **6**, 209–23.

Gleeson, C. and Gray, N. F. (1997). *The Coliform Index and Waterborne Disease: Problems of Microbial Drinking Water Assessment*. London: E. and F. N. Spon.

Gray, N. F. (2004). *Biology of Wastewater Treatment*, 2nd edn. London: Imperial College Press.

Hibler, C. P. and Hancock, C. M. (1990). Waterbourne giardiasis. In *Drinking Water Microbiology*, ed. G. A. McFeters. New York: Springer-Verlag, pp. 271–93.

HMSO (1989). *The Water Supply (Water Quality) Regulations*, 1989. Statutory Instrument 1989/147. London: HMSO.

Korich, D. G., Mead, J. R., Madore, M. S., Sinclair, N. A. and Sterling, C. R. (1990). Effects of ozone, chlorine dioxide, chlorine and monochloramine on *Cryptosporidium parvum* oocyst viability. *Applied and Environmental Microbiology*, **56**(5), 1423–8.

Lazarova, V., Savoye, P., Janex, M. L., Blatchley, E. R. and Pommepuy, M. (1999). Advanced wastewater disinfection technologies: state of the art and perspectives. *Water Science and Technology*, **40**(4/5), 203–13.

LeChevallier, M. W. and McFeters, G. A. (1990). Microbiology of activated carbon. In *Drinking Water Microbiology*, ed. G. A. McFeters. New York: Springer-Verlag, pp. l04–19.

Logsdon, G. S. (1990). Microbiology and drinking water filtration. In *Drinking Water Microbiology*, ed. G. A. McFeters. New York: Springer-Verlag, pp.120–46.

Scott, K. and Hughes, R. (eds.) (1996). *Industrial Membrane Separation Technology*. London: Blackie Academic and Professional.

Sobsey, M. D., Dufour, A. P., Gerba, C. P., LeChevallier, M. W. and Payment, P. (1993). Using a conceptual framework for assessing risks to human health from microbes in drinking water, *Journal of the American Water Works Association*, **85**, 44–8.

USEPA (1990). *Drinking Water Regulations Under the Safe Drinking Water Act*. Fact Sheet, May. Washington, DC: US Environmental Protection Agency.

Water Authorities Association (1985). *Guide to the Microbial Implications of Emergencies in the Water Services*. London: Water Authorities Association.

WHO (2004). *Guidelines for Drinking Water Quality,* Vol. 1. *Recommendations*, 3rd edn. Geneva: World Health Organization.

PART IV
PROBLEMS ARISING IN THE DISTRIBUTION NETWORK

Chapter 20
The distribution network

20.1 Introduction

After treatment the water has to be conveyed to the consumer. This is done by a network of distribution pipes, also known as water mains, which are laid underground normally under roads and pavements. There is, however, much more to water distribution than just pipes; there are service reservoirs and booster stations as well.

20.2 Service reservoirs

Service reservoirs are needed primarily because the water resource, and often the treatment plant, is usually a considerable distance from the centre of population. Where groundwater resources are used, pumps need to be capable of pumping water at peak demand rates, rather than at the average daily demand rate. The service reservoir ensures that all peak demands for water are met, while smaller pumps and trunk mains can be used to cope with the average daily flow-rates rather than the peak demand rates, which may be 50–80% greater.

Like electricity, the demand for water varies during the day (diurnal variation). Although the peaks (high) and troughs (low) in demand can be calculated fairly adequately from experience, service reservoirs and water towers are needed to ensure that these demands are fully met (Figure 20.1). They also have other important functions such as providing a reserve storage capacity in case of problems at the treatment plant or with the trunk mains. They also compensate for any variation in water quality; for example, when water comes from more than one source.

Although the flow to the service reservoir remains constant, the level in the reservoir will rise and fall according to demand. The size of the reservoir varies according to the size of the area served, although enough storage for 24 or 36 hours is usually selected. While generally constructed of concrete, smaller reservoirs are often of steel or brick. An important design feature is that the tanks should be watertight, not only to prevent water loss, but to prevent contamination from outside the tank. The tanks are often split into a number of separate chambers to allow periodic cleaning, although more recent designs

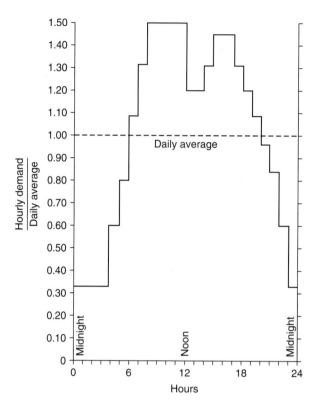

have been circular prestressed concrete tanks. Where this design is used at least
two tanks are required if the supply is not to be interrupted during tank cleaning
or maintenance. The reservoir must ensure an adequate hydraulic head to
produce sufficient pressure within the main to push the water up into the storage
tanks in household attics and even small blocks of flats. A minimum head of
30 m is required by the fire brigade for fire-fighting, whereas a head in excess of
70 m would result in an unacceptably high loss of water via leaks within the
distribution network. High pressure also causes excess wear on household
equipment such as taps, stop-taps (also widely called stopcocks or stopvalves),
washing-machine valves and ball-valves in WCs, all of which are designed to
operate at moderate pressures. Installing a pressure regulator on the supply pipe
can control excessive pressure or problems with pressure surges. Noise from
household plumbing also increases as the water pressure increases. Exceptionally tall buildings will probably require a system of pumps to raise the water to
the storage tanks at the top. The installation, operation and maintenance of such
pumps are the responsibility of the owner.

Where service reservoirs are required in flat areas, sufficient hydraulic head
may be obtained by the construction of a water tower. Water towers serve the
same purpose as a service reservoir except that they are generally much smaller,
serving only small distribution zones. The water usually has to be pumped up to

the tower from the main, thus increasing the cost of supply. Although service reservoirs are often below ground and rectangular, modern water towers have moved away from the traditional large brick or cast-iron tanks that were once familiar to a variety of unusual shapes made from concrete. The most common design is the inverted mushroom shape supported on a remarkably slender tower.

20.3 Water mains

There are two main categories of water main. The trunk mains are the largest and do not have any branch or service pipe connections. They are used for transporting large volumes of water from the source to the treatment plant, from the plant to the service reservoir, and from one reservoir to another. The distribution mains consist of a pipe network of smaller, varying sized pipes, which is highly branched. It is the distribution main that supplies individual houses.

In large towns or cities the distribution main system is usually arranged into pressure zones controlled from specific service reservoirs (Figure 20.2). In practical terms this allows for better leakage control and metering of water usage in each district. Although these zones are independent in terms of supply, they are often interconnected by valves to allow for the transfer of water between zones if the need arises. Within each zone any specific part of the distribution network can be isolated by valves, in order to isolate leaks, make new connections and to carry out maintenance or repairs. Within each pressure zone the distribution network consists mainly of loop (ring) circuits although spurs (dead-ends) are often necessary where the housing pattern does not allow a loop

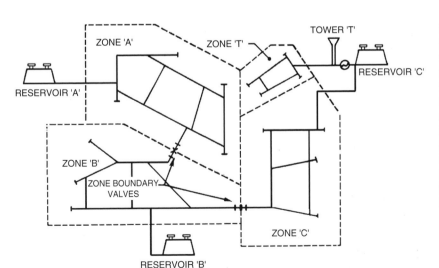

Figure 20.2 Schematic diagram showing that water distribution networks are broken down into operational supply zones each supplied by a service reservoir. Reproduced from Latham (1990) with permission from the Chartered Institution of Water and Environmental Management.

main to be used. In a long spur the water can stay in the main for a considerable time before being consumed, which can adversely affect its quality, especially towards the end of the pipe. Also, if repairs are required, then it is necessary to isolate the entire length of pipe. In contrast, it is possible to isolate a small section of pipework for repairs in a loop main without cutting off the supply to houses either side. This also ensures that the water is not retained for excessively long periods within the distribution system before it is used and that a good even pressure is maintained throughout the loop.

The distribution network consists of rigid pipes of varying sizes ranging from 450 mm diameter and sometimes even bigger, down to 50 mm with 75 mm, 100 mm and 150 mm all common. New mains normally use either 100 or 150 mm pipes as standard. At the end of each system, or in new housing estates or cul-de-sacs, you will often see the flexible plastic pipes, which come in coils, being used. These generally come in two standard diameters, 63 and 90 mm. The pipes come in a variety of materials. The most commonly used are iron (cast, spun or ductile), asbestos cement, uPVC (unplasticized polyvinyl chloride) and also mDPE (medium density polyethylene). There are new draft specifications for all pipe materials published by CEN, the European Committee for Standardization. Owing to the effects on water quality, asbestos cement pipes are no longer installed and are being replaced by plastic pipework whenever possible (Section 22.3). However, the cost of renovation or replacement can be horrendously expensive, so older or less favoured materials are not likely to be replaced very quickly. In fact, existing pipes are only replaced when the size is no longer adequate to supply the amounts of water now required, or if their performance in terms of leakage is no longer acceptable.

Hydrants are important not only for fire-fighting, but for flushing out the main. The location of hydrants is decided by the needs of the fire brigade and the requirements of the water company. In fact, water companies have a legal requirement to locate hydrants where the fire brigade requires them. In the UK and Ireland, hydrants are located in an underground chamber to protect them against frost and general damage. A yellow plaque marked with the letter H painted in black indicates the exact location of the hydrant. The figures above and below the crosspiece of the letter H indicate the diameter of the main and the distance of the hydrant from the plaque, respectively. The optimum diameter for fire-fighting is 100 mm, although 75 mm is satisfactory for a ring main.

20.4 Service pipes

The pipe that conveys the water from the mains to the consumer's house is the service pipe. For single dwellings this is a small pipe of less than 25 mm in diameter, although it is usually only 13 mm. When more than one dwelling is served, for example blocks of flats, institutions such as schools or hospitals, or industrial premises, then the service pipe may be larger and possibly even the same diameter as the main itself when supplying large industrial complexes.

The service pipe is split into two sections, usually by the water company's stop-tap. The communication pipe runs from the water mains to the boundary of the street, which is usually the outer wall or fence of the property to which the supply is going. The boundary stop-tap is situated as close to the boundary as possible, usually just outside the boundary fence (Figure 20.3). This section of the service pipe is owned by the water company, who are responsible for its maintenance and repair. They also own and are responsible for the boundary stop-tap, although the household it serves may operate it if necessary. The service pipe then runs from the boundary stop-tap into the house. The householder usually has his or her own stop-tap within the building, as close as possible to where the supply enters. This section of pipe between the two stop-taps is known as the supply pipe. The property owner is responsible for the maintenance and repair of the supply pipe, not the water company. In fact, all the company's responsibilities end at the boundary stop-tap. This division of responsibility is perhaps the most common cause of bad feeling among customers, who are used to other utilities such as gas, electricity and telephone companies taking responsibility for the service to within the building, or at least to the meter. This is not the case with water. Although water meters may be fitted anywhere on the service pipe, including within the property, this does not alter the division of responsibility for the service pipe, with the supply pipe remaining the responsibility of the property owner, even if it also includes the water meter (Section 26.1).

A common problem in older houses is that the water main can be remote from the property. Although the communication pipe may still be fairly short, just extending to the boundary of the street in which it is laid, the supply pipe may be very long. In these cases the private supply pipe may be laid across

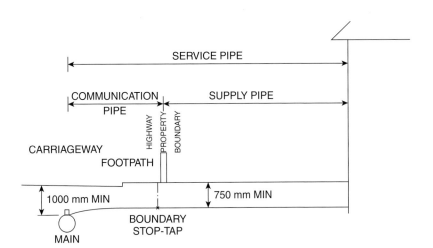

Figure 20.3 Arrangement of service pipes showing minimum depths to which pipes should be buried. Reproduced from Latham (1990) with permission from the Chartered Institution of Water and Environmental Management.

private land, under roads and even occasionally through other buildings on its way to the customer's property. Such complex arrangements can cause appalling problems when the supply pipe requires attention. The problems of locating the pipe and then detecting a leak or blockage is usually compounded by obtaining permission to dig up someone's lawn or driveway. It can also involve very high costs, especially if the repair involves digging up a section of local authority highway.

Before 1974 in the UK, local authorities generally held responsibility for both housing and water supply. During this period many houses were supplied with water using joint supply pipes. These serve a number of properties and are common throughout the UK (Figure 20.4). Currently between 20% and 50% of all service pipes in any supply zone are shared. Like all service pipes only the communication pipe is the responsibility of the water company; the supply pipe in this instance is owned jointly by the properties served. Because these pipes are old they are prone to problems. The most common complaint is that the houses at the end of a joint supply pipe have low pressure or insufficient water. This is often seen when terraces of houses or cottages are modernized and the water usage of the new families is much greater than when the pipe was first laid. Old pipes are much more prone to bursting or blockages, and as with long supply pipes they can be difficult to locate and repair. It was common practice to lay joint supply pipes supplying terraced houses through cellars, so obtaining access for maintenance and repair is not always easy. Interpretation of ownership and the degree of responsibility for service pipes can be very complex and useful guides have been published to assist in sorting out problem installations (Water Research Centre, 1991, 1992).

Service pipes can be made of mild steel, wrought iron, copper, lead, polyethylene or uPVC. Until the widespread introduction of plastic pipes, which are now universally used, the selection of the correct metal for the service pipe was very important. This is because where the water is soft service pipes can be

Figure 20.4 Examples of joint supply arrangements (cp = communication pipe; jsp = joint supply pipe; bst = boundary stop-tap).

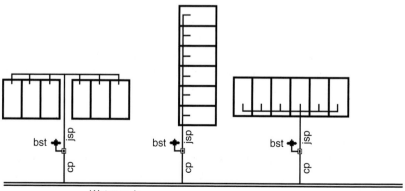

readily corroded. Iron and steel pipes are extremely strong and are jointed by
screwing together or by flanges, so they can withstand high pressures. Mild steel
is generally galvanized to protect it from corrosion. Unlike iron and steel,
copper can be used with very thin walls, which makes it cheap and light to work
with. Copper pipes can be easily jointed and are readily bent into shape, and can
withstand high pressures. The smooth surface in copper pipes results in a very
low resistance to flow so that smaller diameters can be used compared with
other metal pipes. Copper and galvanized steel should not be used together, as
there will be enhanced corrosion due to electrolytic action (Chapter 27). Lead
was the most widely used material for service pipes and for household plumbing
before the Second World War as it is so malleable and was the best material
available at that time. However, justified health concerns have resulted in lead
communication pipes being replaced by the water companies whenever they are
encountered. Where householders replace their lead supply pipes, the water
companies will also replace the communication pipe. The problem of lead pipes
and health is examined in Section 27.3.

All the service utilities now use plastic pipes or plastic ducting. In the UK
these pipes are colour-coded to prevent accidents, with water pipes coloured blue,
gas yellow, electricity black and telephone grey. Plastic water pipes have many
advantages, especially where the soil or water conditions pose problems of
freezing or corrosion. To protect service pipes from frost they should be laid at
least 750 mm below the surface. Polyethylene is tough, light, cheap, flexible, easy
to work and join, frost resistant and a non-conductor of electricity. Unplasticized
polyvinyl chloride, like polyethylene, is not impervious to gas and loses its
strength rapidly at temperatures above 70 °C. Pipes made from uPVC are more
rigid than polyethylene and they have a greater tensile strength, so that more
complex piping systems can be used as a wide variety of joints are available. The
greatest practical problem with using plastic pipes is that their exact location
cannot be detected so readily using electromagnetic surface detectors. This can be
overcome by using metal trace wire, but this is expensive. There are a number of
new location techniques for plastic pipes under investigation. The most promising
appear to be those based on acoustic methods. Sound waves travel readily through
water, so by applying a pulse to a pipe containing water it will be transmitted
along its length. In this way the exact location of the pipe can be identified by
using a suitable detector to pick up the vibration (Godley and Wilcox, 1992). Care
must always be taken to accurately record the exact route of such pipes and to lay
them wherever possible at right angles from the main to the boundary stop-tap
and then in a direct line to the house.

It was common practice in houses constructed before 1970 to earth
household electrical systems to the incoming service pipe. This was phased out
during the 1960s but can still be found occasionally in houses built as late as
1968–9. Plastic pipes are non-conductors of electricity, so an alternative
earthing system will be required.

New service pipe connections are made to the water main without interrupting the flow in the main. A special device is used to drill the pipe and insert a ferrule onto which the communication pipe is fitted. This is carried out at the crown (top) of the pipe with the communication pipe looped over slightly in a snake shape to allow for any ground movement or thermal expansion or contraction without damaging the connection to the main.

20.5 Conclusions

The distribution system is perhaps the least impressive part of the supply chain simply because it is hidden away underground and does not involve the use of the impressive unit processes associated with the treatment plant. However, the distribution network comprising service reservoirs and hundreds of thousands of kilometres of pipework that supply clean, safe water to nearly every household nationwide, each with their own service connection and many with their own meters, is really a great engineering achievement. For example, there are 52.7 million drinking water consumers in England and Wales living in 2284 water supply zones supported by 4691 service reservoirs. This is operated by 26 separate water supply companies who together maintain 326 471 km of water mains (Chapter 1).

The nature of the pipework that makes up the distribution system can affect the nature of the finished water as it travels from the treatment plant to the consumer's tap often altering its aesthetic and health-related quality (Table 3.2). The materials used in the manufacture of the water distribution pipes have the most significant effect on quality. For example, asbestos fibres are released from asbestos cement pipes, polycyclic aromatic hydrocarbons can leach or be eroded from bitumen and coal-tar linings, while iron is released as older cast-iron pipes corrode. Aesthetic problems arise from sediment, discolouration, as well as odour and taste, which are often related to the microbial biofilm that grows on the internal surface of the pipework. This biofilm also supports a range of larger organisms that occasionally appear in tap water, but the biofilm can affect the microbial quality of the water by allowing both pathogens and opportunistic bacteria to regrow.

The constant maintenance and renewal of the distribution system is a major cost factor to the water companies, especially as water resources are becoming increasingly overstretched and consumers are less reluctant to accepted water restrictions when leakage from mains is as high as 40% in some areas (Section 1.5). However, replacement programmes of older sections of pipework need to be accelerated, thereby saving water, reducing the risk of contamination via fractures and improving quality by using more appropriate pipe materials. In terms of water security the distribution network is particularly vulnerable to deliberate interference or contamination, and water utilities are increasingly vigilant to ensure supplies are well protected (Chapter 30). Without water our health and indeed the very basis of our modern society is threatened. So although not seen by consumers, the proper maintenance and investment in the distribution network plays a critical part in bringing adequate supplies of high-quality drinking water into our homes.

References

Godley, A. and Wilcox, P. (1992). *Plastic Pipe Location and Leak Detection on Plastic Pipes*. Report FR 0257. Marlow: Foundation for Water Research.

Latham, B. (1990). *Water Distribution*. London: Institution of Water and Environmental Management.

Open University (1975). *Water Distribution, Drainage, Discharge and Disposal*. Milton Keynes: Open University Press, PT 272–7.

Water Research Centre (1991). *A Guide to Water Service Pipes in Scotland*. Swindon: Water Research Centre.

Water Research Centre (1992). *A Guide to Water Service Pipes*. Swindon: Water Research Centre.

Chapter 21
Aesthetic quality

21.1 Odour and taste

Considerable odour and taste problems develop while the water is in the distribution system due to the material from which the mains are constructed, or the effect of biological growths on the walls of the pipes. However, most problems originate either in the raw water (Chapter 8) or from the treatment plant (Chapter 16). The water supply company is careful to eliminate these as possible sources of odour before testing the water within the distribution system to isolate the problem area (Figure 8.1).

The major complaint arising from the distribution system is that of musty–earthy odour (Figure 8.2). These are due to the development of micro-organisms on the walls of the distribution network. Although actinomycetes and fungi are generally responsible, heterotrophic bacteria can also cause similar problems if present in large numbers. This suggests that a high activity of any micro-organism, regardless of its identity, may result in odours. A number of steps can be taken to reduce microbial growth within the distribution system, the main one being to ensure adequate disinfection with sufficient residual chlorine. Other methods include reducing the amount of organic matter and nutrients in the water by more effective treatment (Chapter 14) and using new pipe materials that do not encourage microbial development. The growth of actinomycetes and fungi is controlled primarily by temperature, with optimum growth occurring at 25 °C. At temperatures below 16 °C growth is so reduced that complaints due to odour are generally eliminated. It is therefore essential to prevent water in distribution systems from standing for long periods and warming up. During warm weather conditions it is possible for many supplies to reach ambient temperatures as a consequence of this long residence time, thus encouraging the unwanted growth of micro-organisms. Long residence times also encourage organic material to flocculate and settle, which then acts as a source of food for micro-organisms and small animals (Chapter 24). The water supply companies are careful not to over-design distribution systems to ensure constant and rapid movement of water, and to avoid spur mains wherever possible (Chapter 20).

Free chlorine will slowly convert ammonia or monochloramine into odorous dichloramine, a process known as disproportionation (Section 14.2). As a consequence dichloramine concentrations tend to increase towards the end of, or the extremities of, the distribution network. The reaction is considerably enhanced by lower pH values so that chloramine odours are often produced where waters from different origins are mixed, which often occurs in larger cities where the water may be coming from several different sources.

Iron is a major problem for a number of the larger water supply companies. It can originate from the raw water, from the use of iron salts as coagulants during treatment, or more commonly due to the corrosion of old iron mains. Iron imparts an unpleasant taste to the water and has a mean taste threshold concentration of 3 mg l^{-1}, although for the most sensitive 5% of the population this threshold concentration decreases to just 0.04 mg l^{-1}. Clearly what is potable to the majority will not be to those who are more sensitive to tastes and odours. In response to this particular problem there is a major programme of mains replacement and renovation usually by lining with a plastic coating by all UK water utilities. The EC has set maximum permissible concentrations of iron and manganese in drinking water of 200 and 50 μg l^{-1} respectively as measured at the consumer's tap. This value is based on taste and aesthetic criteria rather than health grounds.

Iron mains are not the only material to cause problems. Odours are occasionally released from older lining materials, especially bituminous-based compounds that release naphthalene. This gives the water a strong oily odour (Chapter 23).

21.2 Discolouration and iron

Poor treatment plant operation can result in particulate matter getting into the distribution system and causing discolouration and other problems. The main cause is the use of excessive amounts of coagulant, which results in either iron or aluminium entering the system, although manganese, algae and organic matter can all arise from poor treatment. As the contaminated water passes through the distribution system, deposition and sedimentation take place, thereby reducing the concentration of these substances downstream. Examples are given for three different cast-iron distribution networks for the metals aluminium, manganese and silica (Figure 21.1).

Even the most efficient treatment plant can only remove colloidal and suspended organic matter, with dissolved organic material only slightly reduced (20–25%) by conventional treatment. It would not be uncommon to have 10 mg l^{-1} of organic matter, all in soluble form, entering the distribution system (Chapter 25). When organic matter enters the distribution system it is fairly rapidly reduced. This is due to a number of mechanisms such as co-precipitation with iron, manganese or aluminium, adsorption onto corrosion products

Figure 21.1 Schematic diagram showing variation in quality in three cast-iron distribution water mains networks. Concentrations of aluminium, silica and manganese in mg l^{-1} and pipe diameters in mm. Reproduced from Ainsworth *et al.* (1981) with permission from WRc Plc.

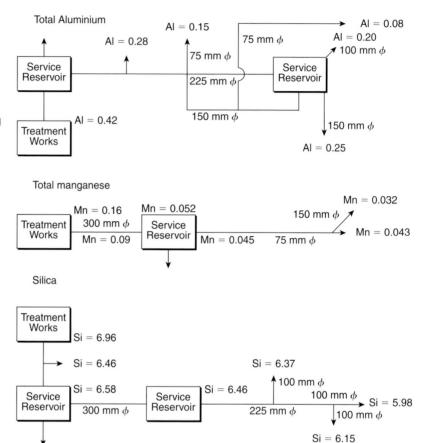

or utilization by micro-organisms. Whichever mechanism is responsible, the organic matter is removed from solution and concentrated in the deposits forming a thin layer of loose debris; this is easily flushed out at higher flow-rates. This debris is the food source for a variety of animals and micro-organisms that find a home in the water mains (Chapter 24). The actual amount of organic matter will depend on the water source. For example, groundwaters and very pure surface waters will contain very little organic matter in solution so there will be little or no related activity in the mains. Upland sources from peaty moorland areas are rich in humic compounds, which give the water a moderate organic matter concentration. This water shows a slight decrease in organic matter content as it passes through the distribution system, suggesting some activity. Lowland surface waters contain high concentrations of dissolved organic matter that is significantly reduced as it passes through the mains. If this organic matter supports excessive microbial growth, normally as a slime that covers the walls of the pipework, discolouration as well as taste and odour problems can result (Section 21.1). Micro-organisms are likely to cause discolouration if counts exceed $10^3 \, ml^{-1}$, although this is unlikely in either groundwaters or high-quality surface waters. The main source of bacteria will be lowland surface water supplies, where counts can exceed $10^6 \, ml^{-1}$. This high level of bacterial activity can cause a number of other related problems (Chapter 25).

Although iron may originate from the use of coagulants, increases in the total iron content of a water supply as it passes through the mains is generally caused by corrosion. Mechanical activity and disturbing existing deposits cause temporary increases in iron, while corrosion is a major source of consumer complaints. Galvanic corrosion is explained in Section 27.2 and occurs where different metals or alloys are coupled together in water. Most water mains are made of cast-iron, and these are also susceptible to corrosion, although this proceeds more slowly. As it corrodes, iron (Fe^{2+}) is released into solution (De Rosa and Parkinson, 1986).

The major factors controlling corrosion rates are those related to the nature of the water itself. These include:

1. *Increases in alkalinity or hardness* (concentration of calcium and magnesium ions) of the water, which results in a decrease in the corrosion of iron.
2. *Increases in the concentration of chloride or sulphate*, which increases the corrosion rate, although in the case of chloride the presence of hydrogencarbonate (HCO_3^-) significantly reduces its corrosivity.
3. *The deposition of sediment and the growth of microbial organisms* on the pipe, leading to oxygen depletion, which results in localized corrosion under such deposits and growths.
4. *Corrosion caused by bacterial action* (e.g. sulphate-reducing bacteria) although this is more common in lowland surface supplies, which contain the significant amounts of organic matter needed to maintain such populations of micro-organisms.
5. *Water with a low dissolved oxygen concentration* ($<4 \, mg \, l^{-1}$) can also produce enhanced corrosion.

Soft moorland water is therefore more corrosive than hard lowland water. This is supported by measurements of the depth that corrosion penetrates into iron mains. The mean depth that soft, acidic water penetrates is $60 \, \mu m$ per annum, which is four times greater than for hard groundwaters. Lining materials protect most iron pipes, with bituminous materials such as coal tar commonly used in the past at a thickness of about $100 \, \mu m$. However, if the water is very aggressive, then this material only offers temporary protection, so plastic liners are now used. Any holes in the lining, or any cracks in the inner skin of iron oxides, silicates or aluminosilicates, which is formed during casting and acts as a protective barrier against corrosion of the metal underneath, or any abraded areas, will become areas of corrosion.

Three types of water resources have been identified as being more likely to corrode metallic pipes and therefore cause discolouration problems. These are:

1. Soft waters that are normally derived from upland catchment areas and have a hardness and alkalinity of $<50 \, \text{mg} \, \text{l}^{-1}$ which currently represent a third of the total volume of UK supplies.
2. Chloride and sulphate-rich waters, due to interactions with other ions, especially hydrogencarbonate. No fixed concentration can be set for corrosivity although $50 \, \text{mg} \, \text{l}^{-1}$ of either chloride or sulphate should be considered as possibly corrosive. Consequently this is a particular problem where saline intrusion occurs in boreholes in coastal areas or where water is being reused.
3. Anaerobic waters – water with less than $4 \, \text{mg} \, \text{l}^{-1}$ dissolved oxygen can become corrosive and cause discolouration complaints. Some groundwaters may fall into this category.

If discolouration is not being caused by problems at the water treatment plant or by microbial action within the distribution mains, discolouration due to rusty deposits is most likely to be due to corrosion where the above types of water supplies are involved. However, most supplies suffer from corrosion and even hard waters will corrode iron mains if they contain high levels of organic matter. Chloride is not removed during sewage or water treatment. Every packet of salt we buy ends up down the drain because chloride is not retained to any significant extent by the body. Urine contains 1% chloride, so as the water is used and reused for drinking, the level of chloride builds up significantly, which can increase the corrosion potential of the water.

Population centres in mainland Europe are generally more dispersed, with water distribution systems not as extensive or integrated as in Britain. In northern Europe in particular, the distribution systems are much newer than those in the UK, so that mains causing the discolouration of water due to iron are more common in Britain. This problem does not occur naturally, so while high natural iron concentrations in drinking water may be allowed through derogation of drinking water regulations, problems caused by the distribution system are not.

21.3 Sediment and turbidity

Consumers often find fine sediment in the water drawn from their taps. It is normally only noticed when a bath has been run or a glass of water left to stand. Sediment can consist of either organic matter, including micro-organisms, or insoluble material, mainly iron and manganese. Sediment is often the extreme problem of particulate matter being present in water, as usually only a slight cloudiness of the water (turbidity) will be noticed. The origins of these sediments have already been discussed elsewhere in this book, so only a summary is given here.

Particulate inorganic and organic matter, micro-organisms and algae in the raw water are generally removed at the treatment stage. However, due to operational problems it is possible for unfiltered water to occasionally pass into the distribution system. The presence of fine clay or silt particles and algae will increase the turbidity. If the turbidity exceeds 5 FTU then it will be clearly visible in a glass of water and is usually rejected by the consumer on aesthetic grounds. Although inorganic particles such as clay do not usually affect health, taste may be affected. High turbidity may also warn the consumer of a treatment plant malfunction, which may mean that the water has not been fully treated or disinfected. Turbid waters should always be treated with suspicion. In some instances, such as with groundwaters, soluble iron and manganese may be allowed into the mains; they will then form insoluble particulate iron and manganese as the water is aerated and the pH increases. This especially occurs within the distribution system where water from another source is mixed, thus altering the chemical nature of the water and causing precipitation of the metals. An increase in the turbidity of groundwater after rain may indicate that surface runoff is getting into the borehole.

The solids, which are normally either brown or orange in colour, either settle in the distribution mains forming a deposit or are carried on to the consumer. The settled deposit in the mains will eventually be resuspended and also reach the consumers. Corrosion of water mains results in soluble ferrous iron (Fe^{2+}) entering the water. This may rapidly precipitate out in the mains adding to the pipe deposits, or more likely it will only form the insoluble ferric (Fe^{3+}) form as it enters the consumer's water storage tank or comes out of the tap. Corrosion may result in thick encrustations on the surface of the iron pipes from which small flakes and particles are constantly being broken off.

Soluble organic matter is not fully removed at the treatment plant and is adsorbed by the existing sediment in the distribution mains and also by micro-organisms living on the walls of the mains and in the debris itself. This source of material is a particular problem in organically enriched lowland rivers, which are often used for supplies (Section 20.3). Sediment problems tend to occur at the end of distribution systems, especially at dead-ends, in fact, anywhere that the solids can settle and accumulate. This results in brownish sediment in the water not unlike fine soil, and occasionally animals may also be associated with the sediment. This problem is exacerbated when repair and maintenance work is carried out on the distribution system, or if sections are cleaned due to low-pressure problems. Cleaning is often carried out to remove sediment and slime accumulations on the walls of the pipes. It can be carried out using a variety of methods, including flushing out debris by opening hydrants and allowing large flows to scour out any material inside, by pushing a tight-fitting swab through the pipe by water pressure to wipe the pipe surfaces clear of slime and accumulated solids, or by air-scouring. The latter is achieved by pumping air into the water stream, which causes so much turbulence that all the debris is scoured from the pipe. The dirty water is allowed to run to waste at the next hydrant. Corrosion encrustations require a different and more erosive technique to scrape the surface of the pipe clean. This is usually only carried out before relining the pipe with epoxy resin, cement mortar or other permissible materials.

Within household plumbing systems sediment only really occurs from corrosion or by soluble iron and manganese being oxidized into their insoluble forms. These problems are more common in households receiving private groundwater supplies. Occasionally water can have a sandy sediment present. This is usually associated with the corrosion of galvanized storage tanks or galvanized pipework (Chapter 27).

21.4 Conclusions

Aesthetic problems are the most frequent causes of complaint about drinking water quality. This is not surprising as consumers are only able to judge water quality using the criteria of clarity, smell and taste. From the producer's point of view aesthetic quality is also important, not only in terms of compliance with water quality standards, but as an indicator of problems either with the raw water, treatment or distribution. The EC Drinking Water Directive (98/83/EEC) (EC, 1998) requires that drinking water must be acceptable to consumers in terms of colour, odour, taste and turbidity, with no abnormal changes recorded. The EC has also indicated that member states should strive to achieve a turbidity of <1.0 nephelometric turbidity units (NTU) where surface waters are used. This has been adopted as the national standard in the UK.

The Water Supply (Water Quality) Regulations 2000 requires water to be wholesome, which is defined in Part III of the Statutory Instrument. Wholesomeness can be summarized here as 'water that does not contain any substance, micro-organism or parasite (whether or not a parameter listed in the Drinking Water Directive or a National Standard) at a concentration or value which, either on its own or in conjunction with any other substance, micro-organism or parasite it contains (whether or not a parameter) would constitute a potential danger to human health'. For an exact definition you must read the Regulations itself, but it is clearly stated that while there are mandatory standards, the consumer's rights in relation to safe drinking water are guaranteed. The definition of wholesomeness does not include palatability, although the indicator parameters listed in Part C of the EC Drinking Water Directive (Appendix 1) do ensure the organoleptic quality of the water.

References

Ainsworth, R. G., Calcutt, T., Elvidge, A. F. *et al.* (1981). *A Guide to Solving Water Quality Problems in Distribution Systems*. Technical Report 167. Medmenham: Water Research Centre.

De Rosa, P. J. and Parkinson, R. W. (1986). *Corrosion of Ductile Iron Pipe.* Technical Report 241. Water Research Centre: Swindon.

DETR (2000). *The Water Supply (Water Quality) Regulations 2000.* Statutory Instrument No. 3184. London: Department of Environment, Transport and the Regions, Water Supply and Regulation Division, HMSO.

EC (1998). Council Directive 98/83/EEC of 3 November 1998 on the quality of water intended for human consumption. *Official Journal of the European Communities*, **L330** (5.12.98), 32–53.

Chapter 22
Asbestos

22.1 The nature of asbestos

Asbestos is the name given to a class of naturally occurring fibrous silicate minerals containing iron, magnesium, calcium or sodium. Its name is derived from the Greek meaning unquenchable or inextinguishable, which refers to the fact that it cannot be destroyed by fire. It is a widely used material, with over 5×10^6 tonnes mined annually during the 1980s, although over the past 15 years demand for asbestos has been slowly declining. Among its important properties are its resistance to fire, its thermal and electrical insulation properties and its binding capacity, which provides strength and stability when it is incorporated with cement or other materials or used in the formulation of brake linings and even floor tiles (Commins, 1988).

There are a number of different types of asbestos; the most important are listed in Table 22.1. Those most widely used are chrysolite (white asbestos) (Figure 22.1), amosite (grey) and, in the past, crocidolite (blue). These are the forms most commonly used in asbestos cement products including water pipes.

Asbestos occurs naturally worldwide, although the largest mined deposits are found in Canada, Russia, Africa, Australia and the USA. In Britain outcrops of chrysolite and tremolite are found in Snowdonia, Caernarvonshire, Cornwall, Hereford and Worcester, Aberdeenshire, Banffshire, Inverness-shire, Skye, Sutherland, Jersey and Guernsey. Although none of these are mined commercially, where asbestos is mined on a large scale up to 1200×10^6 fibres l^{-1} can find their way into drinking waters. In the UK the major outcrops occur at Lochnager in Scotland and on the eastern slope of Moel Yr Ogof in Snowdonia, North Wales. Neither of these areas drain into water supply catchments, and so no studies have been carried out to determine if there is any significant natural runoff of asbestos.

Asbestos fibres in water supplies originate from a number of sources; for example, from the dissolution of natural asbestos mineral, industrial effluents, atmospheric fallout and the effects of aggressive and corrosive water on asbestos cement water distribution pipes and storage tanks. Once asbestos is released into the atmosphere it is inevitable that it will find its way into water, which acts as an irreversible sink for the fibres.

Table 22.1 *Most important types of asbestos mined and used, and the main cations linking the silicate structure*

Mineral	Main cation(s) linking silicate structures
Serpentine group	
Chrysolite (white asbestos)	Mg^{2+}
Amphibole group	
Amosite	Fe^{2+}, Mg^{2+}
Crocidolite (blue asbestos)	Na^{+}, Fe^{2+}, Mg^{2+}
Anthophyllite	Mg^{2+}, Fe^{2+}
Tremolite	Ca^{2+}, Mg^{2+}
Actinolite	Ca^{2+}, Mg^{2+}, Fe^{2+}

Figure 22.1 Chrysolite asbestos fibre from surface water in the Piedmont district of Northern Italy. Reproduced from Buzio *et al.* (2000) with permission from Elsevier Ltd.

Interest in waterborne asbestos was triggered in the USA and Canada by the discovery of relatively high levels of fibres in Lake Superior and in a number of water supplies in mining areas (Sigurdson *et al.*, 1981). This coincided with studies on the inhalation and ingestion of asbestos dust in occupational situations. Asbestos fibres found in water are generally very small. For example, chrysolite fibres are typically 0.5–2.0 µm long and only 0.03–0.1 µm in diameter. Optical (light) microscopy is therefore inadequate for identifying or counting fibres in water. Asbestos analysis is complex and time consuming, involving a combination of transmission electron microscopy with either selected area electron diffraction or energy dispersive X-ray analysis. Routine monitoring of water supplies for asbestos is not carried out, so the specific problems of monitoring asbestos in drinking water or automated methods of analyzing fibre concentrations in water have not received much attention. The problems of sampling, sample preparation and analysis are reviewed by Commins (1988).

Since asbestos is so ubiquitous it can enter the water cycle by a number of different routes resulting in most ground and surface water supplies containing fibres. Concentrations vary between background concentrations of 10^4 fibres l^{-1} to 10^{11} fibres l^{-1} in waters receiving runoff from mining areas. Higher concentrations are usually found near cities and industrial centres (Millette *et al.*, 1983). Studies carried out in the UK, the Netherlands and Germany have indicated that levels in drinking water range from 0.2 to 2.0×10^6 fibres l^{-1}, with an average concentration of 1.0×10^6 fibres l^{-1} (Conway and Lacey, 1984). This is similar to findings in the USA and Canada, but in these countries asbestos is widely mined and more widely used in allied industries. This results in concentrations in excess of 10×10^6 fibres l^{-1} not being uncommon (Chatfield and Dillon, 1979; Millette *et al.*, 1979; Conway and Lacey, 1984; Hardy *et al.*, 1992).

22.2   Health effects and standards

Comparatively little work has been carried out on the health effects of asbestos in drinking water, with most studies performed on inhalation effects. The effects of inhaling asbestos fibres are associated with lung disorders such as asbestosis, which is incapacitating, causing cancer of the lung and pleura, and cancer of the gastrointestinal tract. The lung cancer takes the form of bronchial lung carcinoma, whereas cancer of the pleura, which is the membrane surrounding the lung, produces a tumour known as mesothelioma. Gastrointestinal cancers are usually in the form of peritoneal tumours, the peritoneum being a membrane associated with the gastrointestinal region. All types of asbestos are potentially carcinogenic, although it is considered by most experts that crocidolite (blue asbestos) is the most dangerous. In fact, the importation of crocidolite into the UK has been banned since 1970, although it is still commonly encountered. Owing to the risk to health, atmospheric quality standards have been set for chrysolite at 2 fibres ml^{-1} of air and for crocidolite at 0.2 fibres ml^{-1} of air. Asbestos is ubiquitous in the environment and almost everyone is exposed to airborne asbestos, albeit in very small amounts compared with those employed in mining or associated industries where asbestos is used. Relatively few people were developing mesothelioma up to the mid 1990s, about 200 each year in the UK; but there has been a rapid increase in reported cases annually over the past decade, often in people with little known contact with asbestos.

Inhaled fibres can pass through the lung tissue into the bloodstream to other parts of the body such as the gastrointestinal tract. Likewise, ingested asbestos, usually short fibres, is capable of penetrating the intestinal wall and accumulating in various tissues and organs or being eliminated in urine (Webber and Covey, 1991). Inhaled fibres can also be subsequently ingested as they are brought up by ciliary activity from the lungs. While elevated concentrations of asbestos fibres have been detected in colon cancers of asbestos workers, there is no consensus in the medical profession as to whether it is inhaled or ingested fibres that are responsible for the increased incidence of gastrointestinal cancer

(Commins, 1988; Webber and Covey, 1991). This makes the present position on the health effects of asbestos in drinking water inconclusive, which is why the EU has set no drinking water standards to date. In contrast, the US Environmental Protection Agency (USEPA) has included asbestos in the National Primary Drinking Water Regulations. The current maximum contaminant level (MCL) is set at 7×10^6 fibres l^{-1} for fibres $>10\,\mu$m in length, which approximates to $0.2\,\mu$g l^{-1} of asbestos (Appendix 2). However, the scientific basis for the selection of this limit is unclear. In the 1993 revised World Health Organization (WHO) guidelines, asbestos was included for the first time. It was categorized under 'chemicals not of health significance at concentrations normally found in drinking water'. It is grouped with silver and tin, for which no guide values have been set. This remains the current situation (WHO, 2003).

Cantor (1997), who reviewed the evidence of asbestos in drinking water and cancer, demonstrated the inconsistency between the various epidemiological studies. He concluded that only for cancers of the stomach, pancreas and kidney were elevated risks demonstrated with any degree of consistency. This inconsistency is reflected in two recent studies. In Woodstock, New York state, asbestos cement water distribution pipes were installed during the mid to late 1950's. Due to the corrosiveness of the water the asbestos cement pipes began to decay releasing chrysolite and crocidolite fibres, but the problem of asbestos contamination of the water was not identified until 1985. Of the five sampling sites within the distribution network, four of the sites had $>10 \times 10^6$ fibres l^{-1}, the maximum being 301×10^6 fibres l^{-1}. Extensive epidemiological studies have failed to find any link between the asbestos contamination and cancer within the community at Woodstock (Browne *et al.*, 2005). In contrast, a significant risk of gastrointestinal cancer, and stomach cancer in particular, has been reported in Norwegian lighthouse keepers, whose drinking water comes from water harvested from asbestos roofs installed in 1945 (Kjærheim *et al.*, 2005). However, Varga (2000) concluded that there is no unequivocal evidence that ingested asbestos is carcinogenic. However, as the fibres are known to be able to penetrate the gut wall, the high adsorption capacity of the fibres creates the possibility of cocarcinogenic or cogenotoxic action with adsorbed organic compounds, especially micro-pollutants in chlorinated drinking waters. So the evidence for the health risks associated with drinking water remains unclear.

22.3 Asbestos cement distribution pipes

Asbestos cement has been widely used to make low-cost, high-strength concrete products, especially pipes and tanks. The fibres act as a binder or filler in the cement and have been widely used for water distribution mains and in the construction of service reservoirs. It is not understood exactly what factors cause the fibres to be released from the pipes (i.e. exfoliation), whether it is mechanical or chemical, but increased levels of fibres have been recorded by a study carried out in the UK (Conway and Lacey, 1984 Hardy *et al.*, 1992). Eighty-two potable supplies were examined after they had passed through asbestos cement pipes. All the waters were considered to be either aggressive or moderately aggressive (i.e. soft and acidic). The pipe lengths examined ranged from 375 to 10 500 m, with the age of the pipes ranging from 7 to 35 years. After passing through the pipes, 17 samples contained fibre concentrations greater than 1×10^6 fibres 1^{-1}, and 4 in excess of $3 \times 10^6 1^{-1}$. As the fibre levels could not be attributed to natural sources it showed conclusively that they were from the concrete pipes. The study showed that pipe length, age and the aggressiveness of the water were all important factors affecting the concentration of asbestos fibres in drinking water. After flushing the mains by opening hydrants, or if the fire brigade used large amounts of water, then fibre density could increase up to 14×10^6 fibres 1^{-1}. The problem was worst in spur mains where fibres tended to accumulate along with other debris. In the UK about 10% of the water mains are made from asbestos cement, containing on average between 10% and 15% asbestos by weight. Asbestos is also widely used in gaskets, packing for pumps, glands and joints. In a national survey of asbestos cement distribution networks in Canada, chrysolite was the predominant type of fibre recorded. Concentrations varied from $<0.1 \times 10^6$ fibres 1^{-1}, which was below the detection limit, to 2000×10^6 fibres 1^{-1}; the medium length of fibres was 0.5–0.8 µm. In terms of supplies tested, 25% of the population served had water containing $>1 \times 10^6$ fibres 1^{-1}, 5% $>10 \times 10^6$ fibres 1^{-1} and 0.6% $>100 \times 10^6$ fibres 1^{-1} (Chatfield and Dillion, 1979).

Some asbestos fibres are much smaller than bacteria and can readily pass through internal membranes within the body. They are transparent in water, so even when the water is grossly contaminated, as has happened in some regions of Canada, the USA and Africa, where asbestos is mined, the water supplies remain clear to the naked eye or only slightly turbid (cloudy). Research on the use of contaminated water in showers and portable home humidifiers have shown that the fibres can be readily converted to an aerosol and inhaled. Of particular concern is the use of ultrasonic humidifiers, which produce airborne fibre concentrations directly proportional to the concentration of asbestos in water. In some homes these units were found to produce airborne asbestos concentrations well in excess of the exposure level of 0.2 fibres cm^{-2} of air (Hardy *et al.*, 1992).

22.4 Conclusions

The WHO (2003) concludes that 'although asbestos is a known carcinogen by the inhalation route, available epidemiological studies do not support the hypothesis that an increased cancer risk is associated with the ingestion of asbestos in drinking water'. So, 'in the absence of consistent and convincing evidence that ingested asbestos is hazardous to health', the WHO concludes there is no need to establish a guideline for asbestos in drinking water. In contrast, the USEPA (2006) states that asbestos, after a lifetime exposure above the MCL, has the potential to cause lung disease and cancer of the lungs and other internal organs, although its inclusion in the National Primary Drinking Water Standards is based on the increased risk of developing benign intestinal polyps (USEPA, 2005). There has been a rapid increase in the once rare lung cancer mesothelioma over the past decade, with people affected often only having been exposed to very low levels of asbestos. There is increasing evidence that white asbestos, chrysolite, is far more dangerous than previously thought, bringing millions of people who had no idea that they had been exposed to potentially dangerous levels of asbestos into the increased risk category. For this reason any exposure to asbestos must be considered hazardous. In relation to drinking water the evidence is far from clear, leaving several unanswered questions. Is there a causal link between asbestos in drinking water and cancer in humans? At what levels can exposure to asbestos in drinking water be considered safe? How significant is aerosol formation in the inhalation of asbestos? How much asbestos is there in drinking water supplies, especially at the end of distribution pipes, or where supplies come from areas served by asbestos cement pipes? Which supplies are receiving high levels of asbestos from fallout or erosion of natural deposits?

The size, length to diameter ratio, shape and crystalline structure of asbestos fibres are all known to be as important as concentration in terms of carcinogenesis; However, current water analysis does not differentiate between the types of fibres. The physicochemical processes fibres are exposed to in either water or the gut can modify these characteristics; however, little is still known about their fate or action within the body. The indications of elevated risks for gastric, kidney and pancreatic cancers from asbestos in drinking water in many studies, although not currently supported by the inadequate epidemiological database that currently exists, merits further research.

Asbestos is chemically very stable, does not evaporate or dissolve, it is non-biodegradable and not affected by photolytic process, making it very persistent in the environment. The small size of fibres ensures that it is distributed widely within the aquatic environment. Asbestos is partially removed by coagulation and filtration (Buzio et al., 2000), but in most areas it is primarily a distribution problem with the removal of fibres only achieved by home treatment systems (Chapter 29). Since the discovery of elevated asbestos fibres in drinking water, the use of asbestos cement for water mains and tanks has largely been phased out in the developed world. Although, in general, existing asbestos cement pipes are not replaced unless they show significant signs of deterioration, when they are replaced pipes made from modern safe materials are used.

With contradictory epidemiological evidence, asbestos in drinking water must still be considered as potentially hazardous. Asbestos cement pipes within distribution networks should be identified and monitored regularly to ensure that they are not shedding fibres. Spur mains receiving water from asbestos cement distribution systems should receive special attention. The action of flushing on fibre concentrations also needs further study. Where asbestos is found naturally in the environment or where supplies are exposed to contamination from mining areas, then consumers should consider installing suitable point-of-use treatment systems (Chapter 29), especially where these supplies are not receiving water treatment employing coagulation and filtration. Water harvesting from asbestos roofs should be avoided, as should the use of asbestos cement water storage tanks. Without epidemiological evidence it is impossible to predict an acceptable limit for the number of fibres in drinking water, but clearly consumers want drinking water with as few asbestos fibres present as possible. Indeed the majority will shrink away from the concept that water containing the US MCL of 7×10^6 fibres ($>10\,\mu m$ in length) l^{-1} is safe. This is compounded by the problem that drinking water retains a high clarity even with high concentrations of fibres present. However, with asbestos largely unmonitored and excluded from drinking water standards for most of the world, including Europe, few have any idea what levels of fibres are in their drinking water supplies or that it may even be a problem. Currently hampered by poor detection limits it will be some time before technology is sufficiently developed to allow suppliers and consumers alike to estimate with any degree of accuracy low-level asbestos contamination in drinking water.

References

Browne, M. L., Varadarajula, D., Lewis-Michl, E. L. and Fitzgerald, E. F. (2005). Cancer incidence and asbestos in drinking water, Town of Woodstock, New York, 1980–1998. *Environmental Research*, **98**, 224–32.

Buzio, S., Pesando, G. and Zuppi, G. M. (2000). Hydrogeological study on the presence of asbestos fibres in water of Northern Italy. *Water Research*, **34**, 2817–22.

Cantor, K. P. (1997). Drinking water and cancer. *Cancer Causes and Control*, **8**, 292–308.

Chatfield, E. J. and Dillon, M. J. (1979). *A National Survey for Asbestos Fibre in Canadian Drinking Water Supplies*. Report 79-EHD-34. Ottawa, Ontario: National Health and Welfare.

Commins, B. T. (1988). *Asbestos Fibres in Drinking Water*. Scientific and Technical Report 1. Maidenhead: Commins Associates.

Conway, D. M. and Lacey, R. F. (1984). *Asbestos in Drinking Water*. Technical Report 202. Medmenham: Water Research Centre.

Hardy, J., Highsmith, V. R., Costa, D. L. and Krewer, J. A. (1992). Indoor asbestos concentrations associated with the use of asbestos-contaminated tap water in portable home humidifiers. *Environmental Science and Technology*, **26**, 680–6.

Kjærheim, K., Ulvestad, B., Martinsen, J. I. and Andersen, A. (2005). Cancer of the gastrointestinal tract and exposure to asbestos in drinking water among lighthouse keepers (Norway). *Cancer Causes and Control*, **16**, 593–8.

Millette, J. R., Clark, P. J. and Pansing, M. F. (1979). *Exposure to Asbestos from Drinking Water in the United States*. EPA-600-79-028. Cincinnati: US Environmental Protection Agency.

Millette, J. R., Clark, P. J., Stober, J. and Rosenthal, M. (1983). Asbestos in water supplies of the United States. *Environmental Health Perspectives*, **53**, 45–8.

Sigurdson, E. E., Levy, B. S., Mandel, J. *et al.* (1981). Cancer morbidity investigations: lessons from the Duluth study of possible effects of asbestosis in drinking water. *Environmental Research*, **25**(1), 50–61.

USEPA (2005). *National Primary Drinking Water Regulations*. Washington, DC: Office of Water, US Environmental Protection Agency.

USEPA (2006). *Technical Fact Sheet: Asbestos*. Washington, DC: Office of Water, US Environmental Protection Agency.

Varga, C. (2000). Asbestos fibres in drinking water: are they carcinogenic or not? *Medical Hypotheses*, **55**(3), 225–6.

Webber, J. S. and Covey, J. R. (1991). Asbestos in water. *Critical Reviews in Environmental Control*, **21**(3–40), 331–71.

WHO (2003). *Asbestos in Drinking Water*. Report No. WHO/SDE/WSH/03.04/02. Geneva: World Health Organization.

Chapter 23

Coal-tar linings and polycyclic aromatic hydrocarbons

23.1 Introduction

Polycyclic (polynuclear) aromatic hydrocarbons (PAHs) are organic compounds with two or more fused aromatic rings of carbon and hydrogen atoms. They predominantly enter the atmosphere from combustion processes and while PAHs are ubiquitous in the environment they are not generally found in high concentrations in water resources. Normal concentrations in uncontaminated groundwater resources are in the range of $0.5\,ng\,l^{-1}$. Due to PAHs having low solubilities but a high affinity for particulate matter they are not found at high concentrations in natural waters (Section 6.4). Drinking water can become contaminated by PAHs where coal-tar and bitumen linings have been used to protect distribution pipes from corrosion.

In the UK, 60% by length of the distribution pipes are ductile iron. These were coated internally by immersing the pipe sections in a batch of coal-tar preparation held at a high temperature. The coal tar is intended to protect the pipe. Asbestos cement pipes are also occasionally coated internally with bitumen materials, as are water storage tanks used by the water companies. Finished water is often distributed over considerable distances through ductile iron mains lined with bitumen and related coal-tar products, which are derived from petroleum sources. This means that the water may have a long contact time with such tar-based materials between treatment at the water treatment plant and being used by the consumer. The length of contact may be significantly increased where the water remains static or flows very slowly towards the end of spur mains.

The coal-tar lining usually applied to iron pipes contains massive levels of PAHs and these readily leach into the water supply by diffusing into the water itself. Fluoranthene is the most soluble PAH compound and so comprises the bulk of the total PAHs present in water in contact with coal-tar linings. These compounds can also be found in drinking water due to small fragments of the lining becoming detached because as coal-tar pitch ages small particles can be shredded from the lining by the flow. Exfoliated particles are dense black in colour and are commonly seen in mains deposits and when the mains are flushed. The particles contain high concentrations of all the PAH compounds listed in Table 6.9, up to 50% PAH by weight. Biofilm on the inside of the pipe will also contain PAHs leached from the liner, and together with the exfoliated fragments this will also carry PAHs to consumers' taps in the form of fine particulate matter. Under normal operating conditions there are sufficient PAHs in a coal-tar lining to produce water containing PAHs at $0.2\,\mu g\,l^{-1}$ for well over 100 years.

Under normal circumstances the increase in PAHs will be within the EC limit of $0.1\,\mu g\,l^{-1}$, although studies have shown that increased levels are common in some distribution systems, often exceeding the prescribed limit. Groundwater appears to leach PAHs from linings more readily than surface waters, whereas temporarily high levels are associated with repair work or where new pipes have been laid. Leaching experiments where water was kept in a section of new main pipe for 24 hours resulted in PAH levels of $49\,\mu g\,l^{-1}$ ($49\,000\,ng\,l^{-1}$), which only fell to around the old EC limit level of $0.2\,\mu g\,l^{-1}$ ($200\,ng\,l^{-1}$) after 20 weeks, when they stabilized at that concentration. Details of the guideline values for PAHs and the associated health effects are given in Section 6.4.

Coal tar contains higher concentrations of PAHs than bitumen linings. The National Water Council in 1976 agreed that coal tar was no longer suitable for lining drinking water pipes and recommended alternative linings. Since that time the use of coal tar has been drastically reduced, although there are still many thousands of miles of coal-tar-lined water mains in the UK supplying water to consumers. The alternative bitumen lining materials now used also contain PAHs, although these leach out at far slower rates.

It was common many years ago to line household plumbing pipes, fittings and storage cisterns with coal tar, while repairs to storage tanks were often carried out by plugging the holes and painting over them with coal-tar pitch. It is now illegal in the UK to use coal-tar or any bitumen-based material in household plumbing systems (Water Byelaw No. 8). The presence of PAHs in drinking water is considered in greater detail in Section 6.4 and their toxicity has been reviewed in depth elsewhere (Clement International Corporation, 1990).

23.2 Conclusions

Coal-tar linings are still very common in distribution mains in many parts of the world. Almost all PAHs are carcinogenic although their potency varies, the most hazardous being benzo[*a*]pyrene (BaP). Fluoranthene, being highly soluble, is the commonest PAH associated with these linings and widely found in drinking water, although it is considered not to pose a health risk to consumers at the concentrations normally encountered. For that reason it has been removed from both the World Health Organization (WHO) drinking water guidelines and the recent EC Drinking Water Directive (98/83/EEC). Significant concentrations of BaP in drinking water, in the absence of similar levels of fluoranthene, are indicative of the presence of coal-tar particles exfoliated from the linings. The current WHO guideline for BaP is $0.7\,\mu g\,l^{-1}$ (WHO, 2003), while the EC and US Environmental Protection Agency have set more stringent mandatory values at 0.01 and $0.2\,\mu g\,l^{-1}$ respectively. Since the ban on the use of coal-tar linings in 1977 in the UK, strenuous efforts have been made by water utilities to reline or replace mains that show signs of deteriorating linings. This is reflected by the high degree of compliance by the water utilities for PAHs. Granular activated carbon filters can be used to remove 99.9% of PAHs at point of use (Chapter 29).

References

Clement International Corporation (1990). *Toxicological Profile for Polycyclic Aromatic Hydrocarbons*. Atlanta: Agency for Toxic Substances and Disease Registry, Public Health Service, US Department of Health and Human Services.

WHO (2003). *Polynuclear Aromatic Hydrocarbons in Drinking Water. Background Document for Preparation of WHO Guidelines for Drinking Water Quality*. Report No. WHO/SDE/WSH/03.04/59. Geneva: World Health Organization.

Chapter 24
Animals on tap

24.1 Microbial slimes in distribution pipes

Microscopic organisms such as bacteria and fungi are common in water mains. They grow freely in the water, and more importantly form films or slime growths (biofilms) on the side of the pipe wall, which makes them far more resistant to attack from residual chlorination (Chapter 25). In an operational sense water supply companies find biofilm formation undesirable as it increases the frictional resistance in the pipes, thereby increasing the cost of pumping water through the system. Certain bacteria attack iron pipes increasing the rate of corrosion, and can also affect other pipe materials. In terms of water quality, microbial biofilms can alter the chemical nature through microbial metabolism, reducing dissolved oxygen levels and producing end-products such as nitrates and sulphides. Odour and taste problems have been associated with high microbial activity in distribution systems, as have increased concentrations of particulate matter in drinking water (Chapter 21). The microbial biofilms are also the major food source for larger organisms that are normally found in the bottom or at the margins of reservoirs, lakes and rivers.

Many upland rivers contain water of an exceptionally high quality, with a low density of suspended solids and good microbial quality. Such supplies may only receive rudimentary treatment, so very small animal species and juvenile forms of larger species can enter the mains in large numbers. Some animals are able to penetrate filters, or can enter the distribution system due to operational problems at the treatment plant. This means that populations of small animals are common in distribution systems. Free-swimming (planktonic) species such as *Daphnia* (water flea) are unable to colonize the mains, whereas many of the naturally occurring bottom sediment dwelling species (benthic) such as *Cyclops, Aelosoma* and *Nais* easily adapt to life in the mains and even form breeding populations. The species which give rise to most complaints are not necessarily those species which enter the mains in the greatest numbers from the resource; rather it is those which are able to colonize and reproduce successfully within the distribution system.

So how do these species survive in what is, after all, clean and chlorinated water? Most treated water contains particulate organic or plant material in the form of algae, although most of the organic matter present will be in a dissolved form (>85%). Most of the animals present are invertebrates feeding off either the particulate organic matter comprising the bottom sediment (detritivores) or else in suspension (filter feeders), or the microbial biofilm covering the pipe surface (grazers). Carnivores are rare in distribution mains allowing populations of grazers, filter feeders and detritivores to flourish. Where invertebrate carnivores do occur very basic food chains are created (Figure 24.1). The density or number of animals in the mains will increase if there is an increase in organic matter or algae entering the system from the source, whereas the accumulation of organic material, including dead animals, provides a rich source of food and an ideal habitat for many species. For this reason animals are often associated with discoloured water and often cause discolouration. Where invertebrates such as *Asellus* are found in iron-rich water, then their faeces can contain up to 70% iron oxides by weight increasing discolouration. Also, as organic particles tend to accumulate towards the end of spur mains, in cul-de-sacs for example, the problem of discolouration, sediment and animals tends to increase towards the end of the distribution pipe. The ecology of water mains has been studied by Smart (1989) who carried out doctoral research at the University of Leicester.

The source of water is important in promoting animal growth. Groundwaters contain less dissolved and particulate organic matter than surface waters, being removed by filtration as the water moves through the aquifer. Also, far fewer animals naturally occur in such waters to replenish or inoculate the mains.

Figure 24.1 Within the distribution main a simple food web develops with different trophic levels associated with the biofilm, bottom sediment or both.

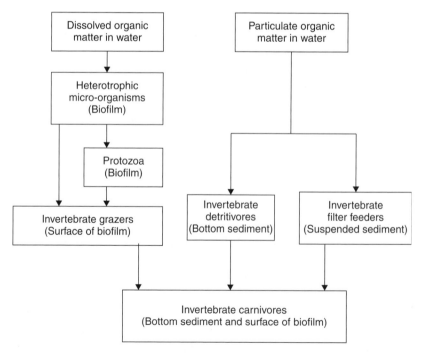

The lack of light in aquifers prevents algal development, which is an important source of organic matter in many other resources. It is the eutrophic lowland reservoirs or rivers that have the greatest potential for supporting large animal populations, but these particular sources usually receive considerable and often complex treatment before entering the mains (Chapter 4). In practice it is upland sources, which receive basic treatment only, which give rise to major consumer complaints of animals in drinking water. The greater the level of organic matter in supplies, the greater the problems from excessive biofilm growth in distribution systems, often made worse by not having animals present to graze and control the development of the biofilm.

24.2 Water supplies

One of the common complaints about water quality is the presence of animals. Although surprisingly large animals miraculously plop out of the tap on rare occasions, such as fish (e.g. the slim ten-spined stickleback), tadpoles and even baby frogs, the vast majority of complaints are about smaller species. Animals in tap water are due to two possible causes. If it is a mains supply then the animals are living in the distribution system; if it is a private water supply then the animals may be making their way straight down the pipe.

24.2.1 Private supplies

Problems occur mainly because the water is being taken either directly from a stream or spring, or from a wide borehole or old hand-dug well. In my experience the larger beasts come from these types of sources. Where wells are used then covering them to exclude light will ensure that no algae will develop and, of course, will ensure that no frogs can enter to lay eggs. In streams where the water is taken from a small chamber built into the stream-bed it is inevitable that some animals will be sucked into the pipe when the water is being used in the house. In both instances it is advisable to fit a series of mesh screens over the inlet, starting with a broad grade and ending with a narrower grade to ensure that animals and silt are both excluded. Mesh screens on the inlet pipes of wide shallow wells are also advisable. These screens will require periodic cleaning, but they will ensure that the larger species are kept out of the pipework. Periodic examination of the water storage tank should also be made for animals and, if present, it should be drained and carefully cleaned out, ensuring that all the silt and any surface slime growths in the tank are also removed using a weak solution of bleach. There should also be a mesh screen on the outlet pipe from the storage tank. Elvers, baby eels, migrate up all major rivers each year, especially those on the west coast. They swarm in their millions against the flow of water until they reach the smaller upland streams from which isolated dwellings often take their water supply. Unless there is a fine screen over the intake, then the occasional elver will get through. Elvers often get into domestic water storage tanks and occasionally may grow into fully grown eels, but only if the tank is uncovered, living off whatever they can find. At worst they will starve to death and decay, which is the fate of most of the other creatures.

24.2.2   Mains supplies

All water mains contain animals and to date about 150 different species have been identified in British water distribution networks (Smalls and Greaves, 1968). The invertebrates commonly found in USA distribution systems are reviewed by Levy (1990). Most of these species are aquatic and enter the mains from the treatment plant, either during construction or maintenance work, or during a temporary breakdown. Other species that are either terrestrial such as earthworms and slugs, or are aquatic for only part of their life cycle, enter the mains during construction, maintenance or via poorly fitting manholes and vents. Obviously terrestrial species, or those that are not aquatic for the whole of their life cycle will eventually die out, so problems arising from these animals are transient. The major sources of complaints are due to truly aquatic species that are able to colonize the mains and breed successfully, although flying insects may enter service or storage reservoirs via unprotected vents and lay eggs at the water surface. This results in a seasonal problem of aquatic fly larvae. This is normally overcome by the use of 0.5 mm mesh screens placed over all access points. Animals are almost ubiquitous in water mains. In a survey of 35 water treatment plants in the Netherlands, van Lieverloo *et al.* (1997) recorded animals in all the distribution systems they served, although the number of species recorded was far less than that recorded by Smalls and Greaves (1968) in their UK study.

24.3   Common species and remedial measures

So what kind of animals can we expect to find? The largest beasts in British water mains will be the water louse (*Asellus* sp.) and the water shrimp (*Gammarus* sp.). Once the animal has been identified, then the water supply company may take the following action depending on the species. The isolated appearance of a single organism, although potentially distressing, should not be taken as a serious indication of a problem. Often animals such as *Asellus* or mussels are present in very large numbers in water mains without ever appearing at the consumer's tap and without causing any reduction in water quality. Conversely, the creature that falls into the sink from the tap may have been the only one. Animals usually appear at the kitchen tap, or less often in the storage tank. If animals appear in taps not directly connected to the rising main, then the water storage tank should be examined immediately. Whenever possible invertebrates should be eliminated before they pass into the distribution system. Once in the distribution network several control options are available. Chemical control includes disinfectants

(chlorine and chloramines) and pesticides such as pyrethrins and copper sulphate. Physical control includes hydrant flushing, cleaning and relining pipes, and the elimination of dead-ends and areas of low flow. The main groups of animals found in tap water in the British Isles are described below together with control options.

24.3.1 Isopoda (e.g. *Asellus aquaticus*)

These are the largest animals that most consumers encounter emerging from their taps. They are up to 15 mm long and quite obvious (Figure 24.2). They usually live among the bottom debris in streams and lakes feeding on organic debris and therefore readily adapt to the conditions within the distribution system. They cling tightly to the pipe wall and can withstand adverse conditions such as low dissolved oxygen concentrations, and therefore usually survive once they get into the mains and can reproduce fairly successfully. Most complaints occur when the adults die after reproduction and are swept away in the flow of water. For this reason few will be seen during the winter, although they are still in the mains. If they have to be removed then the water supply company will use an aggressive flushing technique in mains up to 200 mm in diameter (water- or air-scouring) to dislodge them, or even swabs. Swabs are made from polyurethane foam and are forced by water pressure through the mains, gently scouring the sides of the pipe as they move forward. In practice swabbing is not effective if the pipework is heavily encrusted.

In extreme cases the mains may have to be treated with a pesticide such as pyrethrin, which would involve shutting the system down (Sands, 1969). This is particularly the case where the mains are old and heavily encrusted, allowing *Asellus* a better surface on which to hold. Pyrethrins have a low toxicity to mammals and are extracted from the flowers of *Chrysanthemum cinerariae-folium*. There are now a number of synthetic analogues of natural pyrethrins,

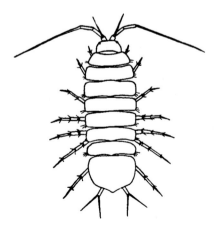

Figure 24.2 *Asellus* sp., an isopod. The invertebrate is on average 8–10 mm long. Reproduced from Macan (1970) with permission from Pearson Education.

called pyrethroids, that appear to have an equally low toxicity to humans and yet have a high insecticidal activity. The most widely used analogue is permethrin (Mitcham and Shelley, 1980; Crowther and Smith, 1982) and is used at an optimum concentration of $10\,\mu g\ l^{-1}$, which is not considered to be a risk to consumers (Fawell, 1987), with the World Health Organization (WHO) guideline value for the compound in drinking water at $20\,\mu g\ l^{-1}$ (WHO, 2004). Treatment takes place in controlled sections of the distribution system by closure of boundary stop-valves to prevent reinfestation from adjacent untreated sections. An optimal contact time of 24 hours is required followed by flushing to remove dead animals. The pesticide is readily adsorbed by the pipework and organic material present, so the concentration must be constantly checked along the entire length of the treated main to ensure adequate pesticide is present. In many countries the addition of any pesticide to drinking water is prohibited. However, all natural pyrethrins and synthetic pyrethroids are extremely toxic to freshwater organisms, including fish, so treated water must not be discharged directly to surface waters under any circumstances.

24.3.2 Amphipoda (e.g. *Gammarus pulex*)

These are similar in habit to *Asellus*, although slightly smaller, reaching 13 mm (Figure 24.3). They are unable to cling onto the sides of the water main and so actively swim, resting in crevices and the debris at the base of the pipe. *Gammarus* is easier to remove because it is unable to cling to the pipe surface, so flushing should be adequate. Swabbing is particularly successful in new mains, but in older mains dosing with pyrethrins may have to be considered.

24.3.3 Insecta (e.g. chironomid larvae)

Many flying insects have freshwater larval stages. The most frequently encountered in water mains are the chironomid (gnat) larvae (Figure 24.4). These can be up to 25 mm in length and may be highly coloured, although most are considerably smaller, 5–10 mm, and white in colour and so are inconspicuous. One or two species of chironomid are parthenogenic. This means that the

Figure 24.3 *Gammarus pulex*, an amphipod. This species varies considerably in length but is usually <12 mm in length when recovered from tap water. Reproduced from Macan (1970) with permission from Pearson Education.

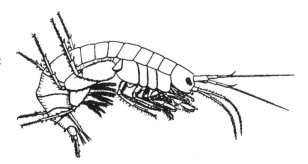

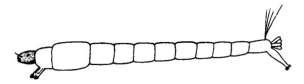

Figure 24.4 Chironomid larvae. These dipteran larvae vary widely in length. Reproduced from Macan (1970) with permission from Pearson Education.

females can reproduce in the absence of males and so can complete their life cycles within the distribution system (e.g. *Paratanytarsus grimmii*). These species have resulted in serious infestations in distribution systems (Williams, 1974; Berg, 1995). However, most occurrences are due to small larvae getting through the treatment plant, or adults entering vents and laying eggs in service reservoirs. Control requires the use of a pesticide, usually permethrin.

24.3.4 Oligochaeta (segmented worms) (e.g. *Nais communis*)

Earthworms sometimes appear, usually after maintenance or repair work. The most common group of worms in the water mains are very small and inconspicuous in comparison, 5–7 mm long and only 0.3 mm in diameter. They are almost transparent and so usually go unnoticed by consumers, but when they wriggle or swim then they may catch the eye, especially when the water is in a clear glass. These species are usually controlled by the residual chlorine in the water and so can be eliminated by raising the free chlorine concentration to between 0.5 and 1.0 mg l^{-1} throughout the system (Sands, 1969). This has to be kept at that level for several weeks and will itself be the cause of consumer complaints. Usually the free chlorine concentration of 0.2 mg l^{-1} throughout the distribution system will prevent a recurrence of this species. Flushing and swabbing can also be used, but infestations quickly reappear unless chlorination is adequate.

24.3.5 Hirudinea (leeches) (e.g. *Erpobdella octoculata*)

Leeches are very unusual in water mains, but if they do appear then complaints are certain and justified. The maximum size you can expect to encounter from your tap will be 35–40 mm in length. They are difficult to dislodge from the mains and require the consumers to be disconnected while the system is heavily dosed with chlorine (12 mg l^{-1}) for at least six hours before being flushed out.

24.3.6 Mollusca (snails) (e.g. *Potamopyrgus jenkinsi*)

This is a very common animal in water mains, reaching 5 or 6 mm in length, although much larger specimens will be found. They cling to the wall of the pipe

and are not readily removed by flushing. Swabbing is effective in newer mains. In practice, consumers rarely come across them in the water; however, if they have an aquarium the snails rapidly colonize it and grow, eventually taking over the tank, encrusting the sides and all the water circulation pipework. Like most of the animals in your water, snails are harmless to consumers in temperate areas such as the British Isles and do not act as intermediate hosts for metazoan parasites.

One particular species of non-native freshwater bivalve that is causing growing concern both in the USA and Europe is the zebra mussel (*Dreissena polymorpha*). Originally from the Caspian and Black Sea areas where they live in lakes and slow-moving rivers, the pea-sized molluscs have been transported via ballast water and on the hulls of ships to become established in many river systems. They can quickly become established excluding the natural fauna and covering the substrate and any submerged objects up to a density of 300 000 animals m^2. They slowly find their way into other catchments and once in water resources little can be done to control them.

24.3.7   Nematoda (roundworms or eelworms)

Nematodes are the most common organism in drinking water, although they are too small to be seen without the aid of a microscope. They live off bacteria and so are found in areas of the distribution system that are rich in organic matter, especially dead-ends. These organisms can be effectively controlled by flushing or swabbing.

24.3.8   Smaller crustacea (e.g. *Cyclops, Chydorus*)

Another common group is the small crustaceans (water fleas), which have a thin, hard outer shell (carapace), which they shed as they grow. The most common species is *Cyclops*, which are transparent and only 1.5 mm long. They are only noticed in water when they move with their characteristic darting action. *Chydorus* are even smaller and cannot be seen with the naked eye. However, they do occur in very large numbers, and as they cast their carapaces, which adsorb iron from the water and so become coloured, the water can become severely discoloured. These are usually successfully controlled by flushing and good residual disinfection. It is necessary to flush a sufficient amount of water through the pipe section to ensure that all the debris is removed, as solid particles move at a much lower rate than the water.

24.3.9   Terrestrial animals (e.g. slugs and earthworms)

Terrestrial animals in the mains usually die rapidly and so do not require control as such. However, the potential points of entry of these animals should be checked, especially service reservoirs. Normal entry is during repair and

maintenance work close to the affected consumers' homes, and so are transient problems.

24.4 Sampling water mains for animals

Water supply companies usually take samples of water from the distribution systems when flushing to check on the number and type of animals present. This is performed by allowing a standard volume of water up to $2.5\,m^3$ (2500 litres) to be taken from a hydrant. Special nets are used which have an aperture size of $142\,\mu m$, and so retain all the animals large enough to cause complaints. A flow gauge is fitted to the standpipe along with a special net (Figure 24.5). The hydrant is then turned on full. Once the required volume of water has passed through the net, it is rolled up with all the animals inside, sealed in a polythene bag and sent to the company biologist for analysis (Department of the Environment, 1985). The abundance of species is important; for example, hundreds of thousands of water fleas per 2500 litre sample will give rise to dirty water complaints, whereas hundreds or even several thousands per sample are

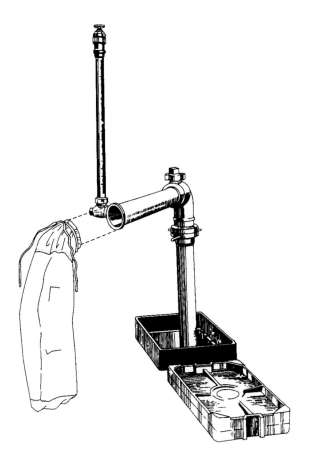

Figure 24.5 Vernon-Morris flow gauge and sampling net used to collect macro-invertebrates from the mains supply. Reproduced from Department of the Environment (1985) with permission from Defra.

unlikely to result in any complaints. In contrast, more than 50 *Gammarus* or *Asellus* per sample would be a cause for concern.

Van Lieverloo *et al.* (2004) examined sampling techniques to quantify invertebrates from water distribution mains in the Netherlands. They compared the use of 500 and 100 μm mesh plankton nets and found that the smaller size was effective in quantifying all but the smallest animals (e.g. *Turbellaria*, Nematoda and Copepoda). Although the numbers of small, less abundant or sessile species were not accurately assessed using their method, they concluded that these species should not be the primary focus of monitoring by water utilities, as consumer complaints were not generally associated with the smaller species. In order to improve the accuracy of quantifying small invertebrates they advise filtering the 100 μm filtrate with a 30 μm plankton gauze filter. The number of invertebrates in unchlorinated water supplies has been studied in the Netherlands (van Lieverloo *et al.*, 1997). Samples collected by flushing water mains has shown that *Asellus* represents about 75% of the total invertebrate biomass with maximum densities varying between <1 and 1000 m^{-3}. Maximum cladocera and copepod densities were very high at 1000–10 000 m^{-3} compared to nematode and oligochaete numbers, which ranged from <10 to 100 m^{-3}.

24.5   Microbial pathogens and invertebrates

None of the invertebrates found in water distribution systems in the UK are associated with metazoan parasites, although this is not true of tropical regions where aquatic invertebrates act as intermediate hosts to a number of parasites. For example the small crustacean *Cyclops* is the intermediate host of the guinea worm (*Dracunculus medinensis*). Although restricted to tropical and sub-tropical countries, there is no evidence that transmission occurs via treated water distribution systems.

Invertebrates feeding on the biofilm and sediment in the distribution system will harbour and concentrate the micro-organisms that comprise or are associated with these food resources. In the USA it has been shown that the animals found in distribution mains can harbour a wide range of opportunistic bacteria, which are protected from disinfection (Table 24.1) (Levy *et al.*, 1986). Non-pathogenic enterbacteriaceae have also been recovered from nematodes collected from distribution systems (Lupi *et al.*, 1995). Wolmarans *et al.* (2005) isolated bacteria from all the invertebrates recovered from distribution mains in South Africa, including several bacterial genera and species that are either pathogenic or opportunistic pathogens of humans. They confirmed that diarrhoea, meningitis, septicaemia and skin infections were among the diseases associated with these organisms. While there appears to be little risk as such to public health from the presence of invertebrates, a potential contaminant route does exist. Therefore they conclude that invertebrates in drinking water should be controlled at levels as low as technically and economically feasible.

Table 24.1 *Bacteria associated with four categories of common invertebrates collected from a drinking water distribution network in the USA. Reproduced from Levy et al. (1986) with permission from the American Water Works Association*

	Invertebrate type			
Bacterium	Amphipod	Copepod	Fly larva	Nematode
Acinetobacter sp.	+			
Achromobacter xylooxidans	+			
Aeromonas hydrophila			+	
Bacillus sp.	+			
Chromobacter violaceum	+			
Flavobacterium meningosepticum		+		
Moraxella sp.	+	+		
Pasteurella sp.	+	+		
Pseudomonas diminuta		+		
Pseudomonas cepacia			+	+
Pseudomonas fluorescens	+		+	
Pseudomonas maltophilia	+			
Pseudomonas paucimobilis	+	+		
Pseudomonas vesicularis	+			
Serratia sp.	+			
Staphylococcus sp.	+			

24.6 Conclusions

There are three main groups of animals found in drinking water, permanent species that complete their whole life cycle within the distribution system (e.g. water fleas, *Asellus*, *Gammarus*), temporary species whose life cycle is partially aquatic (e.g. insect larvae) and transient species that are terrestrial (e.g. slugs and earthworms). Significant numbers of animals in distribution systems are associated with the presence of biofilms and low flow rates. Houses connected to the end of spur mains are the most likely to suffer from animals in their water. However, where larger areas are affected this may be due to the area being supplied from a poorly managed service reservoir. Low pressure can also cause a build-up of organic material in some sections of the mains with a resultant increase in the animal population.

So, apart from consumers seeing the animals because of their size, or when they are small because of their active movement, animals can also cause other complaints. The presence of animals can cause discoloured water due to the faeces of larger animals or the exoskeletons cast off by small crustaceans. Also, as the animals die and decay, they result in intermittent taste and odour problems. It has been concluded that while the presence of invertebrates poses little or no health risk to consumers, including gastrointestinal infection, some groups, such as the very young or those with impaired immune systems, could be at risk (FWR, 1996). Thus the problem of animals in drinking water is primarily aesthetic and should not be a cause for concern to consumers who should, however, report all incidents to their supplier. Operationally, invertebrates in drinking water should be controlled at levels as low as technically and economically feasible, with infestations dealt with quickly using flushing where possible. The use of pesticides, such as permethrin, should be used as a last resort where serious infestations are present. The removal of soluble and particulate organic matter from water entering the distribution system that provides both the habitat and food for these animals is the best preventative action (van Lieverloo et al., 1997). This is reflected by the reduction in reported infestations by animals with increasing efficiency of water treatment.

Those on private supplies should ensure animals are unable to enter the pipework at source. The use of a point-of-use filter to remove animals is unwise as they quickly decay on the filter causing unpleasant tastes and odours and encourage bacterial development. Filters should only be used to remove occasional small organisms and where higher numbers are suspected, and the density of small animals tends to be seasonal, then filters must be replaced more frequently during these periods. The use of an activated carbon filter fitted after the physical filter will remove any tastes and odours, but where surface waters or shallow wells are used then point-of-use ultraviolet disinfection should always be seriously considered.

References

Berg, M. B. (1995). Infestation of enclosed water supplies by chironomids (Diptera: Chironomidae): two case studies. In *Chironomids: From Genes to Ecosystems*, ed. P. S. Cranston. East Melbourne, Australia: CSIRO, pp. 241–6.

Crowther, R. F. and Smith, P. B. (1982). Mains infestations control using permethrin. *Journal of the Institution of Water Engineers and Scientists*, **36**(3), 205–14.

Department of the Environment (1985). *Methods of Biological Sampling: Sampling of Macro-Invertebrates in Water Supply Systems 1983*. Methods for the Examination of Waters and Associated Materials. London: HMSO.

Fawell, J. K. (1987). *An Assessment of the Safety in Use of Permethrin for Disinfestations of Water Mains*. Report PRU 1412-M/1. Medmenham: Water Research Centre.

FWR (1996). *The Health Significance of Animals in Water Distribution Systems*. Final report to the Drinking Water Inspectorate, Report DWI0756. Medmenham: Foundation for Water Research.

Levy, R. V. (1990). Invertebrates and associated bacteria in drinking water distribution lines. In *Drinking Water Microbiology*, ed. G. A. McFeters. New York: Springer-Verlag, pp. 225–48.

Levy, R. V., Hart, F. L. and Cheetham, R. D. (1986). Occurrence and public health significance of invertebrates in drinking water systems. *Journal of the American Water Works Association*, **78**(9), 105–10.

Lupi, E., Ricci, V. and Burrini, D. (1995). Recovery of bacteria in nematodes from a drinking water supply. *Journal of Water Supply: Research and Technology – Aqua*, **44**, 212–18.

Macan, T. T. (1970). *A Guide to Freshwater Invertebrate Animals*. London: Longman.

Mitcham, R. P. and Shelley, M. W. (1980). The control of animals in water mains using permethrin, a synthetic pyrethroid. *Journal of the Institution of Water Engineers and Scientists*, **34**, 474–83.

Sands, J. R. (1969). *The Control of Animals in Water Mains*. Technical Paper 63. Medmenham: Water Research Association.

Smalls, I. C. and Greaves, G. F. (1968). A survey of animals in distribution systems. *Water Treatment and Examination*, **17**, 150–86.

Smart, A. C. (1989). An investigation of the ecology of water distribution systems. Unpublished Ph.D. thesis, University of Leicester, UK.

Van Lieverloo, J. H., van Buuren, R., Veenendaal, G. and van der Kooij, D. (1997). How to control invertebrates in distribution systems: by starvation or by flushing? In *Proceedings of the American Water Works Association Water Quality Technology Conference, 9–12 November, 1997*. Denver, CO: American Water Works Association.

Van Lieverloo, J. H., Bosboom, D. W., Bakker, G. L. *et al.* (2004). Sampling and quantifying invertebrates from drinking water distribution mains. *Water Research*, **38**, 1101–12.

WHO (2004). *Guidelines for Drinking Water Quality*, 3rd edn. Geneva: World Health Organiziation.

Williams, D. N. (1974). An infestation by a parthenogenic chironomid. *Water Treatment and Examination*, **32**(2), 215–31.

Wolmarans, E., du Preez, H. H., de Wet, C. M. and Venter, S. N. (2005). Significance of bacteria associated with invertebrates in drinking water distribution networks. *Water Science and Technology*, **52**(8), 171–5.

Chapter 25
Pathogens in the distribution system

25.1 Introduction

Water that is bacteriologically pure when it enters the distribution system may undergo deterioration before it reaches the consumers' taps. Contamination by micro-organisms can occur through air valves, hydrants, booster pumps, service reservoirs, cross-connections, back-syphonage or through unsatisfactory repairs to plumbing installations. Invertebrates found in distribution networks can also harbour and concentrate a range of micro-organisms (Section 24.5) (Table 24.1). Further problems can also arise within the domestic plumbing system (Chapter 28).

The main danger associated with drinking water is the possibility of it becoming contaminated during distribution by human or animal faeces. This was the case in the Bristol outbreak of giardiasis in 1985, when contamination occurred through a fractured main. A major outbreak of typhoid fever occurred in 1963 in Switzerland when sewage seeped into the water mains through an undetected leak in the pipe. There are many more examples indicating that the microbial quality of water is potentially at risk while in the distribution system. Uncovered service reservoirs can also be a major source of contamination, especially from birds, and there is growing concern about the safety of drinking water while in the distribution system from terrorism (Section 25.4).

25.2 Microbial contamination

Usually the mains is under considerable positive pressure; however, back-syphonage can occur in the distribution system if the water pressure drops and there are faulty connections or fractures in the pipe. In this way contaminants can be sucked into the distribution system. The problems are more likely to occur when water supply pipes are laid alongside sewerage systems. They should be as remote as possible from each other, although in practice this is very difficult.

On passing through the distribution system, the microbiological properties of the water will change. This is due to the growth of micro-organisms on the walls of the pipes, which form a thin bacterial layer, and in the bottom sediments and debris. Although not usually the source of serious health problems, non-pathogenic bacteria can result in serious quality problems, causing discolouration and a deterioration in taste and odour (Chapter 21).

Bacterial growth within the distribution system is known as regrowth, and it is defined as an increase in the heterotrophic plate count (HPC) (Section 19.4) in treated water as it moves along the distribution system. Increased heterotrophic bacteria within the distribution system are due to a number of factors, usually the absence of a residual disinfectant combined with either contamination from outside the distribution network, or more commonly from bacterial regrowth. The growth of bacteria in the water and on pipe surfaces is limited by the concentration of essential nutrients in the water. Organic carbon is the limiting nutrient for bacterial growth, with the growth of bacteria directly related to the amount of assimilable organic carbon (AOC) in the water. There are a number of different methods for calculating AOC concentrations in water. These fall into two broad categories: (1) those that measure the concentration of organic carbon that can be converted into microbial biomass using a parameter such as ATP as the determinant (Van der Kooij et al., 1982); and (2) those that measure the potential mineralization of dissolved organic carbon (DOC) by micro-organisms, by assessing the decrease in organic carbon concentration over time. This is commonly referred to as the biodegradable dissolved organic carbon (bDOC) (Huck, 1990: Gibbs et al., 1993).

Among the major genera found in distribution systems are *Acinetobacter, Aeromonas, Listeria, Flavobacterium, Mycobacterium, Pseudomonas* and *Plesiomonas*. Some of these organisms can be considered as opportunistic pathogens (Table 25.1). The type of micro-organism and the number depend on numerous factors such as the water source, type of treatment, residual disinfectant, corrosion, presence of sediments and bDOC levels in the treated water. The development of slimes, or biofilms as they are known, leads to the survival of other bacteria. Certainly biofilms protect bacteria from disinfectants. *Legionella* in particular is able to survive within biofilms, as are *Pseudomonas* and *Aeromonas* spp. all of which are potential pathogens (Stout et al., 1985). *Mycobacterium avium, Mycobacterium intracellulare* and other mycobacteria have been associated with lung infections (Von Reyn et al., 1994) and are all highly resistant to chlorine. They readily multiply within the distribution system in the sediment that builds up at the end of spur mains and in biofilms (Taylor et al., 2000; Falkinham et al., 2001). The development of non-pathogenic coliforms such as *Klebsiella* is also possible in biofilms (LeChevallier et al., 1996), but when detected it is important for the water operator not to assume a non-faecal cause such as regrowth in the distribution system, even in the absence of *Escherichia coli*. Iron bacteria of the genera *Lepthothrix* and *Gallionella* can form biofilms on the surface of corroding iron mains resulting in taste and discolouration problems at the consumer's tap. Fungi and actinomy-cetes may also be present in low numbers and can significantly affect the taste

Table 25.1 *Examples of the HPC bacteria isolated from distribution and raw waters. Reproduced from Bitton (1994) with permission from John Wiley and Sons Ltd*

Organism	Distribution water		Raw water	
	Total	Percentage of total	Total	Percentage of total
Actinomycete	37	10.7	0	0
Arthrobacter spp.	8	2.3	2	1.3
Bacillus spp.	17	4.9	1	0.6
Corynebacterium spp.	31	8.9	3	1.9
Micrococcus luteus	12	3.5	5	3.2
Staphylococcus aureus	2	0.6	0	0
S. epidermidis	18	5.2	8	5.1
Acinetobacter spp.	19	5.5	17	10.8
Alcaligenes spp.	13	3.7	1	0.6
Flavobacterium meningosepticum	7	2.0	0	0
Group IVe	4	1.2	0	0
Group MS	9	2.6	2	1.3
Group M4	8	2.3	2	1.3
Moraxella spp.	1	0.3	1	0.6
Pseudomonas alcaligenes	24	6.9	4	2.5
P. cepacia	4	1.2	0	0
P. fluorescens	2	0.6	0	0
P. mallei	5	1.4	0	0
P. maltophilia	4	1.2	9	5.7
Pseudomonas spp.	10	2.9	0	0
Aeromonas spp.	33	9.5	25	15.9
Citrobacter freundii	6	1.7	8	5.1
Enterobacter agglomerans	4	1.2	18	11.5
Escherichia coli	1	0.3	0	0
Yersinia enterocolitica	3	0.9	10	6.4
Group IIk biotype 1	0	0	1	0.6
Hafnia alvei	0	0	9	5.7
Enterobacter aerogenes	0	0	1	0.6
Enterobacter cloacae	0	0	1	0.6
Klebsiella pneumoniae	0	0	0	0
Serratia liquefaciens	0	0	1	0.6
Unidentified	65	18.7	28	17.8
Total	347	100	157	99.7

and odour of the water. Nitrification can also increase the HPC count when ammonia is present in the treated water (Skadsen, 1993).

25.3 Control

With good treatment and adequate disinfection with chlorine, ensuring a residual disinfection to control regrowth within the distribution system and to deal with any contamination, serious problems should not arise. When supplies have a high chlorine demand due to the presence of organic matter and humic acids, for example, then it is difficult to maintain sufficient residual chlorine in the system. It is also necessary to strike a balance so that those closest to the treatment plant on the distribution network do not receive too large a dose of chlorine in their water and those at the end too little. Research has shown that many bacteria, viruses and protozoan cysts are far more resistant to chlorination than the indicator bacteria used to test disinfection efficiency. Currently the residual concentration of free chlorine leaving the treatment plant is less than $1.0 \, \text{mg} \, \text{l}^{-1}$ and usually nearer to $0.5 \, \text{mg} \, \text{l}^{-1}$. Most treated waters continue to exert some chlorine demand, which reduces this free chlorine residual even further. Further chlorine is lost by its interaction with deposits and corrosion products within the distribution system, so that free chlorine residuals do not often persist far into the network, leaving the water and the consumer at risk.

The problem stems from treatment processes being primarily designed to remove particulate matter rather than dissolved organic matter. Chlorination and ozonation lead to increases in the AOC concentration in water, whereas coagulation and sedimentation can remove up to 80% of the AOC present. The AOC is also removed by biological activity during filtration, although residence times are far too short to allow significant reductions. Success has been achieved by chemically oxidizing the AOC into simpler compounds to make them more readily removable by biological filtration. There is evidence that the micro-organisms living on granular activated carbon filters can also significantly reduce AOC concentrations (Huck et al., 1991).

The type and stability of the material from which the water mains are made have a significant influence on the density of biofilm development. Iron pipes support 10 to 45 times more growth than plastic pipes (Niquette et al., 2000). So, as there are so many factors involved with controlling regrowth it is probably impossible to completely eliminate it within the distribution system. However, regrowth can be limited by: (1) minimizing the concentration of organic matter entering the distribution system; (2) ensuring the material from which the pipework and fittings are made are both chemically and biologically stable; (3) maintenance of a disinfection residual through the distribution system; and (4) prevention of water stagnation and sediment accumulation within the mains.

25.4   Drinking water security

Water supplies are uniquely vulnerable to terrorism, whether it is aimed at humans, livestock or crops. The key actions that can be taken against supplies are (1) physical damage to water treatment and distribution systems, or the computer systems used to operate them, interrupting the supplies or preventing adequate treatment; (2) deliberate chemical contamination; and (3) bioterrorism using either micro-organisms or biotoxins. Access to water supplies and to water after treatment via service reservoirs and the distribution system by way of fire hydrants, makes water particularly at risk (Denileon, 2001) (Chapter 30).

Chemical contaminants are not very effective due to the volume of chemical required and the relative toxicity of most chemicals. In contrast, biological agents have been widely developed for warfare but rarely employed (Hawley and Eitzen, 2001). Although most are airborne many of these organisms and toxins are equally effective via water. The infective dose of the disease agent varies significantly as does the effect on the target organism or population. Microbial agents are infectious, can be subsequently spread from person to person in some cases, or via contaminated food, are stable within the environment, colourless and odourless and have delayed response times, unlike chemical contaminants that cause an effect in the target organism relatively quickly.

Biological agents are classed into two categories by the US Center for Disease Control and Prevention (Rotz *et al.*, 2002). Category A micro-organisms posing the highest risk with high morbidity and mortality rates include smallpox, anthrax, plague and botulism, while category B agents pose a much lower risk and rates of mortality and morbidity such as brucellosis, typhus fever and cholera. There is a third group, category C, for emerging biological agents such as hantaviruses and tickborne haemorrhagic fever viruses. A wide range of bacteria, fungi and algae produce toxins, mostly potent neurotoxins, posing a particular risk to water supplies. These include aflatoxins, botulinum toxins, microcystins, ricin and saxitoxin.

25.5   Conclusions

The presence of biofilms is currently causing water microbiologists much concern, especially as pathogenic micro-organisms can be protected from inactivation by disinfectants by the biofilm. Two bacteria, *Legionella* and *Mycobacterium* can survive and actively grow in distribution mains causing significant health risks to consumers (Chapter 28). Control of these growths is difficult and research in this area is active. However, as stated earlier when discussing chlorination, the risk from chlorination by-products that are known to be carcinogenic at higher concentrations than those found in water must be balanced against the risk of microbial pathogens from inadequate disinfection. Therefore the main lines of defence must be adequate residual disinfection, reduction in AOC in treated water and the prevention of contamination of distribution mains by potential pathogens. The problem of regrowth in distribution systems is considered further by Van der Wende *et al.* (1989), Maul *et al.* (1991), Van der Kooij (1999) and Bartram *et al.* (2003).

Water treatment and disinfection are the front-line protection for water supplies targeted by terrorists, and most treatment plants employ continuous pollutant and/or toxicity detectors in some form. However, unless an attack is suspected then there is very little that can be done to protect treated waters except higher security and restricted access to the distribution network. Once a threat is suspected then there is a wide range of rapid response kits to assess chemical, biological or radioactive agents (States *et al.*, 2004). Water utilities should carry out vulnerability assessments, and have in place both a security plan to protect supplies as well as an emergency response plan in the event of a terrorist attack (Chapter 30). What is clear is that water supplies are a potential terrorist target, making everyone vulnerable to attack within their own homes. So a new level of security is required to ensure the safety of drinking water.

References

Bartram, J., Cotruvo, J., Exner M. and Glasmacher, A. (eds.) (2003). *Heterotrophic Plate Counts and Drinking Water Safety.* London: IWA Publishing.

Bitton, G. (1994). *Wastewater Microbiology.* New York: Wiley-Liss.

Denileon, G. P. (2001). The who, what, why and how of counter terrorism. *Journal of the American Water Works Association*, **93**(5), 78–85.

Falkinham, J. O., Norton, C. D. and LeChavallier, M. W. (2001). Factors influencing the numbers of *Mycobacterium avium, Mycobacterium intracellulare* and other mycobacteria in drinking water distribution systems. *Applied and Environmental Microbiology*, **67**(3), 1225–31.

Gibbs, R. A., Scutt, J. E. and Croll, B. T. (1993). Assimilable organic carbon concentrations and bacterial numbers in a water distribution system. *Water Science and Technology*, **27**(3–4), 159–66.

Hawley, R. L. and Eitzen, E. M. (2001). Biological weapons – a primer for microbiologists. *Annual Review of Microbiology*, **55**, 235–53.

Huck, P. M. (1990). Measurement of biodegradable organic matter and bacterial growth potential in drinking water. *Journal of the American Water Works Association*, **82**(7), 78–86.

Huck, P. M., Fedorak, P. M. and Anderson, W. B. (1991). Formation and removal of assimilable organic carbon during biological treatment. *Journal of the American Water Works Association*, **83**(12), 69–80.

LeChevallier, M. W., Welch, N. J. and Smith, D. B. (1996). Full-scale studies of factors related to coliform regrowth in drinking water. *Applied and Environmental Microbiology*, **62**(7), 2201–11.

Maul, A., Vagost, D. and Block, J. C. (1991). Microbiology of distribution networks for drinking water supplies. In *Microbiological Analysis in Water Distribution Networks*, eds. A. Maqul, D. Vagost and J. C. Block. Chichester: Ellis Horwood, pp. 11–31.

Niquette, P., Servais, P. and Savoir, R. (2000). Impacts of pipe materials on densities of fixed bacterial biomass in a drinking water distribution system. *Water Research*, **34**(6), 1952–6.

Rotz, L. D., Khan, A. S., Lillibridge, S. R., Ostroff, S. M. and Hughes, J. M. (2002). Public health assessment of potential biological terrorism agents. *Emerging Infectious Diseases*, **8**(8), 225–30.

Skadsen, J. (1993) Nitrification in distribution systems. *Journal of the American Water Works Association*, **85**(7), 95–103.

States, S., Newberry, J., Wichterman, J. *et al.* (2004). Rapid analytical techniques for drinking water security investigations. *Journal of the American Water Works Association*, **96**(1), 52–64.

Stout, J., Yu, V. L. and Best, M. G. (1985). Ecology of *Legionella pneumophilia* within water distribution systems. *Applied and Environmental Microbiology*, **49**, 221–8.

Taylor, R. H., Falkinham, J. O., Norton, C. D. and LeChavallier, M. W. (2000). Chlorine, chloramines, chlorine dioxide and ozone susceptibility of *Mycobacterium avium*. *Applied and Environmental Microbiology*, **66**(4), 1702–5.

Van der Kooij, D. (1999). Potential for biofilm development in drinking water distribution systems. *Journal of Applied Microbiology Symposium Supplement*, **85**, 39S–44S.

Van der Kooij, D. (2003). Managing regrowth in drinking-water distribution systems. In *Heterotrophic Plate Counts and Drinking-water Safety*, ed. J. Bartram, J. Cotruvo, M. Exner and A. Glasmacher. London: IWA Publishing, pp. 199–232.

Van der Kooij, D., Visser, D. A. and Hijnen, M. A. M. (1982). Determining the concentration of easily assimilable organic carbon in drinking water. *Journal of the American Water Works Association*, **74**, 540–5.

Van der Wende, Characklies, W. G. and Smith, D. B. (1989). Biofilms and bacterial drinking water quality. *Water Research*, **23**, 1313–22.

Von Reyn, C. F., Maslow, J. N., Barber, T. W., Falkenham, J. O. and Arbeit, R. D. (1994). Persistent colonization of potable water as a source of *Mycobacterium avium* infections in AIDS patients. *Lancet*, **343**, 1137–41.

PART V

PROBLEMS IN HOUSEHOLD PLUMBING SYSTEMS

Chapter 26
Household plumbing systems

26.1 Entry to the home

Houses are connected to the mains by the service pipe (Figure 20.3). The connection from the mains to the boundary stopcock is known as the communication pipe, and this is owned and maintained by the water supply company. The boundary stopcock, as the name suggests, is located just outside the boundary of the property served, with the stopcock being a control valve that turns off the water supply to the house. It is usually buried 1 m below ground, at the bottom of an access chamber known as the guard pipe; which can be a short piece of any suitable pipe, the top access being protected by a small metal cover. The stopcock can be turned on or off using a long-handled key, the exact nature of which depends on the type of handle on the stopcock itself. It is switched off by turning the key in a clockwise direction. From here the service pipe, which is now known as the supply pipe, carries the water to the house. It rises slightly to ensure that all air bubbles escape. From the boundary stopcock onwards all the pipework and the appliances are the householder's responsibility. The supply pipe must be at least 750 mm below the ground at all times to protect it from frost. It usually enters the house at the kitchen and rises up from the floor underneath the kitchen sink. The pipe (the rising main) continues directly upwards either to the cold-water tank (often referred to as the cold-water cistern because it is operated by a ballcock valve), or to feed the other cold-water draw-off points around the house. There should be an indoor stopcock positioned where the pipe enters the house and before the connection from the rising main to the sink to be able to cut off the supply to the house. The kitchen tap should always be connected directly to the rising main. If the supply pipe enters the house at another point, for example through a cloakroom or garage, then the stopcock will be fitted there. To close a stopcock it should be turned clockwise. In some houses, generally newer ones, there may be a drain cock positioned just above the house stopcock to allow the rising main to be drained if necessary (Figure 26.1).

26.2 Plumbing systems

The layout of the water system in the house depends on a number of factors including whether there is non-electrical central heating. Essentially houses

Figure 26.1 Definition and
location of service pipe,
boundary stop valve and
supply pipe. Reproduced
from White and Mays
(1989) with permission
from the Water
Regulations Advisory
Scheme.

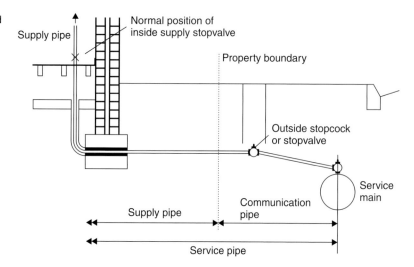

Figure 26.1 Definition and location of service pipe, boundary stop valve and supply pipe. Reproduced from White and Mays (1989) with permission from the Water Regulations Advisory Scheme.

have either a high-pressure direct system, or a low-pressure indirect system that requires a cold-water storage cistern in the attic or in another elevated position.

High-pressure direct water systems are common in the USA, Canada and most of Europe. The water comes into the building and directly feeds all the cold taps and any appliances that also use cold water. This includes the WC (lavatory), hot-water storage cylinder and even the washing-machine and dishwasher. These systems are much simpler and cheaper to install than the indirect system and avoid having lots of pipework and a storage cistern for the cold water in the attic, or occasionally in the airing cupboard. In the UK high-pressure systems are only found in older properties built before 1939 and new buildings. In many countries such as Sweden, the Netherlands, Germany, Denmark and Belgium storage tanks or cisterns are considered to be unhealthy, and a potential health hazard, and so are prohibited.

The vast majority of houses in the UK and Ireland have indirect or low-pressure systems, with only the cold-water tap that serves the kitchen, and possibly the cold-water tap to the washing-machine, directly connected to the rising main (Figure 26.2). The rising main goes up into the attic to discharge into a cold-water storage cistern. From this cistern distribution pipes feed all the other cold taps, the lavatory cistern and the hot-water system. Most water supply companies insist on the use of a storage cistern to ensure that at times of peak usage all demands are satisfied, which would not be the case if everyone had direct systems. This is why on occasions the water tank may start to fill even though water has not been used recently. This is because, due to peak demand, the pressure has been too low to refill the tank. A major advantage of having a low-pressure system is that there is less noise, especially from lavatory systems. Burst pipes can be awful, but under mains pressure they can be extremely serious. Also, when the water supply is cut off for some reason, for example

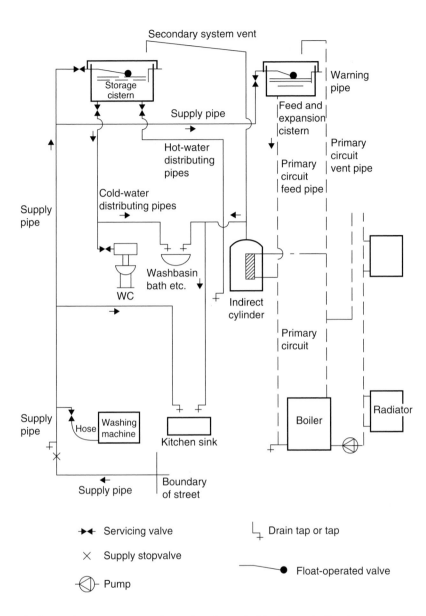

Figure 26.2 Schematic diagram of typical household plumbing system. Reproduced from White and Mays (1989) with permission from the Water Regulations Advisory Scheme.

during maintenance or repair work, a household with an indirect system still has plenty of water stored in the attic for essential use.

The rising main feeds the storage tank or cistern. Older tanks are made of galvanized iron, which is very heavy and liable to corrode. Modern tanks are made of plastic, which is much lighter, and are generally 50 gallons in size (227 litres), although different sizes are available. Cold-water storage tanks in older houses are often open. This allows access for dust, insects, birds, bats, rodents and, of course, their droppings (Section 28.1), and fibres from the insulation (Section 28.2). It is therefore important that the tank is covered. Water tanks now come with closely fitting lids; however, existing tanks can be effectively covered by placing a thick slab of expanded polystyrene insulation on the top. It may be a good idea to weigh the cover down so that it does not move or fall into the tank due to excessive air currents in the attic, especially if it is cross-ventilated.

The rising main will be a 15 mm diameter pipe, although with low pressures a slightly larger pipe (22 mm) may be used. The supply of water enters at the top of the storage tank via a float (ball) valve similar to that in the WC. It is operated by a lever that is opened and shut by a float made of copper or plastic which floats on the surface of the water. Also positioned at the top of the tank is the overflow pipe to carry away the excess water if the ball-valve fails for some reason, or the washer needs replacing so that the supply of water from the rising main cannot be fully shut off. This is to prevent the tank overflowing and pouring through the ceiling. A leaking overflow needs fairly quick attention; not only is it a nuisance to neighbours due to the noise it makes, especially late at night, but it is an offence in many countries, including the UK, to allow water to be wasted in this way. During cold weather, if the overflow freezes up, the water storage tank will overflow and flood the house. The WC operates in exactly the same way as the water storage tank, using a float valve, so the same applies to the overflow for the WC cistern. The overflow pipes are usually located above the back door of the house, to ensure maximum attention (Section 1.5).

The water from the tank is distributed to the rest of the house by two 22 mm diameter pipes fitted about 75 mm from the base of the tank. This is to allow debris to settle so that it is not carried into the plumbing system. One of the pipes will supply the cold taps and the WC. Usually the pipe goes directly to the bath, with the cold taps of the washbasins and the WC cistern taken off this pipe using 15 mm branches. The water flows through these pipes by gravity and so is at a lower pressure than that from the rising main. There is usually another stopcock on the rising main above the connection to the cold-water tap in the kitchen. This allows the water supply to the water storage tank, and thus the whole house, to be shut off while retaining the supply for drinking at the cold-water tap in the kitchen. The second pipe from the tank feeds into the bottom of a hot-water cylinder. This feed pipe also has a stopcock fitted to allow the flow of water to the hot-water cylinder to be shut off during maintenance. The cylinder can be drained using the drainage valve at the base of the cylinder. The hot-water cylinder is made of copper and can hold about 30 gallons (100–150 litres) of water. The water is heated either by an electric element heater and/or by a heating coil from a boiler. The hot taps are then fed from the hot-water cylinder. If the water overheats in the cylinder it expands and the excess escapes into the cold-water cistern in the attic (Figure 26.2).

Water is heated either directly or indirectly depending whether it comes into contact with a heater. With direct heaters the water is heated within the hot-water cylinder by one or more immersion heater elements and the hot water is then supplied to the hot taps. If a boiler is used the water is heated by circulation through the boiler and the heat is transferred to the water in the hot-water cylinder and stored there. An immersion heater is often used in the cylinder as well. Neither of these systems can supply central heating radiators. Indirect heating is usually combined with the central heating system. The water in the cylinders is heated indirectly by a separate water circuit that goes through the boiler, whereas a separate hot-water circuit supplies the radiators. The central heating system is fed by a separate water circuit, supplied by a small feed and expansion cistern that is also supplied by the rising main via a small ball-valve (Figure 26.3). It is also located in the attic and is needed only to top up the system due to evaporation losses. If the feed and expansion tank is being refilled by a constant drip, check that the ball-valve does not need adjusting. If there is no water in the overflow pipe then the dripping means that there is a leak somewhere in the central heating system and the water is escaping either from the enclosed pipework or a radiator.

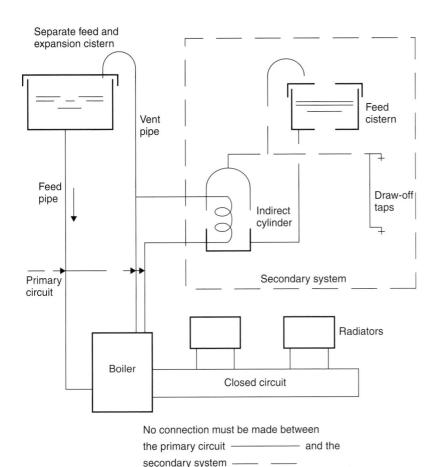

Figure 26.3 Schematic diagram of an indirect heating system. Reproduced from White and Mays (1989) with permission from the Water Regulations Advisory Scheme.

Hot water always rises above cold water because as it heats up it becomes lighter (less dense). In the hot-water cylinder the hot water therefore collects at the top, so cold water is fed in at the base to replace any hot water drawn. When a boiler is fitted the heat exchange pipe runs from the top to the bottom of the cylinder, so that the hot water enters at the top ensuring the most efficient exchange of heat. The hot water is taken from the top of the cylinder to feed the hot-water taps around the house (Figure 26.3). A branch from the hot-water outlet pipe from the hot-water cylinder goes back up to the cold-water cistern in the attic. This hangs over the edge of the tank and is not immersed. It allows for expansion and for the escape of air from the cylinder. The pipe from the boiler does not discharge its hot water, it runs through the cylinder in a sealed coil. This allows heat exchange.

It is possible, if there is a drop in pressure in the mains due to a burst or perhaps due to the fire brigade using large amounts of water, that, if the plumbing has been installed incorrectly, back-suction or back-syphonage will occur. Water is syphoned back through the plumbing system to the rising main and into the water supply distribution main. For example, hose extensions are commonly fitted to cold-water taps and if the hose is immersed in water and back-syphonage occurs, all the water in the sink is sucked back into the mains until the syphon is broken. The example often quoted is the garden hose left in a bucket of weedkiller or insecticide that is drawn back up the hose and into the public supply. It is an unlikely scenario, but not impossible. If the bucket of weedkiller is replaced by a water butt containing lots of murky water, or a fishpond that is being filled, then the scenario becomes more plausible. Back-syphonage is considered in detail in Section 28.1. Devices are now available to prevent back-syphonage and there is now a byelaw in the UK that a back-flow prevention device (or pipe interrupter) should be attached to any tap onto which a hosepipe is connected. These are widely available and simply screw onto the tap, or can be fitted into a section of the hose (Figure 28.1).

26.3   Water Byelaws and regulations

Water Byelaws have been in force for many years in the UK and cover the design, installation and maintenance of plumbing systems, water fittings and appliances connected to the public water supply and are used to prevent waste, undue consumption, misuse, or the contamination of water supplied by the water undertakers. Based on the Model Water Byelaws (White and Mays, 1989) they have been replaced by national regulations, i.e. the *Water Supply (Water Fittings) Regulations 1999* in England and Wales (HMSO, 1999) or the *Water Supply (Water Fittings) Byelaws 2000* in Scotland. If any changes or alterations are made to a consumer's supply system then the Water Regulations Inspector must be notified so that he or she can inspect it, regardless of whether the householder or a plumber carries it out. Simple repairs or renewals are exempt.

However, water regulations are legal requirements and if they are not adhered to then enforcement, and possible prosecution, could follow. All new connections to the mains will require the plumbing to be inspected before the connection is made to ensure that it is satisfactory. The regulations apply to systems in all types of premises from the point where water enters the property's service pipe. Premises without a public water supply connection are not governed by these regulations. The UK water supply companies operate the Water Regulations Advisory Scheme (WRAS) that provides detailed information and help about the new regulations (http://www.wras.co.uk). They publish a number of documents including the *Water Regulations Guide*, which also includes the Scottish Byelaws, and the *Water Fittings and Materials Directory*. Nearly all countries have either specific water or building regulations that ensures plumbing is both safe for the consumer and will not contaminate the supply network.

26.4 Conclusions

The majority of aesthetic supply problems arise from the householder's own plumbing, so installation and maintenance must be carried out by a registered plumber. The major problems arising from plumbing are leaching of metals, in particular lead, due primarily to galvanic corrosion (Chapter 27), and the development of micro-organisms within the plumbing system or contamination of the water via open storage tanks (Chapter 28).

References

HMSO (1999). *The Water Supply (Water Fittings) Regulations 1999*. Statutory Instrument 1999/1506. London: HMSO.

White, S. F. and Mays, G. D. (1989). *Water Supply Bylaws Guide*, 2nd edn. Chichester: Water Research Centre and Ellis Horwood.

Chapter 27

Corrosion and metal contamination from pipework and fittings

27.1 Introduction

Problems with iron are most likely to arise at the treatment plant or within the distribution network rather than the home plumbing system. However, corrosion can result in a number of metals or alloys from which pipework or plumbing fittings are made contaminating drinking water. The most important of these are lead, copper, zinc and iron.

27.2 Corrosion

Most types of corrosion involve electrochemistry. For corrosion to occur the components of an electrochemical cell are required, i.e. an anode, a cathode, a connection between the anode and cathode (external circuit) and finally a conducting solution (internal circuit), in this instance drinking water. The anode and cathode are sites on the metal pipework that have a difference in potential between them, which may be the same metal or different metals. When this occurs then oxidation (the removal of electrons) occurs at the cathode, which is negatively charged. In practice metal dissolution occurs at the anode, releasing the metal into solution (Figure 27.1).

Corrosion cells form on the same piece of metal where there are adjacent anodes and cathodes. These are due to the non-uniformity of the surface and can be caused by minute differences in the pipe surface formed during manufacture, or by stress imposed during installation. Any imperfections in the pipe will create tiny areas of metal with different potentials.

Corrosion is more rapid where two different metals are coupled together. This is known as galvanic corrosion. The most serious situation is where a copper pipe is used to replace a section of lead pipe (Section 27.3). The conducting solution (electrolyte) is the water, the copper pipe is the cathode and the lead pipe is the anode. The lead pipe slowly corrodes, releasing the metal into solution. Metals can be listed in order of their tendency to corrode or go into solution, i.e. the galvanic series. The metals and alloys most commonly used in water treatment and distribution can be placed in electrochemical order, which is the decreasing order in which they tend to corrode when connected

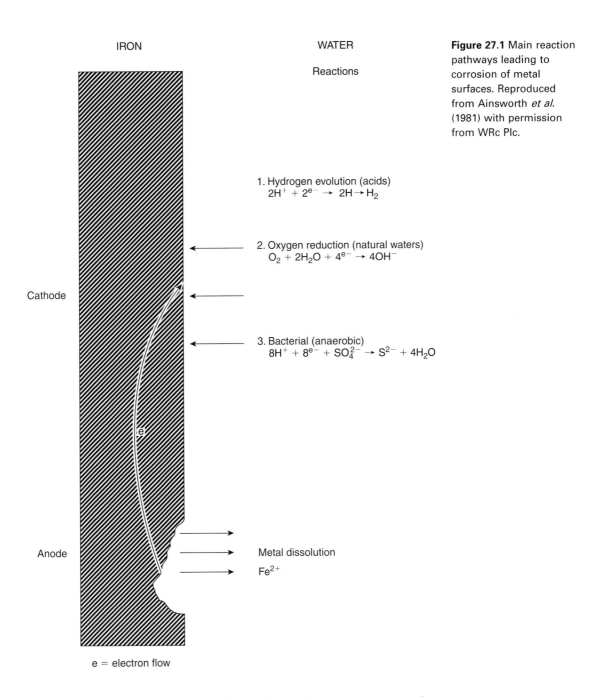

IRON   WATER

Reactions

Figure 27.1 Main reaction pathways leading to corrosion of metal surfaces. Reproduced from Ainsworth *et al.* (1981) with permission from WRc Plc.

1. Hydrogen evolution (acids)
$$2H^+ + 2e^- \rightarrow 2H \rightarrow H_2$$

2. Oxygen reduction (natural waters)
$$O_2 + 2H_2O + 4e^- \rightarrow 4OH^-$$

Cathode

3. Bacterial (anaerobic)
$$8H^+ + 8e^- + SO_4^{2-} \rightarrow S^{2-} + 4H_2O$$

Metal dissolution

Anode Fe^{2+}

e = electron flow

under ordinary environmental conditions. The order is: manganese, zinc, aluminium, steel or iron, lead, tin, brass, copper and bronze. The further apart two metals are on this scale, the greater the potential difference between them and so the greater the rate of corrosion. Iron pipes will corrode if connected by brass fittings, and of course lead pipes will corrode if connected to copper fittings. In household plumbing it is therefore best to use all the same metal, e.g. copper pipes and copper fittings, or metals very close on the electrochemical scale, e.g. copper pipes and brass or bronze fittings.

Galvanized steel, steel coated with zinc, was widely used for water storage tanks and is still available. Zinc is more likely to corrode than steel so any crack or blemish on the surface coating of zinc will result in a cell being formed with zinc as the anode and zinc being released into the water (Section 27.5).

27.2.1   Factors affecting corrosion

The rate of corrosion increases rapidly at higher temperatures and more acidic pH values. The corrosion of metals is therefore more pronounced in softer water and in the hot-water circuit. The hot-water cylinder is usually made from copper, which is a particular target for corrosion. Differences in dissolved oxygen or the hydrogen ion concentration (pH) can also set up a difference in potential within a pipe. This occurs most often under rivets, washers or in crevices where dissolved oxygen may be lower. The area of the metal in contact with the higher dissolved oxygen concentration will become the cathode and the area with the lower dissolved oxygen concentration becomes the anode and so corrodes. The same occurs with differences in hydrogen ion concentration (Section 21.2).

Once the surface of the pipework has been corroded, it provides a habitat for micro-organisms. These are able to attach themselves to the pipe and are not flushed away in the water flowing past. As the micro-organisms build up they increase the resistance to flow of water through the pipework. Some of these micro-organisms, in particular iron and sulphur bacteria, can also corrode the pipework (both metal and non-metal). This is mainly a problem in the mains and at the treatment works as it is mainly buried metal pipes that are affected, although concrete pipes can also be attacked.

When corrosion occurs it is in fact a chemical reaction accompanied by the generation of a very small amount of electrical energy. This results in metal ions leaving the site of corrosion (the anode) and entering the water (Figure 27.1). This dissolution of the metal ion leaves behind excess electrons that migrate towards the cathode, hence the electric current. These excess electrons are consumed at the cathode in a balancing reaction. There are in essence three types of balancing reaction. The first, which is most common in drinking water, is the reduction of oxygen:

$$O_2 + 2H_2O + 4^{e^-} \rightarrow 4OH^-$$

where e are electrons. This means that where the corrosion is due to differences in the dissolved oxygen concentration there is a tendency for the oxygen concentration to equilibrate and the corrosion will stop. However, if the cathode is situated where there is plenty of water movement and the anode is in an area where the water is stagnant (e.g. under a rivet), then equilibrium will not occur and corrosion will continue.

Where the water is acidic, and this becomes increasingly important below pH 6, and where oxygen is absent from the surface of the metal such as occurs beneath deposits, then hydrogen ions are released to form hydrogen:

$$2H^+ + 2^{e-} \rightarrow 2H \rightarrow H_2.$$

This is the reaction seen when charging a car battery, with hydrogen bubbles being formed at the cathode.

The final reaction involves a type of bacteria, which is only found under anaerobic conditions, which reduces sulphate to sulphide

$$8H^+ + 8^{e-} + SO_4^{2-} \rightarrow S^{2-} + 4H_2O.$$

In practice these reactions ensure the assimilation of electrons at the cathode, thus driving the rate of corrosion at the anode, regardless of whether the water is aerobic or anaerobic (De Rosa and Parkinson, 1986).

27.2.2 Prevention

Corrosion can be controlled by choosing the correct metal and alloys, or by using special corrosion-resistant metals or using plastic pipes and tanks. Surfaces can be coated or lined with special protective films. Lining is more commonly carried out in water mains to protect iron pipes, but is also possible in smaller pipes. Corrosion can also be prevented by connecting the pipework to another piece of metal that will act as an anode and so will be deliberately corroded instead of the existing pipework. This is known as cathode protection and the metal insert, a sacrificial anode, should be installed in metal storage tanks to protect them from aggressive water, and also in the hot-water cylinder, which is particularly at risk. Sacrificial anodes are generally made of magnesium alloy fitted onto a steel rod for use with galvanized steel tanks, whereas copper cylinders are protected using aluminium rods. Although there are many other ways of protecting metal pipework from electrochemical corrosion, the selection of the correct materials and fitting a sacrificial anode are all that is normally required.

Chemical corrosion, simply dissolving the metal, can also occur in soft water areas with lead the most readily corroded. The only effective way to control this is to increase the pH of the water to between 8.0 and 8.5. This is done at the water treatment plant by adding various alkali hydroxides, silicates or borates. Under alkaline conditions these form an impervious layer on the wall of metal pipes, usually calcium carbonate, protecting them from further corrosion. Excessive scaling can reduce the diameter of the pipe significantly, seriously reducing the flow rate of water through the pipe. The whole question of the formation of scale in pipes to offer protection, and the measurement of the

Table 27.1 *Summary of types I and II pit corrosion of copper pipes*

Feature	Type I corrosion	Type II corrosion
Pit shape	Broad and hemispherical	Steep sided and narrow
Pit contents	Copper(II) oxide and chloride	Crystalline copper(II) oxide
Corrosion product covering pit	Thick hard green layer of calcium and copper carbonates	Small black mound of copper(II) oxide and basic copper sulphate
Cause	Cold hard waters, presence of carbon (graphite) film produced during manufacture	Hot ($>60\,°C$), soft water (<40 mg $CaCO_3$ l^{-1}), low pH, hydrogencarbonate: sulphate ratio >1
Typical time for pipe failure	3 years	15–20 years
Control	Removal of carbon film during manufacture	pH control to ≥ 7.4

corrosiveness of water, is under debate at present. Because of the widespread use of copper pipework, much attention has been paid to the forms of corrosion of this metal. Corrosion causes pitting of the inner surface of the pipes leading to eventual perforation. Two forms are well documented (Table 27.1), although other forms have been described.

27.2.3   Problems arising from corrosion

In household plumbing the corrosion of lead, galvanized steel and copper pipework leads to discoloration and occasionally taste problems. Lead is also toxic, and the corrosion of lead and copper is dealt with in the following sections. The corrosion of galvanized steel is slightly different. The zinc layer is there to protect the steel underneath by corroding preferentially, ensuring that any exposed steel becomes the cathode rather than the anode. As the zinc corrodes it forms zinc hydroxycarbonate, which then forms a protective layer on the steel. However, in soft waters where the alkalinity is low (less than 50 mg l^{-1} as $CaCO_3$), or in water with a high carbon dioxide concentration, which is common in groundwaters from private wells and boreholes, the corrosion product tends to be zinc carbonate. This does not adhere to the metal surface and so is lost into the water, giving the appearance of sand in the water. It is advisable not to use galvanized tanks or pipework for such aggressive waters. The corrosion of galvanized steel is increased significantly if it is coupled with copper tubing, especially if it is used upstream of the galvanized tank and the water is cuprosolvent (i.e. able to corrode copper pipework). Even a small amount of copper can significantly increase the rate of corrosion. So for domestic purposes, stay clear of galvanized steel.

Water byelaws do not permit dissimilar metals to be connected unless either effective measures are taken to prevent deterioration (i.e. the installation of a sacrificial anode) or where deterioration is unlikely to occur through galvanic action. Metals can be safely mixed as long as the sequence of metals is kept as follows: in downstream order of use – galvanized steel, uncoated iron, lead and copper. Rigid and flexible plastic pipes are now widely used in preference to metal pipes and can be used in conjunction with any other materials fairly safely. A common problem is that even if this sequence has been strictly adhered to, the water return (overflow) pipe from the copper hot-water cylinder may be returned to a galvanized tank. If the water supply byelaws are carefully followed then corrosion will be minimal (White and Mays, 1989). A test method to assess the potential of metallic products to contaminate water supplies has been published by the British Standards Institution (1991).

27.3   Lead

For centuries, lead has been known to be toxic, so there is perhaps more known about the acute and chronic adverse effects of lead on humans than about any other poison. Although the effects of high concentrations of lead on the body are known, research over the past 35 years has highlighted the problems caused by much lower levels of exposure that were previously thought to be harmless. Since its removal from petrol, drinking water has become the largest controllable source of human exposure to lead both in Europe and the USA.

The effects of long-term low-level exposure are mainly neural, affecting the brain, causing behavioural changes and deficits in intelligence levels. The problem has been brought to light by a number of studies in Britain and the USA, although it is the specific studies carried out at Glasgow and Edinburgh Universities which have shown fairly conclusively that exposure to low levels of lead primarily from drinking water has adverse effects on the learning ability of children (Lansdown and Yule, 1986; Thornton and Culbard, 1987; WHO, 2003).

27.3.1   Sources of lead

Lead is ubiquitous in the environment. It is found in food, especially canned foods, air, paint, glazes, dust, fumes and, of course, drinking water. Although the exposure to lead from some of these sources has occurred for decades and in some instances for centuries, there are clear indications that exposure levels are decreasing. Legislation to limit the use of lead in paint and industrial sources of lead has had a significant effect. The introduction of lead-free petrol has contributed enormously to reducing the levels in fumes and dust, especially in our towns, with reduced levels in gardens and garden produce as a result. The main sources of lead are food, water and air, and although the contributions from food and air are decreasing, the contribution from drinking water has remained largely unaltered and in many instances is now the major single source of the metal in our diet.

Raw waters, both ground and surface waters, rarely contain lead in concentrations in excess of $10 \, \mu g \, l^{-1}$. Lead salts are not used in water treatment and there is little chance of lead levels increasing in the mains. The problem arises from the use of lead in the connection of water supplies to the house, and the widespread use of lead pipes and fittings in the home.

Throughout Europe and the UK houses built before 1964 widely employed lead piping and fitments. Lead-lined storage tanks were replaced in the 1930s but are still in use in some areas of Scotland. Houses built before 1939 have extensive lead piping fitted. All water supply companies used lead for service pipe connections, mainly because lead is so malleable and thus will not fracture if the mains, or another part of the pipe, move slightly. The use of such connections was phased out many years ago, but there are still many millions of houses connected to the mains by lead service pipes, even though there is no lead piping in the house itself. The problem arises where the water is plumbosolvent, which means that it can dissolve lead. The most important factor affecting plumbosolvency is the pH of the water. The rate of leaching increases dramatically below pH 8.0, so that soft acidic waters (pH 6.5), which are so common in certain areas, will rapidly dissolve the lead from the pipes into solution. Thus, those most at risk are those living in older houses which still have lead pipes, and which have acidic water. On this basis the worst affected areas in the UK include most of Scotland, the north of England, Wales and the south-west of England. Problems are found in some major British cities including Glasgow, Edinburgh, Birmingham, Liverpool, Manchester and Hull. However, the softness of water is not the only factor related to plumbosolvency, and even in hard water areas lead can still be a problem. Other factors include the water temperature, the contact time between the water and the source of lead and the amount of lead piping, which affects the surface area of lead exposed to the water. These are all considered below.

There has been an enormous amount of research into the factors affecting plumbosolvency in the hope that a simple solution might be found to this major problem. Among the factors affecting lead leaching into drinking water are vibrations in pipes, scouring by high water velocities, thermal expansion effects, the age and nature of the pipes, the presence of particulate lead deposits and electrochemical reactions caused by mixing metal pipes (Section 27.2). Because of the complexity of all these factors interacting together, there is a significant variation in the lead concentrations in water within individual water supplies over time and between supplies to different houses in the same area. This has led to much research and debate over how to actually take samples for the determination of lead in drinking water.

A number of interesting studies have been carried out. One of the most fascinating is the Renfrew lead pipe survey, which was carried out between

1976 and 1977, in which every house in the Renfrew district was checked for lead piping. Lead piping was not used in Renfrew after 1968, so although the total housing stock in the district at the time of the survey was 77 806, only those built before 1968 were inspected, about 55 393 in all. As many of the houses shared a communication pipe or supply pipe (Figure 20.4), it was found that there were only 21 070 communication pipes, of which 96% were lead. Of the 21 000 supply pipes identified over 80% were lead. Although the average length of the communication pipe was 19 m, some houses receiving private water supplies had supply pipes in excess of 1.5 km. Of the houses inspected 1592 had lead-only pipes from the point of entry into the house to the kitchen cold tap. A further 13 349 homes had the usually worse situation of having a combination of lead and copper pipes. The majority, about 40 452 (73%) had no lead pipes running to the cold-water tap, most of these being made of copper. Therefore 27% (14 941) of homes had some lead leading from the mains to the cold-water tap in the kitchen. It was important to identify those houses where the kitchen cold-water tap was not connected directly to the supply pipe, but was fed from a cold-water storage tank. In Renfrew, 875 houses (1.6%) fell into this category. The percentage of these storage tanks that were lead-lined is unknown, but it is estimated to be about 45%. The survey also noted that just over 2% of houses had lead pipework leading to the hot tap and that 6% of bathroom taps were connected with lead pipework. The survey, the first of its kind, was enormously useful. Each household was made aware of the presence of lead pipes and was given a leaflet explaining the problems and the measures that could be taken to avoid excessive lead levels in drinking water. The council determined the extent of the problem and had the opportunity to identify those most at risk (Britton and Richards, 1981).

27.3.2 Intake of lead and health effects

The intake of lead from food and water in the UK is between 70 and 150 $\mu g \, d^{-1}$ for adults and between 70 and 80 $\mu g \, d^{-1}$ for children between the ages of two and four years. This is only an approximation and may in fact be unhelpful. It is perhaps more pertinent to consider how much lead is being taken in by consumers in their tap water. The Department of the Environment carried out a survey of lead in drinking water during 1975–6. Samples were taken of tap water from 3000 households throughout the UK. The results are shown in Table 27.2, and it is clear that although most houses were receiving drinking water with less than 10 $\mu g \, l^{-1}$ lead, significant numbers were receiving drinking water in excess of 50 $\mu g \, l^{-1}$: 7.8% in England, 8.8% in Wales and a staggering 34.4% in Scotland. These samples were taken during the daytime and although the water was not flushed before sampling the water had not been standing in the pipes overnight when much more dissolution of lead would have occurred. These results therefore do not tell the true extent of the problem. Although the

Table 27.2 *Lead concentrations in households in the UK during a survey in 1975–6. Reproduced from Department of the Environment (1977) with permission from Defra*

Lead concentration in daytime sample (μg 1⁻¹)	Percentage of households			
	England	Scotland	Wales	Total
0–9	66.0	46.4	70.5	64.4
10–50	26.2	19.2	20.7	25.3
51–100	5.2	13.4	6.5	6.0
101–300	2.2	16.0	1.5	3.4
≥301	0.4	5.0	0.8	0.9
Total	100.0	100.0	100.0	100.0

contribution of lead from drinking water to the diet of most people will be small, it contributes in excess of 50% when households have lead pipes and receive plumbosolvent water. Of course, this percentage increases dramatically in individuals who drink a lot of water, which children often do, especially infants fed with formula milk feeds made up with tap water.

Lead is odourless, tasteless and colourless when in solution, which makes even fairly high levels in drinking water undetectable unless chemically analyzed. Another problem is that lead is reasonably soluble, so the body readily absorbs it. A considerable proportion of lead in water may be in particulate form rather than in solution and this may affect its uptake (Hulsmann, 1990). Absorption varies widely between individuals, with children absorbing lead far more readily than adults. This is especially true of young children, infants and fetuses, who are all growing rapidly; in fact, they absorb lead five times faster than adults.

Most studies have related symptoms to blood lead levels with a critical effect threshold for lead in adults of between 40 and 50 μg per 100 ml (μg dl⁻¹). It is extremely difficult to relate lead levels in water directly to lead levels in blood, although this has been done for bottle-fed infants in Glasgow (Lacey, 1983). In the USA, blood lead levels above 15–20 μg dl⁻¹ are considered by the US Environmental Protection Agency (USEPA) to cause concern, although there is still much debate over what is a safe blood level, and subsequently a safe lead concentration in drinking water (Mullenix, 1992). Renal disease and kidney damage, increased hypertension, inhibition of biosynthesis of haem causing anaemia, and interference of calcium metabolism are key health effects of lead. Central and peripheral nervous systems are also primary targets for lead, causing neurological and behaviour problems. Reproductive problems are also recorded including depressed sperm counts. Acute effects of lead poisoning include dullness, restlessness, irritability, poor attention span, headaches, muscle tremor, abdominal cramps, kidney damage, hallucinations and loss of memory. Chronic symptoms include tiredness, sleeplessness, irritability, headaches, joint pain, gastrointestinal symptoms, lower scores at psychometric tests, mood swings and peripheral neuropathy (USEPA, 1986). Pregnant women are particularly at risk from lead, including increased risk of malformation, which doubles at blood levels of just 7–10 µg dl^{-1}, and the risk of pre-term delivery, which is four times higher at 14 µg dl^{-1} than at 8 µg dl^{-1} (McMichael *et al.*, 1986).

Lead attacks the nervous system and can result in mental retardation and behavioural abnormalities in the young and unborn. This has been supported by studies carried out by a number of research teams in Scotland and the USA. A study by Edinburgh University in 1987 found that even children who were not exposed to high levels of lead and who had blood lead levels 50% less than that considered safe were showing signs of learning difficulties. The study, which looked at the mental aptitude of 500 children, concluded that they performed on average 6% worse. The research team concluded that there is perhaps no safe level of lead exposure for children (Thornton and Culbard, 1987).

Lead is a cumulative poison and the World Health Organization (WHO) have identified a number of especially vulnerable groups. These are the fetus, infants up to six years of age, pregnant women and patients with specific disorders that require an increased intake of water or the need for renal dialysis. Also, it is important to remember that children more readily take up lead, that it causes them more serious damage in the early years when the brain is developing. Research by Glasgow University in the city, which is one of the worst affected areas in Europe for elevated lead concentrations in drinking water, discovered a number of worrying facts about the risks that unborn children face from lead. For example, they found that infants whose mothers drank water with a high lead concentration during pregnancy were twice as likely to be mentally retarded. The rate of stillbirths also increased within this group. When placentas were examined only 7% of the placentas from normal births had high levels of lead compared with 61% of those from infants who were stillborn or died shortly after birth. In a city where in the late 1970s and early 1980s it was common to find tap water with 10, 15 and even 20 times more lead than the 100 µg l^{-1} considered safe at that time, it is perhaps not surprising that more than 10% of all the infants born had lead levels in their blood at the time of birth which would have been considered unsafe for fully grown adults.

27.3.3 Standards

The WHO (2004b) reduced its 1984 guideline for lead of 50 µg l^{-1} to the current value of 10 µg l^{-1} set in 1993. The EC responded to this by also reducing the maximum limit from 50 µg l^{-1} to 10 µg l^{-1} in the revised Drinking Water Directive (98/83/EEC). However, given the problems facing some member states in achieving this new standard a 15-year transition period was allowed, although an interim value of 25 µg l^{-1} had to be achieved by 25 December 2003. The USEPA also reduced its maximum contaminant level (MCL) with a new action level target value in drinking water of $15\,\mu g\,l^{-1}$, so that if more than 10% of tap water samples exceed this level then the water supply company is forced to take immediate action to reduce the corrosiveness of their water. So lead levels are primarily managed by appropriate water treatment controlling dissolution of the metal in household plumbing. A similar approach has been adopted in the UK. However, it should be a priority of all householders to replace household plumbing and supply pipes at the earliest opportunity, as reducing water corrosiveness as lead piping and fittings age cannot solely control elevated lead.

27.3.4   Levels in water

Lead is not commonly found in normal supplies at elevated concentrations ($>10\,\mu g\,l^{-1}$). Lead contamination is almost exclusively due to dissolution of the metal from household pipework and fittings. It is recognized that soft acidic waters are plumbosolvent, and therefore areas where substantial plumbosolvency problems will occur can easily be predicted by carrying out simple water quality measurements. Although it is also known that some hard waters can also give rise to high lead levels, it is not so easy to identify such sources accurately. In 1980 it was estimated that between 7 and 10.5 million of the total 18.5 million households in the UK had lead pipes somewhere between the connection with the mains and the kitchen tap. It was thought that of the 5 million homes in areas where water is significantly plumbosolvent about 50% had lead in the connection pipes and that many more had lead pipework within the house.

In the UK, some of the worst affected areas outside Scotland, where the situation remains a cause for concern, include parts of Lancashire and Greater Manchester, which are served by reservoirs located in the Lake District. Other areas of concern where the water is potentially very plumbosolvent are Preston, Hull, Blackburn and Blackpool. Two extreme examples were found by Friends of the Earth (1989) in tap water in Dilmorth near Preston and in Blackburn, which contained 3600 and $7750\,\mu g\,l^{-1}$ of lead, respectively. Such high levels are generally due to old and damaged lead pipes where particulate lead is scoured from the surface of the pipe. Many houses are served by very long lead service pipes and this can result in long pipe–water contact times, resulting in high lead levels where the water is soft and plumbosolvent. In a survey of 2000 houses in the UK the lead concentration in tap water was shown to be related to the pH and the length of the service pipe (Table 27.3). Service pipes shorter than 10 m were shown to have significantly lower lead levels (Pocock, 1980). Most of the Midlands and London are free from problems. The situation in other developed countries is similar to the UK (Gendlebien et al., 1992). Mean lead concentration in drinking water in the USA is $2.8\,\mu g\,l^{-1}$ (Levin et al., 1989), although this depends on the plumbosolvency of the water and the number of

Table 27.3 *Percentage of houses, based on length of service pipe and water pH, with a first flush of lead in excess of 100 μg l^{-1}. Reproduced from Department of the Environment (1977) with permission from Defra*

Length of lead household and supply pipes (ft)	pH of mains water			
	<6.8	6.8–7.5	7.6–8.3	>8.4
<16	60	22	4	10
16–32	62	16	6	10
32–65	66	28	15	4
≥65	74	30	20	7

houses with lead pipes included in the study. For example, the median concentration of lead in five Canadian cities was 2.8 μg l^{-1} but a study in Ontario (Canada) that studied the intake of lead over a 7-day period showed a range of 1.1–30.7 μg l^{-1} with a median value of 4.8 μg l^{-1} (DNHW, 1992). In Glasgow (Scotland) 40% of samples tested in the early 1980s exceeded 100 μg l^{-1} (Sherlock and Quinn 1984).

27.3.5   Factors affecting lead levels

Many factors affect the concentration of lead in water and these have been studied extensively. The solubility of lead is temperature dependent, so much so that marked differences are seen over the year, with summer levels often double those recorded in the winter. As a guideline, an increase of 10 °C doubles the rate of leaching through oxidation. This is also seen when samples are taken from hot and cold taps, with levels significantly higher in hot water drawn from taps. This may cause problems in bathrooms, where the hot tap is often used to rinse teeth. Hot-water taps should never be used for drinking. In Scotland, a survey found that it was widespread practice to fill the kettle from the hot-water tap to make it boil faster (Table 27.4). It is important to wash out the kettle each time it is used if a lead problem is suspected. This is because boiling water increases the lead concentration. Table 27.4 shows the difference between first draw samples, where the water has been stored in the pipe overnight and so the lead has had plenty of time to solubilize; and samples taken during the day, which show the normal background level expected from the plumbing. Levels of lead taken during the day from the hot-water tap show a three- to ten-fold increase in lead concentration. The range of lead concentrations in the kettle from 14 households had a maximum over 20 times higher than the EC limit at that time of 50 μg l^{-1}. In a survey conducted in 1979 on water usage, it was discovered that only one-third of the population emptied their kettle before refilling it each time (Hopkins and Ellis, 1980). It is important for consumers never to top up kettles, but to always empty them before refilling.

Table 27.4 *Lead concentrations in the cold water from a number of different households taken first thing in the morning (first draw) and during the day, from the hot tap during the day and in the kettle. Reproduced from Britton and Richards (1981) with permission from the Chartered Institution of Water and Environmental Management*

	Lead concentration in samples (mg 1^{-1})			
Household No.	Cold water first draw	Cold water random daytime	Hot water (random)	Kettle (range)
1	0.19	0.04	0.41	0.03–1.03
2[a]	0.06	0.04	0.28	0.26–0.37
3[a]	0.03	0.02	0.43	0.04–0.53
4	0.28	0.15	0.88	0.18–0.30
5[a]	0.05	0.03	0.71	0.03–1.05
6[a]	0.55	0.07	0.39	0.16–0.65
7[a]	0.48	0.15	0.47	0.23–0.61
8	0.04	<0.01	0.46	0.03–0.13
9[a]	0.12	0.05	0.43	0.12–0.32
10	0.68	0.10	0.31	0.38–1.09
11	0.21	0.04	0.15	0.34–0.51
12[a]	0.16	0.09	0.37	0.06–0.39
13[a]	0.50	0.26	0.34	0.23–0.45
14[a]	0.07	0.07	0.55	0.07–0.56

[a] Households where hot water is known to be drawn regularly or occasionally when filling kettle.

Apart from the contact period, temperature and pH, there are many more secondary factors that contribute to the overall lead concentration. The velocity of water flowing through lead pipes has an important influence. At very low flows there is an increased contact time, so that the lead concentrations are high (Figure 27.2). This is why the water first drawn from taps after it has been standing in the pipes overnight is rich in lead and why it is always best to flush this lead-rich water from the pipework before using the water. The first water drawn from the kitchen tap in the morning will invariably be to fill the kettle, so this, combined with the concentration effect of boiling, will mean that the first cup of tea of the day will contain the highest concentrations of lead of the day as well! The rule for consumers, if they have lead pipes, is to always flush taps before drinking the water. The period of flushing depends largely on the length of the lead service pipe to the house, which if it is a joint supply pipe or is quite a distance from the boundary stop-tap (i.e. more than 30 m) will take much longer. This is considered in more detail at the end of this section.

As expected, as the velocity of flow in the pipe increases, the lead concentration begins to decrease in the water. However, at high flow velocities

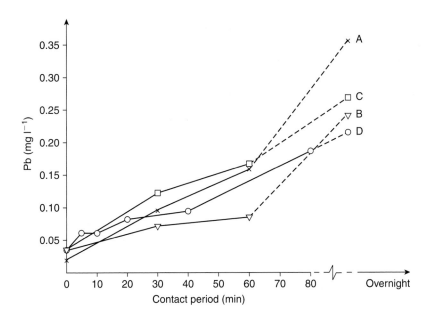

Figure 27.2 The effect of contact time between drinking water and lead pipes in four homes (A to D). Reproduced from Britton and Richards (1981) with permission from the Chartered Institution of Water and Environmental Management.

the lead concentration can, in older pipes, rapidly increase again due to lead deposits being physically scoured from the pipe. This may be a problem with lead joint supply pipes that serve a number of households and where the water usage is high at certain times of the day or week. The lowest concentrations of lead are found at intermediate flow rates. What is important is not to drink water from the tap without flushing first, if the water has been standing in the tap for a long period. As can be seen in Figure 27.2, 30 minutes standing in the pipe can double the lead concentration in water and in some instances it can treble in 60 minutes. This is, of course, dependent on the plumbosolvency of the water, but illustrates the point that consumers must be careful.

The age of lead pipes is also critical in terms of the amount of lead leached. As they age the plumbosolvent action increases due to a greater surface area being exposed, with localized corrosion accelerating at sites where there are casting faults and impurities. In Glasgow the average lifespan of a communication pipe is 58 years. Britton and Richards (1981) have calculated the lifespan of such pipes to be between 43 and 55 years, although with effective water treatment to reduce plumbosolvency this may be increased to between 135 and 170 years.

Lead can be a short-term problem in modern buildings. Tin–lead soldered joints are widely used to connect copper pipework. There is always a danger when different metals are used in plumbing that a galvanic interaction will occur, in this instance between the lead solder and the copper pipe, resulting in electrochemical corrosion (Section 27.2). The rate of corrosion is much higher under these circumstances than that which occurs naturally, it is enhanced by the flow of water and will be much higher in hot-water pipes. This source of lead does

Table 27.5 *Illustration of the problem with plumbosolvency after the removal of lead plumbing at a school. Reproduced from Britton and Richards (1981) with permission from the Chartered Institution of Water and Environmental Management*

Sampling point[a]	pH	Copper concentration (mg 1^{-1})	Lead concentration (mg 1^{-1})
Water supply to school	6.9	0.01	0.01
Cold tap: sink unit (single) (classroom unit)	6.4	0.55	0.12
Cold tap: standpipe (old building)	6.3	0.78	0.01
Cold tap: wash-hand basin (boys' toilets)	5.8	0.99	0.09
Cold tap: wash-hand basin (girls' toilets)	6.3	1.57	0.12
Cold tap: sink unit (primary 7 class)	6.6	1.44	0.28
Servery: drinking tap	6.5	2.20	0.26
Cold tap: sink unit (servery)	6.6	1.89	0.14
Cold tap: sterilizer (servery)	6.7	1.70	0.08
Cold tap: wash-hand basin (foyer of servery) right hand	6.6	1.44	0.60
Cold tap: wash-hand basin (foyer of servery) left hand	6.6	1.78	0.17
Servery: drinking tap	6.8	0.56	0.70
Servery: drinking tap	6.3	1.73	0.23
Servery: drinking tap	6.3	1.58	0.11

[a] All are random daytime samples taken in May and June 1980. All lead replaced with copper in 1977–8.

not usually result in serious increases in the lead intake, as it is generally only the internal pipework that is affected and not that to the cold-water tap in the kitchen (Oliphant, 1983). However, soldered connections of copper pipes in new homes can result in up to 390 μg 1^{-1} of lead in the water (Cosgrove *et al.*, 1989). High lead concentrations have also been reported in newly constructed hospitals and schools with pipework made entirely from copper pipes. Table 27.5 shows such an example for a school on Arran in Scotland. These samples were taken three years after new copper pipes had replaced the lead plumbing. Leaching of lead from soldered joints and brass fittings decrease over time, whereas lead pipes continue to release the metal. Plumbers should clearly avoid the over-zealous use of lead solder. The corrosion of tin–lead solder has been reviewed by Gregory (1990).

In 1986 the US Congress banned the use of solder containing lead in potable water supply and plumbing systems. It did this by classifying it as a hazardous substance and requiring warning labels to be used stating that it was prohibited for use with potable water systems. Only lead-free materials can now be used. Solder is considered to be lead-free if it contains no more than 0.2% lead, whereas pipes and fittings must not exceed 8.0% lead (USEPA, 1988). Water coolers (drinking fountains) are commonly used in schools in the USA, and many were found to have lead-lined water reservoirs and lead-soldered parts leading to elevated lead concentrations in the water. Legislation, in the form of the *Lead Contamination Control Act 1988*, amended the *Safe Drinking Water Act* to protect children from this exposure by banning the sale or manufacture of coolers that are not completely lead-free.

Lead pipe, because of its age, is far more likely to burst and need repair. Often a section of copper pipe is inserted. This causes a significant increase in the water lead concentration in this section by direct electrochemical corrosion of the adjacent lead surfaces. Copper being taken up in the water and deposited onto the lead pipe as the water flows through the rest of the pipework exacerbates the problem. At each site where copper is deposited, a new galvanic cell is produced causing further corrosion. In the Renfrew study it was shown that much higher lead concentrations were recorded after copper inserts had been made either in the service pipe connection with the main or in the household plumbing. If property owners need to replace a section of lead pipe then they should, if possible, replace the whole pipe. Otherwise they should use plastic pipe, and special connections. In the past they would have been recommended to replace sections with new lead pipe as a last resort, rather than to use a section of copper tubing if they were unable to replace all the pipework. Since the introduction of the water byelaws in the UK it is now illegal to install any lead pipework, fittings, storage tank or a tank lined with lead, or to use lead in any repair in the home plumbing system. Copper pipes are no longer permitted for use in repairs of lead pipework unless galvanic corrosion is prevented. In practice this is almost impossible and so plastic pipe should be used to repair lead pipes using special fittings. Some PVC (polyvinyl chloride) pipes also contain lead compounds that can contaminate drinking water so only certified plastic piping should be used. Remember that removing a section of metal pipe and replacing it with plastic pipe may affect the earthing of any electrical system.

27.3.6 Remedial measures

There are a number of possible remedial measures that can be taken to reduce lead concentrations in drinking water. These are: (1) to reduce the level of lead in the resource although this is rarely a problem; (2) to alert the consumer of the possibility of high lead levels and to identify who is at risk and what action should be taken to reduce exposure; (3) to reduce the corrosivity of water by neutralization; (4) to line lead service pipes; or (5) to replace lead service pipes.

An education programme is only an interim solution. All consumers receiving increased lead levels should flush their tap before using the water for drinking or preparing food. For most people about $40 \, l \, d^{-1}$ would have to be flushed to be effective, which is about $14.3 \, m^3 \, yr^{-1}$. As much as 100 litres may have to be flushed for those with joint supply pipes or longer supply pipes, which should be effective in 95% of cases, which is equivalent to $36.5 \, m^3 \, yr^{-1}$. The total water usage each year is given to provide some idea of the annual cost involved if supplies are metered. However, it is still a cheap option for consumers, although it is a significant waste of water. Water companies would clearly be unhappy about widespread flushing in the long term because of the extra water demand and the extra wastewater generated which will require treatment at the sewage treatment plant. Many people now install a household treatment system. These are examined in detail in Chapter 29, but in reality none are 100% effective, although those using reverse osmosis or ion-exchange are capable of removing 90–95% of the lead present under optimum conditions. It is undesirable to rely on a household system to remove a toxic substance such as lead in the long term.

The usual remedial action is to treat all the water supplied to consumers at the water treatment plant to reduce its plumbosolvency, but this is expensive and causes secondary problems. This often involves relatively simple changes in water treatment practice to ensure that the water reaches the consumer at a pH of between 8.0 and 8.5. However, in practice operational problems can make this difficult. The required pH is achieved by adding alkalis such as lime, caustic soda or soda ash (Chapter 14). Lime is the cheapest, but is more difficult to handle resulting in excessive costs in terms of capital equipment and labour at smaller treatment plants. Some small plants use a filter containing semi-calcined (slaked) dolomite, which is simple to operate. Orthophosphate is also known to help control lead concentrations in waters with high alkalinity, although there may be long-term operational, quality and environmental problems associated with its use (Lee *et al.*, 1989; Colling *et al.*, 1992).

An example of water neutralization is the water supply to Glasgow from Loch Katrine, which had a pH of 6.3. This resulted in random daytime water samples containing on average more than $100 \, \mu g \, l^{-1}$ lead. From April 1978 to April 1980, the pH was increased to 7.8 so that by the time it reached the consumer the water had a pH of between 6.9 and 7.6. This resulted in 80% of the random daytime samples being below the $100 \, \mu g \, l^{-1}$ limit. Since April 1980 the mains have been dosed to a pH of 9, so that samples at the tap now have a pH in excess of pH 8.5, which has resulted in 95% of the homes treated now having lead levels below $100 \, \mu g \, l^{-1}$, the remainder having lead-lined tanks or long service pipes.

Lead pipes can be lined with a plastic coating to prevent corrosion, but the best option, and the most successful and reliable, is to replace the lead pipes with plastic or copper. It is desirable, but not necessary, to replace all the household plumbing. Just the service pipe from the mains to the drinking water tap, which is normally in the kitchen, needs to be replaced. The new regulations that implement the EC Drinking Water Directive in the UK require the water undertakers (either a company or a council) to replace their part of the service pipe when requested to do so in writing by a consumer who wishes to replace his or her part of the lead service pipe in water supply zones considered to be at risk. It is difficult to estimate a cost for replacing a supply pipe as this depends on exactly where it runs. This is normally achieved by simply pulling a new smaller-bore plastic pipe through the old service pipe, removing the need for costly excavations. Plastic pipes are always used to replace service pipes in the UK. If the lead pipework in the rest of the house is to be retained, it is essential to use plastic for the new supply pipe as copper should never be used upstream of any lead pipe for the reasons already outlined (Section 27.2).

If the cost of replacing lead pipes is compared with the cost of adjusting the pH of the finished water at the treatment plant, then pipe replacement is far more expensive in terms of capital costs. For the water supply companies, replacing all the lead service pipes, which they are in fact legally required to do if requested, would probably cost the same as identifying and treating all suspect supplies. If they did replace all the communication pipes then they would certainly save in the long term, as the extra operating costs of treating the water would be very high indeed and this would be a continuing annual cost. The cost of replacing the household supply pipes and associated plumbing would cost almost double what it would cost the water supply companies to replace their lead communication pipes. There are still many thousands of lead-lined tanks supplying water for all domestic purposes, including drinking. Their use is restricted to dwellings, mainly tenement properties, in the Edinburgh and Glasgow areas. These are often exposed to the atmosphere and although the water serving Glasgow from Loch Katrine, for example, is treated to keep its pH well above 8, the pH in these tanks falls to less than 8 on storage and results in the leaching of lead.

To achieve the new EC standard of $10\,\mu g\,l^{-1}$, pipe replacement is seen as the only long-term solution. In France it is estimated that it would take 20 years to replace all the lead pipes currently in use. In Belgium the problem is particularly acute in the eastern region where soft moorland waters are used for supply. Also, in some areas the raw water contains natural concentrations of lead and cadmium that have to be removed by treatment at high pH. It is estimated that to conform to the new standard would require 50 000 pipe replacements, which could take up to 20 years to complete. In most European countries lead is now banned. However, in Finland, where lead has been banned for over 100 years, there are no known problems with lead in water supplies.

27.3.7 Babies and lead

People who live in an older property and who are planning to start a family, or indeed already have children, should be advised to have their water tested for lead. If the water contains high levels, and during the vital time of pregnancy anything over 5 to $10\,\mu g\,l^{-1}$ is probably too high, the mother should drink an alternative supply of water. The WHO specifies lead concentrations in excess of

$10\,\mu g\,l^{-1}$ are too high for young children and especially for fetuses. Personally, I would prefer women not to drink water containing more than $5\,\mu g\,l^{-1}$ while they are pregnant, similarly for very young children. After six years of age the higher limit of $10\,\mu g\,l^{-1}$ is probably safe. At a concentration of $5\,\mu g\,l^{-1}$ the total intake of lead from drinking water is approximated at $3.8\,\mu g\,d^{-1}$ for infants and $10\,\mu g\,d^{-1}$ for adults (WHO, 2003). Ideally, the lead service pipe and any lead plumbing should be replaced or, alternatively, bottled water should be used. However, in new homes with copper pipework there is also a significant risk of elevated lead concentrations in the water. If in doubt ask your local water undertaker to check the lead concentration of the water from the cold-water tap in the kitchen. If the mother is going to breast-feed, she should continue to drink bottled water. If the baby is to be bottle-fed and there is an unacceptable concentration of lead in the kitchen tap water, then bottled water should be recommended for making up the feeds, but **ONLY IF THE FOLLOWING ADVICE IS CAREFULLY ADHERED TO**:

(1) **DO NOT** use sparkling mineral water, use only still water.
(2) Check that the bottled water has a low nitrate and sulphate concentration, and has as low a mineral content as possible.
(3) If it does not give details of the composition of the water on the label, then it should not be used.
(4) If it tastes salty or has a discernible mineral taste, then it should not be used.
(5) Before use, empty the kettle fully and then **BOIL** the bottled water before making up the feed.

All bottled waters are fine for older children, except for the very salty or strong mineral waters (Section 29.2). For children below the age of five years always use a non-sparkling, low-nitrate and low-sulphate water. As you can see, I am not too happy about recommending the use of bottled water to feed babies. This is because there are real risks involved, as explained in Section 29.2. Only a few bottled waters, in my opinion, are currently suitable for bottle-feeding babies and infants, and some may pose a serious threat to health if used for making up formula feeds. Extreme care must therefore be taken. Finally babies and young children up to the age of six should be not be allowed to swallow bath water as hot water contains much higher concentrations of lead, or to swallow tap water when cleaning their teeth, if the home has lead piping. Children should be taught not to drink from hot-water taps or from the cold-water tap in the bathroom from the earliest age regardless whether lead is a problem or not.

27.4 Copper

Copper is an essential element for human health. The average western diet will contain between 2 and 5 mg of copper each day; it is also widely used in drugs with concentrations up to 20 ppm common. Although it can be toxic at high concentrations, it rarely causes problems in genetically normal subjects. This is because the body can excrete excessive copper, unlike lead, which is accumulated. Because copper is essential to the body it is bound up in protein complexes, which reduces the concentration of free copper ions throughout the body, thus avoiding toxicity. It has been noted that copper lowers the toxicity of lead, and that lead induces copper deficiency symptoms. Tissue concentrations of copper remain fairly constant, except for changes due to age, regardless of exposure. This was shown by a study on copper miners in South America who were exposed to an atmosphere containing 2% copper ore dust for up to 20 years. Liver and blood analysis showed their copper levels to be normal (WHO, 2004a).

Concentrations in water resources are generally low. For example, the median value in surface waters reported in the USA is 0.01 mg l^{-1} (ATSDR, 2002). Copper is a common element in sewage, resulting primarily from the corrosion of household plumbing, so surface waters may contain up to four times more copper downstream of sewage treatment plants (IPCS, 1998). Copper is naturally present in elevated concentrations in ground and surface waters in metalliferous areas, but more often its presence in water is due to the attack of copper plumbing. Where copper sulphate has been used as an algicide in reservoirs, increased copper levels have occasionally been reported in drinking water, although this is rare. If acidic drinks or food are stored in copper vessels for long periods, then high concentrations can result, which will lead to mild toxicity, the symptoms of which are nausea, vomiting, diarrhoea and general malaise. The taste threshold for copper is between 2.4 and 2.6 mg l^{-1} (Zacarias et al., 2001).

Copper pipes are widely used in houses for the hot- and cold-water systems. There are also in excess of 6.5 million copper service pipes in the UK, which represents about 42% of all service pipe connections. The corrosion of copper pipework is known as cuprosolvency and results in copper being leached into solution, giving rise to blue-coloured water and occasionally taste problems. These problems occur with soft acidic waters or with private supplies from boreholes where free carbon dioxide in the water produces a low pH. Apart from pH, other important factors in cuprosolvency are hardness, temperature, the age of the pipe and oxygen availability. The inside of the pipe is protected by a layer of oxides, but water with a pH less than 6.5 and a low hardness less than 60 mg l^{-1} as $CaCO_3$ will result in corrosion. Hot water accelerates the corrosion rate, although oxygen must be present. If oxygen is absent, corrosion will be negligible regardless of the temperature. The corrosion of pipes can be severe causing bursts in the pipework, usually the hot-water pipes, the central heating system or the hot-water cylinder. Under such conditions some form of corrosion control is recommended. The corrosion of copper pipework is considered in more detail in Section 27.2.

Where the water is cuprosolvent an astringent, bitter taste may be produced and copper levels may increase enough to cause nausea. This is why water should not be drunk from hot-water taps, as the water heated in the copper cylinder will have been in contact with copper for some time at increased temperatures, allowing maximum corrosion to occur. Copper is a relatively soft metal and corrosion due to turbulent or excessive flows can occur. Where the water is also soft with a low pH and warm, this will seriously aggravate the situation. Pitting of copper pipes is associated with hard to moderately hard waters, often occurring in the cold-water pipes where carbon dioxide exceeds 5 mg l^{-1} and dissolved oxygen is high. This also gives rise to copper in the water. Raising the pH of the water to 8.0–8.5 overcomes most corrosion problems. Where water is running or the pipework has been flushed by allowing the tap to run for several minutes, then elevated copper will not occur. Where water has been left standing in the pipe overnight, or for longer, then concentrations up to 30 mg l^{-1} have resulted.

Using the premise that copper concentrations above 2.5 mg l^{-1} will cause taste problems, and above 1 mg l^{-1} discoloration and corrosion will occur, the EC has set a revised maximum value for copper in the current Drinking Water Directive (98/83/EEC) of 2.0 mg l^{-1}. In the USA the approach taken by the USEPA is different. The National Primary Drinking Water Regulations set an action level of 1.3 mg l^{-1}. The level of copper in the water at the consumer's tap is maintained by controlling the corrosiveness of the finished water. If more than 10% of the tap water samples exceed this action level then more stringent steps are required at the treatment plant to prevent corrosion. The National Secondary Drinking Water Standards, which are not mandatory, sets a maximum concentration of 1.0 mg l^{-1} to prevent staining of laundry and sanitary ware. The WHO has set a guideline value for copper of 2 mg l^{-1}.

Copper rarely causes health problems, although householders can have problems with taste, especially if they are particularly sensitive. If a glass of water containing 3.0 mg l^{-1} of copper is drunk at one go it will normally result in nausea, however, if drunk over a couple of hours then the effect is reduced. Accidental ingestion of large doses of copper causes gastrointestinal bleeding, haematuria and acute renal failure amongst other symptoms; lower doses have a similar effect to food poisoning causing headache, nausea, vomiting and diarrhoea (Agarwal *et al.*, 1993).

The most common problem relating to copper will be blue water and the staining of sanitary ware. A particular problem associated with new plumbing is increased concentrations of copper. In these circumstances children should not be allowed to drink from bathroom taps or drink their bath water, as the levels of copper can be high enough to make them feel nauseous. If the water is to be used to make a drink the cold-water taps should be flushed for a few minutes in the morning or if the water has been standing in the pipes for over an hour. Unless the water is acidic the problem should quickly resolve itself as the pipework becomes coated with a layer of oxides. Children should be taught only to drink water from the cold-water tap in the kitchen. Copper begins to impart a blue colour to water at concentrations above 4–5 mg l^{-1}. Such water should not be drunk or used in cooking or used for washing hair that has been highlighted or bleached.

27.5 Zinc

Zinc, like copper, is essential to human health but if ingested in gross amounts from water it will have an emetic effect. The inhalation of zinc-containing fumes is much more serious, but in water fairly high levels are permissible, the limiting factor being its taste threshold. The EC has not included zinc in the Drinking Water Directive, although the USEPA has set a non-mandatory National Secondary Drinking Water Standard of 5 mg l^{-1}. The WHO originally set a guideline value of 5 mg l^{-1} based on aesthetic quality and this was revised downwards to 3 mg l^{-1} in the 1993 standards based on taste. In the latest revision the WHO (2004b) have not set any guideline value for zinc as it does not pose a health threat but acknowledge that taste will be impaired at higher concentrations.

Zinc concentrations in surface and ground waters are normally below 0.01 and 0.05 mg l^{-1} respectively. Elevated concentrations can arise in metalliferous areas where zinc ore has been mined. The prime source of zinc in most homes will be from the corrosion of galvanized steel tanks and pipework. This is caused by low pH, so increasing the pH will help. When copper and galvanized zinc are used together then corrosion of galvanized steel occurs due to galvanic action. This is particularly important if the water is cuprosolvent (acidic) and the copper piping is located upstream of the galvanized steel. Just a small amount of copper (0.1 mg l^{-1}) is enough to increase the corrosion rate of the galvanized steel considerably. The UK water byelaws do not permit different metals to be connected where there is a chance of corrosion occurring through galvanic action. If copper is used downstream of galvanized steel, no problem will occur. The corrosion of galvanized steel is discussed fully in Section 27.2, and is the source of the sandy water often noticed when running water in the bath.

The use of galvanized steel pipework, tanks or fittings in plumbing systems is no longer recommended, especially with other metals such as copper. New plastic water tanks are more effective than the old galvanized tanks, are considerably lighter and are less likely to burst.

27.6 Conclusions

Corrosion in household plumbing is the single most important source of metals in drinking water, with lead one of the most hazardous contaminants found in drinking water. Those most at risk from short-term exposure to lead are pregnant women and infant children, with the unborn child particularly at risk from even low-level exposure (5 µg l^{-1}) of this dangerous contaminant. Exposure to lead has been shown to slow a child's physical growth and mental development, and can also lead to behaviour problems as well as measurable mental retardation. The kidneys, nervous system and liver can all be damaged by lead, with blindness and even death possible outcomes.

Drinking water standards universally require low lead concentrations in drinking water ($\leq 15\,\mu g\,l^{-1}$). To achieve this action has been taken in all countries to reduce the corrosiveness of water by buffering the finished water as it leaves the treatment plant. However, this can only be seen as a short-term solution, because as lead pipes age so the rate of leaching increases regardless of the pH of the water. The only safe scenario is the complete replacement of all lead pipework and fittings from homes, schools and other buildings, in particular lead service pipes. This is a major undertaking for water utilities and will take many years to complete, so a system of prioritization must be established to ensure those most at risk are identified and their pipes replaced first. At best the service pipe and the connection to the kitchen tap must be replaced with plastic pipe, while grants should be made available for the subsequent replacement of lead pipes and tanks where applicable. In the USA lead pipes and solder continued to be used in domestic buildings up to 1986, so this may be affecting enormous numbers of householders, many of whom will have no idea that they have lead pipes or are potentially at risk. I would strongly recommend that if you move into a new house that was built before 1970 in the UK or 1980 in the USA and you are planning a family, are pregnant or have young children, that you contact your water utility and have your water checked for lead. Ensure that you take a non-flushed sample, first thing in the morning (i.e. first draw). Also teach your children not to drink water from any tap other than the kitchen cold tap, and only use water from this tap for making drinks and cooking. Always check that when you are having work done on your plumbing that lead-free materials are being used. Finally point-of-use activated carbon filters, sand filters, standard cartridge and micro-filters do not remove metals including lead. Only reverse osmosis and distillation units are effective.

References

Agarwal, S. K., Tirwari, S. C. and Dash, S. C. (1993). Spectrum of poisoning requiring hemodialysis in a tertiary care hospital in India. *International Journal of Artificial Organs*, **16**(1), 20–2.

Ainsworth, R. G., Calcutt, T., Elvidge, A. F. *et al.* (1981). *A Guide to Solving Water Quality Problems in Distribution Systems*. Technical Report 167. Medmenham: Water Research Centre.

ATSDR (2002). *Toxicological Profile for Copper (Draft)*. Atlanta, GA: Agency for Toxic Substances and Disease Registry, US Department of Health and Human Services, Public Health Service.

British Standards Institution (1991). *Specification of Requirements for Suitability of Metallic Materials for Use in Contact with Water Intended for Human Consumption with Regard to Their Effects on the Quality of the Water*. BS DD201:1991. London: BSI.

Britton, A. and Richards, W. N. (1981). Factors influencing plumbosolvency in Scotland. *Journal of the Institution of Water Engineers and Scientists*, **35**, 349–64.

Colling, J. K., Croll, B. T., Whincup, P. A. E. and Harwood, C. (1992). Plumbosolvency effects and control in hard waters. *Journal of the Institution of Water and Environmental Management*, **6**, 259–68.

Cosgrove, E., Brown, M. J. and Madigan, P. (1989). Childhood lead poisoning: case study traces sources to drinking water. *Journal of Environmental Health*, **52**, 346–9.

De Rosa, P. J. and Parkinson, R. W. (1986). *Corrosion of Ductile Iron Pipe*. Technical Report 241. Swindon: Water Research Centre.

Department of the Environment (1977). *Lead in Drinking Water: a Survey in Great Britain 1975–76*. Pollution Paper 12. London: HMSO.

DNHW (1992). *Guidelines for Canadian Drinking Water Quality: Supporting Documentation: Lead*. Ottawa, Canada: Department of National Health and Welfare.

Friends of the Earth (1989). Poison on tap. *Observer Magazine*, 6 August, 15–24.

Gendlebien, A., Jakson, P. and Agg, R. (1992). *Lead in Drinking Water: the Impact of Existing and Future Standards on other EC Member States*. Report FR 0343. Marlow: Foundation for Water Research.

Gregory, R. (1990). Galvanic corrosion of lead solder in copper pipework. *Journal of the Institution of Water and Environmental Management*, **4**(2), 112–18.

Hopkins, S. M. and Ellis, I. C. (1980). *Drinking Water Consumption in Great Britain*. Technical Report 137. Medmenham: Water Research Centre.

Hulsmann, A. D. (1990). Particulate lead in water supplies. *Journal of the Institution of Water and Environmental Management*, **4**(1), 19–25.

IPCS (1998). *Copper*. Environmental Health Criteria: 200, International Programme on Chemical Safety. Geneva: World Health Organization.

Lacey, R. F.(1983). Lead in water, infant diet and blood: the Glasgow Duplicate Diet study. *Science of the Total Environment*, **41**, 235–57.

Lansdown, R. and Yule, W. (1986). *The Lead Debate: Environment, Toxicology and Child Health*. London: Croom Helm.

Lee, R. G., Becker, W. C. and Collins, D. W. (1989). Lead at the tap: sources and control. *Journal of the American Water Works Association*, **81**(7), 52–62.

Levin, R., Schock, M. R. and Marcus, A. H. (1989). Exposure to lead in US drinking water. In *Proceedings of the 23rd Annual Conference on Trace Substances in Environmental Health*. Cincinnati, OH: US Environmental Protection Agency.

McMichael, A. J., Vinpani, G. V., Robertson, E. F., Baghurst, P. A. and Clark, P. D. (1986). The Port Pirie cohort study: maternal blood lead and pregnancy outcome. *Journal of Epidemiology and Community Health*, **40**, 18–25.

Mullenix, P. I. (1992). Can safe lead levels in drinking water be deduced from current scientific evidence? In *Regulating Drinking Water Quality*, ed. C. E. Gilbert and E. I. Calabrese. Boca Raton, FL: Lewis, pp. 37–46.

Oliphant, R. I. (1983). *The Contamination of Potable Water by Lead from Soldered Joints*. External Report 125E. Medmenham: Water Research Centre.

Pocock, S. I. (1980). Factors influencing household water lead: a British National Survey. *Archives of Environmental Health*, **35**, 45.

Sherlock, J. C. and Quinn, M. J. (1984). Relationship between blood lead concentrations and dietary lead intake in infants: the Glasgow Duplicate Diet Study 1979–1980. *Food Additives and Contaminants*, **3**, 167–76.

Thornton, I. and Culbard, E. (eds.) (1987). *Lead in the Home Environment*. Northwood: Science Reviews.

USEPA (1986). *Air Quality Criteria for Lead*. EPA-600/8-83/028F, US Environmental Protection Agency. Durham, North Carolina: Research Triangle Park, NC.

USEPA (1988). *Handbook for Special Public Notification for Lead for Public Drinking Water Supplies*. Report 570/9-88-002. Washington, DC: Office of Water, US Environmental Protection Agency.

White, S. F. and Mays, G. D. (1989). *Water Supply Bylaws Guide*, 2nd edn. Chichester: Water Research Centre and Ellis Horwood.

WHO (2003). *Lead in Drinking Water*. Report No. WHO/SDE/WSH/03.04/09. Geneva: World Health Organization.

WHO (2004a). *Copper in Drinking Water*. Report No. WHO/SDE/WSH/03.04/88. Geneva: World Health Organization.

WHO (2004b). *Guidelines for Drinking Water Quality*, Vol. 1. *Recommendations*, 3rd edn. Geneva: World Health Organization.

Zacarias, I., Yanez, C. G., Araya, M. *et al.* (2001). Determination of the taste threshold for copper in water. *Chemical Senses*, **26**(1), 85–9.

Chapter 28
Micro-organisms, fibres and taste

28.1   Micro-organisms in plumbing systems

We have seen that the contamination of drinking water by pathogenic and non-pathogenic micro-organisms occurs mainly at source (Chapter 13), although contamination can also occur during treatment or within the distribution systems (Chapters 19 and 25). The contamination of otherwise potable water can also occur within the consumer's premises. This is generally due to the type of plumbing system installed, a lack of basic maintenance and the careless use of appliances.

The householder is responsible for the maintenance and repair of the supply pipe, which runs from the water supply company's connection pipe at the boundary stop-tap to the house. All the water entering the house passes through this pipe so any fracture will allow possible contamination to enter. Sewerage pipes should be located well away from the supply pipe, preferably on the other side of the building.

Back-syphonage in plumbing systems is more of a problem in older buildings as modern building regulations and water byelaws incorporate measures to prevent it. It generally occurs when a rising main supplying more than one floor suffers a loss of pressure at a low point in the system, causing a partial vacuum in the rising main. Atmospheric pressure on the surface of, for example, a bath full of water on an upper floor in which a hose extension or a shower attachment from an open tap has been left, will push the contents of the bath back up through the hose and tap into the plumbing system to fill the partial vacuum. Plumbing systems suffer from constant changes in pressure, so care should always be taken when hoses are left to run water into containers. Farmers, who often need to fill large pesticide sprayers with water, need to take special care to ensure that the end of the hosepipe is never allowed to fall below the liquid surface in case of back-syphonage. This is extremely important if private supplies are pumped by a submersible pumping system from an underground storage tank. Once the pump is switched off any water left in the pipework connected directly to the pump will drain back into the tank. If the pump has been used to fill a watering can or a water trough, for example, and if the end of the hose is not removed before the pump is switched off, then the contents of the container will be drained by back-syphonage into the storage tank resulting in contamination. Mechanical back-flow prevention devices are available to prevent water flowing back into the household plumbing system or even the mains distribution system from the consumer's premises. There are a number of back-flow devices, both mechanical and non-mechanical including pipe interrupters which are now mandatory for all hosepipes in the UK (Figure 28.1). These are fully explained by White and Mays (1989) and more technically by Halford and Bond (1986).

Water storage tanks must always be kept covered. When left open they are susceptible to being fouled by birds and vermin, and so can become a breeding ground for a wide variety of micro-organisms. Many houses have mice in the attic, a favourite place for overwintering, under or in the six inches of insulation. The only source of water in the attic is the cold-water storage tank, so fouling by vermin is fairly common. In 1984 there was an outbreak of *Shigella sonnei* at a boarding school in County Dublin, which caused widespread gastrointestinal illness among the pupils. The source was identified as the water storage tank which was supplying all their drinking water and which had been fouled by pigeons.

Contamination by rats can also lead to leptospirosis although entry of the bacteria into the body is normally via small cuts or the mucous membranes from where they enter the bloodstream and affect the kidneys, liver and central

Figure 28.1 Typical back-flow prevention device used with hosepipes. Reproduced from White and Mays (1989) with permission from the Water Regulations Advisory Scheme.

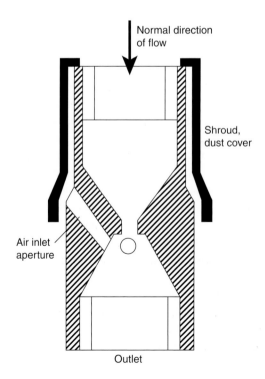

nervous system. Nearly all those affected work in high-risk occupations, e.g. handling animals, meat processing, sewage workers, fish-farmers, river inspectors and water scientists. Infection of people in low-risk categories comes almost exclusively from swimming in contaminated water, although all those engaged in water sports on both inland and coastal waters are at risk. The rapid increase in reported cases in the UK since 1986 has been linked with the increasing popularity of water sports and of windsurfing in particular. The chance of infection from drinking water is slight, but tanks should be covered to be on the safe side.

28.1.1 Legionnaires' disease

Legionnaires' disease was first reported in 1976 when there was a major epidemic among the residents of a hotel in Philadelphia. The American Legion was holding its annual conference at this hotel, hence the name of the disease. Since then numerous outbreaks of the disease have been reported, including many in the UK. In 1985, 75 000 cases were reported in the USA with 11 250 deaths. The bacteria survive and grow within phagocytic cells, multiplying in the lungs causing bronchopneumonia and tissue damage. Although over 20 strains of *Legionella pneumophila* are known, only one serotype is thought to be a serious threat to health.

The bacterium has been associated with domestic water systems, especially hot water that is stored between 20 and 50 °C. It appears that a long retention time and the presence of key nutrients such as iron provides ideal conditions for the bacterium to develop. Therefore iron storage cisterns subject to corrosion will be susceptible. In addition, water pipes that are installed alongside hot-water pipes or other sources of heat may also allow the bacteria to develop. The bacterium is widespread in natural waters so any water supply can become contaminated. It is resistant to residual disinfection and can actively grow within the distribution system (Section 25.2). The bacterium survives in the biofilm that grows on the surface of the plumbing system, with the protozoa that graze on this biofilm acting as hosts for *Legionella* (Abu Kwaik *et al.*, 1998). The main mode of transmission appears to be from bathing, especially showers (Tobin *et al.*, 1980), although heat exchangers, condensers in air-conditioning units, cooling towers as well as showerheads have all been found to be common niches for the bacteria leading to human infection. Legionella are commonly found in hospital water systems, and hospital-acquired (nosocomial) legion-naires' disease is now a major health problem (Joseph *et al.*, 1994). Prevention of legionella infection is normally achieved by either hyperchlorination or thermal eradication of infected pipework, although these methods are not always successful or the effect permanent. The only effective preventative method has been shown to be the use of ultraviolet (UV) sterilizers as close to the point of use as possible.

28.1.2 *Mycobacterium*

Mycobacterium avium, Mycobacterium intracellulare and other mycobacteria are opportunistic pathogens of considerable significance that cause a wide range of diseases in humans, including pulmonary disease, cervical lymphaclemopathy as well as localized and soft tissue infections (Von Reyn *et al.*, 1994). Mycobacteria are ubiquitous in all types of aquatic environments, including groundwaters, surface waters and distribution systems (Jenkins, 1991). The majority of waterborne mycobacterial outbreaks are attributable to treatment deficiencies such as inadequate or interrupted chlorination, although the bacteria are particularly resistant to chlorination (Taylor *et al.*, 2000). Other factors may also influence the growth of this organism in water supplies, such as pitting and encrustations found inside old water pipes, which protect bacteria from exposure to free chlorine (Section 25.2). Mycobacteria also colonize areas where water is moving slowly, as in water distribution systems in large buildings such as blocks of flats, offices and hospitals, thus continuously seeding the system (Du Moulin and Stottmeier, 1986). Disease associated with these bacteria is steadily rising, particularly among patients with AIDS (Von Reyn *et al.*, 1994).

28.1.3   Other sources

Non-pathogenic bacteria can develop in rubber extension pipes on taps and in mixer taps, leading to taste and odour problems. Long rubber extension pipes that extend into contaminated water (during dish or clothes washing at the kitchen sink) will quickly develop thick microbial contamination inside the pipe. Such pipes should be avoided whenever possible because they can cause a significant reduction in water quality, and can also result in back-syphonage. If consumers need to use these pipes, and many elderly or disabled people find them especially useful, then they should be removed and cleaned regularly.

People using home treatment systems should ensure that they are regularly cleaned or the cartridges replaced as stated in the instructions. Micro-organisms will grow on activated carbon granules held in cartridges and may, ironically, cause taste and odour problems. If consumers are not changing the cartridges regularly, then it is best to not use them at all (Chapter 29).

28.2   Fibres, including asbestos

Most of the fibres found in drinking water are asbestos (Chapter 22). These originate either from the resource, or more likely from asbestos cement pipes used in the water distribution network. Asbestos can also originate in the home plumbing system from the water storage tank. This is only a problem where open-topped tanks are located in the roof, which can receive asbestos fibres from atmospheric fallout. This may be significantly higher in cities and industrial areas, so open-air tanks in particular need to be sealed. In many countries isolated dwellings that do not have a natural water supply close at hand collect rainwater from their roofs for use in the home. Rainwater harvesting is especially common on farms, where roof water is usually collected for watering the animals. Where this water drains off asbestos sheeting or asbestos roof tiles, then significant levels of asbestos fibres can be found in the drinking water (Section 22.2). Rainwater collected from roofs is fine for most household purposes, but cannot be recommended for drinking without treatment. In practice it usually tastes awful, as organic matter from the roof quickly decays anaerobically in the bottom of the storage tank producing sulphides, which smell and taste dreadful. Iron levels can also be high if the water is collected from an old corrugated roof and may stain clothes during washing. If rainwater has to be used, then consumers must be advised to regularly clean out any debris from the storage tank, to keep the tank covered to prevent algal growth and colonization by dipteran larvae and other insects and, if possible, prevent any organic debris, especially leaves, from entering the tank. Rainwater is naturally acidic and so acts aggressively on the roof surface and the tank. If the roof is made of asbestos-based material, galvanized steel or has been bitumen-sealed, then it should not be used for collecting drinking water (Section 29.4).

Glass, carbon and other mineral fibres may also be carcinogenic to humans. Animal experiments have confirmed that they can produce tumours. Most attics are now insulated, and indeed the water tank is also insulated, often with glass fibre. Attics have an airflow that can at times be fairly turbulent, so fibres are constantly being blown around the attic space. If the water storage tank is open these fibres will fall into and collect in the tank. It is therefore advisable to cover it to prevent fibres, and other material, from falling into the tank. Asbestos fibres can accumulate in the tank and if the house is situated at the end of a cul-de-sac or if the distribution main is a spur, then the tank should be checked periodically to see if there is any accumulated debris. If there is a deep deposit (greater then 10 mm) then the tank should be drained and all the accumulated debris carefully cleaned out. This should only be necessary at very long intervals.

If sediment in the tank is a serious problem then the distribution system should be suspected and the water supply company contacted to flush out the mains. After the company has completed flushing the consumer should shut off the supply to the storage tank, and then run the kitchen tap, which is connected directly to the supply pipe, until the water is absolutely clear. The consumer will usually only need to run the tap for 10 minutes or so, but where there is a long service pipe it may take a long time, perhaps an hour or more, before all the debris is finally flushed out.

28.3 Odour and taste

Household plumbing systems are in many ways ideal places for microbial growth to occur. The many connections leave areas of stagnant water where sediment can accumulate and micro-organisms can survive. The system may contain organic-based pipes, tanks and even washers where micro-organisms can live and obtain nutrients. The problem is acute in plumbing systems that have a low turnover of water and where water is warmed by running cold-water pipes close to those carrying hot water. Warming also occurs in large blocks of flats where the cold-water tap may be fed from a storage tank, which in turn is fed by an auxiliary pump because the pressure in the water mains is insufficient to reach the top of the building. In large buildings and offices it is important that the pipes carrying drinking water are insulated by lagging and that they are distributed through the building via separate ducts away from central heating systems and hot-water pipes. Under warm conditions actinomycetes in particular thrive, so that musty and earthy odours will be produced (Chapter 21). In large buildings the central storage tanks should be examined periodically for microbial growth and, if necessary, cleaned out. The materials and fittings used in plumbing systems should be those tested and certified as not supporting microbial growth.

An astringent taste in water can be caused by the corrosion of pipes and fittings containing copper or zinc. Where corrosion is severe due to water acidity, the water may become undrinkable. Standard taste threshold concentrations for copper vary from $3\,\text{mg l}^{-1}$ for the most sensitive 5% of the population to $7\,\text{mg l}^{-1}$, which is the mean taste threshold concentration. This compares with 5 and $20\,\text{mg l}^{-1}$, respectively, for zinc. These figures may in practice be too high, especially in soft water areas.

Polyethylene pipes, especially low-density polyethylene, are permeable to a wide range of compounds, especially hydrocarbons and phenols. Although convenient for use with cold water in household plumbing systems, polyethylene can occasionally be the unexpected source of an odour problem. Careful consideration should be given to where these pipes are to be laid. For example, contamination could occur if the pipework is in contact with bituminous materials or the solvents that are now so widely used in the construction and building industries. The diffusion of materials from gas pipes can also lead to problems. An interesting example is a polyethylene pipe that ran along the walls of a cellar. The walls had been painted with bitumen paint to damp-proof the cellar and the gas supply entered the house through the cellar and ran parallel against the water pipe to the meter. As a result of the permeability of the polyethylene pipe the water occasionally had a strong tarry taste.

28.4 Conclusions

Microbial development within household plumbing can be potentially very serious due to the development of *Legionella* and *Mycobacterium* spp. For that reason special attention should be paid to any replacement or renovation of household pipework to ensure that suitable materials are used and that the cold-water system is protected from hot-water pipes. Storage tanks in the attic must also be covered to prevent contamination, a common entry point for pathogens carried by birds and vermin. If the water byelaws or building regulations are closely followed then no biofilm development should occur and no problems arise.

References

Abu Kwaik, Y., Gao, L., Stone, B. J., Venkataraman, C. and Harb, O. S. (1998). Invasion of protozoa by *Legionella pneumophila* and its role in bacterial ecology and pathogenesis. *Applied and Environmental Microbiology*, **64**(9), 3127–33.

Du Moulin, G. C. and Stottmeier, K. D. (1986). Waterborne mycobacteria: an increasing threat to health. *American Society of Microbiology News*, **52**, 525–9.

Halford, I. and Bond, K. (1986). *Devices with Moving Parts for the Prevention of Backflow in Water Installations*. Technical Report 245. Swindon: Water Research Centre.

Jenkins, P. A. (1991). Mycobacterium in the environment. *Society for Applied Bacteriology Symposium Series*, **20**, 137S–41S.

Joseph, C. A., Watson, J. M., Harrison, T. G. and Bartlett, C. L. R. (1994). Nosocomial legionnaires' disease in England and Wales 1980–92. *Epidemiology and Infection*, **112**, 329–46.

Taylor, R. H., Falkinham, J. O., Norton, C. D. and LeChavallier, M. W. (2000). Chlorine, chloramines, chlorine dioxide and ozone susceptibility of *Mycobacterium avium*. *Applied and Environmental Microbiology*, **66**(4), 1702–5.

Tobin, J. O., Beare, J., Dunnill, M. S. *et al.* (1980). Legionnaires' disease in a transplant unit: isolation of the causative organism from shower baths. *Lancet*, **2**(8186), 118–21.

Von Reyn, C. F., Maslow, J. N., Barber, T. W., Falkenham, J. O. and Arbeit, R. D. (1994). Persistent colonization of potable water as a source of *Mycobacterium avium* infections in AIDS patients. *Lancet*, **343**, 1137–41.

White, S. F. and Mays, G. D. (1989). *Water Supply Bylaws Guide*, 2nd edn. Chichester: Water Research Centre and Ellis Horwood.

PART VI
THE WATER WE DRINK

Chapter 29
Alternatives to tap water

29.1 Introduction

The vast majority of us have our water delivered to our homes fully treated and ready to consume. We turn on our taps and expect all the water to be safe and palatable to drink at any time during the day or night. Similarly we expect water, at the correct pressure, to operate our washing- and dishwashing-machines, our showers and other household appliances. In fact it is only until there is a water shortage or a problem with quality that any of us actually even think about water. That is not the case for those on private supplies who have to maintain their own borehole and pump. For this group water quality is often a continuous cause of concern. So what are the options for those on public and private supplies who are concerned about their water quality? Bottled waters offer a short-term alternative to mains supply for consumptive purposes (Section 29.2), while home, or point-of-use, treatment systems offer a long-term solution to problematic water (Section 29.3). Harvesting the water from roofs and recycling water from washing-machines, showers and baths has become increasingly popular for supplying water for external uses such as car-washing and watering the garden, although this water is normally unsuitable for drinking without treatment (Section 29.4).

29.2 Bottled water

The increase in the consumption of bottled water worldwide is phenomenal with estimated global sales around 1.55×10^{11} litres in 2004. Compound annual growth rates (1999–2004) vary between 4.2% in France, which already has a high per capita consumption rate, to 20.9% in China, where per capita consumption is still comparatively low. The average growth rate worldwide is 9.4% (Table 29.1a). In 1980 just 30 million litres of bottled water were drunk in the UK, rapidly rising over the intervening period to between 200 and 250 by 1991 and to 1770 million litres by 2002. This rapid increase in bottled water consumption continues although the per capita consumption in the UK at 29 litres per annum is still low compared with over 100 litres per person each year in other European countries (Italy 184, France 142, Spain 137, and Germany 125 litres) (Table 29.1b). While the market for sparkling water is more

Table 29.1 *Comparison of national consumption of bottled drinking water between 1999 and 2004. Reproduced with permission from the International Bottled Water Association*

(a) Consumption by top ten countries

Country	Consumption of bottled water (million litres)		Compound annual growth rate (%)
	1999	2004	
USA	17 337	25 766	8.2
Mexico	11 572	17 671	8.8
China	4607	11 887	20.9
Brazil	5655	11 591	15.4
Italy	8919	10 654	3.6
Germany	8307	10 306	4.4
France	6943	8545	4.2
Indonesia	3434	7357	16.5
Spain	4075	5502	6.2
India	1681	5123	25.0
Sub-total	72 530	114 402	9.5
All other	25 868	39 879	9.0
Total	98 398	154 281	9.4

(b) Annual per capita consumption by top fifteen countries

Country	Litres consumed each year per capita	
	1999	2004
Italy	155	184
Mexico	117	169
United Arab Emirates	110	164
Belgium-Luxembourg	122	148
France	117	142
Spain	102	137
Germany	101	125
Lebanon	68	102
Switzerland	90	100
Cyprus	67	92
USA	64	91
Saudi Arabia	75	88
Czech Republic	62	87
Austria	75	82
Portugal	70	80
Global average	16	24

or less static, being traditional in many countries, the sales of still water continues to see major growth and represent the bulk of all sales (e.g. >95% in the USA). One of the largest developments in Europe is the increased use of bottled water coolers, which has always been widely appreciated in the USA (Finlayson, 2005). The USA remains the largest single market for bottled water, which increased by 8.6% from 2003 to 6.8 billion gallons (27.7×10^9 litres) in 2004, which represents an annual per capita consumption rate of 24 gallons (91 litres), and an expenditure in that year alone of US\$ 9 803 000 000, making bottled water one of the fastest growing and valuable sectors in the food industry (www.ibwa.com). Consumption of bottled water in the USA is set to exceed 8.5 billion gallons (32×10^9 litres) by 2007.

Bottled water is a generic term that describes all water sold in containers. There are other terms such as spring, table or natural mineral water, which are specific terms defined by legislation. Many different types of water sources are used in the bottling industry including natural springs, boreholes and previously treated municipal supplies.

In Europe bottled water labelled as natural mineral waters come from recognized and licensed sources. They are regulated separately from other types of bottled waters under the EC Mineral Waters Directive (80/777/EEC) and have no restriction on mineral content being exempt from the normal guidelines and limits laid down by the EC Drinking Water Directive (98/83/EEC). All other types of bottled waters are regulated under the Drinking Water Directive, and so must conform to all the prescribed standards, including the limits on minerals (e.g. calcium, sodium, magnesium, sulphate, hydrogencarbonate, chloride). However in 1996 bottled waters labelled 'spring water' became subject to many of the requirements that applied only to natural mineral waters such as being bottled at source, but spring waters must still satisfy the prescribed chemical, physical and microbiological limits of the Drinking Water Directive. Bottled waters are therefore of three types: natural mineral waters, spring waters and other types of bottled water. The latter two are just as pure, if not purer, than the mineral waters and unlike the mineral waters conform to the higher standards as laid down for drinking water. The exemption for mineral waters was to allow for personal taste differences. For example, some people like the very strong mineral waters such as Vittel, others prefer the medium mineral content found in brands such as Perrier.

29.2.1 Standards

European Union

During the mid 1970s the EC identified a number of concerns relating to bottled and natural mineral waters. These were: (1) the national requirements of individual member states constituted barriers to free trade; (2) doubtful claims were being made about the medical or health benefits of such waters; (3) microbiologically unfit products were on the market; (4) some sources were contaminated or were at serious risk from contamination; and (5) the standard of bottling practice left much to be desired. This led to the development and finally the introduction of the Natural Mineral Waters Directive in August 1980 (80/777/EEC), which has since been amended twice (Directives 96/70/EEC; 2003/40/EC). The Directive only applies to waters described as natural mineral waters and, since 1996, spring waters that are sold within the EC. It put into place a scheme for the recognition and exploitation of natural sources of mineral water implemented by national authorities. The Directive stresses that only the purest of groundwaters with no organoleptic defect may be bottled and marketed as natural mineral water. Only those products that have been recognized by the national authority can be called a natural mineral water, even though their purity and composition may be excellent. In the Directive there is a long list of requirements that have to be investigated and fulfilled before an application for natural mineral or spring water status is even considered.

The original Directive only required natural mineral waters to be free from pollution and not contain toxic substances. The later amending Directives have introduced limit values for a range of potentially toxic elements. There is no maximum permissible concentration for individual anions (e.g. chloride, sulphate, nitrate) or for total dissolved solids (TDS) in the Directive. However, limits have been set for a range of ions that may pose a threat to public health in Annex: I of Directive 2003/40/EC (Table 29.2). The source must be free from pathogenic micro-organisms and parasites, with maximum total bacteria content $<20\,\text{ml}^{-1}$ at 20–22 °C in 72 hours and less than $5\,\text{ml}^{-1}$ at 37 °C in 24 hours. As disinfection or sterilization of mineral waters is prohibited, colony counts in still waters, although low at source, may rise to $10^5-10^6\,\text{ml}^{-1}$ within the first few days of bottling but will slowly decline during storage over the following months. These bacteria are part of the normal bacterial flora of the water and are normally harmless. Indeed, it is generally thought that they prevent the development of undesirable bacteria (Edberg, 2005). The Directive requires that the source is free and protected from pollution. Although no chemical standards are specified, most member states have included standards for a range of volatile organic compounds, polycyclic aromatic hydrocarbons and pesticides. The chemical quality of groundwaters is also protected under the EC Water Framework Directive. Spring waters must conform, like other bottled waters, to the Drinking Water Directive quality standards.

Natural mineral waters can be further classified into three groups: (1) natural mineral water, which is still; (2) carbonated mineral water, which is still water with added carbon dioxide; and (3) naturally carbonated natural mineral water, which is water that is carbonated in its natural form. Among the restrictions in the Natural Mineral Waters Directive is the prohibition of transport of water to bottling plants, so that the water must be bottled at source. Prohibition on treatment of the water is included, although a number of treatments are allowed particularly those relating to the bacteriological quality (e.g. simple filtration and

Table 29.2 *Maximum limits in mineral waters as specified in Annex I of the EC Mineral Waters Directive (2003/40/EC)*

Constituents	Maximum limits (mg 1^{-1})
Antimony	0.005
Arsenic	0.01 (as total)
Barium	1.0
Boron	To be set
Cadmium	0.003
Chromium	0.05
Copper	1.0
Cyanide	0.07
Fluorides	5.0
Lead	0.01
Manganese	0.5
Mercury	0.001
Nickel	0.02
Nitrates	50.0
Nitrites	0.10
Selenium	0.01

carbonation), as long as no change is made to the essential characteristics of the water. Bottles must be securely closed with tamper-proof seals and there are restrictions on the claimed health benefits. Labelling requirements include the name and location of the source; a list of the major anions, cations and the TDS; and information on any treatments (e.g. use of ozone-enriched air for the removal of iron, manganese, sulphur or arsenic).

Other bottled waters are not required to be bottled at source or to have a stable or fixed composition, which are mandatory requirements for natural mineral waters although spring waters are also exempt from the latter requirement. Unlike natural mineral and spring waters any treatment is permissible for other bottled waters to achieve drinking water directive standards, and labelling requirements are left to individual member states.

USA

Bottled water in the USA is subject to both federal and state regulations. At the federal level, water is treated as a packaged food product and is regulated by the US Food and Drug Administration (FDA) primarily under the Federal Food, Drug and Cosmetic Act and sections of Title 21 of the Code of Federal Regulations. These include (1) standard of identity (i.e. type of bottled water); (2) standards of quality; (3) good manufacturing practices; (4) labelling requirement; and (5) misbranding, adulteration and recall provisions. At the state level, regulation is based on either the FDA model, so it is treated solely as a food, or alternatively using the environmental model based on the US Environmental Protection Agency (USEPA) Safe Drinking Water Act. The former model is normally adopted by states although the FDA standards must be as stringent as the USEPA Primary Drinking Water Standards.

The Code of Federal Regulations (CFR) (21CFR: Part 165) includes Standard of Identity definitions. This provides specific definitions for different types of bottled water based largely on source and treatment. So in the USA all bottled waters must include the appropriate name on its label reflecting the applicable Standard of Identity definition. The classifications are bottled, drinking, artesian, ground, distilled, deionized, reverse osmosis, mineral, purified, sparkling, spring, sterile and well. For example, the term 'mineral water' applies only to bottled waters that have a TDS concentration of $\geq 250\,mg\ l^{-1}$, comes from a hydrogeologically protected source comprising one or more boreholes or springs and contains no added minerals (21CFR: 165.110).

Limits for microbiological, physical, chemical and radiological substances for both source water and finished bottled water products are defined under the Standard of Quality for bottled waters (21CFR: Part 165). The FDA has established standards for more than 75 parameters; however, where bottled waters contain excessive amounts of one of these parameters then a statement to that effect must be clearly made on the label. Bottled water is one of few food products with its own specific Good Manufacturing Practices that also specifies quality parameters and testing frequency (21CFR, Part 129).

There are also industry-based standards, administered in the USA by the International Bottled Water Association (IBWA), which is the trade association representing the bottled water industry. Founded in 1958, IBWA's membership includes US and international bottlers, distributors and suppliers. They have their own model regulations (available at www.bottledwater.org) that have been adopted by over a dozen states as the standard for regulation of bottled water. All bottled waters imported into the USA must comply with the Federal and State regulations.

29.2.2 Mineral content

The mineral content of bottled water is dependent solely on the rocks that the water comes into contact with and the length of time it has been in contact with them. Some highly mineralized waters have very complicated hydrogeological origins, and some are naturally carbonated due to the upwelling of carbon dioxide. The sparkle in sparkling water is caused by carbon dioxide, which can occur naturally but more often than not it is added later. Perrier, for example, has found itself under criticism for labelling its waters as naturally carbonated. In the past the carbonated water was bottled directly, but problems were encountered with fluctuating carbon dioxide concentrations. It now extracts the water and gas separately and mixes them under controlled conditions to ensure a consistent product.

Table 29.3 *Indications and criteria for labelling and describing mineral waters specified in Article 9 of EC Mineral Water Directive (2003/40/EC)*

Indications	Criteria
Low mineral content	Mineral salt content, calculated as a fixed residue, not greater than 500 mg l^{-1}
Very low mineral content	Mineral salt content, calculated as a fixed residue, not greater than 50 mg l^{-1}
Rich in mineral salts	Mineral salt content, calculated as a fixed residue, greater than 1500 mg l^{-1}
Contains bicarbonate	Bicarbonate content greater than 600 mg l^{-1}
Contains sulphate	Sulphate content greater than 200 mg l^{-1}
Contains chloride	Chloride content greater than 200 mg l^{-1}
Contains calcium	Calcium content greater than 150 mg l^{-1}
Contains magnesium	Magnesium content greater than 50 mg l^{-1}
Contains fluoride	Fluoride content greater than 1 mg l^{-1}
Contains iron	Bivalent iron content greater than 1 mg l^{-1}
Acidic	Free carbon dioxide content greater than 250 mg l^{-1}
Contains sodium	Sodium content greater than 200 mg l^{-1}
Suitable for the preparation of infant food	–
Suitable for a low-sodium diet	Sodium content less than 20 mg l^{-1}
May be laxative	–
May be diuretic	–

Europeans traditionally drank mineral waters as an aid to health, a tradition that is well known and still preserved at many European spa towns. Highly mineralized waters such as Contrexeville (France), Fiuggi (Italy) and Radenska (the former Yugoslavia) are renowned for curing urinary disorders and helping to break up kidney stones, Other waters with a high hydrogencarbonate concentration are used to ease indigestion. Evian was supposed to soothe skin diseases, whereas Apollinaris helped bronchial complaints. These were the traditional reasons for drinking mineral waters. The Mineral Waters Directive no longer allows any therapeutic claims to be made for bottled waters attributing properties relating to the prevention, treatment or cure of human diseases. However, it does allow specific indications to be used on labels (Table 29.3). Some idea of the contents of a range of bottled waters widely available in the British Isles is given in Table 29.4. The key parameters to look for on the label of bottled water are discussed below, although in the EU this only applies to natural mineral and spring waters, as other bottled waters conform to the limits for anions and cations set out in the Drinking Water Directive (98/83/EEC).

Calcium (Ca^{2+}) Calcium concentrations range from less than 10 to over 520 mg l^{-1} (e.g. Crodo Valle D'oro contains 519 mg l^{-1}), although most bottled waters contain between 100 and 120 mg l^{-1}. There is no danger to health at

Table 29.4 Chemical constituents (in mg l^{-1}) of some major bottled waters marketed in the British Isles. The list is by no means comprehensive and the selection of water included is purely arbitrary. Waters are listed according to country of origin

Country	Trade name	Ca^{2+}	Mg^{2+}	K$^+$	Na$^+$	HCO$_3^-$	SO$_4^{2-}$	NO$_3^-$	Cl$^-$	pH	Dry solids
Belgium	Bru	23	22.6	1.8	10.0	209	–	1.0	4.0	–	–
	Spa Reine	3	1.3	0.5	2.5	11	5	1.9	2.7	–	–
France	Evian	78	24.0	1.0	5.0	357	10	3.8	4.5	7.2	309
	Perrier	140	3.5	1.0	14.0	348	51	–	30.9	–	500
	Vichy	100	9.0	60	1200	3000	130	–	220	–	–
	Vittel	505	110	4.0	14.0	403	1479	0.7	11.0	7.0	2580
	Volvic	10	6.0	5.4	8.0	64	7	4.0	7.5	7.0	110
Ireland	Ballygowan	117	18.0	3.0	17.0	400	15	2.0	28.0	6.9	450
	Carlow Castle	117	15.4	5.3	13.1	355	61	8.6	10.2	7.4	560
	Glenpatrick	112	15.0	1.1	12.2	400	19	–	26.0	7.5	–
	Kerry Spring	76	9.6	–	24.0	244	8	2.1	46.0	6.9	–
	Slievenamon	112	15.0	1.1	12.2	400	19	–	20.0	7.5	–
	Tipperary	37	23.0	17.0	25.0	282	10	2.3	19.0	7.7	272
Italy	Aqua Fabia	129	4.0	0.9	11.2	384	–	–	19.5	–	–
	Clavdia	96	30.0	88.0	66.0	561	49	4.5	58.0	–	–
	Crodo lisiel	53	6.7	2.8	4.7	103	79	3.2	2.5	–	–
	C. Valle D'oro	519	49.0	5.0	1.8	69	1398	–	1.8	–	–
	San Pellegrino	208	56.4	3.0	41.1	226	539	1.0	71.0	7.1	11204
	S. Antonio	61	8.9	0.7	3.9	196	10	18.7	9.0	7.7	218
	Sorgente Panna	31	6.1	0.9	6.5	104	19	1.8	8.9	7.4	136
Sweden	Ramlosa	2	0.5	1.8	220	535	14	1.9	24.0	8.7	515

UK										
Ashbourne	102	24	3.0	10.0	–	60	6.0	30.0	–	420
Buxton Spring	56	20	1.0	24	123	–	0.01	38.0	7.4	–
Cerist	3	1.5	0.4	3.0	–	7	0.8	8.1	6.3	19
Highland Spring	45	20	–	12.5	181	10	0.2	16.0	–	–
Malvern Spring[a]	76	32	3.0	9.0	–	41	6.0	16.0	7.1	360
Pentre Nant Spring	35	12	1.9	50	–	28	–	34.0	8.2	–
Prysg	11	4.3	0.1	8.2	40	4	–	–	–	–
Snowdonia Spring	7	6.1	1.0	5.5	6	6	–	17.7	–	–
Strathglen Spring	24	5.0	0.3	7.5	95	10	1.8	11.0	7.3	119
Stretton Hills	41	7.2	1.6	10.8	106	33	12.6	20.0	–	–

[a] The Malvern Springs Pure Water Company Ltd bottled at Aston Manor Brewery in Birmingham. There is another Malvern brand produced by Schweppes Ltd, which is a mineral water.

these levels as calcium is an essential element for building bones and teeth, especially in children. Calcium contributes to water hardness, which is known to reduce cardiovascular disease (Chapter 10), and high levels of calcium make the water palatable.

Magnesium (Mg^{2+}) Along with calcium, magnesium is a major constituent of hardness and a major dietary requirement for humans. Levels are generally low in bottled water, although there are exceptions. Vittel contains 110 mg l^{-1} and Apollinaris 122 mg l^{-1}. When present as magnesium sulphate it makes an effective laxative, so avoid high concentrations unless you need help in this direction. It is best not to give bottled waters with high magnesium concentrations to children under seven years of age.

Potassium (K^+) Although it is an essential element, the body finds it difficult to deal with excess potassium, resulting in kidney stress and possible kidney failure. Although potassium is not considered to be toxic, long-term exposure to high potassium concentrations should be avoided. It is advisable not to select any bottled water with more than 12 mg l^{-1} potassium for regular drinking. Check waters carefully as this is one element that is often exceeded. Typical low-potassium waters are Buxton Spring, Glenpatrick, Perrier, Evian, Aqua Fabia and Spa Reine.

Sodium (Na^+) This is a key element in bottled waters and is essential for health. In healthy adults excess sodium is excreted, but in sensitive adults such as those with hypertension and heart weakness, infants with immature kidneys and the elderly, high sodium levels may cause problems and so a low-sodium diet is often advised. The EC limit for sodium in drinking water is 200 mg l^{-1}, and many mineral waters exceed this. For those identified as being at risk, all high-sodium waters should be avoided and those with low sodium levels such as Evian, Volvic and Cardo used instead. Certainly a water with a maximum of 20 mg l^{-1} should be sought. High-sodium waters such as Vichy, Vichy Calalan, San Narciso and the Russian Borjomi, which contains nearly 1500 mg l^{-1} of sodium, could be injurious to people at risk if taken regularly. For routine family drinking look for a water with less than 200 mg l^{-1}, and preferably less than 50 mg l^{-1}, such as Glenpatrick, Buxton Spring, Volvic or Evian. Those on a low-sodium diet should look for bottled waters containing a maximum of 20 mg l^{-1}. The others should be kept for the occasional glass of something interesting and refreshing, not drunk regularly. Sodium can act as a laxative when present as sodium sulphate (i.e. Glauber salts).

Deaths of babies have occurred from drinking formula feeds made up with high-sodium mineral waters, so use the water with the lowest sodium content you can get. Remember that other minerals apart from sodium, which may be found in excess, can also cause potentially serious problems when used to make up babies' feeds. The use of mineral waters for bottle-feeding babies is considered in Section 29.5.

Sulphate (SO$_4{}^{2-}$) Magnesium and sodium sulphate are both strong laxatives, so although the human body can adapt to moderately high levels of sulphate in water, sudden increases will lead to a purgative effect. This could be extremely serious in young infants and sensitive adults, so families with young children should avoid mineral waters with high sulphate levels. The EC limit in drinking water is 250 mg l^{-1}, but many mineral waters contain much more than this. Avoid any mineral waters with a sulphate level in excess of 30 mg l^{-1} if you want to use them on a regular basis. For very young children about 10 mg l^{-1} seems a sensible maximum.

Nitrate (NO$_3{}^-$) and nitrite (NO$_2{}^-$) Many aquifers are showing increasing nitrate levels and there is a trend in many countries for bottled water companies not to give details of nitrate. If they do not, then avoid them. Excessive nitrate can give rise to two problems: infantile methaemoglobinaemia and the formation of nitrosamines (Chapter 5). The nitrate itself is not a direct toxicant, but it is the conversion of nitrate to nitrite that causes the problems. This can occur readily in the digestive tract under certain conditions, so low nitrate levels are desirable. The EC set a maximum nitrate limit in drinking water of 50 mg l^{-1}, but there is no real safe limit, it is a case of the lower the better. Of those companies that do give information on nitrate, all have low concentrations, with Buxton Spring having less than 0.01 mg l^{-1}. Any bottled water containing more than 0.05 mg l^{-1} nitrite should be avoided and only nitrite-free bottled water should be used to make up formula feeds for babies.

Chloride (Cl$^-$) Chloride is not dangerous at the concentrations found in bottled waters. It has a taste threshold of about 200 mg l^{-1}, but levels are generally much lower than this in mineral waters. If bottled water is to be used for making up drinks such as tea, coffee and fruit juices, then the lower the chloride concentration the better. Natural surface waters contain between 15 and 35 mg l^{-1} chloride, whereas seawater contains 35 000 mg l^{-1}. Concentrations up to 1500 mg l^{-1} are probably drinkable without any adverse effects.

Fluoride (F$^-$) Fluoride levels should be given on the side of bottles. In general they are not, but when they are, they are usually <1 mg l^{-1}. Many bottled waters contain more than the EC maximum of 1.5 mg l^{-1}; for example, Ramlosa contains 2.2 mg l^{-1}. Such levels are fine for the occasional drink, but if consumers are going to use a bottled water as a complete replacement for tap water, for both drinking and cooking, then the fluoride level must be 1.0 mg l^{-1} or less to avoid dental fluorosis and other associated problems (Chapter 17).

Hydrogen ion concentration (pH) In still mineral waters the pH ranges from 6.0 to 8.0, which is fine. In carbonated waters, however, the pH can decrease in some brands to 5.0 or even 4.5. Although this keeps the water bacteriologically pure, it does make the water aggressive and can lead to problems such as the leaching of minerals from teeth if drunk regularly. Bear in mind, however, that many canned carbonated soft drinks are equally aggressive.

Total dissolved solids (TDS) This is a gross measure of all the solid material left after the water has been evaporated at 180 °C. This will include mainly dissolved minerals and so the higher the TDS, the saltier it will taste. Generally water with more than about 500 mg l^{-1} TDS has a distinctive mineral taste, and above 1000 mg l^{-1} it has a strong mineral taste. Above 1500 mg l^{-1} it will taste salty, although this will vary according to personal palates.

All aspects of bottled water has been reviewed by Dege and Senior (2005).

29.3   Point-of-use water treatment

Over the past 25 years the number of manufacturers and distributors of home water treatment systems, more commonly referred to as point-of-use treatment systems, has mushroomed. The diversity of systems available has also expanded. The systems range from small simple jug filters, which produce about two litres at a time and just remove sediment and fine solids from the water, to in-line systems producing such pure water that it can be used in a car battery instead of distilled water. Most of the systems are based on physical filtration, ion-exchange, activated carbon, reverse osmosis or ultraviolet (UV) radiation. Often a combination of these processes is used with physical filtration normally always required to prolong the use of other units (Table 29.5). The basis of all these processes has already been described in relation to full-scale water treatment plants. The same technology has been scaled down for use in the home.

29.3.1   Physical filtration and activated carbon

Point-of-use water treatment starts with water filters. The simplest systems are physical filters that remove anything larger than the pores in the filter material. They are normally used to remove sediment and iron precipitate, although they can also be used to remove pathogens such as protozoan cysts and even bacteria when very small pore sizes are employed. Filters are used in two ways. In-line cartridge filters are fitted into the plumbing system, usually under the kitchen sink so that all the water to the kitchen tap is treated. Alternatively, jug filters are small containers of about two litres capacity fitted with a filter reservoir that is also used as the lid. Water is poured into the reservoir and passes through the filter into the jug.

Cartridge filters are usually plumbed into the mains connection to the kitchen cold-water tap (Figure 29.1). This is very simple and can be done by any DIY enthusiast. Of course, not all the water from the cold-water tap needs to be treated so it is common to take a branch connection from the mains pipe and then have a separate tap at the sink solely for drinking water (Figure 29.2). Most manufacturers offer easy to install systems using a simple self-piercing connection, and all that is required is to tighten it onto the incoming mains water pipe. They supply a simple to fit small-bore tap and all the other pipe

Table 29.5 *Most effective home treatment methods for the removal of selected chemical and physical problems*

Problem material or substance	Fibre filter	Activated carbon filter	Reverse osmosis	Deionization	Distillation	Ultraviolet radiation
Sodium	−	−	++	++	++	−
Arsenic	−	−	++	++	++	−
Lead	−	−	++	++	++	−
Cadmium	−	−	++	++	++	−
Potassium	−	−	++	++	++	−
Sulphate	−	−	++	++	++	−
Hardness (Ca)	−	−	++	++	++	−
Hardness (Mg)	−	−	++	++	++	−
Nitrate	−	−	++	++	++	−
Chloride	−	−	++	++	++	−
Faecal bacteria	−	−	++	−	++	++
Viruses	−	−	++	−	++	++
Protozoan cysts[a]	++	−	++	−	++	+
Organics	−	++	++	−	+	−
THM, TCE[b]	−	++	++	−	+	−
Dioxins	−	+	++	−	+	−
Chlorine	−	++	++	−	+	−
Pesticides	−	++	++	−	++	−
Sediment	++	+/−	++	−	++	−
Taste/odour	−	++	++	−	−	−
Asbestos[a]	++	−	++	−	++	−

Key: −, Non-effective; +/−, some reduction; +, good to moderate reduction; ++, excellent reduction.
[a] Requires a filter with a pore size of < 1 μm to be effective.
[b] THM = Trihalomethanes; TCE = trichloroethylene.

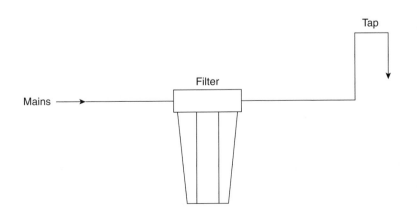

Figure 29.1 Typical plumbing arrangement for a water filter fitted to treat all the water to the kitchen cold-tap.

Tap

Filter

Mains

Figure 29.2 Alternative method of plumbing a water filter into the cold-water supply in the kitchen. This only treats a proportion of the water to the sink and so requires a separate tap. The smaller volume passing through the filter unit, however, will prolong the life of the cartridge.

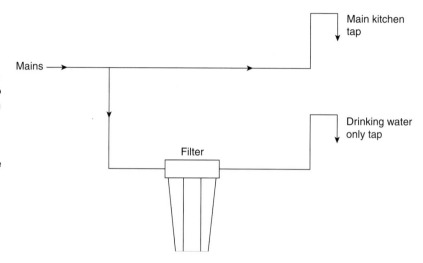

Figure 29.3 Typical layout of a DIY filter kit.

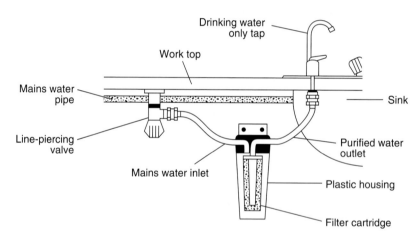

connections are push-fit. The plastic housing for the filter cartridge is simply screwed into the wall (Figure 29.3). The water from the mains enters the plastic filter housing and then passes under pressure through the cartridge filter, the purified water passing up the centre of the housing flowing onwards to the tap. This is the most common type of system and there are a wide variety of cartridge filters that can be used. Fine particulate matter, including sediment and iron precipitate, are removed by cartridges made out of single strands of yarn, which are wound spirally to form the filter, or consist of pads containing felted or bound fibres. Spirally wound filters are made from cotton, wool, rayon, fibreglass, polypropylene, acrylic, nylon and other polymers. The yarns are brushed to raise the fibres to create a filtering medium. These cartridges come in three general grades depending on the type of yarn used. Coarse woven filter cartridges are used to pretreat household waters to remove particles $>50\,\mu m$ before further finer filtration or treatment by another process. Medium woven filter cartridges are usually adequate for drinking waters and remove particles down to $20\,\mu m$ in diameter and last longer than the finer filters. Fine woven filter cartridges are generally used in laboratories as they remove all particles down to $5\,\mu m$ or less. They provide very good quality drinking water and may be needed if the turbidity in the water is made up of very fine particles. Pathogens are often adsorbed onto particulate matter so filters that remove solids as small as $5\,\mu m$ will produce water that has most of the pathogens removed, although this will not be satisfactory for those with compromised immune systems.

Apart from the spirally wound filters, cartridges can also be packed with loose felted fibres of cellulose, wool, glass fibre or polypropylene. These can remove particles as small as $0.5\,\mu m$, but will quickly become blocked if the water is visibly dirty or turbid. Other materials, such as ceramics, can also be used to achieve the same level of filtration. If the pore size is small enough even bacteria may be removed and fibre cartridge filters with a pore size of $1\,\mu m$ effectively remove protozoan cysts of *Giardia* and *Cryptosporidium*.

The larger the pore size, the longer the life of the filter cartridge. Therefore, when fine filters are used, prefiltering the water using a coarse grade of filter will significantly increase the life of the second filter. As filtration continues the pores become progressively clogged with retained material, and so the average pore size gradually becomes smaller. All particles will be retained, including any substances adsorbed onto the particles. Even a physical filter will remove aluminium or pesticide residuals that are adsorbed onto particulate matter.

Physical filters are widely employed with groundwaters to remove turbidity and red water caused by iron particles and rust. These sediment filters are designed to treat all the water coming into the house and so should be fitted on the rising main before the junction feeding the kitchen tap. If a water softener is installed it must be installed after the filter. If the water is seriously cloudy or rusty, a coarse filter cartridge must be used before a finer filter cartridge (in series) with both cartridges changed frequently.

Taste and odours caused by chlorine, decaying organic matter and metals can also be removed by using simple cartridge systems. These consist of a hollow cartridge filled with granular activated carbon instead of fibres. Activated carbon is also excellent at removing organic chemicals such as chlorinated hydrocarbons including trihalomethanes and trichloroethylene, pesticides, industrial solvents and naturally occurring humic substances. These humic substances also cause colour, which can also be removed by activated carbon. They operate under normal pipe pressures, as do fibre cartridges, and although the fibre cartridge becomes discoloured indicating that it needs to be replaced, it is not possible to tell when activated carbon cartridges need to be replaced. Their effective lifespan depends on the concentration of the organic material and residual chlorine in the water, so advice must be sought from the supplier about how often they should be replaced. With in-line systems prefiltration using a fibre filter is usually necessary unless the water is exceptionally clear and free from solids. Like all point-of-use treatment units, the activated carbon cartridge system needs only to be placed on the supply to the kitchen cold-water tap. Special units are available that only treat the water required for drinking via a special tap (Figure 29.3). This considerably prolongs the life of the filter cartridge.

When using cartridge filters the system should always be flushed through for at least 10 seconds before using the water for drinking or cooking. This is especially important if the tap has not been used for 24 hours or longer. The cartridges should be replaced if the water pressure drops or if taste and odours begin to reappear. Fibre filters and activated carbon cartridges should be changed fairly regularly to maintain maximum performance. Certainly do not use any cartridge filter for longer than six months because foul tastes and odours can originate from old cartridges. As with full-scale granular activated carbon filters, heterotrophic bacteria can colonize the surface of activated carbon, leading not only to taste and odour problems but also to a risk of colonization by opportunistic pathogens (Reasoner *et al.*, 1987). Therefore, four months is probably a maximum period for most in-line activated carbon cartridges, and four weeks for the small activated carbon units used in jug filters.

29.3.2   Reverse osmosis

Reverse osmosis removes dissolved impurities such as manganese and sulphate, and is achieved by using a semi-permeable membrane. The water is forced through the membrane under pressure while dissolved and particulate materials are left behind ensuring that there is a constant concentration gradient with the liquid passing through the membrane from the concentrated side. This means that there is a production of wastewater in which all the various contaminants are concentrated and which has to run to the drain. Typical removal (or rejection as it is referred to in reverse osmosis) characteristics for a reverse osmosis system are given in Table 29.6, with on average between 94% and 98% of the TDS removed. The most critical factor in point-of-use units is the type of membrane used. Different membranes are used depending on whether the water has been chlorinated or not.

The performance of the membrane depends very much on the quality of the raw water. For example, its life is prolonged if chlorine is removed by activated carbon pretreatment and if iron is removed using a catalyzed manganese (II) dioxide filter, which oxidizes the iron to an insoluble form that can then be removed by a simple fibre filter. The membrane may need to be flushed occasionally and bacterial development can also become a problem. Reverse osmosis is ideal for removing aluminium, copper, nickel, zinc and lead. It can also remove 85–90% of all organic compounds including trihalomethanes, polychlorinated biphenyls, pesticides and benzene. When used downstream of an activated

Table 29.6 *Nominal retention characteristics of reverse osmosis membrane units*

Chemical	Symbol	Rejection (%)
Sodium	Na^+	87–93
Calcium	Ca^{2+}	94–97
Magnesium	Mg^{2+}	96–98
Potassium	K^+	87–94
Iron	Fe^{2+}	95–98
Manganese	Mn^{2+}	95–98
Aluminium	Al^{3+}	98–99
Ammonium	NH_4^+	86–92
Copper	Cu^{2+}	98–99
Nickel	Ni^{2+}	98–99
Zinc	Zn^{2+}	98–99
Strontium	Sr^{2+}	96–98
Cadmium	Cd^{2+}	96–98
Silver	Ag^+	93–96
Mercury	Hg^{2+}	96–98
Barium	Ba^{2+}	96–98
Chromium	Cr^{3+}	96–98
Lead	Pb^{2+}	96–98
Chloride	Cl^-	87–93
Hydrogencarbonate	HCO_3^-	90–95
Nitrate	NO_3^-	60–75
Fluoride	F^-	87–93
Silicate	SiO_2^{2-}	85–90
Phosphate	PO_3^{4-}	98–99
Chromate	CrO_2^{4-}	86–92
Cyanide	CN^{4-}	86–92
Sulphite	SO_3^{2-}	96–98
Thiosulphate	$S_2O_3^{2-}$	98–99
Ferrocyanide	$Fe(CN)_6^{3-}$	98–99
Bromide	Br^-	87–93
Borate	$B_4O_2^{2-}$	30–50
Sulphate	SO_4^{2-}	98–99
Arsenic	As	94–96
Selenium	Se^{2-}	94–96

Also removes 85–90% of all organics, i.e. THMs, PCBs, pesticides, herbicides, benzene, etc.

carbon system, organic compounds are essentially eliminated. Nitrate is not completely removed, however, and so to remove all the nitrate present (99%) an ion-exchange system should be used upstream of the reverse osmosis unit. The installation of ion-exchange systems will increase the capacity of the reverse osmosis unit ten fold, also making it much more efficient. Some problems have been reported due to excessive water pressure or microbial degradation of the membrane material, causing membrane fracture and thus treatment failure.

29.3.3   Ion-exchange

Ion-exchange systems use a resin onto which certain undesirable ions are adsorbed and replaced by different ions. The usual ion-exchange reaction exchanges calcium and magnesium ions for sodium ions, which reduces hardness. Most domestic water softeners work on this principle. Softening does not reduce the alkalinity or TDS, but scale formation caused by calcium, magnesium, carbonate and hydrogencarbonate is eliminated.

Nitrate can be removed by using a strongly basic anion resin for the preferential removal of nitrate and sulphate anions. These are replaced by chloride ions thus:

$$CaNO_3 + RCl \rightarrow RNO_3 + CaCl$$

$$MgNO_3 + RCl \rightarrow RNO_3 + MgCl$$

where R is the resin within the unit. Exchange only continues as long as there are sufficient ions in the resin to be exchanged. The resin is regenerated by flushing with concentrated brine (Section 14.2). Ion-exchange systems are used to treat the water for the entire home, whereas reverse osmosis is used for the kitchen tap only.

Ion-exchange does not remove bacteria or viruses, sediment, taste, odours or organisms. Apart from softening water, such systems need to be used with fibre and activated carbon filters for full treatment (Table 29.5). There is concern that there is a higher rate of cardiovascular disease in soft water areas; also, the corrosion of certain metals and materials are increased in soft waters. This option should therefore be considered carefully. If a consumer is receiving a sodium-free diet, then remember that base exchange removes calcium and magnesium at the expense of increasing the sodium concentration. If this is a problem there are salt-free resin ion-exchange units now available.

29.3.4   Disinfection

One of the most effective ways of ensuring that your water is microbially safe is to install your own disinfection system. Ultraviolet radiation is extremely effective and is able to kill almost all known micro-organisms found in water. The systems available are all generally similar. A UV lamp is housed in a transparent quartz sleeve. The water enters one end of the unit and flows around the protected lamp ensuring adequate exposure. The greater the volume of water required, the larger the lamp that is used to ensure an adequate exposure time. These units are plumbed into the incoming water main and are generally designed to treat all the water, although smaller units are available to fit just the supply to the kitchen cold-water tap. If all the water is supplied from a storage tank, the UV sterilizer can be fitted directly to the incoming water main in the attic. The lamp is housed in a metal box with a warning light and also an audible warning signal, if required, to indicate lamp failure. The lamp indicator and control switch can be remote from the actual unit if the sterilizer itself is in a cellar, attic, or otherwise not visible. The lamp runs for six to nine months on a continuous basis using either 18 or 35 W. In terms of running costs it is equivalent to having a 100 W light bulb on for five to eight hours each day. Even at high flows removal of 99.7% of faecal coliforms has been achieved in trials and so at the lower flows expected in the average household even higher levels can be achieved. Bacteria may survive, if they are protected by particulate matter, and the efficiency of UV sterilizers can be increased by installing a fibre filter upstream of the UV lamp. A word of caution; the units come with full instructions and usually special safety devices to ensure that the lamp is switched off before the protective housing is removed. Ultraviolet light can severely damage your eyesight so under **no, absolutely no**, circumstances should you allow the light to be activated if the protective cover is not in place.

An alternative method of disinfection is to install a chlorinator. Most piped water supplies are already chlorinated and so it is not advisable to chlorinate the water a second time. However, those served by private supplies, especially surface streams, springs or shallow wells, should consider either an UV system or a chlorinator. Chlorine tablets (calcium hypochlorite) are held in a tall tube, the base of which is perforated to allow the water to flow past the tablets, slowly dissolving them and releasing chlorine. The dosing chamber is a simple box that is plumbed into the pipework, although it is best to have the chlorinator as far upstream of your system as possible to reduce chlorinous tastes and odours. In terms of aesthetic quality, a fibre filter followed by an UV sterilizer is preferable. However, where there is no power, or the power is intermittent, passive chlorination is the only answer. Physical filtration using a pore size of $< 0.5\,\mu m$ will remove bacteria but not all viruses.

29.3.5   Other systems

There are a number of more specific treatment systems also available. Acid water neutralizers reduce the corrosion of copper and lead pipes. They neutralize the water by passing it through a special medium that increases the pH. The water supply flows upwards through the unit and does not require any electricity, backwashing or drainage. Other systems add soda ash or lime to the water. There are also a number of iron removal systems available; including one used for private groundwater supplies that are rich in iron. The water is aerated to bring the iron from its soluble iron (II) state into the insoluble iron (III) state, which is then filtered out.

29.4   Water harvesting

There has been much interest over the past decade in the harvesting of rainwater for water supply purposes, and indeed for many isolated situations rainwater is the only reliable source of drinking water. Increasing water charges and water shortages have all increased interest in capturing this free water resource that would otherwise go to waste. Some of the water companies in England and Wales, where water supplies are particularly stretched during the summer, offer discounted water collection systems and storage containers for garden use. However, the quality of rainwater is quite variable and is easily contaminated during collection and storage. So in practice rainwater harvesting is limited to non-consumptive purposes such as garden watering, flushing the toilet and perhaps washing clothes, with the pipework connected directly to appliances so that the water cannot be inadvertently drunk. The reuse of water by recycling bath and shower water is only suitable for flushing the toilet and watering the garden (Konig, 2001).

Rainwater harvesting has focussed mainly on surfaces such as roofs although driveways and other paved areas can also be used. The volume collected each year is a function of the collection area, the annual rainfall and water storage capacity. It is estimated that for a medium-sized home up to $100 \, \text{m}^3$ per annum could be collected in all areas of the British Isles.

For agricultural and some industrial applications rainwater harvesting is certainly cost effective. However, for the individual householder where a complete rainwater harvesting system could cost anything from £2500 to £10 000 rainwater harvesting may not be cost effective except in the very long term, but will allow unrestricted use of water during water shortages when hosepipe bans have been introduced. Reduced water usage has environmental benefits, but any financial savings depends on (1) whether the water supply is metered; (2) the rainfall pattern and collection area for harvesting being sufficient; (3) whether or not a reduction in metered water usage translates into a reduced wastewater treatment charge; and (4) whether you are prepared to maintain and service the system yourself.

About 45% of the water used in the home could be replaced with harvested rainwater and assuming a daily per capita water usage of 150 litres, then this is equivalent to $270 \, \text{l} \, \text{d}^{-1}$ in a household of four people or $100 \, \text{m}^3$ per annum. In practice a full harvesting system is estimated to save the average household between 10% and 20% of the water normally used; although a keen gardener currently using mains water to water plants would save considerably more (Table 1.14; Section 1.5).

At its simplest, rainwater can be collected directly from the guttering into water butts and then used on the garden. Where rainwater is to be used in the home then a rainwater harvesting system has to be installed. This consists of a number of different parts (Figure 29.4). Rainwater is collected from the roof or

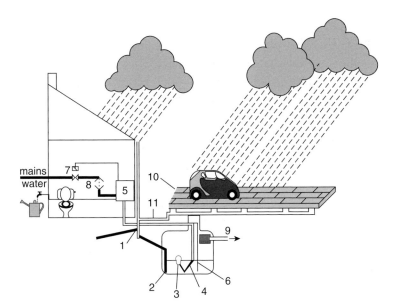

Figure 29.4 Standard rainwater harvesting system. (Environment Agency, 2003). Where: 1 = filter, 2 = smoothing inlet, 3 = pump filter, 4 = pump, 5 = control unit, 6 = water level monitor, 7 = automatic water change over, 8 = AA airgap, 9 = overflow trap, 10 = permeable pavement, 11 = oil trap.

other paved areas and passes through a vertical collection pipe to an underground storage tank. A filter positioned on this vertical collection pipe prevents any leaves or other coarse solids from entering the tank. The water is dispersed on entry into the storage tank by a smoothing inlet, which prevents any accumulated fine sediment at the bottom of the tank being disturbed. Water is removed via a fine mesh filter attached to either a submersible or suction pump. The pressure created by the pump feeds the water into the house via a separate network of pipes to the various appliances. In some houses the water is fed into a separate storage tank in the attic to which the WC and outdoor tap is connected. Control units are widely employed to monitor the water level in the storage tank and that can also automatically change over the supply to mains water when the harvested water is exhausted. This type of arrangement require the use of a type AA airgap to prevent any possibility of back flow of the harvested water into the mains supply. A simple monitor measures the level in the storage tank, and when full excess water overflows into a soakaway. The overflow contains a trap to remove any floatable material that may be in the tank. It is important that a back-flow prevention device is attached to the overflow to prevent any contamination. Storage tanks come in a wide variety of sizes and materials. A widely used method for household water use is to size the tank at 5% of the rainwater supply or of annual demand, using the smallest of the figures. Size will usually be a factor of cost and need to store sufficient water over the driest period. This is reviewed in detail by the Environment Agency (2003).

If water is being harvested for drinking purposes then point-of-use treatment will be required. Bird droppings, wind-blown debris and plant material such as moss and lichen can contaminate water collected from the roof making the water unsuitable for drinking due to pathogens and trace chemical contaminants. Also, depending on the roofing material, a variety of hazardous compounds can be leached into the water including: (1) asbestos fibres from asbestos cement tiles and asbestos sheeting may pose a health risk from aerosols and also lead to filters being blocked; (2) polycyclic aromatic hydrocarbons from flat bitumen-coated roofs and felt repairs may also cause discolouration and odour problems; (3) lead from lead flashing; (4) zinc and iron from galvanized roof sheets and copper from copper sheeting may also cause green discolouration; and (5) grass roofs can lead to discolouration from soil and the loss of fine sediment to the storage tank.

All rainwater is mildly acidic due to the presence of carbonic acid (pH 5.6), although due to the presence of oxides of sulphur and nitrogen most European rainwater has a pH<4.5 making it highly corrosive to normal metal plumbing pipework and fittings, as well as roofing materials. So in practice, rainwater should only be used for drinking as a last resort.

Water collected from paved areas will be of much poorer quality than that collected from roofs and may well be contaminated with oil and animal faeces. Porous paved areas are better than surface drains for collection, but this type of water may need extra treatment before entry into the storage tank, including the use of an oil trap and passive disinfection.

The reuse of grey water in the home is discussed in Section 1.5.

29.5 Conclusions

Bottled water is expensive because it is heavy to transport, so foreign companies are at a disadvantage compared with local bottled water suppliers. However, as the cost margin is small, many of the major companies have brought or developed new companies in target countries rather than exporting water direct.

Bottled waters are covered by specific and general legislation, and so should be of excellent quality. However, there is little evidence that they are better than most tap waters in developed countries, in fact, there has been some disquiet in recent years that some parameters are rarely, or even never, tested in natural mineral waters, for example, radon (Section 12.2). Complex organic compounds can be found at high concentrations in some of the groundwaters from where these mineral waters are abstracted and the chemical quality can be variable. Some companies have over-pumped their groundwater reserves in an attempt to satisfy the growing demand and have found that the quality has altered. Problems at bottling plants have led to numerous breaches of chemical and bacteriological quality standards. Factors such as shelf-life, water is like any other food and so once opened will begin to deteriorate even when kept in a refrigerator, all make the quality of bottled water every bit as suspect as tap water.

If bottled water is required to supplement or substitute piped drinking water on a long-term basis, then identify why. It will be cheaper to correct the problem with the tap water by correcting plumbing problems, or at worst by having to install a point-of-use treatment system. If, however, the consumer needs to use bottled water to supplement tap water, then non-carbonated water should be used. Use water with as little in it as possible and select water that is low in sodium, fluoride, potassium, sulphate and nitrate. Mineral-rich water should not be consumed continuously over a long period of time.

If consumers wish to use bottled water for making up formula milk for a baby, then they must ensure that it is very low in sodium and sulphate ($10\,\text{mg}\,1^{-1}$), with low nitrate ($5\,\text{mg}\,1^{-1}$ as nitrate) and no nitrite. **Carbonated water should not be used**, only still water. Suitable bottled waters in Europe are Evian and Volvic. **The water must be boiled** before making up the feed. Remember, bottled water has not been disinfected, and even though it is normally free from pathogenic micro-organisms at source, there are other normally harmless bacteria present. Contamination can occur, so it must be boiled to be safe. Finally, opened bottles of water should not be stored for longer than 36 hours if they are being used to feed an infant, unless they are in a special sterilized bulk container with a tap. Even so, this should be stored in a refrigerator.

There has been growing concern about the use of PVC (polyvinyl chloride) to bottle water due to concerns over the leaching of endocrine-disrupting substances from the plastic that can mimic the action of the hormone oestrogen in the body, which has been linked with various adverse effects on the reproductive systems of animals and humans. This has led to a significant shift to polyethylene terephthalate (PET) packaging, which is considered much safer. However, studies suggest that with repeated use, toxic chemicals in PET plastic can break down and migrate into the liquid inside. Of major concern is di (2-ethylhexyl) adipate, a carcinogen that has also been shown to cause liver damage as well as reproductive problems. So care should be taken in reusing such bottles over prolonged periods. The bottles used in water coolers are normally made from polycarbonate, which is made from bisphenol-A, which is also a hormone-disrupting chemical. Although leaching is known to occur it is not considered a threat to health.

While point-of-use water treatment systems offer a long-term solution to water quality problems, they require regular maintenance and the occasional replacement of the active units, such as cartridges, etc. Systems that are not looked after properly may result in much poorer quality water than before, so they do require intelligent management. Also, apart from the initial cost, service arrangements, replacement cartridges and other parts can be expensive. There is also considerable concern about the suitability for purpose of many units that are offered for sale, especially over the Internet, with exaggerated and false claims often made. While point-of-use treatment systems are increasingly popular there are currently no British Standards covering home treatment systems. As long ago as 1990, the Drinking Water Inspectorate highlighted in its annual report that extravagant claims for point-of-use treatment systems were often made by suppliers and some manufacturers, and that some devices could even lead to the deterioration of water quality by, for example, encouraging microbial growths. The Inspectorate stated that it did not see a need on health grounds for these devices to be used on water supplies drawn directly from the public mains. However, it recognized the right of consumers who may object to the taste or appearance of their water, or the presence of a particular substance in the supply, to use such a device. It stresses that the installation of these devices must comply with the water byelaws. For consumers with private supplies, the use of such devices may be the only way to obtain water of reasonable quality. In 1991 the Department of the Environment set up an expert group to devise national test procedures for point-of-use treatment systems, which they prefer to call 'point of use devices'. The protocols produced by the Committee on Point of Use Device Test Protocols, in conjunction with WRc, aim to provide potential users with accurate information on the capability of the available systems. Currently protocols for ultraviolet (UV) disinfection units, activated carbon filter units, ceramic and cartridge filters, *in-situ* regenerated ion-exchange nitrate removal units and reverse osmosis units are available. Each protocol represents a standard of good practice although compliance with a protocol does not confer immunity from relevant legal requirements.

In the USA the federal government has appointed the National Sanitation Foundation (NSF) to establish standards for water filters. They provide an on-line directory of filters that they have approved and certified (www.nsf.org). Certification is based on a number of criteria including (1) ability to achieve manufacturer's treatment claims, (2) constructed out of safe materials that will not contaminate the water it is treating, (3) able to operate under expected pressures and use, and (4) must be accompanied by accurate literature. The UK protocols were produced in close collaboration with the NSF. The USEPA (2006) recently produced a set of guidelines for point-of-use and point-of-entry systems. Point-of-use devices, such as reverse osmosis filters, are usually installed under a kitchen sink and can comply with drinking water standards for such contaminants as arsenic, lead and radium. Point-of-entry devices are installed outside the home or business and can treat an even wider variety of contaminants. Further details of the effectiveness of point-of-use treatment systems are given by Bell *et al.* (1984) and Geldreich and Reasoner (1990).

Drinking water from streams and lakes is not advisable when hiking or camping. Surface waters are also used by animals and readily become contaminated. While bacteria and viruses are found in upland streams, the protozoan parasites *Giardia* and *Cryptosporidim* are also commonly recorded and pose a real threat to health. The fact that giardiasis is also known as back-packers disease in the USA reflects the risk from protozoan parasites. They produce cysts and oocysts respectively that can be removed using portable filters developed for the very purpose. Bacteria and viruses can be destroyed by boiling, although cysts and oocysts may resist boiling for up to 10 minutes! An alternative to boiling are iodine or chlorine sterilization tablets. These require very long contact times of up to eight hours to be fully effective with only 90–95% inactivation recorded after 60 minutes. Sterilization tablets are ineffective against protozoan parasites. The most reliable method of treating drinking water in outdoor and emergency situations is the use of ceramic filter elements housed in simple gravityfed units. For example, the Doulton Sterasyl filter candle can remove 100% of cysts and 99.99% of pathogenic bacteria from water.

References

Bell, F. A., Perry, D. L., Smith, J. K. and Lynch, S. C. (1984). Studies on home treatment systems. *Journal of the American Water Works Association,* **76**(4), 126–30.

Dege, N. and Senior, D. A. G. (eds.) (2005). *Technology of Bottled Waters.* Oxford: Blackwell.

Edberg, S. C. (2005). Microbiology of bottled water. In *Technology of Bottled Water,* ed. D. Senior and N. Dege. Oxford: Blackwell Publishing, pp. 388–402.

Environment Agency (2003). *Harvesting Rainwater for Domestic Uses: An Information Guide.* Bristol: Environment Agency.

Finlayson, D. (2005). Market development of bottled waters. In *Technology of Bottled Water,* ed. D. Senior and N. Dege. Oxford: Blackwell Publishing, pp. 6–27.

Geldreich, E. E. and Reasoner, D. J. (1990). Home treatment devices and water quality. In *Drinking Water Microbiology,* ed. G. A. McFeters. New York: Springer-Verlag, pp. 147–67.

Konig, K. (2001). *The Rainwater Technology Handbook: Rain Harvesting in Building.* Dortmund, Germany: Wilo-Brain.

Reasoner, D. J., Blannon, J. C. and Geldreich, E. E. (1987). Microbiological characteristics of third faucet point-of-use devices. *Journal of the Water Works Association,* **79**, 60–6.

USEPA (2006). *Point-of-Use or Point-of-Entry Treatment Options for Small Drinking Water Systems.* EPA: 815-R-06-010. Washington, DC: US Environmental Protection Agency.

Chapter 30
Water security in the twenty-first century

30.1 Introduction

Water supplies are uniquely vulnerable to terrorism, whether it is aimed at humans, livestock or crops. Access to water supplies, and treated water, via service reservoirs and the distribution network places water at particular risk (Denileon, 2001). The key actions that can be taken against supplies are (1) physical damage to water treatment and distribution systems, or the computer systems used to operate them, interrupting the supplies or preventing adequate treatment; (2) deliberate chemical contamination; and (3) bioterrorism using either micro-organisms or biotoxins.

Chemical contaminants are not very effective due to the volume of chemical required and the relative toxicity of most chemicals. In contrast, biological agents have been widely developed for warfare but rarely employed (Hawley and Eitzen, 2001). Although most are airborne many of these organisms and toxins are equally effective via water. The infective dose of the disease agent varies significantly as does the effect on the target organism or population. Microbial agents are infectious, in some cases can be subsequently spread from person to person or via contaminated food, are stable within the environment, colourless and odourless, and have delayed response times unlike chemical contaminants that cause an effect in the target organism relatively quickly.

Biological agents are classed into two categories by the US Center for Disease Control and Prevention (Rotz et al., 2002). Category A micro-organisms posing the highest risk with high morbidity and mortality rates include smallpox, anthrax, plague and botulism, while category B agents pose a much lower risk and rates of mortality and morbidity such as brucellosis, typhus fever and cholera. There is a third group, category C, for emerging biological agents such as hantaviruses and tickborne haemorrhagic fever viruses. A wide range of bacteria, fungi and algae produce toxins, mostly potent neurotoxins, posing a particular risk to water supplies. These include aflatoxins, botulinum toxins, microcystins, ricin and saxitoxin (Section 11.2).

Water treatment and disinfection are the front-line protection for targeted water supplies and most treatment plants employ continuous pollutant and/or toxicity detectors in some form. However, unless an attack is suspected then there is very little that can be done to protect treated waters except higher security and restricted access to the distribution network. Constant surveillance of water quality is imperative but once a threat is suspected then there is now a wide range of rapid response kits that can be employed to detect most chemical, biological or radioactive agents (States *et al.*, 2004). Water utilities should carry out vulnerability assessments, and have in place both a security plan to protect supplies as well as an emergency response plan (ERP) in the event of a terrorist attack. What is clear is that water supplies are a potential terrorist target, making everyone vulnerable to attack within their own homes. So a new level of security is required to ensure the safety of drinking water.

Legislation dealing with water security has been introduced in the USA. The Homeland Security Presidential Directives (HSPDs) and the Public Health Security and Bioterrorism Preparedness and Response Act (Bioterrorism Act) of 2002 require a number of specific actions to be taken in relation to water supplies. These include (1) assessing vulnerabilities of water utilities; (2) developing strategies for responding to and preparing for emergencies and incidents; (3) promoting information exchange among stakeholders; and (4) developing and using technological advances in water security. The US Environmental Protection Agency (USEPA) has been appointed as the agency responsible for identifying, prioritizing and co-ordinating infrastructure protection activities for the nation's drinking water and water treatment systems.

Title IV of the Bioterrorism Act specifically deals with drinking water security and safety. It requires operators of supplies serving more than 3300 persons to carry out vulnerability assessments and to develop ERPs. Vulnerability assessments had to be completed and submitted to the USEPA by 31 March 2004 (systems serving >1 000 000 people), 31 December 2003 (>50 000) or 30 June 2004 (>3300). The ERP must subsequently be agreed within six months of the completion of the vulnerability assessment. Under the Act the USEPA is also required to provide (1) information on potential threats to water systems; (2) strategies for responding to potential incidents; (3) information on protection protocols for vulnerability assessments; and (4) carry out research into water security. Further details can be obtained on-line at http://cfpub.epa.gov/safewater/watersecurity/index.cfm. Although it is not a requirement under the Act, water utilities are strongly encouraged to review and update their vulnerability assessments and emergency response plans at regular intervals. There is a wide range of tools commercially available to help carry out vulnerability assessments and in the formulation of emergency response plans. In other parts of the world, including Europe, water utilities are constantly reviewing the security situation but to date no specific legislation has been introduced. For obvious reasons security plans for individual water supplies or utilities are rarely published.

30.2   Vulnerability assessment

The USEPA defines vulnerability assessment as the susceptibility of a water system to a terrorist attack or other intentional acts intended to substantially disrupt the ability of the system to provide a safe and reliable supply of drinking water or otherwise present public health concerns. Thus drinking water is vulnerable at two levels, from contamination and by the interruption or cessation of supply. The latter is equally serious as the loss of water supply quickly renders manufacturing and business non-operable, fire-fighting and medical care are both severely affected. While agriculture and food production are also severely affected the greatest threat for individuals comes from the potential risk to health.

In the USA, the Bioterrorism Act requires that vulnerability assessments should include, but not necessarily be limited to, water supplies, water collection, pretreatment, treatment, storage and distribution facilities; electronic, computer or other automated systems utilized by the water system; the use, storage, or handling of various chemicals; and the operation and maintenance of such systems. Assessment of vulnerabilities within the system should be formally reviewed annually, although it should be seen as a continuous process, with adjustments made as circumstances alter (USEPA, 2002).

The National Drinking Water Advisory Council, a federal advisory committee that supports the USEPA, has published guidance on the basic elements comprising sound vulnerability assessments; these elements are: (1) character-ization of the water system, including its mission and objectives; (2) identification and prioritization of adverse consequences to avoid; (3) determination of critical assets that might be subject to malevolent acts that could result in undesired consequences; (4) assessment of the likelihood (qualitative probability) of such malevolent acts from adversaries; (5) evaluation of existing countermeasures; and (6) analysis of current risk and development of a prioritized plan for risk reduction (www.epa.gov/safewater/ndwac/council.html). There are now a number of standard approaches to vulnerability assessment, including assessment software such as the Vulnerability Self Assessment Tool (VSAT) developed by the Association of Metropolitan Sewerage Agencies (AMSA) (www.vsatusers.net/).

30.3   Developing security plans

Water safety plans have been used for many years to improve water quality control strategies, in conjunction with excreta disposal and personal hygiene, to deliver sustainable health gains within the population (WHO, 2004) (Section 2.4). These plans use a combination of risk assessment and risk management techniques such as a multi-barrier approach to pathogen control and hazard analysis critical control point (HACCP) principles that are employed primarily by the food industry (Rasco and Bledsoe, 2005). While water safety plans were principally introduced to help achieve health-based targets in developing countries, they equally apply to good water management practice and quality assurance systems (e.g. ISO 9001:2000) already used in developed counties (DWI, 2005). The underlying principles used in the development of water safety plans are very similar to those employed in water security plans.

The National Drinking Water Advisory Council (NDWAC) established a Water Security Working Group (WSWG) in the autumn of 2003 to identify key actions that water utilities could take to maximize security (NDWAC, 2005). Fourteen such actions were identified to ensure continuity and safety of supplies and these are used to formulate water security plans. Similar to normal Environmental Management Systems many of the recommended actions are consistent with procedures already necessary to maintain the technical, managerial and operational requirements to produce water of drinking quality standard and to maintain adequate supplies to householders during operational problems. The actions are broken down individually below using four functional categories i.e. organizational, operational, infrastructure, and external.

30.3.1 Organizational

Action 1: Make an explicit and visible commitment of the senior leadership to security. This includes incorporating security into mission statements; raising awareness of security throughout the organization; making the commitment to security visible to all employees and customers; developing a strong culture and awareness of security within the organization.

Action 2: Promote security awareness throughout the organization. Developing security policies, inclusion on agenda for all meetings; feedback and suggestions from employees and customers; establishing performance standards for security; inclusion of security into job descriptions; appropriate employee training, making security awareness a routine part of day-to-day operations.

Action 5: Identify managers and employees who are responsible for security and establish security expectations for all staff. Appointment of trained personnel to lead and develop security implementation; clear allocation of responsibilities for security.

30.3.2 Operational

Action 3: Assess vulnerabilities and periodically review and update vulnerability assessments to reflect changes in potential threats and vulnerabilities. Security should be continually reviewed and adjusted to take into account changing circumstances within and outside the organization.

Action 4: Identify security priorities and, on an annual basis, identify the resources dedicated to security programs and planned security improvements, if any. Adequate resources both in terms of investment and personnel are required; security should be part of the normal budgeting procedure.

Action 7: Employ protocols for detection of contamination consistent with the recognized limitations in current contaminant detection, monitoring, and surveillance technology. Use of on-line contaminant monitoring and surveillance systems; monitoring of known and surrogate physical and chemical contaminants, and pressure change abnormalities.

Action 10: Monitor available threat-level information and escalate security procedures in response to relevant threats. Close involvement with law enforcement agencies; development of information networks; adequate response procedures to increased levels of risk from terrorism or other sources of potential threat to quality and continuation of supply.

Action 11: Incorporate security considerations into emergency response and recovery plans, test and review plans regularly, and update plans to reflect changes in potential threats, physical infrastructure, utility operations, critical interdependencies, and response protocols in partner organizations. Co-operation with emergency services and other emergency response organizations; integration with other local and national emergency response and recovery plans; arrangement for exchanging personnel and physical assets with other agencies in the event of an emergency or disaster that disrupts operation.

Action 14: Develop utility-specific measures of security activities and achievements, and self assess against these measures to understand and document program progress. Monitoring protocols to assess: policies, procedures, training, and surveillance; assess implementation of schedules and plans; review procedure.

30.3.3 Infrastructural

Action 6: Establish physical and procedural controls to restrict access to utility infrastructure to only those conducting authorized, official business and to detect unauthorized physical intrusions. Includes all areas from source water to distribution and through collection and wastewater treatment; physical access controls (e.g. fencing critical areas, locking gates and doors, installation of barriers at site access points); procedural access controls (e.g. changing access codes regularly, security passes for all employees and contactors); alarms (e.g. motion detectors and intruder alarms); security personnel, including liaison with local residents, police etc.

Action 8: Define security-sensitive information; establish physical, electronic, and procedural controls to restrict access to security-sensitive information; detect unauthorized access; and ensure information and communications systems will function during emergency response and recovery. Increased reliance on computer systems and software to operate water treatment and distribution systems, as well as wastewater treatment plants, makes them particularly vulnerable to potential attack. Physical steps include restricting the number of individuals with authorized access to computer systems; preventing access by unauthorized individuals; procedural steps include restricting remote access, safeguarding critical data through back-ups and storage in safe places; restrict security-sensitive information (e.g. maps and blueprints, operational and technical details, details of storage of hazardous chemicals, details of security plans).

Action 9: Incorporate security considerations into decisions about acquisition, repair, major maintenance, and replacement of physical infrastructure; include consideration of opportunities to reduce risk through physical hardening and adoption of inherently lower-risk design and technology options. Inclusion of security should be an important consideration at the design and construction phase of new facilities and the upgrading, repair or maintenance of existing facilities; consideration as to how new security equipment and procedures will affect the day-to-day operation of plants and the safety of workers.

30.3.4 External

Action 12: Develop and implement strategies for regular, ongoing security-related communications with employees, response organizations, rate setting organizations, and customers. Development of effective communication strategies; communication routes, especially during an emergency; also ensuring customer and employee confidence in security plan.

Action 13: Forge reliable and collaborative partnerships with the communities served, managers of critical interdependent infrastructure, response organizations, and other local utilities. Development of collaborative arrangements with agencies and other utilities to ensure rapid and appropriate co-ordinated response to emergencies; clear delineation of roles and responsibilities during the response to and recovery from an emergency; regular collaboration with key staff in other organizations; sharing response and security information; involvement of local community with security (e.g. identification of suspicious behavior or changes in operational practice at sites).

An example of how each of these actions are assessed is given in Table 30.1.

30.4 The emergency response plan

Water emergencies are not only associated with terrorist attack, they also include other man-made and natural events that result in a major disruption of supply. These include power failure, floods, earthquakes, pollution episodes, failure of the distribution system due to trunk main fracture, failure of the prime water supply source, water treatment plant failure, strikes by staff, shortage of

Table 30.1 *Examples of assessment procedures for a water treatment utility security programme. Reproduced from NDWAC (2005) with permission from the US Environmental Protection Agency*

Action	Summary of action	Potential assessment procedure
1	Explicit commitment to security	Does a written, enterprise-wide security policy exist, and is the policy reviewed regularly and updated as needed?
2	Promote security awareness	Are incidents reported in a timely way, and are lessons learned from incident responses reviewed and, as appropriate, incorporated into future utility security efforts?
3	Vulnerability Assessment up to date	Are reassessments of vulnerabilities made after incidents, and are lessons learned and other relevant information incorporated into security practices?
4	Security resources and implementation priorities	Are security priorities clearly identified, and to what extent do security priorities have resources assigned to them?
5	Defined security roles and employee expectations	Are managers and employees who are responsible for security identified?
6	Intrusion detection and access control	To what extent are methods to control access to sensitive assets in place?
7	Contamination detection	Is there a protocol or procedure in place to identify and respond to suspected contamination events?
8	Information protection and continuity	Is there a procedure to identify and control security-sensitive information, is information correctly categorized, and how do control measures perform under testing?
9	Design and construction standards	Are security considerations incorporated into internal utility design and construction standards for new facilities and/or infrastructure and major maintenance projects?
10	Threat-level based protocols	Is there a protocol/procedure of responses that will be made if threat levels change?
11	Emergency Response Plan tested and up to date	Do exercises address the full range of threats (i.e. physical, cyber, and contamination) and is there a protocol or procedure to incorporate lessons learned from exercises and actual responses into updates to emergency response and recovery plans?
12	Communications	Is there a mechanism for utility employees, partners, and the community to notify the utility of suspicious occurrences and other security concerns?
13	Partnerships	Have reliable and collaborative partnerships with customers, managers of independent interrelated infrastructure, and response organizations been established?
14	Utility-specific measures and self-assessment	Does the utility perform self-assessment at least annually?

chemicals and chemical accidents. Minor operational problems are normally excluded. Therefore, in the event of an emergency each water supply system must have a contingency plan in place to protect public health by maintaining a water supply sufficient for potable use and fire-fighting.

The USEPA have prepared guidance for the formulation of an ERP based on eight core elements. These are:

1. *System-specific information*: Population served and number of service connections; key information about critical system components, including plans and engineering drawings, of the source water, treatment plants, water and chemical storage, water supply zones, distribution system; how to isolate parts of the system.
2. *Roles and responsibilities*. Appointment of an emergency response leader and deputy; definition of command structure; allocation of roles and responsibilities.
3. *Communication procedures*. Internal list of emergency response and operational personnel; external list of first response teams, agencies, and customers; when and how to communicate to the public and media.
4. *Personnel safety*. Training in the effective use of safety and emergency equipment; evacuation procedures; first aid.
5. *Identification of alternative water sources*. Identify short- and long-term alternative water supplies; agreement of partners for supply of bulk tankers and temporary emergency supplies.
6. *Replacement equipment and chemical supplies*. Inventory of spares and suppliers; store of key machinery replacement parts and chemicals; protocol for exchange of such items, including generators and pumps, with other water utilities.
7. *Property protection*. Access controls; protection of perimeters; lock down procedures; protection of equipment; and finally;
8. *Water sampling, analysis and subsequent monitoring*. Emergency protocol for the identification of chemical problem or biohazard, quantification of limits of problems (spatial and temporal), identification of laboratories for emergency analysis, confirmation of safety of supply.

Full details of how to develop ERPs are given in USEPA (2003, 2004a).

A critical element in emergency response planning is deciding when to implement the ERP and what elements to activate when a terrorist attack is only suspected. In the USA the Department of Homeland Security have introduced a tiered notification system to indicate the nature and degree of a terrorist threat. Known as the Homeland Security Advisory System, it sets potential threat levels or conditions that are colour-coded. Condition Green is declared when there is a low risk from terrorist attack, rising through Guarded Condition Blue (general risk), Elevated Condition Yellow (significant risk), High Condition Orange (high risk), and finally Severe Condition Red (severe risk). Using this scheme, the USEPA has prepared specific guidance for water utilities that includes detection, preparedness, prevention and finally protection advice. The response levels are additive, so as the threat increases the extra levels are added

to the actions already required at lower levels. This guidance document is
given in USEPA (2004b) as appendix B 'guarding against terrorist and security
threats, suggested measures for drinking water and wastewater utilities (water
utilities)' and can be downloaded at www.rpa.gov.safewater/watersecurity/pubs/
small_medium_ERP_guidance040704.pdf.

For obvious reasons an ERP is a highly sensitive document that must be kept
highly confidential. However, other agencies who are involved in the planned
security response must be fully briefed and advised of their exact roles and
responsibilities, even though the exact details of the vulnerability assessment
and ERP are kept secret. However, details of the ERP must be disseminated
widely enough within the organization to ensure that it will still operate even
though one or more critical elements (e.g. the utilities headquarters), have been
rendered inoperative. Once the emergency response has been carried out then
detailed recovery plans will be needed to ensure that normal supplies are
resumed as soon as possible. Emergency response must be practised and tested
regularly and both the ERP and recovery plans reviewed annually.

Practical details of how to respond to a drinking water contamination threat
or accident is given in USEPA (2004b) and can be accessed over the Internet
at www.waterisac.org/epa/Guidance/rptp_response_guidelines.pdf. The Water
Information Sharing and Analysis Center (WaterISAC) has developed a useful
on-line library of emergency response documents (www.waterisac.org/).

30.5 Conclusions

With the increased number of terrorist attacks over the past decade, and in
particular since 11 September 2001, there has been a growing awareness of the
vulnerabilities of countries to terrorist attack. Both food and water were quickly
identified as major targets for terrorists, with areas such as milk storage and
collection, and drinking water treatment and supply seen as high-risk areas for
possible attack. The problem for both farmers and water supply utilities is that in
the past there has been little or no security of the product, with numerous points
along the production and supply chain where it could easily be tampered with.
From the consumers' perspective, the meal or drink that they prepare or
purchase is assumed to be contaminant free. Yet protecting food products,
especially those that are not highly processed such as water, that are consumed
directly, or where there is easy access to the bulk product after treatment and
testing, requires a new approach to quality control and security.

The introduction of water safety plans is an effective way of ensuring drinking
water is safe for human consumption. Such plans are based on risk assessment
and risk management of the entire supply chain from catchment to consumer.
It does this through good management practices to minimize contamination of
water resources, removing contaminants through effective and appropriate
treatment, and preventing further contamination from the distribution network,

and, where possible, from domestic plumbing (DWI, 2005). However water safety plans are unable to protect supplies from deliberate terrorist attack or unforeseen catastrophic natural events. Water security aims to (1) assess and reduce vulnerabilities of the drinking water supply chain to potential terrorist attacks; (2) plan for and practise response to identified emergencies and incidents; and finally (3) to develop new security technologies to detect and monitor contaminants and prevent security breaches.

The introduction of the Public Health Security and Bioterrorism Preparedness and Response Act (Bioterrorism Act) of 2002 by the US Government has ensured that the awareness of the vulnerability of water supplies to terrorist or other malicious attack has been universally recognized. The importance of the catchment and of controlling the entire water supply and wastewater treatment cycles within it has been widely accepted, making water security a potentially enormous task. Water resources are equally vulnerable from attacks on wastewater treatment facilities, especially as water reuse and recycling is increasingly used to deal with water scarcity. By adopting the 14 key actions outlined by the Water Security Working Group of the NDWAC, water and wastewater utilities can significantly reduce risk to public health from terrorist attacks and natural disasters.

To carry out full vulnerability assessments, develop security and ERPs is expensive. Implementing key actions identified within security plans may, for smaller utilities, be prohibitively expensive. However, even with limited financial resources improvements in security can be made. For example, by simply increasing organizational awareness of the need for security will itself reduce vulnerability and the resulting increased alertness amongst staff and customers deter possible attack. While security must now become part of the organizational culture of all water utilities, higher security can also lead to greater efficiency and other cost benefits. Benefits include reducing operating costs and chemical usage, re-evaluation of the suitability of treatment processes and the adoption of newer and more versatile technologies such as membrane filtration, all leading to better water quality and reliability of service. The need to include all stakeholders in security will also have positive effects on how water is perceived and used by customers at a time of dwindling resources.

References

Denileon, G. P. (2001). The who, what, why and how of counter terrorism. *Journal of the American Water Works Association*, **93**(5), 78–85.

DWI (2005). *A Brief Guide to Drinking Water Safety Plans*. London: Drinking Water Inspectorate.

Hawley, R. L. and Eitzen, E. M. (2001). Biological weapons – a primer for microbiologists. *Annual Review of Microbiology*, **55**, 235–53.

NDWAC (2005). *Recommendations of the National Drinking Water Advisory Council to the US Environmental Protection Agency on Water Security Practices, Incentives and*

Measures. Washington, DC: Water Security Working Group, The National Drinking Water Advisory Council.

Rasco, B. A. and Bledsoe, G. E. (2005). *Bioterrorism and Food Safety*. Boca Raton, FL: CRC Press.

Rotz, L. D., Khan, A. S., Lillibridge, S. R., Ostroff, S. M. and Hughes, J. M. (2002). Public health assessment of potential biological terrorism agents. *Emerging Infectious Diseases*, **8**(8), 225–9.

States, S. J., Newberry, J., Wichterman, J. *et al.* (2004). Rapid analytical techniques for drinking water security investigations. *Journal of the American Water Works Association*, **96**(1), 52–64.

USEPA (2002). *Vulnerability Assessment Factsheet*. EPA Report 816-F-02-025. Washington, DC: Office of Water, US Environmental Protection Agency.

USEPA (2003). *Large Water System Emergency Response Plan Outline: Guidance to Assist Community Water Systems in Complying with the Public Health Security and Bioterrorism Preparedness and Response Act of 2002*. EPA Report 810-F-03-007. Washington, DC: Office of Water, US Environmental Protection Agency.

USEPA (2004a). *Emergency Response Plan Guidance for Small and Medium Community Water Systems: to Comply with the Public Health Security and Bioterrorism Preparedness and Response Act of 2002*. EPA Report 816-R-04-002. Washington, DC: Office of Water, US Environmental Protection Agency.

USEPA (2004b). *Response Protocol Tools: Planning for and Responding to Drinking Water Contamination Threats and Incidents*. EPA Report 817-D-04-001. Washington, DC: Office of Water, US Environmental Protection Agency.

WHO (2004). *Drinking Water Guidelines*. Geneva: World Health Organization.

Chapter 31
Final analysis

31.1 Introduction

Under normal circumstances each one of us requires between 1.8 and 2.0 litres of water each day in order to maintain a healthy body. In practice people drink very little water on its own; rather they drink a wide variety of beverages made from water. Tea, for example, contains very high levels of aluminium, fluoride and polycyclic aromatic hydrocarbons (PAHs) as well as elevated levels of many other metals and compounds (Flaten and Ødegård, 1988; Lin *et al.*, 2005; Cao *et al.*, 2006). Canned carbonated drinks can have very high levels of aluminium as well as other compounds and metals (Table 15.2). So our exposure to contaminants is more often related to what we drink and eat rather than drinking water per se.

However, in recent years there has been an enormous upsurge of interest in drinking water, which has been reflected in record sales of bottled water (Table 29.1) and the use of water coolers. While bottled water has become fashionable, the public's perception of tap water remains negative (Section 31.3). Yet in reality tap water in the majority of developed countries is of a higher quality than ever before, and equal in quality to bottled waters. So should consumers be concerned over drinking water quality?

Like other food manufacturers, the water utilities have a unique relationship with their consumers who trust them to provide food and water that is completely safe to consume. However, problems can and do arise.

Problems can be categorized into three areas. (1) Traditional problems that water treatment normally effectively deals with, such as infectious diseases associated with waterborne pathogens. (2) Accidental problems that result from poor operational practice or accidents such as elevated levels of aluminium in water, the breakthrough of protozoan oocycts (e.g. *Cryptosporidium*) or spillages of contaminants into water resources that are not detected early enough. (3) Emerging problems such as newly detected compounds or pathogens in water, including pathogens that have not been identified previously as infectious via water (WHO, 2003) or new compounds found in water usually due to better analytical technology or their increase use within the catchment (e.g. endocrine-disrupting compounds and trace organics) (Table 3.2). Two emerging problems that are currently causing significant concern is the increased risk from deliberate contamination that may be associated with terrorism (Chapter 30), and the problem of sustaining water supplies into the future without causing significant ecological damage.

Water utilities normally have a monopoly in terms of providing water, and as domestic consumers are not able to switch supplier, many have responded by treating the water further using point-of-use or point-of-entry systems, supplementing water by using bottled water, or by avoiding drinking water altogether. There remains a significant gulf between expectation and reality for many consumers even though suppliers in developed countries have by and large reached a peak in terms of quality and quantity of the water that they can supply. Changing the public's perception of water quality is a major challenge for suppliers.

This book has focussed on water quality problems, but the most serious scenario is having no water supply at all. Therefore in the dual issues of sensible use of supplies and increased water security are the joint responsibilities of both the supplier and the consumer. So vigilance, dedication and continuing investment is needed from suppliers. Likewise consumers need to realize that as stakeholders they must take a positive interest in water supply and be realistic about their expectations in relation to water quality and use.

31.2 Complaints

Problems can arise at source through natural or man-made contamination, from chemicals added during water purification, during distribution and, of course, within the consumer's own home. For example, discoloured or dirty water could be caused by corrosion of the iron mains, poor or insufficient treatment, microbiological growths in the mains or even replacement of the natural supply with one of a different composition. It may even be a combination of problems.

In the UK, wholesomeness in relation to drinking water is legally defined in the Water Supply (Water Quality) Regulations 2000 and requires water to be not only safe to drink but also aesthetically acceptable to consumers in terms of taste, odour, colour and clarity (Section 21.4). In fact, consumers tend to assess quality subjectively using these four parameters, and although water may be absolutely safe to drink, if it is cloudy or turbid the consumer will assume that it is in fact contaminated and unfit to drink. A good example is milky water. The water is over-aerated as it leaves the tap, usually due to high pressure in the mains, and takes on a milky appearance due to the presence of millions of tiny bubbles. These gradually rise to the surface and the water becomes totally clear and wholesome.

Consumer complaints have traditionally fallen into a number of categories: discoloured water, with or without particulate matter; particulate matter such as rust, grit or sediment; staining of laundry; staining of bathroom fittings; cloudy, milky or chalky water; tastes and odours; the presence of animals such as worms or insects; scale deposits within domestic plumbing; alleged illness; noise; corrosion; and low pressure. Water utilities are now getting more specific complaints, for example, about aluminium, lead, pesticides and nitrate. Consumers are concerned about the effects of specific contaminants on their children, or their unborn child when pregnant. However, these are more concerns than complaints, as the aesthetic quality of the water, the only thing that consumers can generally evaluate, is generally unaffected. This supports the findings of customer satisfaction surveys conducted by the Office of Water Services in the UK, now the Water Services Regulation Authority (Ofwat), which found that such complaints are often the result of media attention (Section 31.3). However, the Internet has given enormous instant access to water quality data published by the water utilities themselves, and also consumers are now able to purchase a range of test kits for a wide range of possible contaminants. So consumer complaints are becoming more technical and often very challenging for suppliers.

In most countries water utilities believe that if consumers are unhappy about the quality of their water then they should complain. Only in this way are they usually aware that problems actually exist and so can try to rectify them, especially as now in the UK it is a criminal offence to supply water that is unfit for human consumption. As water utilities often carry a legal liability to fulfil their obligations to consumers, if consumers are unhappy, they want to know about it. Unfortunately this is not the case elsewhere, and in many countries complaints about water quality are either ignored or treated in a hostile manner. In order to achieve and maintain good-quality water there needs to be a realistic relationship between the supplier and the consumer. Suppliers must realize that consumers are literally risking their lives in drinking the water that is being supplied to them. Therefore water should be subject to the same level of regulation and scrutiny as any other product, such as food or pharmaceuticals, that is ingested. There is a unique expectation and trust on behalf of the consumer, one which suppliers generally take very seriously, so any feedback about quality is important. Likewise, consumers have an obligation to inform and educate themselves about their water, including about how it is affected by their own plumbing system and use.

31.3 Public perception of drinking water quality

Over the past decade increased public awareness of environmental issues has caused a much wider spectrum of consumers to be concerned about the quality of drinking water. This is reflected by a large increase in the sales of bottled water and home treatment systems; and the general acknowledgement that water should be considered more as a food item. However, owing to a lack of knowledge, consumers will generally interpret their risk and safety in an emotional rather than a scientific way. This has led to considerable pressure being put on the water undertakers to implement often unnecessarily stringent treatment and control measures to reduce the concentrations of particular parameters. Most water companies have an action plan of remedial measures to improve water quality based on appropriate risk–benefit and cost–benefit analysis. However, pressure from consumers, politicians and, of course, the media can result in this structured approach being set aside and inappropriate action being taken, often at considerable cost.

One of the major challenges facing the water industry is to understand how consumers perceive water quality, why they are so unhappy about drinking tap water when clearly for the vast majority the quality is excellent and, finally, educating the consumer about water quality and the regulatory functions of water suppliers. To answer some of these questions Ofwat commissions periodic surveys of consumer attitudes.

The first study, conducted by MORI in 1992 , asked consumers if they thought that their tap water was safe to drink. Over two-thirds were satisfied, whereas 20% expressed concern. Surprisingly, only 16% responded as being very satisfied, so most consumers had some doubts about the safety of their drinking water. Asked about the reason for their dissatisfaction, 49% referred to the taste or smell and the presence of chemicals, 38% indicated that their worries were related to stories in the media, whereas 29% referred to the visual appearance of the water. The results showed that the consumers' view of safety was based heavily on the smell of the water. Seventy per cent of consumers found the taste of their tap water to be acceptable, whereas 20% found it unacceptable. This was due to chlorine (25%), unspecified chemicals (17%), a general dislike of tap water (15%) or a metallic taste (9%). The taste of water was the strongest motivation in the purchase of bottled water at that time. The detection of chlorine in drinking water appears to be a major source of complaint with consumers. Ninety per cent thought it either essential or very important that drinking water should be clear and colourless. Only 18% of consumers were dissatisfied with their water in this respect, reporting occasional cloudiness, 'bits' or colour. A good indicator of true consumer dissatisfaction is the use of alternative water supplies or the use of home treatment systems. The survey showed that 39% of consumers did take some personal action to improve quality. Fourteen per cent of consumers boiled their water before drinking, 21% purchased bottled water and 11% used home treatment devices (9% jug filters and 2% in-line systems). The reasons for boiling water or the use of home treatment systems were primarily concerns over the safety of their supplies, although an improvement in taste was also an important factor. The reasons for purchasing bottled water were more complex. Most used bottled water due to taste considerations (74%), although 37% stressed safety reasons. The survey identified a strong belief that bottled water was purer and safer to drink.

Surveys have been carried out every three years since this first major study, and the satisfaction levels and trends have remained more or less constant with summaries of the surveys posted on the Drinking Water Inspectorate (DWI) website (www.dwi.co.uk). The more recent surveys have shown that consumers who are dissatisfied with their drinking water fall into two categories. The major concern is related to the physical properties or attributes of the water, such as taste, appearance, smell, hardness, freshness and temperature. Many in this group clearly felt that these physical problems were an indication of a more serious underlying problem and so attempted to improve the quality by further treatment. The remainder simply refused to drink something that they felt did not taste or look right. While 86% of those surveyed in 2000 did drink tap water, 6% only drank bottled water and 6% were drinking alternatives such as milk and soft drinks (DWI, 2001). The second group had concerns about either the composition and/or the provenance of their water leading to underlying worries about its quality. They had concerns such as the water may have been contaminated, been recycled, or have had undesirable chemicals added during treatment. However, the 2000 survey showed that while such concerns stopped consumers drinking the water directly from the tap, many were still happy to use it in other ways, such as for cooking, making tea and coffee, cleaning their teeth, and even, in some cases, using it to dilute squash for their children. While such consumer concerns and responses may appear difficult to understand to the water engineer, they do demonstrate real anxieties by consumers, often involving considerable expense in their efforts to overcome them. Ofwat have reported that one-third of all consumers, when prompted, think there is a danger to health from things in tap water. Chemicals and bacteria or bugs were identified to be the main concerns.

The Ofwat surveys are unique in that they show that the public's perception of drinking water is largely based on the physical characteristics of their own water supply, rather than its quality in relation to the prescribed standards. The lack of knowledge about water supply and quality is highlighted, as is the desire to know more. The media is also a significant influence on attitudes towards water and the assessment of risk. With up to 60% of quality problems in some areas related to householders' own plumbing systems, it is clear that companies will have to look carefully at the education of consumers if they are to overcome some of the basic aesthetic-based problems highlighted by the surveys.

31.4 The quality of drinking water

The overall quality in drinking water in England and Wales has probably reached the best that it is operationally possible to achieve when supplying water to >53 million people living in 2249 water supply zones. This is the same for most supplies in the USA, Canada and Australia.

In the UK there has been a steady increase in quality since the 1980s although the rate of increase has steadily declined to the current (2005) compliance level of 99.96% (Table 31.1). In 2003, a total of 2 045 473 microbial and chemical tests, for the 55 individual parameters that have a numerical standard, were carried out

Table 31.1 *Overall compliance in water supply zones in England and Wales. (Reproduced from DWI (2006) with permission from the Drinking Water Inspectorate.) New drinking water standards were introduced in 2004 to comply with the improved standards in the Drinking Water Directive (Table 2.7)*

Year	%
1993	98.62
1994	99.05
1995	99.29
1996	99.63
1997	99.71
1998	99.73
1999	99.78
2000	99.81
2001	99.83
2002	99.85
2003	99.88
2004	99.92
2005	99.96
2006	99.96

Figure 31.1 Percentage contraventions of standards by parameter in water supply zones in England and Wales during 2003. Reproduced from DWI (2004) with permission from the Drinking Water Inspectorate.

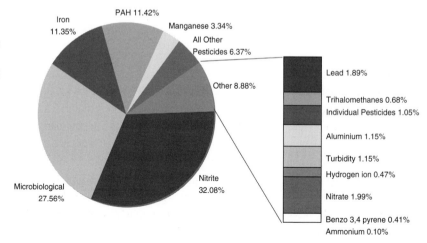

Table 31.2 *Water quality compliance in water supply zones in England and Wales 2001 to 2003 prior to the introduction of the revised drinking water standards (Table 2.7) in 2004. Reproduced from DWI (2004) with permission from the Drinking Water Inspectorate*

Parameter	Total number of tests taken in 2003	Tests not meeting the prescribed concentration or value		Numbers of zones not complying with the standards (numbers in 2003 = 2249)[a]		
		Number	%	2003	2002	2001
Coliforms	148 508	770	0.52	8	12	5
Faecal coliforms	148 523	46	0.03	44	59	60
Colour	42 144	0	0.00	0	2	2
Turbidity	58 245	34	0.06	31	23	42
Odour	14 910	8	0.06	8	6	5
Taste	14 697	5	0.04	5	3	2
Hydrogen ion	54 738	14	0.03	13	10	8
Nitrate	36 463	59	0.16	20	15	28
Nitrite	34 074	950	2.79	141	131	153
Aluminium	38 085	34	0.09	33	23	27
Iron	59 734	336	0.56	259	308	349
Manganese	42 267	99	0.23	78	84	94
Lead	23 192	56	0.24	50	80	106
PAH	11 081	338	3.05	200	185	169
Trihalomethanes	11 644	20	0.17	9	18	22
Total pesticides	35 007	0	0.00	0	6	4
Individual pesticides	808 154	31	<0.01	26	59	71
All others	464 007	159	0.03	100	30	50
TOTAL	**2 045 473**	**2959**	**0.15**	**1025**	**1054**	**1197**

[a] 2289 zones in 2002; 2305 zones in 2001

with only 2959 (0.15%) failing the EC limits. This compares with 1.38% of the 2 529 485 samples failing in 1993, a 90% reduction over that period (DWI, 2003, 2006).

The key problems in relation to compliance in England and Wales during 2003 are nitrite and nitrate (32% of total quality failures), coliforms (28%), iron (11%), PAHs (11%), pesticides (6%), manganese (3%) and lead (2%) (Figure 31.1). The number of tests and zones failing the EC parameter standards are summarized in Table 31.2. Where non-compliance is reported, the Drinking Water Inspectorate investigates and, where necessary, requires the companies to give undertakings to carry out improvements in the distribution system or at treatment works to remedy the problems (Section 1.4). The concept of using a single index to assess water quality is quite new. The DWI uses two such indices, the overall quality index (OQI) and the operational performance index (OPI) to compare

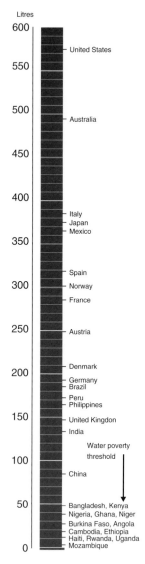

Figure 31.2 Comparison of the average volume of water used per person per day in 2002, and those below the water poverty threshold of $50 \, \text{l ca}^{-1} \text{d}^{-1}$.

overall performance trends within companies over time. The OQI is calculated by companies by averaging the mean zonal percentage compliance for 17 key parameters, giving each parameter an equal weighting, and comparing that with the average for England and Wales. The key parameters are: aluminium, coliforms, colour, faecal coliforms, pH, iron, lead, manganese, nitrate, nitrite, odour, PAHs, taste, total pesticides, individual pesticides, trihalomethanes and turbidity. The OPI is used to measure the operational performance of treatment works and distribution systems. This is calculated by averaging the mean zonal percentage compliance for six parameters: iron, manganese, aluminium, turbidity, faecal coliforms and trihalomethanes.

31.5 Water usage and conservation

Each one of us uses on average between 150 and 580 litres each day, with the UK having one of the lowest per capita water consumption rates in the developed world at 150 litres, although less than 5% of this will be consumed or used for cooking (Figure 31.2). With leakage rates as high as 40% in some distribution systems, it seems strange that we have to treat all this water when so little needs to be of the very best quality. How has this come about?

Early settlements were always close to a source of water, but as society developed, centres of populations rapidly grew so that local water resources became exhausted or polluted. Distribution systems were developed to bring water from rural areas to towns and cities, reservoirs were constructed to ensure sufficient supplies and as the quality of water resources deteriorated so treatment was required. As technology has developed so our per capita water usage has grown requiring more and more treated water to satisfy an ever-increasing demand. So our current water supply system has slowly evolved with a continuous massive investment over the centuries. To alter it to a new radical system, using for example a dual supply system so that only a small fraction of water needs to be fully treated, is simply not feasible financially. Yet we seem to be in a spiral of increasing demands, dwindling supplies, exacerbated in some areas by climate change, and increasing quality problems that require ever more expensive technology to solve. The answer in part is our water usage.

Leaks have always been a factor in supply and as the distribution system ages then increasing losses are inevitable. Huge improvements and investment have been and continue to be made, and as the housing stock continues to rise, so the problems of ever increasing the distribution network to supply these new homes creates new strain on supplies. In a period of less than 30 years we have changed from a society that bathed and washed our clothes once a week and washed the dishes by hand, to one where people shower several times a day, use the washing machine several times a day, often without a full load, and use a dishwasher as standard. Water privatization has in part encouraged this increase in water use, as they are in the business of selling water and in some cases in treating your wastewater as well. So the more you use the more money they make. Of course this depends on whether you are metered or pay a flat charge. The problem now is that while water is a renewable resource, and in theory is still plentiful, in practice there are areas of every country where supplies are inadequate resulting in the choice of reducing water usage or finding more (Section 1.5).

31.6   Water conflict

Water is becoming increasingly scarce not only in poorer arid countries but throughout the developing and developed world. This has led to increasing interest in the link between water resources and international security and conflict. Peter Gleick of the Pacific Institute has made this his particular study and has produced a definitive list of conflicts involving water supply (www.worldwater. org/chronology.html) (Gleick, 2006). He has categorized such conflicts and those involved under six areas: (1) *Control of Water Resources* – water supplies or access to water are at the root of tensions (state and non-state actors); (2) *Military Tool* – water resources or water supply systems are used by a nation or state as a weapon during a military action (state actors); (3) *Political Tool* – water resources or water supply systems are used by a nation, state, or non-state actor for a political goal (state and non-state actors); (4) *Terrorism* – water resources or water supply systems, are either targets or tools of violence or coercion by non-state actors (non-state actors); (5) *Military Target* – water supply systems are targets of military actions by nations or states (state actors); (6) *Development Disputes* – water resources or water supply systems are a major source of contention and dispute in the context of economic and social development (state and non-state actors). Many of these conflicts have been explored by Ward (2002). Pearce (2006) examines the growing global water crisis in his book *When the Rivers Run Dry* and calls for a new water ethic where the water cycle is managed for the benefit of all.

31.7   Drinking water in developing countries

31.7.1   Water scarcity

A sufficient quantity of clean water is the prerequisite to good health and without it humans become susceptible to a surprising wide range of diseases (Tables 3.3 and 31.3) and health-related problems. Access to adequate and safe drinking water should be a basic human right, yet today there are 1.1 billion people globally who do not have access to sufficient safe drinking water. Many of these are managing on as little as five litres a day for all their drinking, washing and cooking needs. Due to a range of factors such as conflict, over-abstraction and climate change, this number is set to double over the next decade. Drinking water quality, especially in terms of pathogens, cannot be isolated from sanitation, with

Table 31.3 *Estimated global morbidity and mortality of water-related diseases during the early 1990s. Reproduced from WHO (1995) with permission from the World Health Organization*

Disease	Morbidity (episodes/year or [a]people infected)	Mortality (deaths/year)
Diarrhoeal diseases	1 000 000 000	3 300 000
Intestinal helminths	1 500 000 000[a]	100 000
Schistosomiasis	200 000 000[a]	200 000
Dracunculiasis	150 000 (in 1996)	–
Trachoma	150 000 000 (active cases)	–
Malaria	400 000 000	1 500 000
Dengue fever	1 750 000	20 000
Poliomyelitis	114 000	–
Trypanosomiasis	275 000	130 000
Bancroftian filariasis	72 800 000[a]	–
Onchocerciasis	17 700 000[a]; (270 000 blind)	40 000 (mortality caused by blindness)

a total of 2.6 billion people globally currently lacking adequate sanitation facilities. The various health problems created by the lack of access to clean drinking water and proper sanitation is having a daily impact on 50% of the population of developing countries (UNDP, 2006). Figure 31.2 compares the average daily volume of water used per capita in a variety of countries with people in the USA and Australia using up to 40 to 60 times more than people in some water scarce areas. The minimum requirement for water has been estimated as 50 litres per capita per day, the so-called water poverty level. This includes 5 litres for drinking, 20 litres for sanitation, 15 litres for bathing and 10 litres for food preparation.

The amount of water a person has access to should in theory be based on the amount of water potentially available. It is generally accepted by hydrologists that the threshold between a country having adequate water resources and not is 1700 cubic metres per capita per year ($m^3 \ ca^{-1} \ yr^{-1}$) i.e. the water stress threshold. In practice very few countries use that much water. For example in Europe the average per capita usage is $726 \ m^3 \ ca^{-1} \ yr^{-1}$, although in North America it is double this at $1693 \ m^3 \ ca^{-1} \ yr^{-1}$ but they have huge reserves of water to meet this excessive demand.

In many areas there is direct conflict between water users, with the bulk of water resources currently tapped for irrigation. For example, in 1990, water abstracted for domestic purposes in Pakistan was estimated at $26 \ m^3 \ ca^{-1} \ yr^{-1}$ compared to $1226 \ m^3 \ ca^{-1} \ yr^{-1}$ for irrigation. The water table is also being adversely affected in many regions by over-abstraction for irrigation with levels falling by as much as 3–4 m yr^{-1} in important agricultural areas such as the Punjab (India) and the north China Plain. Poorer farmers and communities are unable to drill new deeper wells as groundwater levels continue to fall. How this affects arsenic and fluoride levels is unknown, but what is clear is that the poorer sectors of society are not being protected.

The United Nations Development Programme (UNDP) recently published the Human Development Report *Beyond Scarcity: Power, Poverty and the Global Water Crisis* (UNDP, 2006). The report stresses that while there is sufficient water globally to currently meet the needs for human consumption, agriculture and industry, some 700 million people are living below the water stress threshold. The number of people falling below this minimum threshold is estimated to increase to 3 billion by 2025. The report concludes that shortages are generally due to poor policy decisions rather than absolute scarcity, resulting in resources coming increasingly under threat and over-exploited. Two specific factors that are highlighted are under-pricing of water, leading to overuse, and the effects of climate change on resources. In order to resolve the current water crisis the report identifies five broad policy solutions: (1) the development of national water strategies that monitor water availability, assess the sustainable limits to human use and regulate withdrawals within these limits; (2) the adoption of pricing strategies that reflect the real scarcity value of water while maintaining equity among large and small, rich and poor users; (3) cut perverse subsidies for water overuse while ensuring that polluters pay, and create incentives for reducing pollution; (4) recognize the value of ecological services provided by wetlands and other water-based systems; and finally (5) deal with climate change, both through continuing efforts to cut carbon emissions but also to start developing adaptation strategies (UNDP, 2006). Access to sufficient clean water should be acknowledged as a basic human right and to that end the UN has set a Millennium Development Goal of halving the proportion of people without access to safe drinking water by 2015. The full list of 8 Millennium Development Goals, which comprise of 18 specific targets and 48 indicators can be examined at http://devdata.worldbank.org/gmis/mdg/list_of_goals.htm. Under Goal 7 *Future environmental sustainability*, Target 10 specifically aims to: 'Halve, by 2015, the proportion of people without sustainable access to safe drinking water and basic sanitation'. The indicators (30 and 31) are the proportion of population achieving the target for drinking water and sanitation respectively in both urban and rural areas.

31.7.2   Water quality problems

The most important water-associated health problems in developing countries are waterborne diseases, especially those leading to diarrhoea, which is suspected of being responsible for between 3 to 5 million deaths per year, especially among young children. Control of pathogens in drinking water is comparatively straightforward, but poverty combined with water scarcity is a devastating combination. However, adequate supplies of clean water, combined with adequate sanitation and improved hygiene standards, would significantly reduce the incidence of waterborne disease, and especially diarrhoea, in developing countries. So the reduction of pathogens in drinking water has been the priority for many decades. However, in recent years there has been a growing awareness of contamination from naturally occurring chemicals in groundwater and also from anthropogenic activities involving agriculture, industry and urban development.

An interesting example has been the installation of low-cost hand pumps to improve the drinking water quality in rural areas. This has allowed communities to access groundwater that is free from disease-causing microbes resulting in a significant reduction in diarrhoeal diseases. For example in the past 20 years over 4 million tube-wells have been constructed in Bangladesh alone that has brought pathogen-free water to approximately 95% of the population.

What was not known at the time was that much of the groundwater in this and many areas where this strategy was adopted was contaminated with high concentrations of the toxin arsenic. Today more than 20 million people in Bangladesh are exposed to high arsenic concentrations in their drinking water resulting in serious health effects including cancer (Section 9.3). The metal is both tasteless and odourless in water and so unless tested those drinking the water are unaware of its presence or the risk that they are taking. However, as it is difficult and relatively expensive to remove, finding alternative supplies may be the only option, not an attractive scenario in areas where water is very scarce and difficult to access. Currently over 100 million people are affected by elevated arsenic in their drinking water making what should have been a public health success into one of the most serious public health crises on record.

Another major problem in many developing countries is fluoride. While added to drinking water by many countries to prevent dental decay, in excess it results in serious dental and skeletal deformities as well as other health problems. In India, it is estimated that 66 million people drink groundwater containing dangerous concentrations of fluoride, in China a further 10 million people are affected (Sections 17.3 and 17.5).

One of the major challenges facing developing countries is finding appropriate, reliable and cost-effective methods of removing arsenic and fluoride. However, the task of installing these systems at millions of tube-wells appears to be an impossible task. Similarly finding alternative sources of drinking water in the worst affected areas may be equally difficult. One option may be to revert back, where possible, to using faecally contaminated surface water as tackling pathogens through wastewater treatment and water filtration may be more economically effective than removing inorganic contaminants from groundwaters. The reuse of wastewater for irrigation, linked with a more sustainable water management approach, offers a number of possible solutions to many regions affected by water scarcity.

In developed countries serious chemical contamination of drinking water is unusual. This is in contrast to the situation in China, which has experienced an unprecedented period of industrial development over the past 20 years. The state of surface and ground water resources in China is rapidly deteriorating due primarily to industrial pollution, although untreated sewage is also a major factor. It is estimated that >300 million rural Chinese, that is almost 25% of the entire population, lack access to safe drinking water, while the situation in many cities is also critical with almost a third of all drinking water supplied unsuitable for consumption. In November 2005 a chemical explosion at a petrochemical plant released over 100 tonnes of chemicals into the Songhua River that provides the main water supply for Harbin, the capital city of Heilongjiang Province. The water was heavily contaminated with benzene and nitrobenzene forcing the water supplies to the 3.8 million inhabitants to be cut off for 4 days causing widespread panic and suffering. Releasing vast quantities of stored water from upstream reservoirs, which also diluted the contaminants, flushed the chemicals downstream of the city. The water was subsequently treated with activated carbon to remove any remaining contaminants. Currently all surface and ground waters are at risk from contamination in China, and the new State Environmental Protection Agency (SEPA) has set about tackling this enormous task. For example, there are 13 000 petrochemical plants, similar to that which caused the problem at Harbin, built along the Yangtze and Yellow Rivers alone. New drinking water regulations will increase the 35 parameters that are currently monitored to 107 parameters, including organic and inorganic contaminants of industrial origin.

31.8 Conclusions

Investment in adequate water supplies is the single most important investment that can not only save lives but also create future wealth and develop sustainable local economies. According to the UNDP report discussed in Section 31.7, the US$ 10 billion investment required to meet the Millennium Development Goal of halving the proportion of people without access to safe drinking water by 2015 would generate US$ 38 billion a year in economic benefits (UNDP, 2006). However, UNDP argues that poverty, power and inequalities, not scarcity, are at the roots of the problem.

While additive factors of increasing population, wealth and urbanization are resulting in water scarcity throughout the world, climate change is adding to the problem of maintaining adequate supplies of treated water. Climate change and the hydrological cycle are intimately linked with weather patterns and so precipitation, both in terms of volume and timing, is controlled by temperature increase. Climate change will affect every country and region differently. For example, in Canada and Scandinavia it is expected that precipitation will increasingly fall as rain rather than snow. It is also expected that far less of the snow that does fall will remain on the ground for very long but will melt and run off the land. Outside the winter

months more severe storms and periods of drought are forecast, with the overall amount of precipitation not necessarily being less, but falling in more intense events over fewer days. This will significantly reduce groundwater recharge and influence surface water flows. The higher temperatures will also lead to higher plant transpiration rates and greater evaporation from the soil and stored water. Climate change is changing the context for water management and making the possibility of water sustainability very difficult to plan using conventional water management that relies on engineering infrastructural solutions. The future of water supplies relies on adopting a water demand management approach that must be put into place immediately if drinking water quality and adequate supplies are to be maintained into the future (Section 1.5). Currently most urban communities have just enough water to meet demand, but if the ecological status of surface waters is to be adequately protected then there must be a commitment to adopt the policy that all future demands must be met through conservation measures including rainwater harvesting and water reuse within the home (Sections 1.5 and 29.4).

Is my drinking water safe to drink? This is the question that anyone involved in the water supply industry is constantly asked. Accidents and operational problems can and do happen but in most of these cases, although standards are temporarily broken, the water will still be safe to drink. If not, alternative supplies or advice, such as boil water notices are issued. Incidents on the scale of Milwaukee or Camelford (Sections 13.2 and 15.1) are thankfully rare, but the industry is not accident or incident proof. Problems also occur with the quality of bottled water, and accidents such as that at the Perrier factory in February 1990, when an operational problem resulted in contamination of some of their water by trace amounts of benzene, are also thankfully rare. In this case the Company protected its consumers by a rapid but cripplingly expensive recall of all its bottled water stocks.

At the end of the day, however, the regulations and standards relating to piped water ensure that the water is safe to drink. In the UK, for example, there are independent channels to help and advise consumers, to follow up complaints and to check out worries and concerns (Section 1.4). Water companies maintain registers of water quality analyses and these are open for public scrutiny, usually now online. In reality, the public has much less information about many of the suppliers of bottled water, and fewer channels of complaint. The massive investment over the past decades by the water supply companies in the UK has resulted in the highest compliance to EC standards on record. With the revision of the Drinking Water Directive in 1998, water quality in the UK has never been so high or so reliable. The World Health Organization (WHO) water quality guidelines are revised approximately every decade, and since the revision in 1993, all developed countries have implemented their national standards to reflect these WHO recommendations. The latest revision was in 2004 and as a result of this ongoing analysis of health-based drinking water quality by the

WHO, there has been a continuous improvement in drinking water quality that is reflected through Europe, North America and the rest of the developed world. The challenge is sustaining this high level of quality in the future as supplies become less predictable.

A major concern must be the level of organic compounds, and pesticides in particular, in drinking water. Sophisticated treatment technologies such as activated carbon and membrane filtration can only provide the last line of defence. We must, however, look beyond the water industry for solutions, and look towards those who use such chemicals at source, especially the non-agricultural herbicides atrazine and simazine, which are serious endocrine-disrupting compounds (Section 7.3). Water resources are an inseparable part of the environment, so good water quality may only be achievable through the sensitive management and protection of the resources themselves. Better control, application and use of chemicals is often more cost effective than trying to remove them at the water treatment plant. However, for some areas of south-east England where water is coming from chalk aquifers, the levels of pesticides in groundwater are likely to continue to be elevated for decades to come, so treatment in these areas is clearly a priority.

The importance of closer links between suppliers and consumers is self-evident. It is paramount that consumers should be readily able to obtain information about water quality, and help where necessary. The industry needs to invest more in risk–benefit analysis (Gilbert and Calabrese, 1992; Keller and Wilson, 1992) to assess the dangers of certain pollutants to consumers. This is highlighted by the problems surrounding nitrate and aluminium, where the perceived threat was clearly exaggerated and has taken valuable investment capital away from dealing with the more important quality problems, although the treatment of high-nitrate water by bacterial denitrification has led to the further development of that technology for the removal of potentially more dangerous trace organic compounds.

It is time for the consumers themselves to take a share of the huge responsibility involved in achieving high-quality water, and this can only be achieved by openness, education and, above all, a willingness on the behalf of the professionals to interact more closely with consumers. Privatization has to some degree achieved this, and more importantly has put the issue of water quality on the agenda for public debate. Even in countries where the water supply is still in the control of the public sector the consumer should be perceived as the customer, and be treated as such. In the words of the chairman of the Water Service Association in the UK, 'If the consumer is unhappy then we have failed.'

Although I am sure that new issues and problems will come to light in the future, we will continue to see an improvement in water quality throughout Europe. Safeguarding drinking water quality is a shared responsibility, shared between those who use and dispose of chemicals, who treat and supply water, and all of us who use it. A greater understanding of the problems, and an acceptance

of that responsibility by all, will be needed if we are going to preserve one of our greatest assets for future generations: clean, safe drinking water on tap.

References

Cao, J., Zhao, Y., Li, Y. *et al.* (2006). Fluoride levels in various black tea commodities: measurement and safety evaluation. *Food and Chemical Toxicology*, **44**(7), 1131–7.

DWI (2001). *Drinking Water Quality Report of Public Perceptions in 2000*. London: Drinking Water Inspectorate.

DWI (2003). *Chief Inspector's Report 2002: Part 1. Overview of Water Quality in England and Wales*. London: Drinking Water Inspectorate.

DWI (2004). *Chief Inspector's Report 2003: Part 1. Overview of Water Quality in England and Wales*. London: Drinking Water Inspectorate.

DWI (2006). *Chief Inspector's Report 2005: Part 1. Overview of Water Quality in England and Wales*. London: Drinking Water Inspectorate.

Flaten, T. P. and Ødegård, M. (1988). Tea, aluminium and Alzheimer's disease. *Food Chemistry and Toxicology*, **26**, 959–60.

Gilbert, C. E. and Calabrese, E. J. (eds.) (1992). *Regulating Drinking Water Quality*. Boca Raton: Lewis.

Gleick, P. H. (2006). *The World's Water 2006–2007*. Washington, DC: Island Press.

Keller, A. Z. and Wilson, H. C. (1992). *Hazards to Drinking Water Supplies*. London: Springer-Verlag.

Lin, D., Tu, Y. and Zhu, L. (2005). Concentrations and health risk of polycyclic aromatic hydrocarbons in tea. *Food and Chemical Toxicology*, **43**, 41–8.

Pearce, F. (2006). *When the Rivers Run Dry: Water the Defining Crisis of the Twenty-First Century*. Boston, MA: Beacon Press.

UNDP (2006). *Beyond Scarcity: Power, Poverty and the Global Water Crisis*. Human Development Report 2006. New York: United Nations Development Programme.

Ward, D. R. (2002). *Water Wars: Drought, Flood, Folly and the Politics of Thirst*. New York: Riverhead Books.

WHO (1995). *Community Water Supply and Sanitation: Needs, Challenges and Health Objectives*. 48th World Health Assembly, A48/INF.DOC. 2,28 April. Geneva: World Health Organization.

WHO (2003). *Emerging Issues in Water and in Infectious Diseases*. Geneva: World Health Organization.

Appendix 1
EC Drinking Water Directive (98/83/EEC) quality parameters

Parameter	Parametric value
Part A: Microbiological parameters	
Escherichia coli	0/100 ml
Enterococci	0/100 ml
In water offered for sale in bottles or containers	
E. coli	0/250 ml
Enterococci	0/250 ml
Pseudomonas aeruginosa	0/250 ml
Colony count at 22 °C	100/ml
Colony count at 37 °C	20/ml
Part B: Chemical parameters	
Acrylamide	$0.1 \mu g \ l^{-1}$
Antimony	$5.0 \mu g \ l^{-1}$
Arsenic	$10 \mu g \ l^{-1}$
Benzene	$1.0 \mu g \ l^{-1}$
Benzo[*a*]pyrene	$0.01 \mu g \ l^{-1}$
Boron	$1.0 \ mg \ l^{-1}$
Bromate	$10 \mu g \ l^{-1}$
Cadmium	$5.0 \mu g \ l^{-1}$
Chromium	$50 \mu g \ l^{-1}$
Copper	$2.0 \ mg \ l^{-1}$
Cyanide	$50 \mu g \ l^{-1}$
1,2-Dichloroethane	$3.0 \mu g \ l^{-1}$
Epichlorohydrin	$0.1 \mu g \ l^{-1}$
Fluoride	$1.5 \ mg \ l^{-1}$
Lead	$10 \mu g \ l^{-1}$
Mercury	$1.0 \mu g \ l^{-1}$
Nickel	$20 \mu g \ l^{-1}$
Nitrate	$50 \ mg \ l^{-1}$
Nitrite	$0.5 \ mg \ l^{-1}$

Parameter	Parametric value
Pesticides[a,b]	$0.1\,\mu g\,l^{-1}$
Pesticides (total)[a]	$0.5\,\mu g\,l^{-1}$
Polycyclic aromatic hydrocarbons[a]	$0.1\,\mu g\,l^{-1}$
Selenium	$10\,\mu g\,l^{-1}$
Tetrachloroethene plus trichloroethane	$10\,\mu g\,l^{-1}$
Trihalomethanes (total)[a]	$100\,\mu g\,l^{-1}$
Vinyl chloride	$0.5\,\mu g\,l^{-1}$

[a] Relates to specified compounds in Directive 98/83/EEC.
[b] For aldrin, dieldrin, heptachlor and heptachlor epoxide the parametric value is $0.03\,\mu g\,l^{-1}$.

Parameter	Parametric value	Notes
Part C: Indicator parameters		
Physico-chemical and microbiological		
Aluminium	$200\,\mu g\,l^{-1}$	
Ammonium	$0.5\,mg\,l^{-1}$	
Chloride	$250\,mg\,l^{-1}$	Water should not be aggressive
Clostridium perfringens	0/100 ml	From, or affected by, surface water only
Colour		Acceptable to consumers and no abnormal change
Conductivity	$2500\,\mu S\,cm^{-1}$	Water should not be aggressive
Hydrogen ion concentration	pH ≥ 6.5, ≤ 9.5	Water should not be aggressive. Minimum values for bottled waters not $\leq$ pH 4.5
Iron	$200\,\mu g\,l^{-1}$	
Manganese	$50\,\mu g\,l^{-1}$	
Odour		Acceptable to consumers and no abnormal change
Oxidizability	$5.0\,mg\,O_2\,l^{-1}$	Not required if TOC used
Sulphate	$250\,mg\,l^{-1}$	Water should not be aggressive
Sodium	$200\,mg\,l^{-1}$	
Taste		Acceptable to consumers and no abnormal change
Colony count at 22°C	0/100 ml	
Coliform bacteria	0/100 ml	For bottled waters 0/250 ml
TOC		No abnormal change. Only for flows $> 10\,000\,m^3\,day^{-1}$
Turbidity		Acceptable to consumers and no abnormal change. Normally < 1.0 NTU
Radioactivity		
Tritium	$100\,Bq\,l^{-1}$	
Total indicative dose	$0.1\,mSv\,year^{-1}$	

TOC: total organic carbon; NTU: nephelometric turbidity unit.

Appendix 2
US National Primary and Secondary Drinking Water Standards of the US Environmental Protection Agency (2006)

EPA National Primary Drinking Water Standards

Enforceable standards for contaminants considered potentially harmful to health. Contaminant Categories: OC, organic chemical; R, radionuclides; IOC, inorganic chemical; DBP, disinfection by-product; D, disinfectant; M, micro-organism

Contaminant Category	Contaminant	MCL or TT[1] (mg/L)[2]	Potential health effects from exposure above the MCL	Common sources of contaminant in drinking water	Public Health Goal
OC	Acrylamide	TT[8]	Nervous system or blood problems; increased risk of cancer	Added to water during sewage/ wastewater treatment	zero
OC	Alachlor	0.002	Eye, liver, kidney or spleen problems; anemia; increased risk of cancer	Runoff from herbicide used on row crops	zero
R	Alpha particles	15 picocuries per Liter (pCi/L)	Increased risk of cancer	Erosion of natural deposits of certain minerals that are radioactive and may emit a form of radiation known as alpha radiation	zero

Reproduced with permission from US Environmental Protection Agency.

Contaminant Category	Contaminant	MCL or TT[1] (mg/L)[2]	Potential health effects from exposure above the MCL	Common sources of contaminant in drinking water	Public Health Goal
IOC	Antimony	0.006	Increase in blood cholesterol; decrease in blood sugar	Discharge from petroleum refineries; fire retardants; ceramics; electronics; solder	0.006
IOC	Arsenic	0.010 as of 1/23/06	Skin damage or problems with circulatory systems, and may have increased risk of getting cancer	Erosion of natural deposits; runoff from orchards, runoff from glass & electronics production wastes	0
IOC	Asbestos (fibers >10 micrometers)	7 million fibers per liter (MFL)	Increased risk of developing benign intestinal polyps	Decay of asbestos cement in water mains; erosion of natural deposits	7 MFL
OC	Atrazine	0.003	Cardiovascular system or reproductive problems	Runoff from herbicide use	0.003
IOC	Barium	2	Increase in blood pressure	Discharge of drilling wastes; discharge from metal refineries; erosion of natural deposits	2
OC	Benzene	0.005	Anemia; decrease in blood platelets; increased risk of cancer	Discharge from factories; leaching from gas storage tanks and landfills	zero

Contaminant Category	Contaminant	MCL or TT[1] (mg/L)[2]	Potential health effects from exposure above the MCL	Common sources of contaminant in drinking water	Public Health Goal
OC	Benzo(a)pyrene (PAHs)	0.0002	Reproductive difficulties; increased risk of cancer	Leaching from linings of water storage tanks and distribution lines	zero
IOC	Beryllium	0.004	Intestinal lesions	Discharge from metal refineries and coal-burning factories; discharge from electrical, aerospace, and defence industries	0.004
R	Beta particles and photon emitters	4 millirems per year	Increased risk of cancer	Decay of natural and man-made deposits of certain minerals that are radioactive and may emit forms of radiation known as photons and beta radiation	zero
DBP	Bromate	0.010	Increased risk of cancer	By-product of drinking water disinfection	zero
IOC	Cadmium	0.005	Kidney damage	Corrosion of galvanized pipes; erosion of natural deposits; discharge from metal refineries; runoff from waste batteries and paints	0.005

Contaminant Category	Contaminant	MCL or TT[1] (mg/L)[2]	Potential health effects from exposure above the MCL	Common sources of contaminant in drinking water	Public Health Goal
OC	Carbofuran	0.04	Problems with blood, nervous system, or reproductive system	Leaching of soil fumigant used on rice and alfalfa	0.04
OC	Carbon tetrachloride	0.005	Liver problems; increased risk of cancer	Discharge from chemical plants and other industrial activities	zero
D	Chloramines (as Cl_2)	MRDL = 4.0[1]	Eye/nose irritation; stomach discomfort, anemia	Water additive used to control microbes	MRDLG = 4[1]
OC	Chlordane	0.002	Liver or nervous system problems; increased risk of cancer	Residue of banned termiticide	zero
D	Chlorine (as Cl_2)	MRDL = 4.0[1]	Eye/nose irritation; stomach discomfort	Water additive used to control microbes	MRDLG = 4[1]
D	Chlorine dioxide (as ClO_2)	MRDL = 0.8[1]	Anemia; infants & young children: nervous system effects	Water additive used to control microbes	MRDLG = 0.8[1]
DBP	Chlorite	1.0	Anemia; infants & young children: nervous system effects	By-product of drinking water disinfection	0.8
OC	Chlorobenzene	0.1	Liver or kidney problems	Discharge from chemical and agricultural chemical factories	0.1

Contaminant Category	Contaminant	MCL or TT[1] (mg/L)[2]	Potential health effects from exposure above the MCL	Common sources of contaminant in drinking water	Public Health Goal
IOC	Chromium (total)	0.1	Allergic dermatitis	Discharge from steel and pulp mills; erosion of natural deposits	0.1
IOC	Copper	TT[7]; Action Level = 1.3	Short term exposure: Gastrointestinal distress. Long term exposure: Liver or kidney damage. People with Wilson's Disease should consult their personal doctor if the amount of copper in their water exceeds the action level	Corrosion of household plumbing systems; erosion of natural deposits	1.3
M	*Cryptosporidium*	TT[3]	Gastrointestinal illness (e.g., diarrhea, vomiting, cramps)	Human and animal fecal waste	zero
IOC	Cyanide (as free cyanide)	0.2	Nerve damage or thyroid problems	Discharge from steel/metal factories; discharge from plastic and fertilizer factories	0.2
OC	2,4-D	0.07	Kidney, liver, or adrenal gland problems	Runoff from herbicide used on row crops	0.07
OC	Dalapon	0.2	Minor kidney changes	Runoff from herbicide used on rights of way	0.2

Contaminant Category	Contaminant	MCL or TT[1] (mg/L)[2]	Potential health effects from exposure above the MCL	Common sources of contaminant in drinking water	Public Health Goal
OC	1,2-Dibromo-3-chloropropane (DBCP)	0.0002	Reproductive difficulties; increased risk of cancer	Runoff/leaching from soil fumigant used on soybeans, cotton, pineapples, and orchards	zero
OC	o-Dichlorobenzene	0.6	Liver, kidney, or circulatory system problems	Discharge from industrial chemical factories	0.6
OC	p-Dichlorobenzene	0.075	Anemia; liver, kidney or spleen damage; changes in blood	Discharge from industrial chemical factories	0.075
OC	1,2-Dichloroethane	0.005	Increased risk of cancer	Discharge from industrial chemical factories	zero
OC	1,1-Dichloroethylene	0.007	Liver problems	Discharge from industrial chemical factories	0.007
OC	cis-1,2-Dichloroethylene	0.07	Liver problems	Discharge from industrial chemical factories	0.07
OC	trans-1,2-Dichloroethylene	0.1	Liver problems	Discharge from industrial chemical factories	0.1
OC	Dichloromethane	0.005	Liver problems; increased risk of cancer	Discharge from drug and chemical factories	zero
OC	1,2-Dichloropropane	0.005	Increased risk of cancer	Discharge from industrial chemical factories	zero

Contaminant Category	Contaminant	MCL or TT[1] (mg/L)[2]	Potential health effects from exposure above the MCL	Common sources of contaminant in drinking water	Public Health Goal
OC	Di(2-ethylhexyl) adipate	0.4	Weight loss, liver problems, or possible reproductive difficulties	Discharge from chemical factories	0.4
OC	Di(2-ethylhexyl) phthalate	0.006	Reproductive difficulties; liver problems; increased risk of cancer	Discharge from rubber and chemical factories	zero
OC	Dinoseb	0.007	Reproductive difficulties	Runoff from herbicide used on soybeans and vegetables	0.007
OC	Dioxin (2,3,7,8-TCDD)	0.00000003	Reproductive difficulties; increased risk of cancer	Emissions from waste incineration and other combustion; discharge from chemical factories	zero
OC	Diquat	0.02	Cataracts	Runoff from herbicide use	0.02
OC	Endothall	0.1	Stomach and intestinal problems	Runoff from herbicide use	0.1
OC	Endrin	0.002	Liver problems	Residue of banned insecticide	0.002
OC	Epichlorohydrin	TT[8]	Increased cancer risk, and over a long period of time, stomach problems	Discharge from industrial chemical factories; an impurity of some water treatment chemicals	zero

Contaminant Category	Contaminant	MCL or TT[1] (mg/L)[2]	Potential health effects from exposure above the MCL	Common sources of contaminant in drinking water	Public Health Goal
OC	Ethylbenzene	0.7	Liver or kidneys problems	Discharge from petroleum refineries	0.7
OC	Ethylene dibromide	0.00005	Problems with liver, stomach, reproductive system, or kidneys; increased risk of cancer	Discharge from petroleum refineries	zero
IOC	Fluoride	4.0	Bone disease (pain and tenderness of the bones); Children may get mottled teeth	Water additive which promotes strong teeth; erosion of natural deposits; discharge from fertilizer and aluminium factories	4.0
M	*Giardia lamblia*	TT[3]	Gastrointestinal illness (e.g., diarrhea, vomiting, cramps)	Human and animal fecal waste	zero
OC	Glyphosate	0.7	Kidney problems; reproductive difficulties	Runoff from herbicide use	0.7
DBP	Haloacetic acids (HAA5)	0.060	Increased risk of cancer	By-product of drinking water disinfection	n/a[6]
OC	Heptachlor	0.0004	Liver damage; increased risk of cancer	Residue of banned termiticide	zero
OC	Heptachlor epoxide	0.0002	Liver damage; increased risk of cancer	Breakdown of heptachlor	zero

Contaminant Category	Contaminant	MCL or TT[1] (mg/L)[2]	Potential health effects from exposure above the MCL	Common sources of contaminant in drinking water	Public Health Goal
M	Heterotrophic plate count (HPC)	TT[3]	HPC has no health effects; it is an analytic method used to measure the variety of bacteria that are common in water. The lower the concentration of bacteria in drinking water, the better maintained the water system is.	HPC measures a range of bacteria that are naturally present in the environment	n/a
OC	Hexachlorobenzene	0.001	Liver or kidney problems; reproductive difficulties; increased risk of cancer	Discharge from metal refineries and agricultural chemical factories	zero
OC	Hexachlorocyclop entadiene	0.05	Kidney or stomach problems	Discharge from chemical factories	0.05
IOC	Lead	TT[7]; Action Level = 0.015	Infants and children: Delays in physical or mental development; children could show slight deficits in attention span and learning abilities; Adults: Kidney problems; high blood pressure	Corrosion of household plumbing systems; erosion of natural deposits	zero

Contaminant Category	Contaminant	MCL or TT[1] (mg/L)[2]	Potential health effects from exposure above the MCL	Common sources of contaminant in drinking water	Public Health Goal
M	*Legionella*	TT[3]	Legionnaire's Disease, a type of pneumonia	Found naturally in water; multiplies in heating systems	zero
OC	Lindane	0.0002	Liver or kidney problems	Runoff/leaching from insecticide used on cattle, lumber, gardens	0.0002
IOC	Mercury (inorganic)	0.002	Kidney damage	Erosion of natural deposits; discharge from refineries and factories; runoff from landfills and croplands	0.002
OC	Methoxychlor	0.04	Reproductive difficulties	Runoff/leaching from insecticide used on fruits, vegetables, alfalfa, livestock	0.04
IOC	Nitrate (measured as Nitrogen)	10	Infants below the age of six months who drink water containing nitrate in excess of the MCL could become seriously ill and, if untreated, may die. Symptoms include shortness of breath and blue-baby syndrome.	Runoff from fertilizer use; leaching from septic tanks, sewage; erosion of natural deposits	10

Contaminant Category	Contaminant	MCL or TT[1] (mg/L)[2]	Potential health effects from exposure above the MCL	Common sources of contaminant in drinking water	Public Health Goal
IOC	Nitrite (measured as Nitrogen)	1	Infants below the age of six months who drink water containing nitrite in excess of the MCL could become seriously ill and, if untreated, may die. Symptoms include shortness of breath and blue-baby syndrome.	Runoff from fertilizer use; leaching from septic tanks, sewage; erosion of natural deposits	1
OC	Oxamyl (Vydate)	0.2	Slight nervous system effects	Runoff/leaching from insecticide used on apples, potatoes, and tomatoes	0.2
OC	Pentachlorophenol	0.001	Liver or kidney problems; increased cancer risk	Discharge from wood preserving factories	zero
OC	Picloram	0.5	Liver problems	Herbicide runoff	0.5
OC	Polychlorinated biphenyls (PCBs)	0.0005	Skin changes; thymus gland problems; immune deficiencies; reproductive or nervous system difficulties; increased risk of cancer	Runoff from landfills; discharge of waste chemicals	zero
R	Radium 226 and Radium 228 (combined)	5 pCi/L	Increased risk of cancer	Erosion of natural deposits	zero

Contaminant Category	Contaminant	MCL or TT[1] (mg/L)[2]	Potential health effects from exposure above the MCL	Common sources of contaminant in drinking water	Public Health Goal
IOC	Selenium	0.05	Hair or fingernail loss; numbness in fingers or toes; circulatory problems	Discharge from petroleum refineries; erosion of natural deposits; discharge from mines	0.05
OC	Simazine	0.004	Problems with blood	Herbicide runoff	0.004
OC	Styrene	0.1	Liver, kidney, or circulatory system problems	Discharge from rubber and plastic factories; leaching from landfills	0.1
IOC	Tetrachloroethylene	0.005	Liver problems; increased risk of cancer	Discharge from factories and dry cleaners	zero
IOC	Thallium	0.002	Hair loss; changes in blood; kidney, intestine, or liver problems	Leaching from ore-processing sites; discharge from electronics, glass, and drug factories	0.0005
OC	Toluene	1	Nervous system, kidney, or liver problems	Discharge from petroleum factories	1
M	Total Coliforms (including fecal coliform and *E. coli*)	5.0%[4]	Not a health threat in itself; it is used to indicate whether other potentially harmful bacteria may be present[5]	Coliforms are naturally present in the environment as well as feces; fecal coliforms and *E. coli* only come from human and animal fecal waste.	zero

Contaminant Category	Contaminant	MCL or TT[1] (mg/L)[2]	Potential health effects from exposure above the MCL	Common sources of contaminant in drinking water	Public Health Goal
DBP	Total Trihalomethanes (TTHMs)	0.10; 0.080 after 12/31/ 03	Liver, kidney or central nervous system problems; increased risk of cancer	By-product of drinking water disinfection	n/a[6]
OC	Toxaphene	0.003	Kidney, liver, or thyroid problems; increased risk of cancer	Runoff/leaching from insecticide used on cotton and cattle	zero
OC	2,4,5-TP (Silvex)	0.05	Liver problems	Residue of banned herbicide	0.05
OC	1,2,4-Trichlorobenzene	0.07	Changes in adrenal glands	Discharge from textile finishing factories	0.07
OC	1,1,1-Trichloroethane	0.2	Liver, nervous system, or circulatory problems	Discharge from metal degreasing sites and other factories	0.20
OC	1,1,2-Trichloroethane	0.005	Liver, kidney, or immune system problems	Discharge from industrial chemical factories	0.003
OC	Trichloroethylene	0.005	Liver problems; increased risk of cancer	Discharge from metal degreasing sites and other factories	zero
M	Turbidity	TT[3]	Turbidity is a measure of the cloudiness of water. It is used to indicate water quality and filtration effectiveness	Soil runoff	n/a

Contaminant Category	Contaminant	MCL or TT[1] (mg/L)[2]	Potential health effects from exposure above the MCL	Common sources of contaminant in drinking water	Public Health Goal
			(e.g., whether disease-causing organisms are present). Higher turbidity levels are often associated with higher levels of disease-causing micro-organisms such as viruses, parasites and some bacteria. These organisms can cause symptoms such as nausea, cramps, diarrhea, and associated headaches.		
R	Uranium	30 ug/L as of 12/08/03	Increased risk of cancer, kidney toxicity	Erosion of natural deposits	zero
OC	Vinyl chloride	0.002	Increased risk of cancer	Leaching from PVC pipes; discharge from plastic factories	zero
M	Viruses (enteric)	TT[3]	Gastrointestinal illness (e.g., diarrhea, vomiting, cramps)	Human and animal fecal waste	zero
OC	Xylenes (total)	10	Nervous system damage	Discharge from petroleum factories; discharge from chemical factories	10

Notes
[1] Definitions

- Maximum Contaminant Level Goal (MCLG) – The level of a contaminant in drinking water below which there is no known or expected risk to health. MCLGs allow for a margin of safety and are non-enforceable public health goals.
- Maximum Contaminant Level (MCL) – The highest level of a contaminant that is allowed in drinking water. MCLs are set as close to MCLGs as feasible using the best available treatment technology and taking cost into consideration. MCLs are enforceable standards.
- Maximum Residual Disinfectant Level Goal (MRDLG) – The level of a drinking water disinfectant below which there is no known or expected risk to health. MRDLGs do not reflect the benefits of the use of disinfectants to control microbial contaminants.
- Maximum Residual Disinfectant Level (MRDL) – The highest level of a disinfectant allowed in drinking water. There is convincing evidence that addition of a disinfectant is necessary for control of microbial contaminants.
- Treatment Technique (TT) – A required process intended to reduce the level of a contaminant in drinking water.

[2] Units are in milligrams per litre (mg/L) unless otherwise noted. Milligrams per litre are equivalent to parts per million (ppm).

[3] EPA's surface water treatment rules require systems using surface water or ground water under the direct influence of surface water to (1) disinfect their water, and (2) filter their water or meet criteria for avoiding filtration so that the following contaminants are controlled at the following levels:

- *Cryptosporidium* (as of 1/1/02 for systems serving >10,000 and 1/14/05 for systems serving <10,000) 99% removal.
- *Giardia lamblia:* 99.9% removal/inactivation
- Viruses: 99.99% removal/inactivation
- *Legionella:* No limit, but EPA believes that if *Giardia* and viruses are removed/inactivated, *Legionella* will also be controlled.
- Turbidity: At no time can turbidity (cloudiness of water) go above 5 nephelolometric turbidity units (NTU); systems that filter must ensure that the turbidity go no higher than 1 NTU (0.5 NTU for conventional or direct filtration) in at least 95% of the daily samples in any month. As of January 1, 2002, for systems servicing >10,000, and January 14, 2005, for systems servicing <10,000, turbidity may never exceed 1 NTU, and must not exceed 0.3 NTU in 95% of daily samples in any month.
- HPC: No more than 500 bacterial colonies per millilitre
- Long Term 1 Enhanced Surface Water Treatment (Effective Date: January 14, 2005); Surface water systems or (GWUDI) systems serving fewer than 10,000 people must comply with the applicable Long Term 1 Enhanced Surface Water Treatment Rule provisions (e.g. turbidity standards, individual filter monitoring, *Cryptosporidium* removal requirements, updated watershed control requirements for unfiltered systems).
- Filter Backwash Recycling: The Filter Backwash Recycling Rule requires systems that recycle to return specific recycle flows through all processes of the system's existing conventional or direct filtration system or at an alternate location approved by the state.

[4] No more than 5.0% samples total coliform-positive in a month. (For water systems that collect fewer than 40 routine samples per month, no more than one sample can be total coliform-positive per

month.) Every sample that has total coliform must be analyzed for either fecal coliforms or *E. coli* if two consecutive TC-positive samples, and one is also positive for *E. coli* fecal coliforms, system has an acute MCL violation.

[5] Fecal coliform and *E. coli* are bacteria whose presence indicates that the water may be contaminated with human or animal wastes. Disease-causing microbes (pathogens) in these wastes can cause diarrhea, cramps, nausea, headaches, or other symptoms. These pathogens may pose a special health risk for infants, young children, and people with severely compromised immune systems.

[6] Although there is no collective MCLG for this contaminant group, there are individual MCLGs for some of the individual contaminants:

- Haloacetic acids: dichloroacetic acid (zero); trichloroacetic acid (0.3 mg/L)
- Trihalomethanes: bromodichloromethane (zero); bromoform (zero); dibromochloromethane (0.06 mg/L)

[7] Lead and copper are regulated by a Treatment Technique that requires systems to control the corrosiveness of their water. If more than 10% of tap water samples exceed the action level, water systems must take additional steps. For copper, the action level is 1.3 mg/L, and for lead is 0.015 mg/L.

[8] Each water system must certify, in writing, to the state (using third-party or manufacturers certification) that when it uses acrylamide and/or epichlorohydrin to treat water, the combination (or product) of dose and monomer level does not exceed the levels specified, as follows: Acrylamide = 0.05% dosed at 1 mg/L (or equivalent); Epichlorohydrin = 0.01% dosed at 20 mg/L (or equivalent).

EPA National Secondary Drinking Water Standards

National Secondary Drinking Water Standards are non-enforceable guidelines regulating contaminants that may cause cosmetic effects (such as skin or tooth discoloration) or aesthetic effects (such as taste, odour, or colour) in drinking water. EPA recommends secondary standards to water systems but does not require systems to comply. However, states may choose to adopt them as enforceable standards.

Contaminant	Secondary Standard
Aluminium	0.05 to 0.2 mg/L
Chloride	250 mg/L
Colour	15 (color units)
Copper	1.0 mg/L
Corrosivity	Non-corrosive
Fluoride	2.0 mg/L
Foaming Agents	0.5 mg/L
Iron	0.3 mg/L
Manganese	0.05 mg/L
Odour	3 threshold odor number

Contaminant	Secondary Standard
pH	6.5–8.5
Silver	0.10 mg/L
Sulphate	250 mg/L
Total Dissolved Solids	500 mg/L
Zinc	5 mg/L

Appendix 3
World Health Organization drinking water guide values for chemicals of health significance. Health-related guide values have not been set for a number of chemicals that are not considered hazardous at concentrations normally found in drinking water, although some of these compounds may lead to consumer complaints on aesthetic grounds. These are listed in Table 2.6

Chemical	Guideline value[a] (mg/l)	Remarks
Acrylamide	0.0005[b]	
Alachlor	0.02[b]	
Aldicarb	0.01	Applies to aldicarb sulphoxide and aldicarb sulphone
Aldrin and dieldrin	0.00003	For combined aldrin plus dieldrin
Antimony	0.02	
Arsenic	0.01 (P)	
Atrazine	0.002	
Barium	0.7	
Benzene	0.01[b]	
Benzo[a]pyrene	0.0007[b]	
Boron	0.5 (T)	
Bromate	0.01[b] (A, T)	
Bromodichloromethane	0.06[b]	
Bromoform	0.1	
Cadmium	0.003	
Carbofuran	0.007	
Carbon tetrachloride	0.004	
Chloral hydrate (trichloroacetaldehyde)	0.01 (P)	
Chlorate	0.7 (D)	
Chlordane	0.0002	

Reproduced with permission from the World Health Organization.

Chemical	Guideline value[a] (mg/l)	Remarks
Chlorine	5 (C)	For effective disinfection, there should be a residual concentration of free chlorine of $\geq$0.5 mg/l after at least 30 min contact time at pH <8.0
Chlorite	0.7 (D)	
Chloroform	0.2	
Chlorotoluron	0.03	
Chlorpyrifos	0.03	
Chromium	0.05 (P)	For total chromium
Copper	2	Staining of laundry and sanitary ware may occur below guideline value
Cyanazine	0.0006	
Cyanide	0.07	
Cyanogen chloride	0.07	For cyanide as total cyanogenic compounds
2,4-D (2,4-dichlorophenoxyacetic acid)	0.03	Applies to free acid
2,4-DB	0.09	
DDT and metabolites	0.001	
Di(2-ethylhexyl)phthalate	0.008	
Dibromoacetonitrile	0.07	
Dibromochloromethane	0.1	
1,2-Dibromo-3-chloropropane	0.001[b]	
1,2-Dibromoethane	0.0004[b] (P)	
Dichloroacetate	0.05 (T, D)	
Dichloroacetonitrile	0.02 (P)	
Dichlorobenzene, 1,2-	1 (C)	
Dichlorobenzene, 1,4-	0.3 (C)	
Dichloroethane, 1,2-	0.03[b]	
Dichloroethene, 1,1-	0.03	
Dichloroethene, 1,2-	0.05	
Dichloromethane	0.02	
1,2-Dichloropropene (1,2-DCP)	0.04 (P)	
1,3-Dichloropropene	0.02[b]	
Dichlorprop	0.1	
Dimethoate	0.006	
Edetic acid (EDTA)	0.6	Applies to the free acid
Endrin	0.0006	
Epichlorohydrin	0.0004 (P)	
Ethylbenzene	0.3 (C)	
Fenoprop	0.009	

Chemical	Guideline value[a] (mg/l)	Remarks
Fluoride	1.5	Volume of water consumed and intake from other sources should be considered when setting national standards
Formaldehyde	0.9	
Hexachlorobutadiene	0.0006	
Isoproturon	0.009	
Lead	0.01	
Lindane	0.002	
Manganese	0.4 (C)	
MCPA	0.002	
Mecoprop	0.01	
Mercury	0.001	For total mercury (inorganic plus organic)
Methoxychlor	0.02	
Metolachlor	0.01	
Microcystin-LR	0.001 (P)	For total Microcystin-LR (free plus cell-bound)
Molinate	0.006	
Molybdenum	0.07	
Monochloramine	3	
Monochloroacetate	0.02	
Nickel	0.02 (P)	
Nitrate (as NO_3^-)	50	Short-term exposure
Nitrilotriacetic acid (NTA)	0.2	
Nitrite (as NO_2^-)	3	Short-term exposure
	0.2 (P)	Long-term exposure
Pendimethalin	0.02	
Pentachlorophenol	0.009[b] (P)	
Pyriproxyfen	0.3	
Selenium	0.01	
Simazine	0.002	
Styrene	0.02 (C)	
2,4,5-T	0.009	
Terbuthylazine	0.007	
Tetrachloroethene	0.04	
Toluene	0.7 (C)	
Trichloroacetate	0.2	
Trichloroethene	0.07 (P)	
Trichlorophenol, 2,4,6-	0.2[b] (C)	
Trifluralin	0.02	

Chemical	Guideline value[a] (mg/l)	Remarks
Trihalomethanes		The sum of the ratio of the concentration of each to its respective guideline value should not exceed 1
Uranium	0.015 (P,T)	Only chemical aspects of uranium addressed
Vinyl chloride	0.0003[b]	
Xylenes	0.5 (C)	

[a] P: provisional guideline value, as there is evidence of a hazard, but the available information on health effects is limited; T: provisional guideline value because calculated guideline value is below the level that can be achieved through practical treatment methods, source protection, etc.; A: provisional guideline value because calculated guideline value is below the achievable quantification level; D: provisional guideline value because disinfection is likely to result in the guideline value being exceeded; C: concentrations of the substance at or below the health-based guideline value may affect the appearance, taste or odour of the water, leading to consumer complaints.

[b] For substances that are considered to be carcinogenic, the guideline value is the concentration in drinking water associated with an upper-bound excess lifetime cancer risk of 10^{-5} (one additional cancer per 100 000 of the population ingesting drinking water containing the substance at the guideline value for 70 years). Concentrations associated with upper-bound estimated excess lifetime cancer risks of 10^{-4} and 10^{-6} can be calculated by multiplying and dividing, respectively, the guideline value by 10.

Appendix 4
Major pesticides and their degradation (breakdown) products with their relative toxicity limits in drinking water. The toxicity is based on limits set originally by the Federal Health Authority in the former Federal Republic of Germany (Miller *et al.*, 1990), where category A pesticides should not exceed $1\,\mu g\ l^{-1}$, category B $3\,\mu g\ l^{-1}$ and category C $10\,\mu g\ l^{-1}$

Active ingredient	Category[a]	Degradation products[b]	Category[c]
Alachlor	A	2,6-Diethylaniline	–
Aldicarb	B	Aldicarbsulphone (= Aldoxycarb)	B
		aldicarbsulphoxide	B
		Total concentration of aldicarb and main decomposition products	B
Alloxydim	C		
Amitrole	–		
Anilazine	C	2-Chloroaniline	–
		Dichloro-*s*-triazine	C
Asulam	C	*p*-Aminobenzine sulphonic acid	C
Atrazine	B	Desethylatrazine	B
		2-Chloro-4-Ethylamino-6-amino-1,3,5-triazine	B
		Total concentration of atrazine and main decomposition products	B
Azinphos-ethyl	C		
Benalaxyl	C	2,6-Dimethylaniline	–
Benazolin	C		
Bendiocarb	C		
Bentazone	C		
Bromacil	C		
Carbetamide	C	Aniline	A

Adapted from Miller, D.G., Zabel, T.F. and Newman, P.J. (1990). *Summary Report on Environmental Developments June 1989–March 1990*. Report FR0088. Marlow: Foundation for Water Research with permission from the Foundation for Water Research.

Active ingredient	Category[a]	Degradation products[b]	Category[c]
Carbofuran	C		
Carbosulfan	C		
Chloramben	C		
Chloridazon	C		
Chlorfenvinphos	C		
Chlorthiamid	C	Dichlorobenzamide	B
		Dichlobenil	C
		Total concentration of chlorthiamid and main decomposition product	C
Chlortoluron	C	5-Chloro-p-toluidine	–
Clopyralid	C		
Cyanazine	C		
2,4-D	C	2,4-Dichlorophenol	A
Dazomet	A		
Diazinon	A		
Dicamba	C	3,6-Dichlorophenol	A
		3,6-Dichlorosalicylic acid	B
Dichlobenil	C	2,6-Dichlorobenzamide	B
		Total concentration of dichlobenil and main decomposition product	C
Dichlorprop	C	2,4-Dichlorophenol	A
1,2-Dichloropropane	C		
1,3-Dichloropropene	–		
Dikegulac	C		
Dimefuron	C	3-Chloroaniline	–
Dimethoate	C		
Dinoseb	B	Aromatic amines[d] and nitroaromatics[d]	A[c]
Dinoseb-acetate	B	Aromatic amines[d] and nitroaromatics[d]	A[c]
Dinoterb	C	Aromatic amines[d] and nitroaromatics[d]	A[c]
Diuron	C	3,4-Dichloroaniline	–
DNOC	C	Aromatic amines[d] and nitroaromatics[d]	A[c]
Endosulfan	B	Chlorinated cyclic compounds[d]	A[c]
Ethidimuron	C		
Ethiofencarb	C	Ethiofencarbsulphone	
		Ethiofencarbsulphoxide	C
		Total concentration of ethiofencarb and main decomposition products	C
Ethoprophos	A		
Etrimfos	C		
Fenpropimorph	C		
Flamprop-methyl	B	3-Chloro-4-fluoroaniline	–
Fluazifop	C		
Fluroxypyr	C		
Haloxyfop	A		

Active ingredient	Category[a]	Degradation products[b]	Category[c]
Hexazinone	C		
Isocarbamid	C		
Isoproturon	C	*p*-Isopropylaniline	–
Karbutilate	C		
Lindane	B	Chlorinated cyclohexene[d]	A[c]
Linuron	C	3,4-Dichloroaniline	–
Maleic hydrazide	C		
MCPA	A	*p*-Chlorophenol	A
Mecoprop (= MCPP)	C	*p*-Chlorophenol	A
Mefluidide	C		
Metalaxyl	C	2,6-Dimethylaniline	–
Metham sodium	C		
Metazachldr	C	2,6-Dimethylaniline	–
Methabenzthiazuron	C		
Methamidophos	B		
Methomyl	B		
Methyl bromide	–		
Methyl isothiocyanate	B		
Metobromuron	B	*p*-Bromoaniline	–
Metolachlor	B	2-Methyl-6-ethylaniline	–
Metoxuron	C	3-Chloro-4-methoxyaniline	
		1-Chloro-*p*-aminophenol	–
Metribuzin	C		
Monuron	C	*p*-Chloroaniline	–
Nitrothalisopropyl	C	Nitro-aromatics[d]	A[c]
Oxadixyl	C	2,6-Dimethylaniline	
Oxamyl	C		
Oxycarboxin	C		
Parathion	C		
Pendimethalin	C	Aromatic amines[d] and nitroaromatics[d]	A[c]
Pichoram	C		
Pirimicarb	C		
Pirimiphos-methyl	C		
Propachlor	C	*N*-Isopropylaniline	A[c]
		Aniline	A[c]
Propazine	C	Desethylatrazine	B
		Total concentration of propazine and main decomposition product	C
Propoxur	C		
Pyridate	C	3-Phenyl-6-hydroxy-6-chloropyridazine	C
		Total concentration of pyridate and main decomposition product	C
S 421	A	Chlorinated unsaturated aliphatic compounds	A[d]
Sebuthylazine	C	Desethylsebuthylazine	C

Active ingredient	Category[a]	Degradation products[b]	Category[c]
		2-Chloro-4-ethylamino-6-l,3,5-triazine	B
		Total concentration of sebuthylazine and main decomposition products	C
Sethoxydim	C		
Simazine	C	2-Chloro-4-ethylamino-1,3,5-triazine	B
		Total concentration of simazine and main decomposition products	C
TCA	C		
Tebuthiuron	C		
Terbacil	C		
Terbumeton	C	Desethylterbumeton	C
		2-Methoxy-4-ethylamino-6-amino-1,3,5-triazine	C
		Total concentration of terbumeton and main decomposition products	C
Terbuthylazine	C	Desethylterbuthylazine	C
		2-Chlor-4-Ethylamino-6-amino-1,3,5-triazine	B
		Total concentration of terbuthylazine and main composition products	C
Thiofanox	A	Thiofanoxsulphone	A
		Thiofanoxsulphoxide	A
		Total concentration of thiofanox and main decomposition products	A
Triclopyr	C		
Trifluralin	C	Aromatic amines[d] and nitroaromatics[d]	A[c]

[a] Groups of substances are identified by[d]. Subdegradation products and reaction products are not listed. Their existence must be verified in each individual case.

[b] The substances are classified according to the level of knowledge about chronic toxicity (category A to C) or genotoxic potential (limit values must not be exceeded). When determining the acceptable concentration of chlorophenol it is important that the taste of the drinking water is not tainted.

[c] If decomposition products of these groups of substances exist they must be differentiated and identified. Degradation products with genotoxic potential must not exceed their limit values.

Appendix 5
EC Water Framework Directive (2000/60/EC)
Priority Substances

List of priority substances in the field of water policy[a]

	CAS number	EU number	Name of priority substance	Identified as priority hazardous substance
(1)	15972-60-8	240-110-8	Alachlor	
(2)	120-12-7	204-371-1	Anthracene	(Possible)[b]
(3)	1912-24-9	217-617-8	Atrazine	(Possible)[b]
(4)	71-43-2	200-753-7	Benzene	
(5)	na	na	Brominated diphenylethers[c]	Yes[d]
(6)	7440-43-9	231-152-8	Cadmium and its compounds	Yes
(7)	85535-84-8	287-476-5	C_{10-13}-chloroalkanes[c]	Yes
(8)	470-90-6	207-432-0	Chlorfenvinphos	
(9)	2921-88-2	220-864-4	Chlorpyrifos	(Possible)[b]
(10)	107-06-2	203-458-1	1,2-Dichloroethane	
(11)	75-09-2	200-838-9	Dichloromethane	
(12)	117-81-7	204-211-0	Di(2-ethylhexyl)phthalate	(Possible)[b]
(13)	330-54-1	206-354-4	Diuron	(Possible)[b]
(14)	115-29-7	204-079-4	Endosulfan	(Possible)[b]
	959-98-8	na	(alpha-endosulfan)	
(15)	206-44-0	205-912-4	Fluoranthene[e]	
(16)	118-74-1	204-273-9	Hexachlorobenzene	Yes
(17)	87-68-3	201-765-5	Hexachlorobutadiene	Yes
(18)	608-73-1	210-158-9	Hexachlorocyclohexane	Yes
	58-89-9	200-401-2	(gamma-isomer, Lindane)	
(19)	34123-59-6	251-835-4	Isoproturon	(Possible)[b]
(20)	7439-92-1	231-100-4	Lead and its compounds	(Possible)[b]
(21)	7439-97-6	231-106-7	Mercury and its compounds	Yes
(22)	91-20-3	202-049-5	Naphthalene	(Possible)[b]
(23)	7440-02-0	231-111-4	Nickel and its compounds	
(24)	25154-52-3	246-672-0	Nonylphenols	Yes
	104-40-5	203-199-4	(4-(para)-nonylphenol)	

	CAS number	EU number	Name of priority substance	Identified as priority hazardous substance
(25)	1806-26-4	217-302-5	Octylphenols	(Possible)[b]
	140-66-9	na	(para-tert-octylphenol)	
(26)	608-93-5	210-172-5	Pentachlorobenzene	Yes
(27)	87-86-5	201-778-6	Pentachlorophenol	(Possible)[b]
(28)	na	na	PAHs	Yes
	50-32-8	200-028-5	(Benzo[a]pyrene),	
	205-99-2	205-911-9	(Benzo(b)fluoranthene),	
	191-24-2	205-883-8	(Benzo(g,h,i)perylene),	
	207-08-9	205-916-6	(Benzo(k)fluoranthene),	
	193-39-5	205-893-2	(Indeno(I,2,3-cd)pyrene)	
(29)	122-34-9	204-535-2	Simazine	(Possible)[b]
(30)	688-73-3	211-704-4	Tributyltin compounds	Yes
	36643-28-4	na	(Tributyltin-cation)	
(31)	12002-48-1	234-413-4	Trichlorobenzenes	(Possible)[b]
	120-82-1	204-428-0	(1,2,4-Trichlorobenzene)	
(32)	67-66-3	200-663-8	Trichloromethane (Chloroform)	
(33)	1582-09-8	216-428-8	Trifluralin	(Possible)[b]

[a] Where groups of substances have been selected, typical individual representatives are listed as indicative parameters (in brackets and without number). The establishment of controls will be targeted to these individual substances, without prejudicing the inclusion of other individual representatives, where appropriate.

[b] This priority substance is subject to a review for identification as possible 'priority hazardous substance'. The Commission will make a proposal to the European Parliament and Council for its final classification not later than 12 months after adoption of this list. The timetable laid down in Article 16 of Directive 2000/60/EC for the Commission's proposals of controls is not affected by this review.

[c] These groups of substances normally include a considerable number of individual compounds. At present, appropriate indicative parameters cannot be given.

[d] Only Pentabromobiphenylether (CAS number 32534-81-9).

[e] Fluoranthene is on the list as an indicator of other, more dangerous PAHs.

List I and List II substances covered by the EC Dangerous Substances Directive (76/464/EEC)

List I (black list)

Organohalogen compounds and substances which may form such compounds in the aquatic environment

Organophosphorus compounds

Organotin compounds

Substances, the carcinogenic activity of which is exhibited in or by the aquatic environment (substances in List II which are carcinogenic are included here)

Mercury and its compounds

Cadmium and its compounds

Persistent mineral oils and hydrocarbons of petroleum

Persistent synthetic substances

List II (grey list)

The following metalloids/metals and their compounds:

Zinc, copper, nickel, chromium, lead, selenium, arsenic, antimony, molybdenum, titanium, tin, barium, beryllium, boron, uranium, vanadium, cobalt, thallium, tellurium, silver

Biocides and their derivatives not appearing in List I

Substances which have a deleterious effect on the taste and/or smell of products for human consumption derived from the aquatic environment and compounds liable to give rise to such substances in water

Toxic or persistent organic compounds of silicon and substances which may give rise to such compounds in water, excluding those which are biologically harmless or are rapidly converted in water to harmless substances

Inorganic compounds of phosphorus and elemental phosphorus

Non-persistent mineral oils and hydrocarbons of petroleum origin

Cyanides, fluorides

Certain substances which may have an adverse effect on the oxygen balance, particularly ammonia and nitrites

Appendix 7
The USEPA second Drinking Water Contaminant Candidate List (CCL) Published in February 2005

Microbial contaminant candidates

Adenoviruses
Aeromonas hydrophila
Caliciviruses
Coxsackieviruses
Cyanobacteria (blue-green algae), other freshwater algae, and their toxins
Echoviruses
Helicobacter pylori
Microsporidia (Enterocytozoon & Septata)
Mycobacterium avium intracellulare (MAC)

Chemical contaminant candidates

Chemical contaminant	CASRN[1]
1,1,2,2-tetrachloroethane	79-34-5
1,2,4-trimethylbenzene	95-63-6
1,1-dichloroethane	75-34-3
1,1-dichloropropene	563-58-6
1,2-diphenylhydrazine	122-66-7
1,3-dichloropropane	142-28-9
1,3-dichloropropene	542-75-6
2,4,6-trichlorophenol	88-06-2
2,2-dichloropropane	594-20-7
2,4-dichlorophenol	120-83-2
2,4-dinitrophenol	51-28-5
2,4-dinitrotoluene	121-14-2
2,6-dinitrotoluene	606-20-2

Chemical contaminant	CASRN[1]
2-methyl-phenol (o-cresol)	95-48-7
Acetochlor	34256-82-1
Alachlor ESA & other acetanilide pesticide degradation products	N/A
Aluminum	7429-90-5
Boron	7440-42-8
Bromobenzene	108-86-1
DCPA mono-acid degradate	887-54-7
DCPA di-acid degradate	2136-79-0
DDE	72-55-9
Diazinon	333-41-5
Disulfoton	298-04-4
Diuron	330-54-1
EPTC (s-ethyl-dipropylthiocarbamate)	759-94-4
Fonofos	944-22-9
p-Isopropyltoluene (*p*-cymene)	99-87-6
Linuron	330-55-2
Methyl bromide	74-83-9
Methyl-t-butyl ether (MTBE)	1634-04-4
Metolachlor	51218-45-2
Molinate	2212-67-1
Nitrobenzene	98-95-3
Organotins	N/A
Perchlorate	14797-73-0
Prometon	1610-18-0
RDX	121-82-4
Terbacil	5902-51-2
Terbufos	13071-79-9
Triazines & degradation products of triazines	including, but not limited to Cyanazine 21725-46-2 and atrazine-desethyl 6190-65-4
Vanadium	7440-62-2

[1] CASRN (Chemical Abstract Service Registration Number). Chemical abstract service registry numbers are used in reference works, databases, and regulatory compliance documents by many organizations around the world to identify substances with a standardized name.

Glossary

All the terms, and abbreviations, listed below have been explained fully in the text. A brief summary or explanation is given here as an aide-memoire. A more comprehensive glossary can be accessed at the USEPA website (www.epa.gov/safewater/pubs/gloss2.htm).

Absorption The process by which a substance is taken into the body of another substance (normally a biological cell).

Activated carbon Made from materials such as coal or coconut shells, it has a very highly porous structure which is able to adsorb dissolved organic matter and certain dissolved gases from water.

Acute toxicity A toxic effect coming speedily to a crisis, usually caused by a large dose of poison of short duration.

ADAS Agricultural Development Advisory Service.

ADI Acceptable Daily Intake of a substance to ensure no ill effect on consumer.

Adsorption The process by which a gas, vapour, dissolved material or small suspended particle is attracted to, and attached to, the surface of another material either by physical or chemical forces.

Aeration The vigorous mixing of water to put oxygen into solution to strip out carbon dioxide, remove odorous compounds, and to facilitate oxidative reactions.

Aesthetic quality Water quality parameters linked to the senses of smell, taste and sight.

Algae Small, microscopic plant forms found as single cells, colonies or as filaments.

Algal bloom A prolific growth of algae due to nutrient enrichment, resulting in a serious reduction in water quality.

Alum The common name for aluminium sulphate, which is widely used as a coagulant.

Anaerobic A process (normally biological) performed in the complete absence of oxygen.

Anion Negatively charged ion.

AOC Assimilible organic carbon.

Aquifer An underground water-bearing layer of porous rock.

Artificial recharge An artificial method of replacing the water abstracted from an aquifer at rates in excess of those that occur naturally.

Backwashing The process in which the flow is reversed through a slow sand filter or ion-exchange resin to loosen the bed and to flush out any suspended matter collected.

BaP Benzo[a]pyrene.

BDCM Bromodichloromethane

BHT Butylated hydroxytoluene.

Biofilm A slime or growth comprising microbial populations that grow on the inside of pipes and other surfaces.

Breakpoint chlorination Chlorinating water until the chlorine demand is satisfied, with the excess chlorine remaining in the water as free chlorine.

CAS Chemical Abstracts Service.

Catchment The area of land from which all the rainfall contributes to surface and ground waters. Also known as a watershed or river basin.

Cation Positively charged ion.

CEN Comite Europeen de Normalisation publish European standards.

Chloramination The use of *chloramines* as a means of disinfection that provides a longer lasting residual within the distribution system than free chlorine.

Chloramines Compounds formed by the reaction of chlorine with ammonia in water.

Chlorination The use of chlorine gas or solutions to disinfect drinking water.

Chlorine demand The amount of chlorine consumed by organic matter and other oxidizable compounds in the water without leaving a chlorine residue.

Chronic toxicity A toxic effect that continues for a long time and may be either lethal or sub-lethal, generally caused by a low dose of poison over a long time.

Coagulant Normally a salt of aluminium or iron added to water to form a hydroxide precipitate.

Coagulant aid A compound added to water to promote coagulation. These are normally large organic ions whose charges help to form larger flocs.

Coagulation The process in which small particles of floc begin to agglomerate into larger particles that settle out of solution more readily.

Coliforms A group of bacteria found in vast numbers in faeces, used as indicator organisms for microbial pathogens in water.

Conductivity The power of a liquid to transmit an electrical charge that is related to the concentration of ions.

Consumer Either an individual or organization that uses drinking water.

Cryptosporidium A protozoan pathogen commonly found in surface waters and also occasionally groundwaters as an oocyst that is highly resistant and causes diarrhoeal illness that is life-threatening in individuals with severely weakened immune systems.

CTC Tetrachloromethane or carbon tetrachloride – an industrial solvent.

Cyanobacteria Known as blue-green algae, a bacteria containing chlorophyll and phycobilins, associated with the release of algal toxins.

DALY Disability-adjusted life-year.

DBCM Dibromochloromethane.

DBCP 1,2-dibromo-3-chloropropane.

DCB Dichlorobenzene.

DCM Methylene chloride – an industrial solvent.

DCP 1,2 dichloropropane.

DDT Dichlorodiphenyltrichloroethane.

Defra Department of Environment, Food and Rural Affairs with its headquarters in London.

Denitrification The reduction of nitrate to nitrogen gas.

Desalination The removal of dissolved salts from brackish water in order to make it potable.

Diffuse pollution Pollution emanating from a large undefined area.

Disinfection Reduction of the microbial contamination of water.

Disinfection by-product A compound, often toxic, that is formed by the reaction of chlorine or sometimes ozone with an organic compound (e.g. *THMs*).

Distribution system The network of pipes (mains) and supply reservoirs that transport water from the treatment plant to the consumer's house.

Dose–response The quantitative relationship between the dose of a contaminant and the effect caused by the contaminant.

DWI Drinking Water Inspectorate.

Enteric pathogen A pathogen found in the human gut.

Epidemiology The study of the distribution and causes of disease within a population.

Escherichia coli A common bacterium found in the gut that is used as an indicator of faecal contamination of drinking water.

Eutrophication The excessive development of algae in a surface water due to enrichment by nutrients such as nitrogen and phosphorus, leading to significant deterioration in water quality and problems for water treatment.

Filtration Process to remove particles from water by passing it through a porous layer.

Flocculation A process in which small suspended particles agglomerate to form large fluffy flocs through gentle stirring by hydraulic or mechanical means. Aluminium and iron salts are generally added to water to bring this about.

Fluoride A general compound containing fluorine added to water to reduce the incidence of dental caries within the community (fluoridation).

Fluorosis A condition caused by excess fluoride in drinking water resulting in pitting and discolouration of teeth.

FTU Formazin Turbidity Unit. Widely used unit for expressing turbidity.

GAC Granular activated carbon.

Giardia lamblia A protozoan frequently found in rivers and lakes causing a severe gastrointestinal disease called giardiasis.

Groundwater Water from aquifers or other natural underground sources.

Guideline value (*GV*) The concentration of a water quality parameter that based on current knowledge does not represent a significant risk to the health of the consumer (health-related guideline value) or affect the aesthetic quality or usefulness of the water (aesthetic guideline value).

GV Guide value for substances and parameters listed in the World Health Organization drinking water guidelines.

HACCP Hazard analysis and critical control points.

Hardness Caused by the dissolved salts of calcium and magnesium. Hard waters cause scaling and reduce the effectiveness of soap, while soft waters are generally acidic and corrosive.

Heterotrophic plate count (HPC) Used to estimate the level of heterotrophic bacterial activity in water by measuring the number of colonies of heterotrophic bacteria grown on selected solid media at a given temperature and incubation period. Expressed in number of bacteria per millilitre of sample.

Humic acids This is a general name for a wide variety of organic compounds which are the breakdown products of vegetable matter, including peat.

Hydrogencarbonate Bicarbonate (HCO_3^-).

Hydrological cycle The continuous circulation of water between the atmosphere, land and sea by precipitation, transpiration and evaporation.

IARC International Agency for Research on Cancer.

ICRP International Commission on Radiological Protection.

Indicator organism A specific micro-organism, such as *Escherichia coli*, whose presence is indicative of faecal pollution or of more-harmful pathogenic micro-organisms.

Ion-exchange A process in which ions of like charge are exchanged between a solid resin and the water. Water softeners replace the calcium ions in the water for sodium ions thus reducing the water hardness.

Ions Charged particles in water. They are either atoms (e.g. Ca^{2+}, Cl^-) or groups of atoms (e.g. HCO_3^-).

Iron (II) Ferrous iron.

Iron (III) Ferric iron.

ISO International Organization for Standardization.

ISO 9001:2000 International standard that provides a common framework for quality management systems in the design development, production, installation and servicing of a product.

ISO 14001:1996 International standard providing a common framework for the development and implementation of environmental management systems. Such systems ensure that all business operations minimize their environmental impacts and work towards the broad goals of environmental sustainability.

Langelier index An equation for calculating the corrosiveness of water.

Lime The common name for calcium oxide (CaO).

LOAL Lowest-observed-adverse-effect level.

Log removal Widely used to describe the efficiency of a physical–chemical treatment process. Normally applied to the reduction in the concentration of pathogens (i.e. bacteria, protozoa and viruses) by a disinfection process or membrane filtration: 1-log removal is equivalent to a 90% reduction in density of the target organism, 2-log removal a 99% reduction, 3-log removal a 99.9% reduction, etc.

MAC Maximum admissible concentration. The maximum concentration of a substance listed in the EC Drinking Water Directive.

MCB Monochlorobenzene.

MCLs Maximum contaminant levels are enforceable drinking water standards set by the Office of Drinking Water of the USEPA.

MCLGs Maximum contaminant level goals are non-enforceable health-based goals for drinking water set by the Office of Drinking Water of the USEPA.

MCPA 4-(2-methyl-4-chlorophenoxy)acetic acid.

MCPP Mecoprop or 2-(2-methyl-chlorophenoxy)propionic acid.

Methaemoglobinaemia A condition in babies and infants where the oxygen-carrying capacity of haemoglobin is reduced due to the uptake of nitrite.

mg l^{-1} Milligram per litre ($1000\,mg\,l^{-1} = 1\,g\,l^{-1}$).

Ml d^{-1} Megalitre per day ($1\,Ml\,d^{-1} = 1000\,m^3\,d^{-1} = 1\,000\,000\,l\,d^{-1}$).

MPN Most probable number.

MS Mass spectrometry.

MTBE methyl tertiary-butyl ether.

ng l^{-1} Nanogram per litre ($1000\,ng\,l^{-1} = 1\,\mu g\,l^{-1}$).

NOAL No-observed-adverse-effect level.

NOEL No-observed-effect level.

NTU Nephelometric turbidity units.

Ofwat Formally the Office of Water Services, since April 2006 the Water Services Regulation Authority.

Oocyst The dormant stage of a protozoan parasite such as *Cryptosporidium*.

Oxidation A chemical reaction in which the oxygen content of a compound is increased, or in which electrons are removed from an ion or compound. Soluble iron becomes insoluble when oxidized and precipitates out of solution.

Ozone A gas that is an unstable form of oxygen, with the chemical formula O_3, used as an aggressive disinfectant.

P–A Presence–absence.

PAC Powdered activated carbon.

PAH Polycyclic or polynuclear aromatic hydrocarbons.

Pathogen An organism that is capable of causing disease (e.g. bacteria, viruses and protozoa).

PCBs Polychlorinated biphenyls.

PCE Tetrachloroethene or perchloroethylene – an industrial solvent.

PCP Pentachlorophenol.

PCTs Polychlorinated terphenyls.

Pesticide A pesticide is any chemical used to control animal and plant pests. These include fungicides, herbicides and insecticides.

pH A measure of how acid or alkaline water is in relation to the concentration of hydrogen ions present.

Plumbosolvency Solubility of lead in water.

Point-of-entry treatment device A unit or series of units used to treat the water supplied on entry to a building or household.

Point-of-use treatment device A unit or series of units used to selectively treat the water supplied to a single tap normally located in the kitchen.

Point pollution Pollution from a clearly defined source, usually an outlet pipe.

Primary National Drinking Water Regulations Standards set by the USEPA for parameters considered to be potentially harmful to health.

Private supply Water taken from a private source or supplied by a non-licensed supplier.

Public supply Water supplied by a company licensed or authorized to do so.

PVC Polyvinyl chloride.

Radionuclide An unstable isotope of an element that undergoes radioactive decay.

Rapid sand filter A filter containing coarse sand or other filtration media through which water passes at high rates, often under pressure.

Raw water Natural surface or ground water at time of abstraction prior to entry to the treatment plant.

Reduction A chemical process in which electrons are added to an ion or compound, or in which the oxygen content is reduced.

Reservoir A man-made or artificial impoundment used for the collection and storage of water either for supply or to compensate flows in rivers used for abstraction downstream.

Reverse osmosis A process for the removal of dissolved ions and organic compounds from water that is passed under pressure through a semi-permeable membrane.

Risk assessment The prediction from available data of the frequency of an identified hazard or event occurring (likelihood) and the magnitude of any associated consequences.

Secondary National Drinking Water Regulations Standards set by the USEPA for parameters intended to protect the aesthetic quality of water rather than concerned with public health.

Sedimentation Process for the removal of settleable solids within a tank under semi-quiescent conditions. The settled particles form a sludge, which is removed from the tank at regular intervals.

Service reservoir A water tower or reservoir used for storage of treated water within the distribution system.

Slow sand filters Water passes through a layer of fine sand on top of a layer of gravel. Solids are removed by filtration and nutrients are removed by biological activity.

Softening Process to reduce the hardness of water either by precipitation or ion-exchange processes.

Storage reservoir Either a man-made or natural impoundment used for storing water to improve quality and ensure continuity of supply prior to treatment.

Surface water General term for any water body that is found flowing or standing on the surface such as rivers, lakes or reservoirs.

Taste threshold Minimum concentration at which a compound in water can be tasted.

TBA Terbuthylazine.

TCA Methyl chloroform or 1,1,1-trichloroethane – an industrial solvent.

TCB Trichlorobenzene.

TCE Trichloroethylene – an industrial solvent.

TCM Trichloromethane or chloroform – an industrial solvent.

TDI Tolerable daily intake.

TDS Total dissolved solids is a crude measure of the total concentration of inorganic salts in water. Measured by evaporation.

THMs (Trihalomethanes) Organic chemicals formed during chlorination of water containing natural organic compounds such as humic and fulvic acids.

Turbidity Due to colloidal and suspended matter that imparts a cloudiness to water. It is determined by measuring the degree of scattering of a beam of light that is passed through the water.

µg l^{-1} Microgram per litre ($1000\,\mu g\,l^{-1} = 1\,mg\,l^{-1}$).

USEPA US Environmental Protection Agency.

UV Ultraviolet radiation used to disinfect water.

Water poverty threshold The minimum acceptable volume of water for basic needs is set at 50 litres per capita per day.

Water safety plan A set of guidelines that optimizes water quality and supply through optimal systems and operational management. It comprises three basic elements: a system assessment, effective operational monitoring and management.

Water scarcity The situation where water demand exceeds availability.

Water security (developed countries) The protection of the water supply cycle against terrorism or other deliberate acts leading to *water shortage*.

Water security (developing countries) The reliable and secure access to water over time. Includes measures to be taken at times of water scarcity to avoid *water stress*.

Water shortage Where supplies are insufficient to meet certain minimum defined requirements. The minimum per capita volume of water that defines minimum varies between countries.

Water stress A situation brought about by *water scarcity* often leading to conflict.

Water supply zone These are either a discrete area served by a single source or an area supplying no more than 50 000 people. Used as the basic unit for monitoring water quality in England and Wales.

Water undertaker A company holding a licence to supply drinking water. In England and Wales these are the water service companies and the water-only companies.

Water utility A *water undertaker*.

WHO World Health Organization.

Wholesomeness A concept of water quality that is defined by reference to the standards and other requirements listed in the Water Supply (Water Quality) Regulations for England and Wales 1989 or 2003.

WRc Water Research Centre.

WSC Water supply company.

WSP Water safety plan.

YLD Years of healthy life lost not in full health (e.g. years lived with a disability).

YLL Year of life lost by premature mortality.

Index